Renewable Energy and Utility System Optimization for Sustainable Industries

Renewable Energy and Utility System Optimization for Sustainable Industries

Guest Editors

Cristina Rodriguez
Li Sun

MDPI

Basel • Beijing • Wuhan • Barcelona • Belgrade • Novi Sad • Cluj • Manchester

Guest Editors
Cristina Rodriguez
School of Computing
Engineering and Physical
Sciences
University of the West of
Scotland
Paisley
United Kingdom

Li Sun
School of Computing
Engineering and Physical
Sciences
University of the West of
Scotland
Paisley
United Kingdom

Editorial Office
MDPI AG
Grosspeteranlage 5
4052 Basel, Switzerland

This is a reprint of the Special Issue, published open access by the journal *Sustainability* (ISSN 2071-1050), freely accessible at: https://www.mdpi.com/journal/sustainability/special_issues/RE_USO_SI.

For citation purposes, cite each article independently as indicated on the article page online and as indicated below:

Lastname, A.A.; Lastname, B.B. Article Title. *Journal Name* **Year**, *Volume Number*, Page Range.

ISBN 978-3-7258-4327-5 (Hbk)
ISBN 978-3-7258-4328-2 (PDF)
https://doi.org/10.3390/books978-3-7258-4328-2

Cover image courtesy of Carlos Alberto Coutinho Monteiro

Contents

Amal A. Mohamed, Salah Kamel, Ali Selim, Tahir Khurshaid and Sang-Bong Rhee
Developing a Hybrid Approach Based on Analytical and Metaheuristic Optimization Algorithms for the Optimization of Renewable DG Allocation Considering Various Types of Loads
Reprinted from: *Sustainability* **2021**, *13*, 4447, https://doi.org/10.3390/su13084447 1

Ruth Chinyere Anyanwu, Cristina Rodriguez, Andy Durrant and Abdul Ghani Olabi
Evaluation of Growth Rate and Biomass Productivity of *Scenedesmus quadricauda* and *Chlorella vulgaris* under Different LED Wavelengths and Photoperiods
Reprinted from: *Sustainability* **2022**, *14*, 6108, https://doi.org/10.3390/su14106108 28

Ahmed Fathy, Hegazy Rezk, Seydali Ferahtia, Rania M. Ghoniem, Reem Alkanhel and Mohamed M. Ghoniem
A New Fractional-Order Load Frequency Control for Multi-Renewable Energy Interconnected Plants Using Skill Optimization Algorithm
Reprinted from: *Sustainability* **2022**, *14*, 14999, https://doi.org/10.3390/su142214999 41

Sreekanth Manavalla, Feroskhan M., Joseph Daniel, Sivakumar Ramasamy, T. M. Yunus Khan, Rahmath Ulla Baig, et al.
System Design, Optimization and 2nd Law Analysis of a 100 MWe Double Reheat s-CO_2 Power Plant at Full Load and Part Loads
Reprinted from: *Sustainability* **2023**, *15*, 14677, https://doi.org/10.3390/su152014677 66

Hegazy Rezk, Rania M. Ghoniem, Seydali Ferahtia, Ahmed Fathy, Mohamed M. Ghoniem and Reem Alkanhel
A Comparison of Different Renewable-Based DC Microgrid Energy Management Strategies for Commercial Buildings Applications
Reprinted from: *Sustainability* **2022**, *14*, 16656, https://doi.org/10.3390/su142416656 85

Dilip Kumar, Yogesh Kumar Chauhan and Ajay Shekhar Pandey
Performance Investigation of Renewable Energy Integration in Energy Management Systems with Quantum-Inspired Multiverse Optimization
Reprinted from: *Sustainability* **2025**, *17*, 3734, https://doi.org/10.3390/su17083734 107

Amarnath Krishnamoorthy, Cristina Rodriguez and Andy Durrant
Sustainable Approaches to Microalgal Pre-Treatment Techniques for Biodiesel Production: A Review
Reprinted from: *Sustainability* **2022**, *14*, 9953, https://doi.org/10.3390/su14169953 131

Khaled Obaideen, Abdul Ghani Olabi, Yaser Al Swailmeen, Nabila Shehata, Mohammad Ali Abdelkareem, Abdul Hai Alami, et al.
Solar Energy: Applications, Trends Analysis, Bibliometric Analysis and Research Contribution to Sustainable Development Goals (SDGs)
Reprinted from: *Sustainability* **2023**, *15*, 1418, https://doi.org/10.3390/su15021418 161

Farjana Faisal, Mohammad Golam Rasul, Ashfaque Ahmed Chowdhury and Md Islam Jahirul
Optimisation of Process Parameters to Maximise the Oil Yield from Pyrolysis of Mixed Waste Plastics
Reprinted from: *Sustainability* **2024**, *16*, 2619, https://doi.org/10.3390/su16072619 195

Hegazy Rezk, A. G. Olabi, Mohammad Ali Abdelkareem, Abdul Hai Alami and Enas Taha Sayed
Optimal Parameter Determination of Membrane Bioreactor to Boost Biohydrogen Production-Based Integration of ANFIS Modeling and Honey Badger Algorithm
Reprinted from: *Sustainability* **2023**, *15*, 1589, https://doi.org/10.3390/su15021589 . . . . **219**

Farhat Nasim, Shahida Khatoon, Ibraheem, Shabana Urooj, Mohammad Shahid, Asmaa Ali and Nidal Nasser
Hybrid ANFIS-PI-Based Optimization for Improved Power Conversion in DFIG Wind Turbine
Reprinted from: *Sustainability* **2025**, *17*, 2454, https://doi.org/10.3390/su17062454 . . . . **232**

A. G. Olabi, Khaled Obaideen, Mohammad Ali Abdelkareem, Maryam Nooman AlMallahi, Nabila Shehata, Abdul Hai Alami, et al.
Wind Energy Contribution to the Sustainable Development Goals: Case Study on London Array
Reprinted from: *Sustainability* **2023**, *15*, 4641, https://doi.org/10.3390/su15054641 . . . . **254**

Uchechi Ukaegbu, Lagouge Tartibu and C. W. Lim
Optimization of Solar-Assisted CCHP Systems: Enhancing Efficiency and Reducing Emissions Through Harris Hawks-Based Mathematical Modeling
Reprinted from: *Sustainability* **2024**, *16*, 10694, https://doi.org/10.3390/su162310694 . . . . **276**

Afnan Fayez Eliyan, Mohamed Haouari and Ahmad Sleiti
Decarbonizing Public Transportation: A Multi-Criteria Comparative Analysis of Battery Electric Buses and Fuel Cell Electric Buses
Reprinted from: *Sustainability* **2024**, *16*, 9354, https://doi.org/10.3390/su16219354 . . . . **297**

Qiao Wang, Shiyun Chen and Hongtao Yi
A Two-Stage Evaluation of China's New Energy Industrial Policy Package
Reprinted from: *Sustainability* **2024**, *16*, 8264, https://doi.org/10.3390/su16188264 . . . . **318**

Shuangli Pan, Huiyu Liao, Guijun Zheng, Qian Huang and Maozhuo Shan
Cold Chain Distribution Route Optimization for Mixed Vehicle Types of Fresh Agricultural Products Considering Carbon Emissions: A Study Based on a Survey in China
Reprinted from: *Sustainability* **2024**, *16*, 8207, https://doi.org/10.3390/su16188207 . . . . **339**

sustainability

MDPI

Article

Developing a Hybrid Approach Based on Analytical and Metaheuristic Optimization Algorithms for the Optimization of Renewable DG Allocation Considering Various Types of Loads

Amal A. Mohamed [1], Salah Kamel [1], Ali Selim [1], Tahir Khurshaid [2,*] and Sang-Bong Rhee [2,*]

1 Department of Electrical Engineering, Faculty of Engineering, Aswan University, Aswan 81542, Egypt; amal.amin13@yahoo.com (A.A.M.); skamel@aswu.edu.eg (S.K.); ali.selim@aswu.edu.eg (A.S.)
2 Department of Electrical Engineering, Yeungnam University, Gyeongsan 38541, Korea
* Correspondence: tahir@ynu.ac.kr (T.K.); rrsd@yu.ac.kr (S.-B.R.)

Abstract: The optimal location of renewable distributed generations (DGs) into a radial distribution system (RDS) has attracted major concerns from power system researchers in the present years. The main target of DG integration is to improve the overall system performance by minimizing power losses and improving the voltage profile. Hence, this paper proposed a hybrid approach between an analytical and metaheuristic optimization technique for the optimal allocation of DG in RDS, considering different types of load. A simple analytical technique was developed in order to determine the sizes of different and multiple DGs, and a new efficient metaheuristic technique known as the Salp Swarm Algorithm (SSA) was suggested in order to choose the best buses in the system, proportionate to the sizes determined by the analytical technique, in order to obtain the minimum losses and the best voltage profile. To verify the power of the proposed hybrid technique on the incorporation of the DGs in RDS, it was applied to different types of static loads; constant power (CP), constant impedance (CZ), and constant current (CI). The performance of the proposed algorithm was validated using two standards RDSs—IEEE 33-bus and IEEE 69-bus systems—and was compared with other optimization techniques.

Keywords: radial distribution networks; optimal allocation; different types of DGs; hybrid technique; analytical technique; SSA; static load; voltage stability; power loss

1. Introduction

The electric power distribution network is the part which connects the high voltage (HV) electric power transmission network to low voltage (LV) consumers, which is considered the substantial part of the electric power network because the power delivered from the electric power to the loads is assured by operative and reliable distribution networks. The electric power system is sectioned into three parts: generation, transmission, and the distribution network. The last part is the distribution network, which is used to link between the consumers and transmission. The electric power system suffers from two important factors: a low voltage profile and a high percentage of power losses. However, 70% of the power losses are caused at the distribution level (primary and secondary), and the remaining 30% are caused by the transmission system [1]. Hence, the distribution networks attract the main attention of the researchers.

Exiting grids are mostly characterized by centralized bulk power generation far from load centers, requiring long, costly, and lossy transmission lines for power delivery.

By incorporating distribution generation (DG) units in the radial distribution system (RDS) at appropriate places, the losses can reach the minimum value. DG is known as electricity generation by specific capacity generators associated with the electric distribution system or close to the loads.

In smart networks, there is intense interest in the distribution systems to increase the penetration levels of renewable energy sources that are incorporated close to the point-of-use, which achieve lower losses in energy delivery, and this makes the transactions of the power in the lines of the distribution system bidirectional [2].

Research has demonstrated that the incorporation of DG sources in the electric power distribution networks could create various benefits; some of them are:

- Improved voltages of all busses.
- Decreased power losses.
- Improved security of critical buses.
- Network reinforcement.
- The reduction of the congestion of the distribution system.
- The reduction of operation costs.
- The improvement of system reliability

It is very important to incorporate an auxiliary power system beside the main generation in order to supply the loads with sufficient energy. Hence, the electrical network must achieve reliability, which is defined as the probability of this network to perform its intended function for a specific time [3].

In order to achieve these objectives, it is necessary to specify the optimal sizes and location of DG units in distribution systems; otherwise, it could lead to reverse actions, such as the increase in active power losses and a poor voltage profile.

There are multiple types of DGs, which can be discussed as follows:

- Type I: injects only active power, with a unity power factor.
- Type II: injects only reactive power, with zero power factor.
- Type III: injects active and reactive power.
- Type IV: injects active power and consumes reactive power.

Different techniques have been obtained in order to solve the optimization problems: analytical optimization techniques which based on the mathematical model (such as [4,5]) and metaheuristic techniques such as the Ant lion optimization algorithm [6], Hybrid particle swarm optimization, and the Genetic algorithm (GA-PSO) based on active power losses [7,8]; the Artificial Bee Colony Algorithm (ABC) [9]; the Modified Teaching–Learning-Based Optimization algorithm (MTLBO) [10]; and the Harmony Search Algorithm (HSA) [11]. In [12,13], the researchers used a hybrid evolutionary algorithm to specify the optimal DG placement; the improvement of the voltage profile and voltage stability were the main objectives of the researchers in [14,15]. There are also the Whale Optimization Algorithm (WOA) [16] and the Intersect Mutation Differential Evolution optimizer (IMDE) [17].

The loads of the electric distribution system are permanently changing due to consumer demand variations. In industrial areas, the electric power distribution network is susceptible to voltage collapse due to grievous loading conditions.

In load flow studies, it is acceptable to symbolize the composite load properties, as signified from the power given points. In the transmission networks, because of the voltage regulation, the load can be represented by using a constant power model, while in the distribution networks, the voltage is changed along the network lines because most of the buses are not voltage controlled. As such, the properties of the loads in the distribution network analysis are very important and serious.

Distribution power flow studies have been extremely influenced by load modeling. Furthermore, the load characteristics have a noticeable influence on power flow solutions and convergence ability [18].

There are two important types of load: static and dynamic loads. The dynamic loads are not signifying in power flow studies, while static load modeling expresses the active and reactive power as a function of the voltage magnitude of each bus, so it is appropriate for power flow studies, which are in turn categorized into different types of loads, as follows: constant power (CP), constant current (CI), and constant impedance (CZ).

Multiple papers have studied the effect of the DG units in RDS by considering the static load model [19,20]. In order to check the performance of the hybrid technique in power loss reduction and voltage improvement, it is applied to the three types of static loads.

In this paper, a hybrid analytical and metaheuristic technique is proposed in order to calculate the optimal allocation and sizing of multiple types of DGs with different types of load; the analytical technique was proposed to specify the proper size, while the Salp Swarm Algorithm (SSA) was chosen to specify best locations.

SSA is a new efficient optimization algorithm introduced by Mirjalil in 2017 [21]. The behavior of salps in a swarm for foraging or traveling across oceans is the main inspiration of this technique. Salps are a kind of Salpidae family that have a transparent barrel-shaped body; their tissues are highly similar to a jellyfish's structure. The main reason for this behavior of salps in swarms is unclear as yet, but some of the researchers believe that the swarming is performed in order to achieve the best movement using rapid coordinated alterations and foraging.

The SSA technique is simple and uncomplicated to implement, and includes one controlling parameter C_1; the acclimated mechanism of this algorithm allows it to alert solutions, and ultimately finds an exact estimation of the optimal solutions. The SSA gives strongly satisfying results when it is integrated with other techniques; it is also characterized by its speed in convergence, and is suitable for different optimization problems, while the GA is easy to understand, and deals with multi objective optimization, but one of its important disadvantages is being time consuming and computationally expensive. As such, due to these advantages, SSA is an efficient and excellent technique comparable with recent techniques; nevertheless, this cannot be ensured for all optimization problems. Additionally, the SSA has been used to solve single-objective functions in various optimization areas; the Chaotic Salp Swarm Algorithm (CSSA) is a new optimization technique which was introduced by Sayed et al. [22]. In addition, SSA has been utilized in electrical power engineering problems; in [23], El-Fergany proposed a technique to define the parameters of the polarization curves of polymer exchange membrane fuel cells (PEMFC), which are based on SSA, and in [24], the authors presented a modification and application of SSA called the enhanced salp swarm algorithm (ESSA); in [25], the SSA was employed to solve optimal power flow (OPF) problems.

The rest of the paper is sectioned as follows: the problem formulation, including the load modeling, forward/backward load flow algorithm, objective function, and operational constraints are illustrated in Section 2; the SSA is presented in Section 3; Section 4 illustrates the analytical technique; Section 5 presents the simulation results; finally, the conclusion of the paper is illustrated in Section 6.

2. Problem Formulation

2.1. Load Modelling

In this paper, three types of static loads are studied, as follows:

1. Constant impedance (CZ): the power is proportional to the square of the voltage value.
2. Constant current (CI): the power is proportional to the voltage value.
3. Constant power (CP): the power does not change with the voltage variation.

In this article, the exponential form is used to define the static load as follows:

$$P_A = P_{0_A} * V_A{}^{\sigma} \tag{1}$$

$$Q_A = Q_{0_A} * V_A{}^{\sigma} \tag{2}$$

where P_A and Q_A are the active and reactive power at bus A, V is the voltage magnitude at bus A, $\sigma = 0$ at constant power (CP), $\sigma = 1$ at constant current (CI), and $\sigma = 2$ at constant impedance (CZ).

2.2. Forward–Backward Power Flow Algorithm

RDSs have some special properties, such as a radial structure, a high R/X ratio, and unbalanced loads. Due to the previous features, Gauss–Seidel (GS), Newton Raphson (NR), and the fast decoupled method are inefficient in radial distribution system analysis. Hence, the forward–backward load flow is a modern technique which is used in this paper to solve the load flow in RDS, which is divided into two processes: forward sweep and backward sweep. The voltage in the forward sweep is calculated from sending to the end of the feeder, while the power in the backward sweep is calculated from the end of the feeder to the source. Figure 1 illustrates two bus RDS; the active power P_S and the reactive power Q_S that flow from bus S to T can be calculated from backward sweep as follows.

$$P_S = P_T + P_{L,T} + R\left(\frac{(P_T + P_{L,T})^2 + (Q_T + Q_{L,T})^2}{|V_T|^2}\right). \tag{3}$$

$$Q_S = Q_T + Q_{L,T} + X\left(\frac{(P_T + P_{L,T})^2 + (Q_T + Q_{L,T})^2}{|V_T|^2}\right) \tag{4}$$

and the voltage of each bus can be determined in the forward sweep as:

$$V_T^2 = \left(V_S^2 - 2(P_S R + Q_S X) + \left(R^2 + X^2\right)\frac{({P_S}^2 + {Q_S}^2)}{V_S^2}\right) \tag{5}$$

therefore, the active power loss in the branch from S to T can be determined as:

$$P_{loss(S,T)} = R_{S,T}\left(\frac{P_S^2 + jQ_S^2}{|V_S|^2}\right) \tag{6}$$

and the reactive power loss in the branch from S to T can be determined as:

$$Q_{loss(S,T)} = x_{S,T}\left(\frac{P_S^2 + jQ_S^2}{|V_S|^2}\right) \tag{7}$$

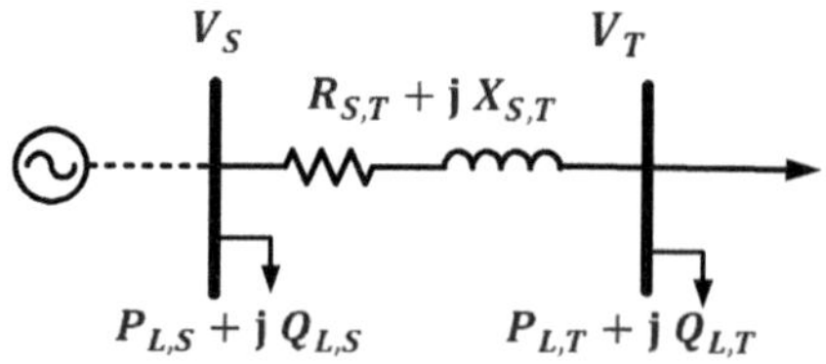

Figure 1. Two bus radial distribution network.

Hence, the total active and reactive power losses in RDS can be estimated as:

$$TPL = \sum_{s,t=1}^{Br} (P_{Loss}(S,T)) \tag{8}$$

$$TQL = \sum_{s,t=1}^{Br} (Q_{Loss}(S,T)) \tag{9}$$

2.3. Objective Function

In general, incorporating DGs into RDS reduces the power losses by reducing the amount of current flowing in the lines; also, it can improve the overall voltage profile, which in turn improves the efficiency of the network.

In the present work, the main objective function is to minimize total active power losses. Therefore, the objective function can be formulated as:

$$P_{Loss} = \sum_{S=1}^{No} \sum_{T=1}^{No} \alpha_{ST} \left(P_S P_T + Q_S Q_T\right) + \beta_{ST} \left(Q_S P_T - P_S Q_T\right) \tag{10}$$

P_S and P_T are the active power at bus S and T, respectively
Q_S and Q_T are the reactive power at bus S and T, respectively.

2.4. Operation Constraints

2.4.1. Equality Constraints

$$P_{Slack} + \sum_{A}^{No} P_{DG(A)} = \sum_{A}^{No} P_{Load(A)} + \sum_{B}^{Br} P_{Loss(B)} \tag{11}$$

$$Q_{Slack} + \sum_{A}^{No} Q_{DG(A)} = \sum_{A}^{No} Q_{Load(A)} + \sum_{B}^{Br} Q_{Loss(B)} \tag{12}$$

$P_{Slack,}$ Q_{Slack} are the active and reactive power of the slack bus, respectively.
No: number of buses; Br: number of branches; P_{Loss} : active power loss.

2.4.2. Inequality Constraints

The voltage at each bus must be between 0.90 and 1.05.

$$V_{min} \le V_A \le V_{max} \tag{13}$$

The total reactive power injected by the DG is less than the total reactive power of loads.

$$\sum Q_{DG} \le \sum_{A}^{No} Q_{Load(A)} \tag{14}$$

The total active power injected by the DG is less than the total active power loads.

$$\sum P_{DG} \le \sum_{A}^{NO} P_{Load(A)} \tag{15}$$

3. Salp Swarm Algorithm (SSA)

The SSA is a type of Salp family, and it is similar to a jellyfish's body. The SSA inhabits oceans, and they derive their movement from water power to look for food, which adjust as swarms, and are known as salp chains.

The SSA typifies the solutions of the given problems by the position number of the salps, which are represented by (salp chain) in the dimension of (n × dim), where n is the search agent and dim is the proposed variable. The salp chain is categorized into two groups; the first one is the leader (in the front of the chain) and the second is the follower (all other salps in the swarm), as shown in Figure 2.

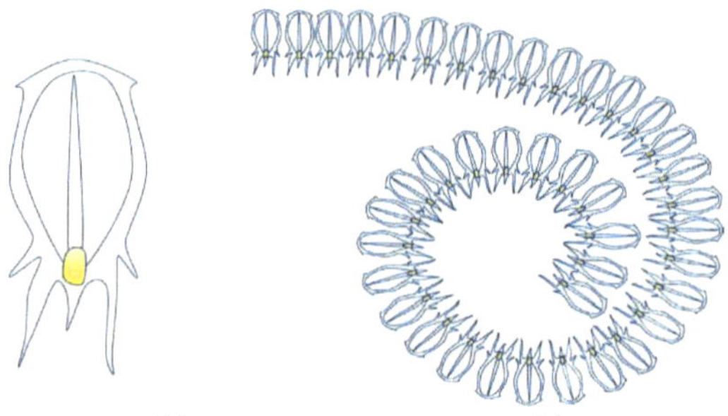

Figure 2. (**a**) Individual salp; (**b**) swarm of salps (salps chain) [19].

The optimal solution (the preferable salp) is considered as the swarm's main target (food), which is followed up by the follower's group. When the leader updates its position according to the best food, it explores the search space carefully to obtain the best solution, and the follower follows the leader.

The leader of the chain updates its place concerning a food source, which is formulated as:

$$A_j^{\,1} = \begin{cases} H_j + c_1\left(\left(U_j - L_j\right)c_2 + L_j\right) & c_3 \geq 0 \\ H_j - c_1\left(\left(U_j - L_j\right)c_2 + L_j\right) & c_3 < 0 \end{cases} \tag{16}$$

$A_j^{\,1}$ is the leader position, H_j is the position of the food source.

L_j and U_j represent the lower and upper boundaries, respectively. c_2 and c_3 are random numbers, where

$$c_1 = 2e^{-\left(\frac{4x}{X}\right)^2} \tag{17}$$

x is the current iteration and X is the maximum number of iterations. By Newton's law of motion, the follower salps update their position:

$$A_j^{\,i} = \frac{1}{2}\left(aw^2 + v_o wi\right) \geq 2 \tag{18}$$

where $A_j^{\,i}$ is the follower salp position, v_o is the initial speed, and w is the number of the iteration.

Then, Equation (18) can be reorganized as:

$$A_j^{\,i} = \frac{1}{2}\left(A_j^{\,i} + A_j^{\,i-1}\right) \tag{19}$$

The application of SSA for the optimal placement of DGs units can be summarized in the following steps:

- Step 1: specify the input variables of the SSA, which include the search agent, the number of iterations, and the lower and upper variable.
- Step 2: start the population of the SSA randomly, using (20).

$$A(n, m) = \text{rand}(n, 1) * (U - L) + L \tag{20}$$

- Step 3: run the forward–backward load flow code and determine the fitness function.
- Step 4: calculate the best position according to the optimal fitness function.
- Step 5: update the position of the leader salp according to (16).
- Step 6: update the position of the follower salp according to (19).
- Step 7: verify the limits of the header and follower of the salp chain.
- Step 8: run the forward–backward load flow code in order to determine the fitness function for the positions which updated, then calculate the optimal position.

- Step 9: repeat the previous steps from step 5 to step 8 until the current iteration equals a maximum number of iterations.
- Step 10: finally, find an optimal position represented by the food source and the associated fitness function.

4. Optimal Sizes

From Figure 1, the total active power losses (P_{Loss}) in the RDS are calculated as:

$$\alpha_{ST} = \frac{R_{ST}}{V_S V_T}\cos(\alpha_S - \alpha_T) \tag{21}$$

$$\beta_{ST} = \frac{R_{ST}}{V_S V_T}\sin(\alpha_S - \alpha_T) \tag{22}$$

V_S and V_T are the voltage magnitude at bus S and T, respectively

$Z_{ST} = R_{ST} + jX_{ST}$, where Z_{ST} bus impedance matrix.

We assume that x is the number of the DG which supplies only active power (Type I) and their places are $M_1, M_2, \ldots M_x$, while $P_{(DG)_{M_1}}, P_{(DG)_{M_2}} \ldots\ldots P_{(DG)_{M_x}}$ are the sizes of the DGs, respectively.

Therefore, the active power from Type I can be:

$[P_{M_1} = P_{(DG)_{M_1}} - P_{(Load)_{M_1}}], \ldots\ldots, [P_{M_x} = P_{(DG)_{M_x}} - P_{(Load)_{M_x}}]$. Furthermore, the number of the DGs supplying reactive power (Type II) is y, and their places are $N_1, N_2, \ldots N_y$, while $Q_{(DG)_{N_1}}, Q_{(DG)_{N_2}} \ldots\ldots Q_{(DG)_{N_y}}$ are the sizes of the DGs, respectively.

The reactive power from Type II can be written as: $[Q_{N_1} = Q_{(DG)_{N_1}} - Q_{(Load)_{N_1}}], \ldots\ldots,$ $[Q_{N_y} = Q_{(DG)_{N_y}} - Q_{(Load)_{N_y}}]$.

The minimum power losses can be obtained if the first derivative of the power losses with respect to the power supplied from the DGs becomes zero; therefore, it can be formed as:

$$\frac{\partial P_{Loss}}{\partial P_{M_1}} = 2\alpha_{M_1M_1}P_{M_1} + 2\sum_{\substack{T=1 \\ T \neq M_1}}^{No}(\alpha_{M_1T}P_T - \beta_{M_1T}Q_T) \tag{23}$$

or

$$\begin{aligned} &\alpha_{M_1M_1}P_{M_1} + \alpha_{M_1M_2}P_{M_2} + \ldots + \alpha_{M_1M_x}P_{M_x} - \beta_{M_1N_1}Q_{N_1} - \beta_{M_1N_2}Q_{N_2} - \ldots \\ &-\beta_{M_1N_y}Q_{N_y} = -\sum_{\substack{T=1 \\ T \neq M_1, M_2, \ldots, M_x \\ T \neq N_1, N_2, \ldots, N_y}}^{No}(\alpha_{M_1T}P_T - \beta_{M_1T}Q_T) \end{aligned} \tag{24}$$

The derivative of the active power loss according to active power injection P_{M_x} can be formulated as:

$$\frac{\partial P_{Loss}}{\partial P_{M_x}} = 2\alpha_{M_xM_x}P_{M_x} + 2\sum_{\substack{T=1 \\ T \neq M_x}}^{No}(\alpha_{M_xT}P_T - \beta_{M_xT}Q_T) \tag{25}$$

or

$$\begin{aligned} &\alpha_{M_xM_1}P_{M_1} + \alpha_{M_xM_2}P_{M_2} + \ldots + \alpha_{M_xM_x}P_{M_x} - \beta_{M_xN_1}Q_{N_1} - \beta_{M_xN_2}Q_{N_2} - \ldots \\ &-\beta_{M_xN_y}Q_{N_y} = -\sum_{\substack{T=1 \\ T \neq M_1, M_2, \ldots, M_x \\ T \neq N_1, N_2, \ldots, N_y}}^{NO}(\alpha_{M_xT}P_T - \beta_{M_xT}Q_T) \end{aligned} \tag{26}$$

Furthermore, the derivative of the active power losses regarding to the reactive power injection Q_{N_1} can be formulated as:

$$\frac{\partial P_{Loss}}{\partial Q_{N_1}} = 2\alpha_{N_1N_1}Q_{N_1} + 2 \sum_{\substack{T=1 \\ T\neq N_1}}^{No} \left(\alpha_{N_1T}P_T - \beta_{N_1T}Q_T\right) \tag{27}$$

or

$$\begin{aligned} &\beta_{N_1M_1}P_{M_1} + \beta_{N_1M_2}P_{M_2} + \ldots + \beta_{N_1M_x}P_{M_x} - \alpha_{N_1N_1}Q_{N_1} - \alpha_{N_1N_2}Q_{N_2} - \ldots \\ &-\alpha_{N_1N_{my}}Q_{N_y} = - \sum_{\substack{T=1 \\ T\neq M_1, M_2, \ldots, M_x \\ T\neq N_1, N_2, \ldots, N_y}}^{No} \left(\alpha_{N_1T}Q_T + \beta_{N_1T}P_T\right) \end{aligned} \tag{28}$$

Similarly, the derivative for Q_{N_y} is

$$\frac{\partial P_{loss}}{\partial Q_{N_y}} = 2\alpha_{N_yN_y}Q_{N_y} + 2 \sum_{\substack{T=1 \\ T\neq N_y}}^{No} \left(\alpha_{N_yT}P_T - \beta_{N_yT}Q_T\right) \tag{29}$$

$$\begin{aligned} &\beta_{N_yM_1}P_{M_1} + \beta_{N_yM_2}P_{M_2} + \ldots + \beta_{N_yM_x}P_{M_x} - \alpha_{N_yN_1}Q_{N_1} - \alpha_{N_yN_2}Q_{N_2} - \ldots \\ &-\alpha_{N_yN_y}Q_{N_y} = - \sum_{\substack{T=1 \\ T\neq M_1, M_2, \ldots, M_x \\ T\neq N_1, N_2, \ldots, N_y}}^{No} \left(\alpha_{N_yT}Q_T + \beta_{N_yT}P_T\right) \end{aligned} \tag{30}$$

After reorganizing the above equations, they can be rewritten as illustrate below:

$$\begin{bmatrix} \alpha_{M_1M_1} & \alpha_{M_1M_2} & \cdots & \alpha_{M_1M_x} & -\beta_{M_1N_1} & \beta_{M_1N_2} & \cdots\cdots\cdots & \beta_{M_1N_y} \\ \alpha_{M_2M_1} & \alpha_{M_2M_2} & \cdots & \alpha_{M_2M_x} & -\beta_{M_2N_1} & \beta_{M_2N_2} & \cdots\cdots\cdots & \beta_{M_2N_y} \\ \vdots & \vdots & \ddots & \vdots & \vdots & \vdots & \ddots & \vdots \\ \alpha_{M_xM_1} & \alpha_{M_xM_2} & \cdots & \alpha_{M_xM_x} & -\beta_{M_xN_1} & \beta_{M_xN_2} & \cdots\cdots\cdots & \beta_{M_xN_y} \\ \beta_{N_1M_1} & \beta_{N_1M_2} & \cdots & \beta_{N_1M_x} & -\alpha_{N_1N_1} & \alpha_{N_1N_2} & \cdots\cdots\cdots & \alpha_{N_1N_y} \\ \beta_{N_2M_1} & \beta_{N_2M_2} & \cdots & \beta_{N_2M_x} & -\alpha_{N_2N_1} & \alpha_{N_2N_2} & \cdots\cdots\cdots & \alpha_{N_2N_y} \\ \vdots & \vdots & \ddots & \vdots & \vdots & \vdots & \ddots & \vdots \\ \beta_{N_yM_1} & \beta_{N_yM_2} & \cdots & \beta_{N_yM_x} & -\alpha_{N_yN_1} & \alpha_{N_yN_2} & \cdots\cdots\cdots & \alpha_{N_yN_y} \end{bmatrix} \begin{bmatrix} P_{M_1} \\ P_{M_2} \\ \vdots \\ P_{M_x} \\ Q_{N_1} \\ Q_{N_2} \\ \vdots \\ Q_{N_y} \end{bmatrix} = - \begin{bmatrix} H_{M_1} \\ H_{M_2} \\ \vdots \\ H_{M_x} \\ I_{N_1} \\ I_{N_2} \\ \vdots \\ I_{N_y} \end{bmatrix} \tag{31}$$

where

$$H_{M_i} = \sum_{\substack{T=1 \\ T\neq M_1, M_2, \ldots, M_x \\ T\neq N_1, N_2, \ldots, N_y}}^{No} \left(\alpha_{M_1T}P_T - \beta_{M_1T}Q_T\right),\ i = 1,2,3\ldots\ldots\ldots\ldots x \tag{32}$$

$$I_{Y_i} = \sum_{\substack{T=1 \\ T\neq M_1, M_2, \ldots, M_x \\ T\neq N_1, N_2, \ldots, N_y}}^{No} \left(\alpha_{N_1T}Q_T + \beta_{N_1T}P_T\right),\ i = 1,2,3\ldots\ldots\ldots\ldots y \tag{33}$$

If we assume the matrix in (31) is equal:

$$\begin{bmatrix} \alpha_{M_1M_1} & \alpha_{M_1M_2} & \cdots\cdots\cdots & \alpha_{M_1M_x} \\ \alpha_{M_2M_1} & \alpha_{M_2M_2} & \cdots\cdots\cdots & \alpha_{M_2M_x} \\ \vdots & \vdots & & \vdots \\ \alpha_{M_xM_1} & \alpha_{M_xM_2} & \cdots\cdots\cdots & \alpha_{M_xM_x} \end{bmatrix} = [A_{11}]_{x\times x}, \text{ and } \begin{bmatrix} -\beta_{M_1N_1} & \beta_{M_1N_2} & \cdots\cdots & \beta_{M_1N_y} \\ -\beta_{M_2N_1} & \beta_{M_2N_2} & \cdots\cdots & \beta_{M_2N_y} \\ \vdots & \vdots & & \vdots \\ -\beta_{M_xN_1} & \beta_{M_xN_2} & \cdots\cdots & \beta_{M_xN_y} \end{bmatrix} = [A_{12}]_{x\times y} \quad (34)$$

$$\begin{bmatrix} -\alpha_{N_1N_1} & \alpha_{N_1N_2} & \cdots\cdots & \alpha_{N_1N_y} \\ -\alpha_{N_2N_1} & \alpha_{N_2N_2} & \cdots\cdots & \alpha_{N_2N_y} \\ \vdots & \vdots & & \vdots \\ -\alpha_{N_yN_1} & \alpha_{N_yN_2} & \cdots\cdots & \alpha_{N_yN_y} \end{bmatrix} = [A_{21}]_{y\times x}, \text{ and } \begin{bmatrix} \beta_{N_1M_1} & \beta_{N_1M_2} & \cdots\cdots\cdots & \beta_{N_1M_x} \\ \beta_{N_2M_1} & \beta_{N_2M_2} & \cdots\cdots\cdots & \beta_{N_2M_x} \\ \vdots & \vdots & & \vdots \\ \beta_{N_yM_1} & \beta_{N_yM_2} & \cdots\cdots\cdots & \beta_{N_yM_x} \end{bmatrix} = [A_{22}]_{y\times y} \quad (35)$$

Then, after reorganization, it becomes:

$$\begin{bmatrix} [A_{11}]_{x\times x} & [A_{12}] \; x \times y \\ [A_{21}]_{y\times x} & [A_{22}]_{y\times y} \end{bmatrix} \begin{bmatrix} [P_{M_i}]_{x\times 1} \\ [Q_{N_i}]_{y\times 1} \end{bmatrix} = \begin{bmatrix} [H_{M_i}]_{x\times 1} \\ [I_{N_i}]_{y\times 1} \end{bmatrix} \quad (36)$$

Next

$$\begin{bmatrix} [P_{M_i}]_{x\times 1} \\ [Q_{N_i}]_{y\times 1} \end{bmatrix} = \text{inverse} \begin{bmatrix} [A_{11}]_{x\times x} & [A_{12}]_{x\times y} \\ [A_{21}]_{y\times x} & [A_{22}]_{y\times y} \end{bmatrix} \begin{bmatrix} [H_{M_i}]_{x\times 1} \\ [I_{N_i}]_{y\times 1} \end{bmatrix} \quad (37)$$

Then

$$\text{Inverse} \begin{bmatrix} [A_{11}]_{x\times x} & [A_{12}]_{y\times x} \\ [A_{21}]_{y\times x} & [A_{22}]_{y\times y} \end{bmatrix} = \begin{bmatrix} [A'_{11}] & [A'_{12}] \\ [A'_{21}] & [A'_{22}] \end{bmatrix} \quad (38)$$

$$[P_{M_i}]_{x\times 1} = [A'_{11}]_{x\times x}[H_{M_i}]_{x\times 1} + [A'_{12}]_{x\times y}[I_N]_{y\times 1} \quad (39)$$

$$[Q_{N_i}]_{y\times 1} = [A'_{21}]_{y\times x}[H_{M_i}]_{x\times 1} + [A'_{22}]_{y\times y}[I_{N_i}]_{y\times 1} \quad (40)$$

Finally, after the installed DG at bus S, the active power obtained can be calculated from the following equation:

$$\left[P_{(DG)_{M_i}}\right]_{x\times 1} = [P_{M_i}]_{x\times 1} + P_{(Load)_i} \quad (41)$$

$P_{(Load)_i}$: active power demand at bus M_i.

The reactive power installed from type II can be determined from the following equation:

$$\left[Q_{(DG)_{N_i}}\right]_{y\times 1} = [Q_{N_i}]_{y\times 1} + Q_{(Load)_i} \quad (42)$$

$Q_{(Load)_i}$: reactive power demand at bus N_i.

Note that:

- For type I of the DG, $M_S \neq N_T \forall\, T$
- For type II of the DG, $N_S \neq M_T \forall\, T$
- For type III of the DG, $M_S = N_T$

The power factor of Type III (injects active and reactive power) is calculated using (41) and (42), which can be evaluated as:

$$PF = \frac{P_{(DG)_i}}{\sqrt{P_{(DG)_i}^2 + Q_{(DG)_i}^2}} \tag{43}$$

5. Results and Discussion

In the present work, two IEEE systems were chosen to check the effectiveness of the proposed technique. The IEEE 33-bus and IEEE 69-bus are shown in Figures 3 and 4; their base voltages and apparent power are 12.66 kV and 100 MVA, respectively.

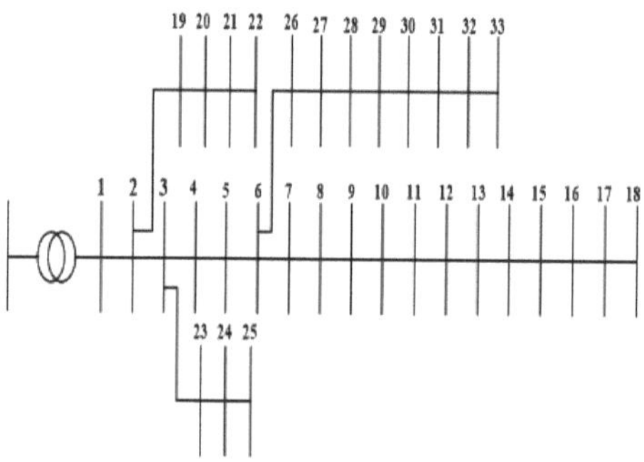

Figure 3. Single line diagram of the 33-bus system.

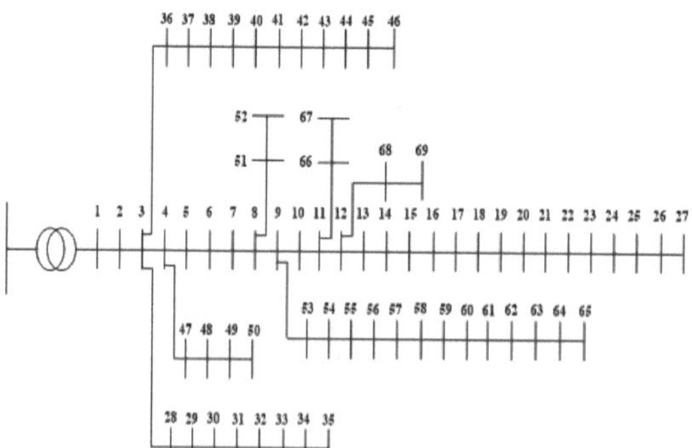

Figure 4. Single line diagram of the 69-bus system.

In order to obtain the optimal size and location of the DGs, the SSA is performed 100 times, and the optimal solutions are reported in tables for three load types. Furthermore, the main parameters of the SSA are specified as several search agents (=30), and the number of maximum iterations = 80.

In the present work, the simulation results of each system are divided to three cases; one DG integration, two DG integrations, and three DG integrations with three DG types and three types of loads: CP, CI, and CZ.

5.1. IEEE 33-Bus Test System

This system consists of 33 buses and 32 branches. The total active power of the load is 3715 kW, and the total reactive power of the load is 2300 kVAR; the total power loss at base case (B.C) is 211 kW, and the minimum voltage is 0.9040 p.u. on bus 18. The convergence characteristics of one, two, and three DGs of type I, type II, and Type III for the three loads are presented in Figures 5–7, respectively. Figures 8–16 describe the voltage profiles of 1DG, 2DG, and 3DG of the three DG types for the three load types. Furthermore, the power loss comparison between the three types of the load is shown in Figures 17–19.

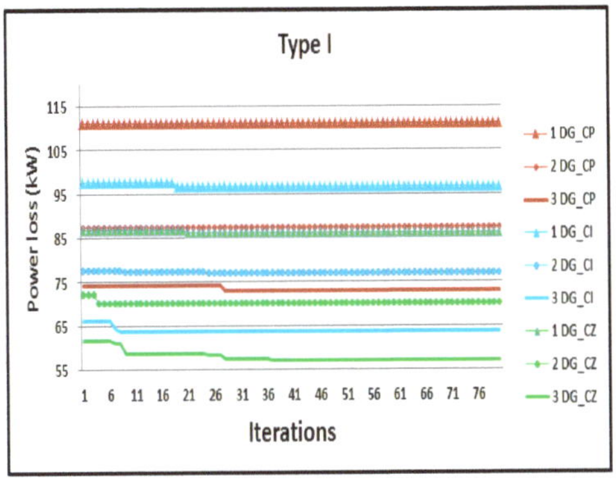

Figure 5. Pl and Iter of DG type I for the three loads.

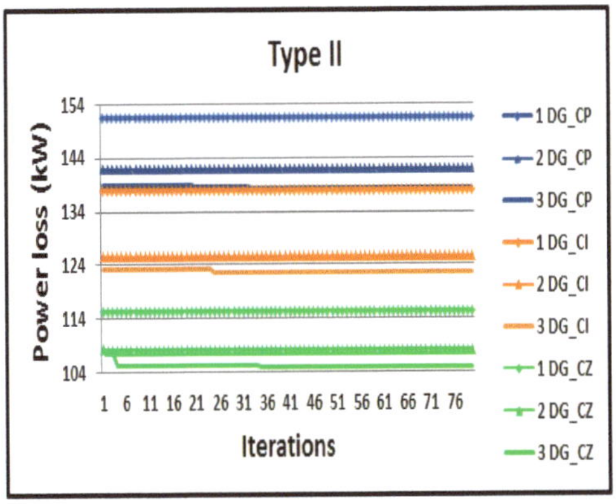

Figure 6. Pl and Iter of DG type II for the three loads.

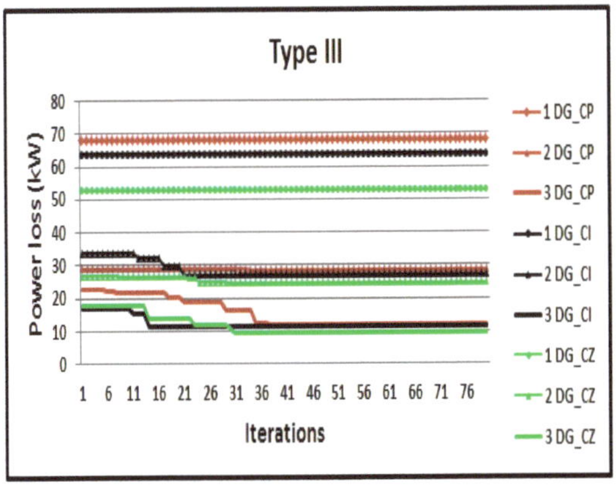

Figure 7. Pl and Iter of DG type III for the three loads.

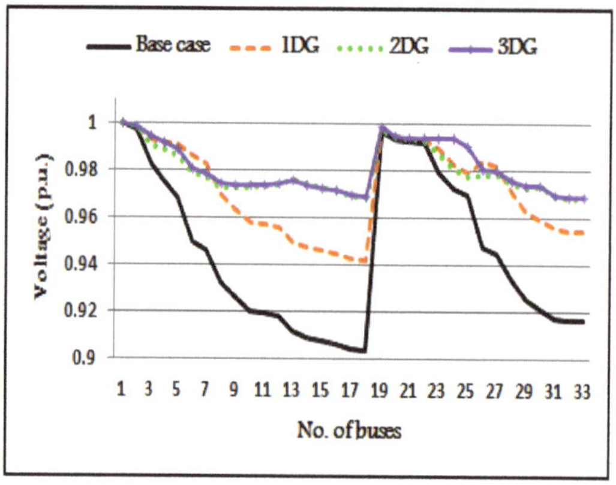

Figure 8. Voltage profile of DG type I for the CP load.

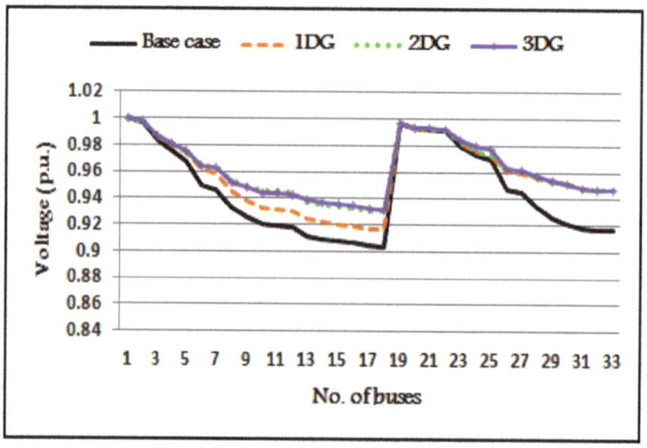

Figure 9. Voltage profile of DG type II for the CP load.

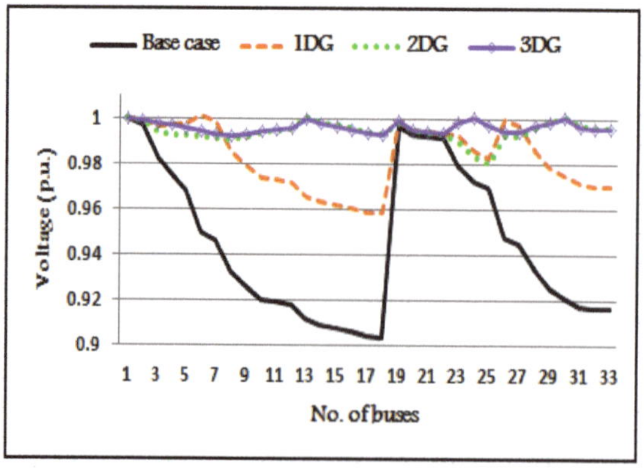

Figure 10. Voltage profile of DG type III for the CP load.

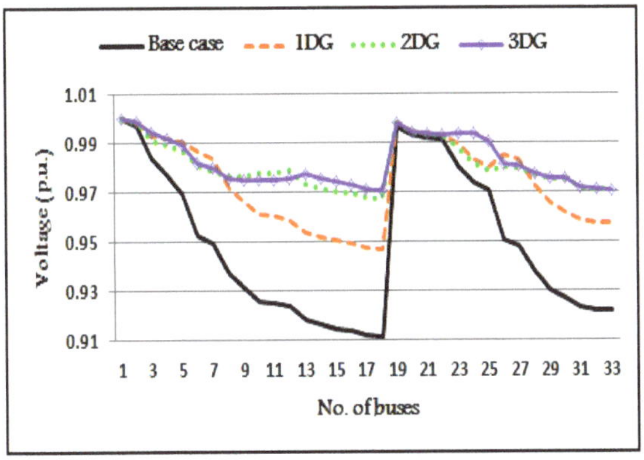

Figure 11. Voltage profile of DG type I for the CI load.

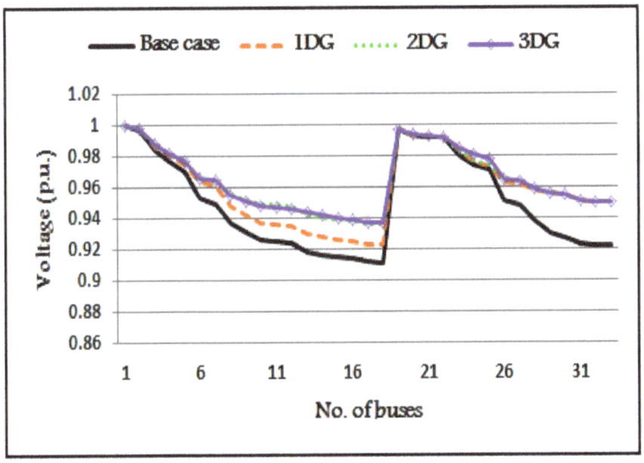

Figure 12. Voltage profile of DG type II for the CI load.

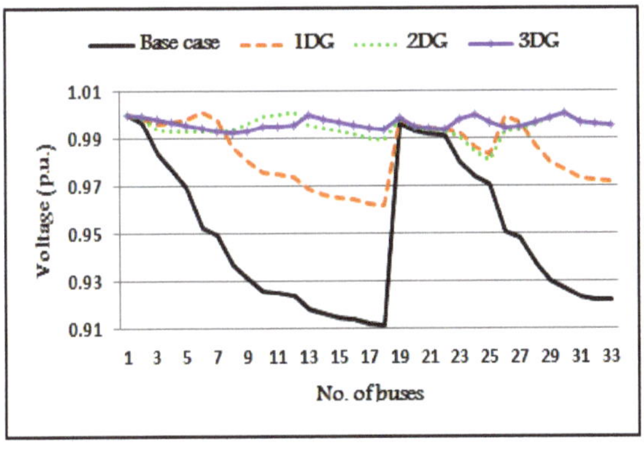

Figure 13. Voltage profile of DG type III for the CI load.

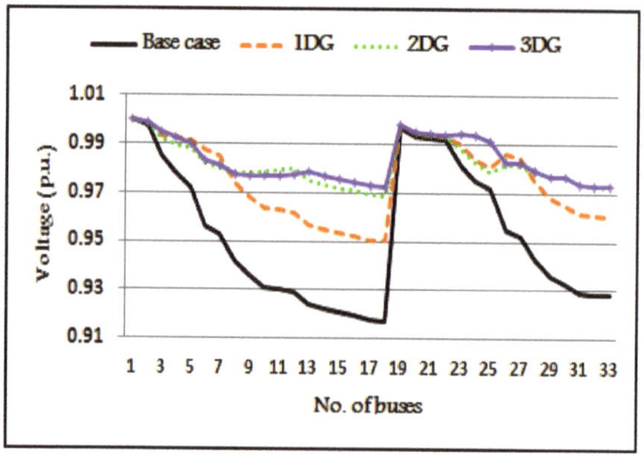

Figure 14. Voltage profile of DG type I for the CZ load.

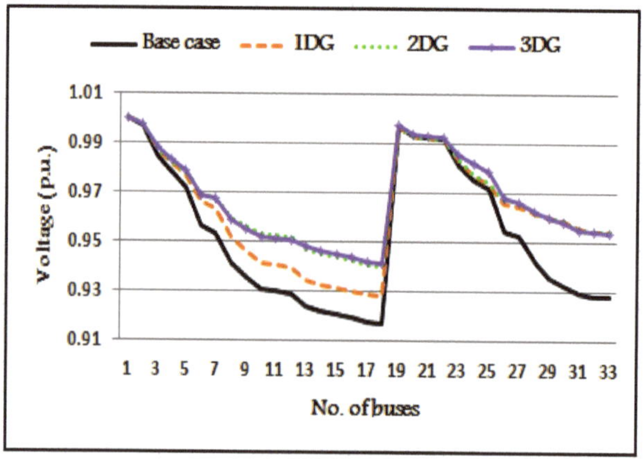

Figure 15. Voltage profile of DG type II for the CZ load.

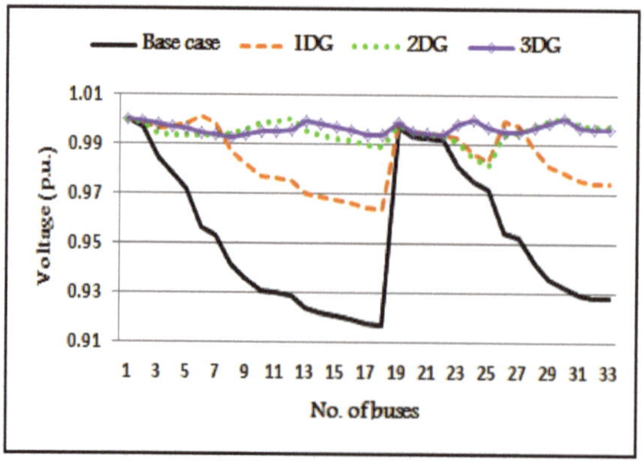

Figure 16. Voltage profile of DG type III for the CZ load.

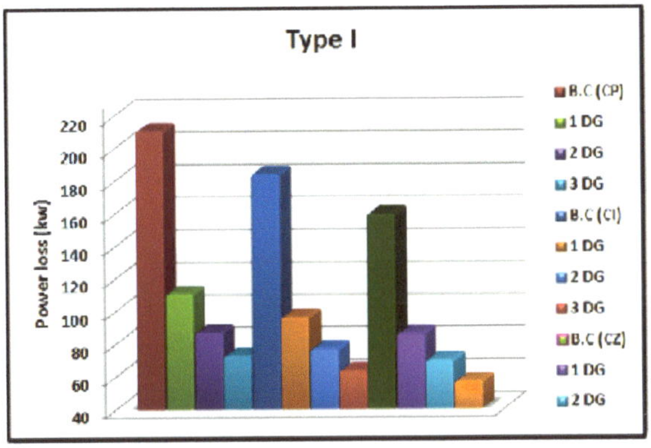

Figure 17. The power loss of DG type I for the three loads.

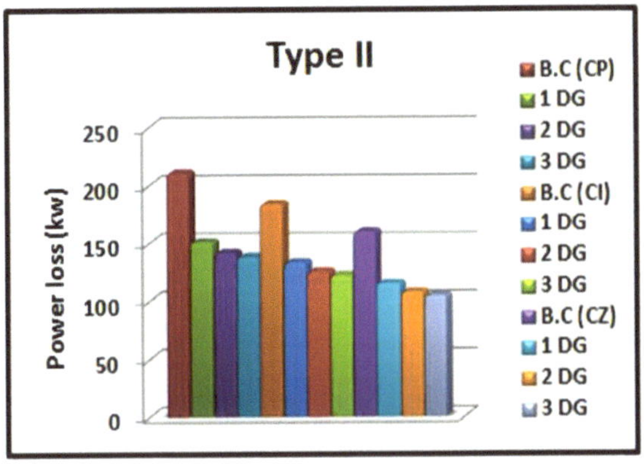

Figure 18. The power loss of DG type II for the three loads.

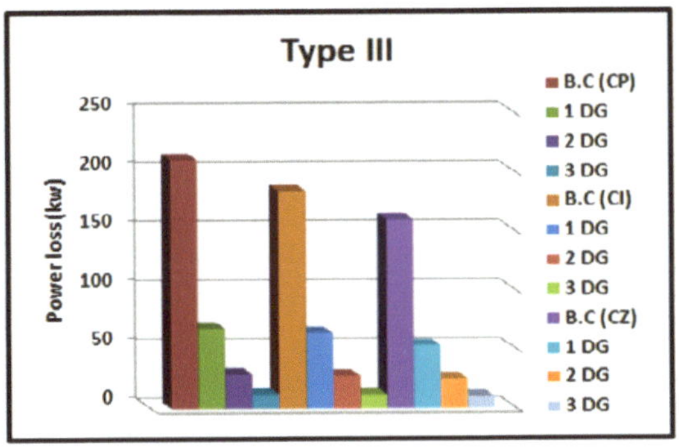

Figure 19. The power loss of DG type III for the three loads.

As shown in the tables, the best results (minimum power losses and maximum voltage profile) are obtained in case of three DGs type III compared with other types of DGs as provide active and reactive power to the system. Furthermore, the results obtained in case of constant impedance is better than the results in other types of loads.

5.1.1. Case1: One DG Integration

For one DG installing, the results are illustrated in Table 1 with the CP, CI, and CZ load types. It is shown from Table 1 that DG type III gives the best simulation results among the other DG types.

Table 1. Optimal results of one DG integration for the three loads on IEEE 33-bus RDS.

Type of Load	**CP Type Load**				**CI Type Load**				**CZ Type Load**			
DG type	B.C	DG type			B.C	DG type			B.C	DG type		
		Type-I	Type-II	Type-III		Type-I	Type-II	Type-III		Type-I	Type-II	Type-III
Location		6	30	30		6	30	6		6	30	6
Size		2.490	1.23	3.028		2.352	1.144	2.844		2.166	1.080	2.648
Total capacity		2.490	1.23	3.028		2.352	1.144	2.844		2.166	1.080	2.648
Power loss (P.L)	210.997	111.17	151.41	67.95	184.3557	96.6	133.6	63.7	159.78	86.2	115.4	52.7
V. min (p.u.)	0.9038	0.9410	0.9162	0.9570	0.9113	0.9460	0.9226	0.9664	0.9173	0.9491	0.9280	0.9632
Min voltage bus	(18)	(18)	(18)	(18)	(18)	(18)	(18)	(18)	(18)	(18)	(18)	(18)
P.L Reduction%		47.31	28.24	67.795		47.07	26.79	65.10		46.53	28.41	67.31

For the CP load: Bus 6 is the optimal bus for DG type I with size 2.490 MW, bus 30 for DG type II, and DG type III with sizes 1.23 MVAR and 3.028 MVA, respectively; the active power loss decreased from 210.997 kW to111.17 kW for DG type I, 151.41 kW for DG type II, and 67.95 kW for DG type III.

For the CI load: Bus 6 is specified as the optimal bus for DG type I with size 2.352 MW, bus 30 for DG type II with size 1.144 MVAR, and bus 6 for DG type III with size 2.844 MVA. The active power loss decreased from 184.355 kW to 96.6 kW at DG type I, 133.6 kW for DG type II, and 63.7 kW for DG type III.

For the CZ load: Bus 6 is specified as the optimal bus for DG type I with size 2.166 MW, bus 30 for DG type II with size 1.080 MVAR, and bus 6 for DG type III with size 2.648MVA. The power loss decreased from 159.78 kW to 86.2 kW for DG type I, 115.4 kW for DG type II, and 52.7 kW for DG type III.

5.1.2. Case2: Two DGs Integration

For two DG installation, Table 2 illustrates the results of optimal allocation of three types of DGs for three load types.

For the CP load: the power loss reduced from 210.997 kW to 87.2876 kW for DG type I, 141.935 kW for DG type II, and 28.56 kW for DG type III.

For the CI load: Bus 12 and 30 were chosen as the best locations for three DG types of DG, and the power loss reduced from 184.3557 kW to 77 kW, 125.6 kW, and 27 kW for DG type I, DG type II, and DG type III, respectively.

For the CZ load: the power loss reduced from 159.78kW to 69.9kW, 108kW, and 24.6kW for DG type I, DG type II, and DG type III, respectively, when the DGs were installed at bus 12 and 30 for the three types of DGs.

Table 2. Optimal results of two DG integration for three types of loads on the IEEE 33-bus RDS.

Type of Load	CP Type Load				CI Type Load				CZ Type Load			
DG type	B.C	DG type			B.C	DG type			B.C	DG type		
		Type-I	Type-II	Type-III		Type-I	Type-II	Type-III		Type-I	Type-II	Type-III
Location		13 30	12 30	13 30		12 30	12 30	12 30		12 30	12 30	12 30
Size		0.832 1.110	0.430 1.044	0.920 1.529		0.869 1.014	0.400 0.971	0.962 1.396		0.806 0.929	0.391 0.911	0.9 1.29
Total capacity		1.942	1.47	2.449		1.883	1.372	2.358		1.734	1.303	2.19
Power loss (P.L)	210.997	87.2876	141.935	28.56	184.356	77	125.6	27	159.78	69.9	108	24.6
V. min (p.u.)	0.9038	0.9667	0.9290	0.9801	0.9113	0.9659	0.9344	0.9836	0.9173	0.9675	0.9394	0.9814
Min voltage bus	(18)	(33)	(18)	(25)	(18)	(18)	(18)	(25)	(18)	(18)	(18)	(25)
P.L reduction%		58.63	32.73	86.46		57.81	31.2	85.21		56.64	33.00	84.74
PF				0.91 0.72		-		0.91 0.72				0.90 0.71

5.1.3. Case 3: Three DGs Integration

For the installation of three DGs, the optimal allocation of the three types of DGs are 13, 24, and 30 for the CP, CI, and CZ load types, as obtained in Table 3; the power loss reduction is at its maximum in the case of the type III DGs in all three types of load among the other types, as they inject active and reactive power.

Table 3. Optimal results of the three DGs' integration for three types of loads on an IEEE 33 bus.

Type of Load	CP Type Load				CI Type Load				CZ Type Load			
DG type	B.C	DG type			B.C	DG type			B.C	DG type		
		Type-I	Type-II	Type-III		Type-I	Type-II	Type-III		Type-I	Type-II	Type-III
Location		13 24 30	13 24 30	13 24 30		13 24 30	13 24 30	13 24 30		13 24 30	13 24 30	13 24 30
Size		0.79 1.07 1.012	0.359 0.52 1.016	0.869 1.189 1.425		0.725 0.958 1.046	0.334 0.490 0.944	0.802 1.175 1.337		0.672 1.015 0.871	0.324 0.490 0.884	0.750 1.128 1.234
Total capacity		2.87	1.890	3.483		2.730	1.769	3.297		2.558	1.699	3.112
Power loss (P.L)	210.997	72.89	138.37	11.77	184.3557	63.6	122.4	11.3	159.78	57	104.9	9.5
V. min (p.u.)	0.9038	0.9670	0.9303	0.9905	0.9113	0.9695	0.9356	0.9918	0.9173	0.9713	0.9405	0.9925
Min voltage bus	(18)	(33)	(18)	(8)	(18)	(33)	(18)	(8)	(18)	(18)	(18)	(8)
P.L reduction%	- -	65.45	34.42	94.42		65.15	32.93	93.81		64.64	34.92	94.11
PF				0.91 0.90 0.71		-		0.91 0.91 0.71				0.90 0.90 0.70

Furthermore, the power losses for the three previous tables with the number of iterations are shown in Figures 5–7.

For the CP load: the power loss reached 72.89 kW in DG type I, 138.37 kW in DG type II, and 11.77 kW in DG type III from B.C 210.997 kW.

In order to further emphasize the strength of the proposed method, Table 4 illustrates a comparison between the proposed method results and other techniques for the CP load case; the maximum benefit can be achieved at this case as the number of the DGs which inject active and reactive power are increased, so that the minimum power losses are achieved and the voltages of all of the buses are improved when the three DGs are placed at the optimal locations, with optimal power factors of 0.91, 0.90, and 0.71, respectively.

Table 4. Optimal results of the three DGs for type III compared with other techniques for CP loads on an IEEE 33 bus.

Case	Technique		Location Size (MVA)			Total Capacity (MVA)	Power Loss	P.f	V min (p.u.)	Loss Reduction %
3DG	Proposed	Bus Size	13 0.869	24 1.189	30 1.425	3.483	11.77	0.91 0.90 0.71	0.9905 (8)	94.42
	GA [26]	Bus Size	14 0.8153	24 1.102	30 1.49	3.407	11.91	0.90 0.89 0.72	NA	94.35
	IA [27]	Bus Size	6 1.098	30 1.098	14 0.768	2.964	22.29	0.82 0.82 0.82	0.99217 (8)	89.45
	PABC [28]	Bus Size	12 1.014	25 0.960	30 1.363	2.889	15.91	0.85 0.85 0.85	NA	92.46

The proposed method achieves a power loss reduction of 94.42%, which is better than other techniques, e.g., 89.45% and 89.43 in Improved Analytical (IA) [27] and Particle Artificial Bee Colony algorithm (PABC) [28].

For the CI load: the B.C power losses are 184.3557 kW; the power losses obtained in the case of inserting three DGs of DG type I is 63.6 kW, DG type II is 122.4kW, and DG type III is 11.3kW.

For the CZ load: the power loss reached 57 kW in DG type I, 104.9 kW in DG type II, and 9.5 kW in DG type III from B.C 159.78 kW.

5.2. IEEE 69-Bus Test System

This system consists of 69 buses and 68 branches, as shown in Figure 6; the total active and reactive power of the load are 3801.49 kW and 2694.6 KVAR, respectively; the losses at base case (B.C) are 224.997 kW, and the minimum voltage is 0.9091 p.u. at bus 65. The convergence characteristics of one, two, and three DGs of DG type I, DG type II, and DG type III for the three loads are presented in Figures 20–22, respectively. Figures 23–31 describe the voltage profiles of 1DG, 2DG, and 3DG of the three DG types for the three load types. Furthermore, the power loss comparisons between the three types of load are shown in Figures 32–34.

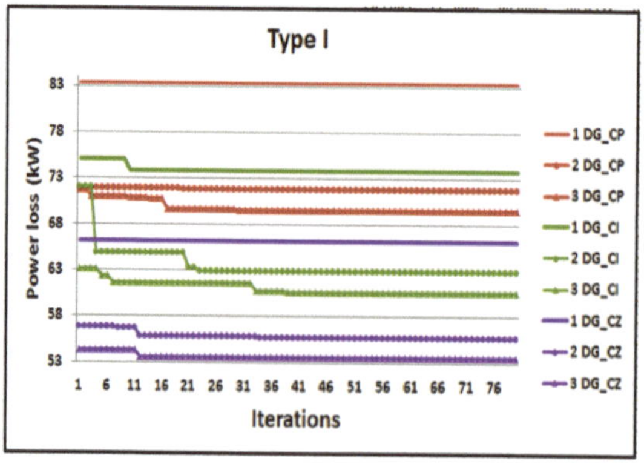

Figure 20. Pl and Iter of DG type I for the three loads.

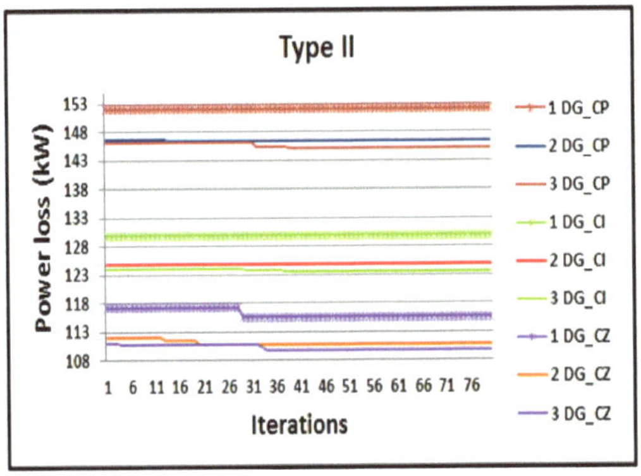

Figure 21. Pl and Iter of DG type II for the three loads.

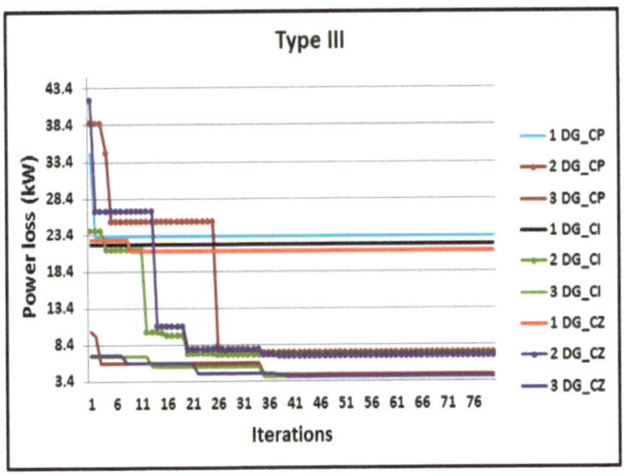

Figure 22. Pl and Iter of DG type III for the three loads.

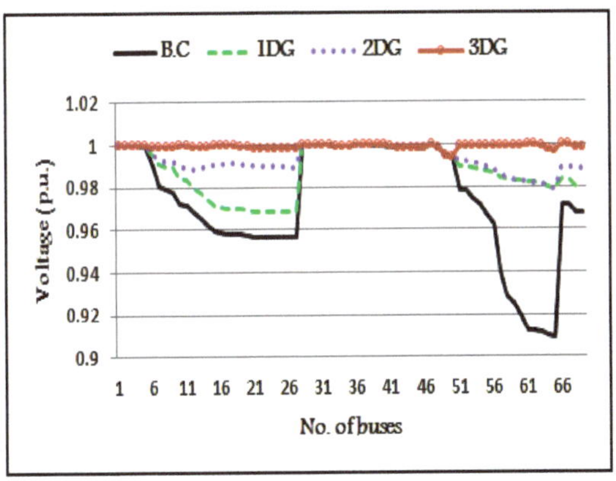

Figure 23. Voltage profile of DG type I for the CP load.

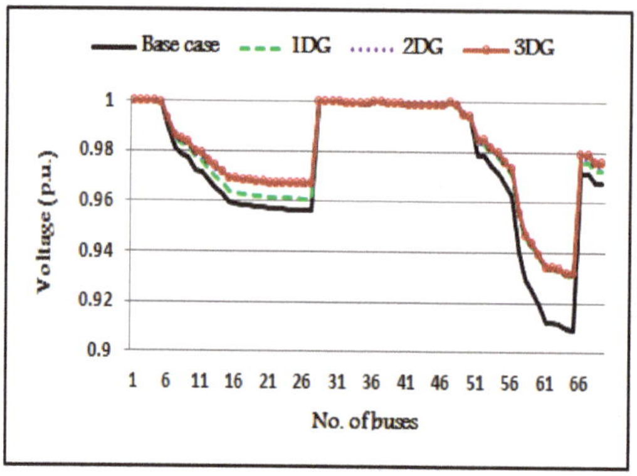

Figure 24. Voltage profile of DG type II for the CP load.

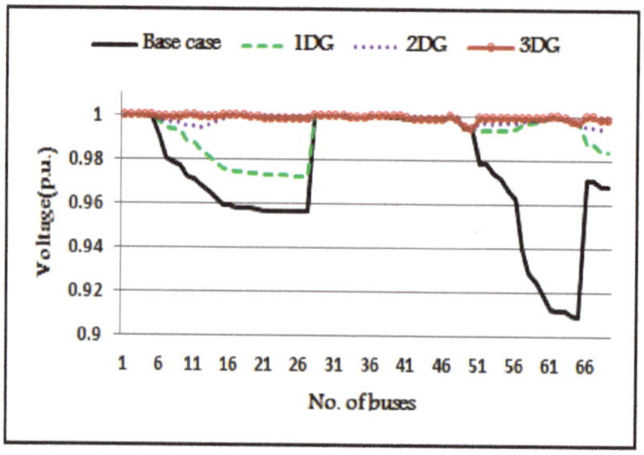

Figure 25. Voltage profile of DG type III for the CP load.

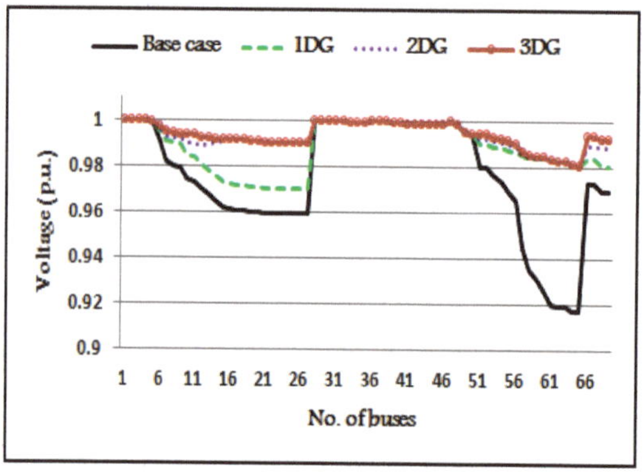

Figure 26. Voltage profile of DG type I for the CI load.

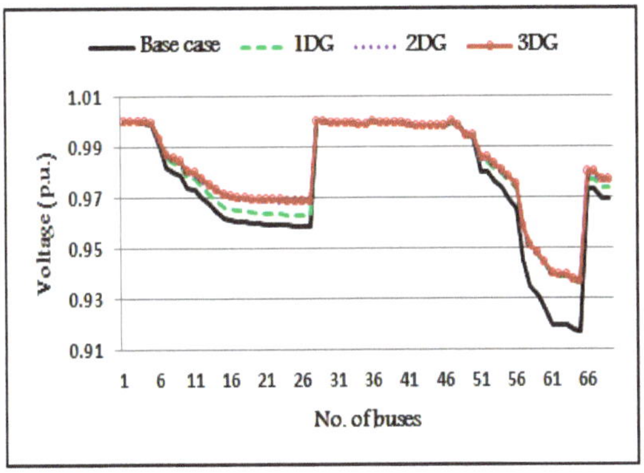

Figure 27. Voltage profile of DG type II for the CI load.

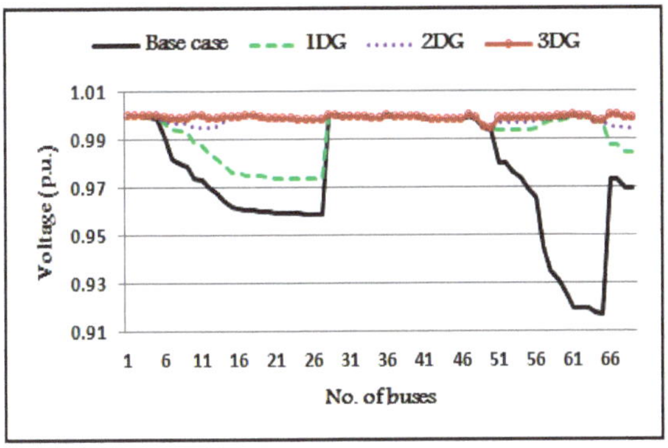

Figure 28. Voltage profile of DG type III for the CI load.

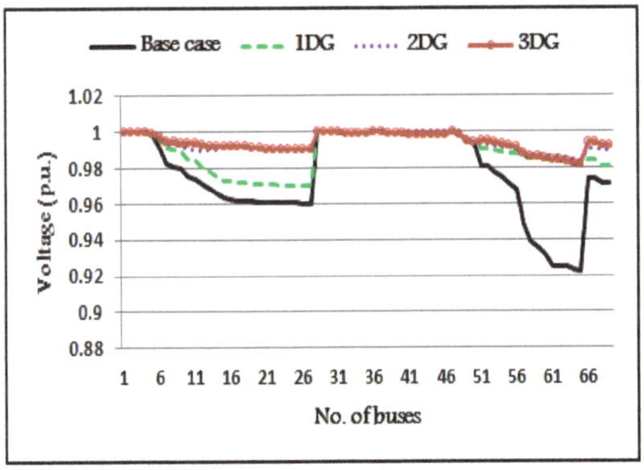

Figure 29. Voltage profile of DG type I for the CZ load.

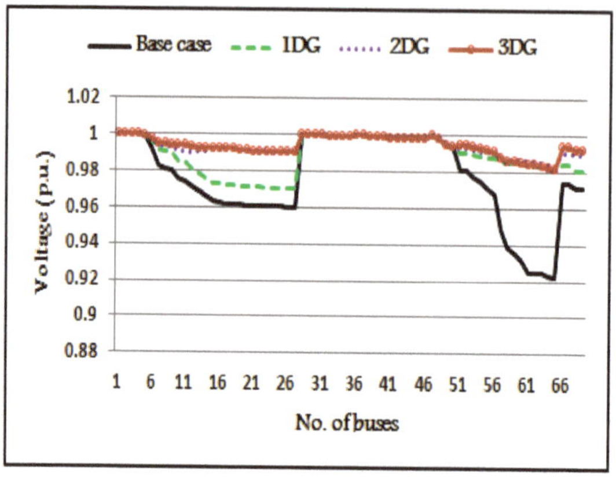

Figure 30. Voltage profile of DG type II for the CZ load.

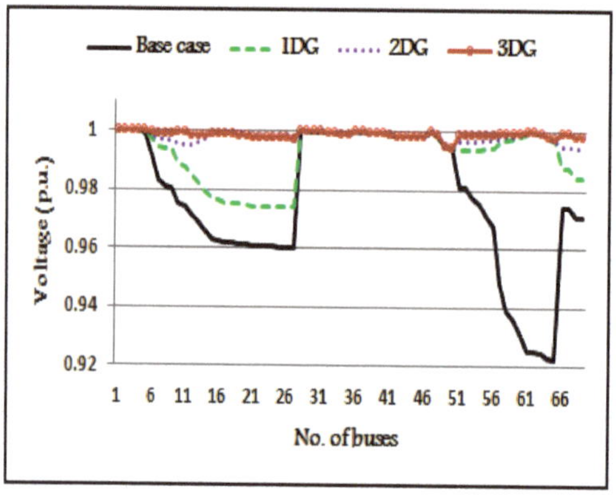

Figure 31. Voltage profile of DG type III for the CZ load.

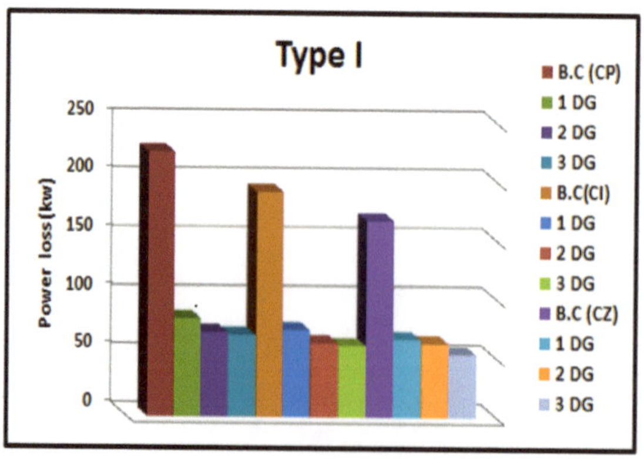

Figure 32. The power loss of DG type I for the three loads.

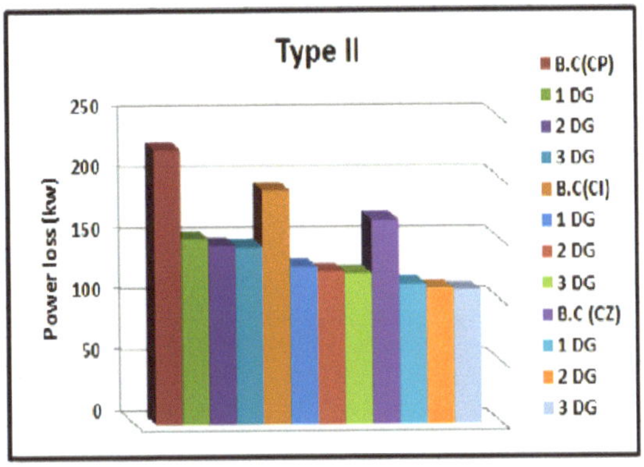

Figure 33. The power loss of DG type II for the three loads.

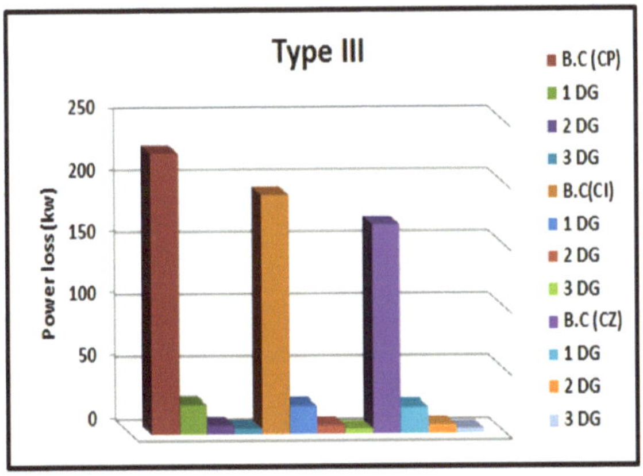

Figure 34. The power loss of DG type III for the three loads.

Similar to the IEEE 33 bus, the minimum power losses in the case of three DGs for type III were constant impedance (CZ) and then constant current (CI), followed by constant power (CP), although most locations of the DGs were the same for the three type of loads. The power factor was improved for three cases.

5.2.1. Case1: One DG Integration

For one DG installation, the simulation results are illustrated in Table 5.

For the CP load: Bus 61 is the optimal bus for three types of DG; the power loss decreased from 224.997 kW to 83.37 kW for DG type I, 152.09 kW for DG type II, and 23.15 kW for DG type III.

For the CI load: Bus 61 is the optimal bus for the three types of DGs; the power loss decreased from 191.5 kW to 73.87 kW for DG type I, 129 kW for DG type II, and 22 kW for DG type III.

For the CZ load: the power losses reduced from 167.2kW at B.C to 66.2kW for DG type I, to 115 kW for DG type II, and to 21 kW for DG type III when the DG was installed at bus 6.

Table 5. Optimal results of one DG integration for three types of loads on an IEEE 69 bus RDS.

Type of Load	CP Type Load				CI Type Load				CZ Type Load			
Case	B.C	DG type			B.C	DG type			B.C	DG type		
DG type		Type-I	Type-II	Type-III		Type-I	Type-II	Type-III		Type-I	Type-II	Type-III
Location		61	61	61		61	61	61		61	61	61
Size		18.1	1.291	2.22		1.655	1.202	2.052		1.594	1.122	1.924
Total capacity		18.1	1.291	2.22		1.655	1.202	2.052		1.594	1.122	1.924
Power loss (PL)	224.997	83.37	152.09	23.15	191.5	73.87	129	22	167.2	66.2	115	21
V. min (p.u.)	0.9091	0.9679	0.9302	0.9734	0.9167	0.9690	0.9361	0.9740	0.9226	0.9700	0.9402	0.9748
Min voltage bus	(65)	(27)	(65)	(27)	(65)	(27)	(65)	(27)	(65)	(27)	(65)	(27)
P.L reduction%		62.95	32.40	89.71		61.43	32.63	88.5		60.40	31.22	87.44
PF				0.81				0.81				0.81

5.2.2. Case2: Two DGs Integration

For two DG installation, the simulation results are illustrated in Table 6.

Table 6. Optimal results of two DG integration for three types of loads on an IEEE 69 bus RDS.

	CP Type Load				CI Type Load				CZ Type Load			
Type of load	B.C	DG type			B.C	DG type			B.C	DG type		
DG type		Type-I	Type-II	Type-III		Type-I	Type-II	Type-III		Type-I	Type-II	Type-III
Location		61 17	61 17	61 17		61 17	61 17	61 17		61 17	61 17	61 17
Size		1.724 0.518	1.236 0.35	2.127 0.626		1.575 0.497	1.147 0.336	1.979 0.606		1.583 0.507	1.069 0.326	1.834 0.559
Total capacity		2.24	1.586	2.752		2.072	1.483	2.585		2.090	1.395	2.393
Power loss (PL)	224.997	71.804	146.51	7.1857	191.5	62.87	124.84	6.87	167.2	62.82	110.9	6.8
V. min (p.u.)	0.9091	0.9769	0.9305	0.9946	0.9167	0.9783	0.9364	0.9946	0.9226	0.9801	0.9405	0.9940
Min voltage bus	(65)	(65)	(65)	(69)	(65)	(65)	(65)	(69)	(65)	(65)	(65)	(69)
PL reduction%		68.09	34.88	96.8		67.17	30.80	96.41		66.69	33.67	95.93
PF				0.81 0.83				0.81 0.83				0.81 0.81

For the CP load: the proposed hybrid method chose buses 61and 17 as the optimal locations for the three types of load, and the power loss reached 71.804 kW for DG type I. 146.51 kW for DG type II, and 7.1857 kW for DG type III from the base case 224.997 kW.

For the CI load: in the constant current load, the power loss reduced from 191.5 kW to 62.87 kW, 124.84 kW, and 6.87 kW for DG type I, DG type II, and DG type III, respectively.

For the CZ load: the power loss was reduced from 167.2kW at B.C to 62.82 kW for DG type I, to 110.9kW for DG type II, and to 6.8 kW for DG type III when the DGs were installed at bus 61 and 17, which are the best locations for the three types of the DG.

5.2.3. Case 3: Three DG Integration

For three DGs installation, the simulation results are illustrated in Table 7.

Table 7. Optimal results of the three DG integration for the three loads on an IEEE 69-bus RDS.

Type of Load	CP Type Load				CI Type Load				CZ Type Load			
DG type	B.C	DG type			B.C	DG type			B.C	DG type		
		Type-I	Type-II	Type-III		Type-I	Type-II	Type-III		Type-I	Type-II	Type-III
Location		11 18 61	11 21 61	11 18 61		11 18 61	11 18 61	11 18 61		11 18 61	11 18 61	11 18 61
Size		0.499 0.377 1.668	0.368 0.231 1.196	0.616 0.452 2.050		0.491 0.359 1.520	0.336 0.241 1.109	0.604 0.438 1.884		0.493 0.358 1.427	0.336 0.231 1.031	0.602 0.425 1.758
Total capacity		2.545	1.795	3.119		2.370	1.686	2.926		2.278	1.597	2.785
Power loss (P.L)	224.997	69.5456	145.21	4.25	191.5	60.70	123.67	4	167.2	53.6	109.7	3.9
V. min (p.u.)	0.9091	0.9770	0.9307	0.9972	0.9167	0.9785	0.9367	0.9974	0.9226	0.9802	0.9408	0.9975
Min voltage bus	(65)	(65)	(65)	(65)	(65)	(65)	(65)	(65)	(65)	(65)	(65)	(50)
P.L reduction%	- -	69.09	35.46	98.11		68.3	35.42	97.91		67.94	34.39	97.67
PF				0.82 0.81 0.81		-		0.82 0.84 0.81				0.82 0.84 0.81

For CP load: the proposed hybrid method specified buses 61, 18, and 11 as the optimal locations for type I and type III, while the buses 11, 21, and 61 are the best for DG type II, and the power losses reached 69.5456 kW for DG type I, 145.21 kW for DG type II, and 4.25 kW for DG type III from B.C 224.997 kW. It was observed that DG type III obtained less power loss compared with the others.

In a 69-bus system, the proposed method achieved the best results compared with the other techniques in the case of three DG integration for type III, as shown in Table 8, when the three DGs are placed at 11, 21, and 61; with an optimal power factor of 0.82, 0.81, and 0.81, respectively, the power loss reduction for the proposed method is 98.11%, which is better than 98.10 in the Hybrid PSO-Analytical [26] and 96.84 in the combined power loss sensitivity (CPLS).

Table 8. Optimal results of three DGs for type III compared with the other techniques for CP loads on an IEEE 69 bus.

Case	Technique		(Location) Size (MVA)			Total Capacity (MVA)	Power Loss	P.f	V min	Loss Reduction%
1 DG	Proposed	Bus Size		61 2.22		2.22	23.15	0.81	0.9734	89.71
	GA [29]	Bus Size		61 2.16		2.16	38.45	NA	NA	82.91
3 DG	Proposed	Bus Size	11 0.616	18 0.452	61 2.050	3.119	4.25	0.82 0.81 0.81	0.9972 (65)	98.11
	Hybrid [30]	Bus Size	18 0.48	61 2.06	66 0.53	3.07	4.30	0.77 0.83 0.82	NA	98.1
	CPLS [31]	Bus Size	21 0.723	61 2.20	64 0.438	3.356	7.1	0.81 0.81 0.81	0.9934 (69)	96.84
	EA [32]	Bus Size	11 0.668	18 0.458	61 2.113	3.239	4.48	0.82 0.83 0.82	NA	NA
	EA-OPF [32]	Bus Size	11 0.611	18 0.456	61 2.067	3.134	4.27	0.81 0.83 0.81	NA	NA

For CI load: the power loss was reduced from 191.5 kW to 60.70 kW, 123.67 kW, and 4 kW for DG type I, DG type II, and DG type III, respectively.

For CZ load: the best voltage profile and minimum power loss were obtained in this type of load when the three types of DGs were placed at 11, 18, and 61 for the three types of loads.

6. Conclusions

This paper presented an efficient hybrid analytical and metaheuristic technique to find the optimal locations and sizes of different and multiple types of DGs in order to minimize active power losses and improve the overall voltage profile. The hybrid technique was applied to three types of static loads in order o check the robustness of the proposed method. A metaheuristic SSA technique was proposed in order to ascertain the optimal allocation of the DG units in the three types of loads—CP, CI, and CZ—while the analytical technique determined the proper sizes. The proposed algorithm was tested on 33-bus and 69-bus systems, and the results illustrated in the tables and figures proved that the power loss reduction reaches the maximum value in the case of the incorporation of three DGs of type III for the three types of loads.

Author Contributions: Conceptualization, A.A.M. and S.K.; data curation, A.S. and A.A.M.; formal analysis, S.K.; resources, A.A.M., T.K. and S.-B.R.; methodology, A.S., T.K. and S.-B.R.; software, A.A.M., S.K. and A.S.; supervision, S.K., T.K. and S.-B.R.; validation, A.S., T.K. and S.-B.R.; visualization, A.A.M., T.K. and S.-B.R.; writing—original draft, A.A.M. and A.S.; writing—review and editing, S.K., T.K. and S.-B.R. All authors together organized and refined the manuscript in the present form. All authors have approved the final version of the submitted paper. All authors have read and agreed to the published version of the manuscript.

Funding: This work was supported by "Development of Modular Green Substation and Operation Technology" of the Korea Electric Power Corporation (KEPCO).

Institutional Review Board Statement: Not applicable.

Informed Consent Statement: Not applicable.

Data Availability Statement: Not applicable.

Conflicts of Interest: The authors declare no conflict of interest.

References

1. Reddy, P.D.P.; Reddy, V.C.V.; Manohar, T.G. Whale optimization algorithm for optimal sizing of renewable resources for loss reduction in distribution systems. *Renew. Wind Water Sol.* **2017**, *4*, 1–13. [CrossRef]
2. Kazerani, M.; Tehrani, K. Grid of Hybrid AC/DC Microgrids: A New Paradigm for Smart City of Tomorrow. In Proceedings of the 2020 IEEE 15th International Conference of System of Systems Engineering (SoSE), Budapest, Hungary, 2–4 June 2020; pp. 175–180.
3. Kemikem, D.; Boudour, M.; Benabid, R.; Tehrani, K. Quantitative and Qualitative Reliability Assessment of Reparable Electrical Power Supply Systems using Fault Tree Method and Importance Factors. In Proceedings of the 2018 13th Annual Conference on System of Systems Engineering (SoSE), Paris, France, 19–22 June 2018; pp. 452–458.
4. Singh, R.K.; Goswami, S.K. Optimum allocation of distributed generations based on nodal pricing for profit, loss reduction and voltage improvement including voltage rise issue. *Electr. Power Energy Syst.* **2010**, *32*, 637–644. [CrossRef]
5. Caponetto, R.; Fortuna, L.; Fazzino, S.; Xibilia, M.G. Chaotic sequences to improve the performance evolutionary algorithms. *IEEE Trans. Evol. Comput.* **2003**, *7*, 289–304. [CrossRef]
6. Ali, E.S.; Abd Elazim, S.M.; Abdelaziz, A.Y. Optimal allocation and sizing of renewable distributed generation using ant lion optimization algorithm. *Electr. Eng.* **2018**, *100*, 99–109. [CrossRef]
7. El-Zonkoly, A. Optimal placement of multi-distributed generation units including different load models using particle swarm optimisation. *IET Gener. Transm. Distrib.* **2011**, *5*, 760–771. [CrossRef]
8. Moradi, M.; Abedini, M. A combination of genetic algorithm and particle swarm optimization for optimal DG location and sizing in distribution systems. *Int. J. Electr. Power Energy Syst.* **2012**, *34*, 66–74. [CrossRef]
9. Abu-Mouti, F.S.; El-Hawary, M.E. Optimal distributed generation allocation and sizing in distribution systems via artificial bee colony algorithm. *IEEE Trans. Power Deliv.* **2011**, *26*, 2090–2101. [CrossRef]
10. García, J.A.M.; Mena, A.J.G. Optimal distributed generation location and size using a modified teaching–learning based optimization algorithm. *Int. J. Electr. Power Energy Syst.* **2013**, *50*, 65–75. [CrossRef]
11. Nekooei, K.; Farsangi, M.M.; Nezamabadi-Pour, H.; Lee, K.Y. An improved multi-objective harmony search for optimal placement of DGs in distribution systems. *IEEE Trans. Smart Grid* **2013**, *4*, 557–567. [CrossRef]
12. Kaur, S.; Kumbhar, G.; Sharma, J. A MINLP technique for optimal placement of multiple DG units in distribution systems. *Int. J. Electr. Power Energy Syst.* **2014**, *63*, 609–617. [CrossRef]

13. Hosseini, S.A.; Madahi, S.S.K.; Razavi, F.; Karami, M.; Ghadimi, A.A. Optimal sizing and siting distributed generation resources using a multiobjective algorithm. *Turk. J. Electr. Eng. Comput. Sci.* **2013**, *21*, 825–850.
14. Sultana, U.; Khairuddin, A.B.; Aman, M.; Mokhtar, A.; Zareen, N. A review of optimum DG placement based on minimization of power losses and voltage stability enhancement of distribution system. *Renew. Sustain. Energy Rev.* **2016**, *63*, 363–378. [CrossRef]
15. Singh, A.; Parida, S. Novel sensitivity factors for DG placement based on loss reduction and voltage improvement. *Int. J. Electr. Power Energy Syst.* **2016**, *74*, 453–456. [CrossRef]
16. Prakash, D.B.; Lakshminaraya, C. Optimal siting of capacitors in radial distribution network using whale optimization algorithm. *Alex. Eng. J.* **2016**. [CrossRef]
17. Khodabakhshian, A.; Andishgar, M.H. Simultaneous Placement and Sizing of DGs and Shunt Capacitors in Distribution Systems by Using IMDE algorithm. *Electr. Power Energy Syst.* **2016**, *82*, 599–607. [CrossRef]
18. Haque, M.H. Load flow-solution of distribution systems with voltage dependent load models. *Electr. Power Syst. Res.* **1996**, *36*, 151–156. [CrossRef]
19. Roy, N.K.; Hossain, M.J.; Pota, H.R. Voltage profile improvement for distributed wind generation using D-STATCOM. In Proceedings of the IEEE PES General Meeting, Detroit, MI, USA, 24–28 July 2011.
20. Dahal, S.; Mithulananthan, N.; Saha, T. Investigation of small signal stability of a renewable energy based electricity distribution system. In Proceedings of the IEEE PES General Meeting, Minneapolis, MN, USA, 25–29 July 2010.
21. Mirjalili, S.; Gandomi, A.H.; Mirjalili, S.Z.; Saremi, S.; Faris, H.; Mirjalili, S.M. Salp Swarm Algorithm: A bio-inspired optimizer for engineering design problems. *Adv. Eng. Softw.* **2017**, *114*, 163–191. [CrossRef]
22. Sayed, G.I.; Khoriba, G.; Haggag, M.H. A novel chaotic salp swarm algorithm for global optimization and feature selection. *Appl. Intell.* **2018**, *48*, 3462–3481. [CrossRef]
23. El-Fergany, A.A. Extracting optimal parameters of PEM fuel cells using salp swarm optimizer. *Renew. Energy* **2018**, *119*, 641–648. [CrossRef]
24. Qais, M.H.; Hasanien, H.M.; Alghuwainem, S. Enhanced salp swarm algorithm: Application to variable speed wind generators. *Eng. Appl. Artif. Intell.* **2019**, *80*, 82–96. [CrossRef]
25. El-Fergany, A.A.; Hasanien, H.M. Salp swarm optimizer to solve optimal power flow comprising voltage stability analysis. *Neural Comput. Appl.* **2019**, *32*, 5267–5283. [CrossRef]
26. Moradi, M.H.; Abedinie, M.; Tolabi, H.B. Optimal multi-distributed generation location and capacity by genetic algorithms. In Proceedings of the 2010 Conference Proceedings IPEC, Singapore, 27–29 October 2010; pp. 614–618.
27. Hung, D.Q.; Mithulananthan, N. Multiple distributed generator placement in primary distribution networks for loss reduction. *IEEE Trans. Ind. Electron.* **2011**, *60*, 1700–1708. [CrossRef]
28. Muthukumar, K.; Jayalalitha, S. Optimal placement and sizing of distributed generators and shunt capacitors for power loss minimization in radial distribution networks using hybrid heuristic search optimization technique. *Int. J. Electr. Power Energy Syst.* **2016**, *78*, 299–319. [CrossRef]
29. Shukla, T.; Singh, S.; Srinivasarao, V.; Naik, K. Optimal sizing of distributed generation placed on radial distribution systems. *Electr. Power Compon. Syst.* **2010**, *38*, 260–274. [CrossRef]
30. Kansal, S.; Kumar, V.; Tyagi, B. Hybrid approach for optimal placement of multiple DGs of multiple types in distribution networks. *Int. J. Electr. Power Energy Syst.* **2016**, *75*, 226–235. [CrossRef]
31. Kumar, A.; Babu, P.V.; Murty, V.V.S.N. Distributed generators allocation in radial distribution systems with load growth using loss sensitivity approach. *J. Inst. Eng. (India) Ser. B* **2017**, *98*, 275–287. [CrossRef]
32. Mahmoud, K.; Yorino, N.; Ahmed, A. Optimal distributed generation allocation in distribution systems for loss minimization. *IEEE Trans. Power Syst.* **2015**, *31*, 960–969. [CrossRef]

sustainability

Article

Evaluation of Growth Rate and Biomass Productivity of *Scenedesmus quadricauda* and *Chlorella vulgaris* under Different LED Wavelengths and Photoperiods

Ruth Chinyere Anyanwu [1], Cristina Rodriguez [1,*], Andy Durrant [1] and Abdul Ghani Olabi [2]

1 School of Computing, Engineering and Physical Sciences, University of the West of Scotland, Paisley PA1 2BE, UK; ruth.anyanwu@uws.ac.uk (R.C.A.); andy.durrant@uws.ac.uk (A.D.)
2 Sustainable and Renewable Energy Engineering Department, University of Sharjah, Sharjah P.O. Box 27272, United Arab Emirates; aolabi@sharjah.ac.ae
* Correspondence: cristina.rodriguez@uws.ac.uk

Abstract: Cultivation has been identified as an essential stage for biofuel production. This research has examined two important parameters for the industrial production of microalgae, namely microalgae growth rate and biomass productivity. *Chlorella vulgaris* and *Scenedesmus quadricauda* were cultivated using a closed photobioreactor (PBR). A novel approach for cultivation and energy input reduction was developed by incorporating periods of darkness during cultivation, as would happen in nature. Three different LED light sources (white, red, and green) were used to determine the conditions that result in the highest growth rate and biomass productivity. *C. vulgaris* and *S. quadricauda* responded differently to lighting conditions. It was found that, depending on the LED source and light period, different growth rates and biomass productivities were obtained. Overall, experimental results obtained in this study indicated that a white LED is more effective than green or red LEDs in increasing microalgae growth rate and biomass productivity. A maximum growth rate of 3.41 d^{-1} and a biomass productivity of 2.369 g $L^{-1}d^{-1}$ were achieved for *S. quadricauda* under a 19 h period of white light alternating with 5 h of darkness. For *C. vulgaris* the maximum growth rate of 3.49 d^{-1} and maximum biomass productivity of 2.438 g $L^{-1}d^{-1}$ were achieved by continuous white light with no darkness period.

Keywords: microalgae cultivation; growth rate; biomass productivity; light intensity; biofuels

1. Introduction

Due to the rapid depletion of fossil fuels and the climate change caused by their use, several natural energy sources such as wind [1], solar [2], and biomass [3,4] have been studied, but these are still limited by the natural environment. According to IEA 2019 reports entitled "tracking transport", bioenergy is currently the only multifunction form of renewable energy that can directly produce electricity as well as fuels used in transportation. Biomass energy is derived from plants; it is considered both a clean energy source [5,6] and a sustainable one [7]. It is sustainable because microalgae biomass does not damage or deplete natural resources. It supports long-term environmental balance; hence, it has been proposed as a potential alternative to petroleum and can provide clean and renewable energy [8]. Microalgae are unicellular organisms with complex and robust cell walls. Among their advantages are the oil yields higher than oil seeds and food crops such as jatropha or corn. Microalga culture does not require pesticides. Residual biomass of microalgae can be used as fertiliser or livestock feed [9]. It does not compete with food crops for arable land, and it is possible to use municipal wastewater as a source of carbon and nutrients to promote growth [10]. In addition, according to the IEA 2019 report "tracking transport", microalgae can be grown on land not suitable for agriculture, thus reducing or eliminating the cost of acquiring land. The high availability of microalgae means that fully commercialised technologies could open the door to significant volumes of advanced biofuels for the transport sector, providing diesel substitutes in sectors that are hard to electrify [11]. Microalgae also have the ability to produce intracellular storage compounds: mainly protein (i.e., 50% to 60%), carbohydrates (10% to 20%), and lipids (20% to 30%).

Lipids are oil-producing materials which represent the carbon and energy reserves in a microalga cell. Lipids include two types: neutral lipids that serve as energy reserves and polar lipids that are constituents of organelles and membranes. Microalgal cells accumulate and store neutral lipids in the form of triacylglycerols (TAGs) [11]. TAG content is the key factor in making biodiesel from microalgae, since the associated fatty acids react with methanol to produce fatty acid methyl esters (FAME)—a reaction known as transesterification. As well as being a feedstock for biodiesel, algal fatty acids are of value as a food supplement [12]. For example, long-chain polyunsaturated fatty acids (PUFAs) are confirmed to be especially beneficial to the human diet, including the essential omega-3 fatty acids [13].

Substantial amounts of TAGs can be accumulated in certain microalgae species such as *Chlorella* and *Spirulina*, both of which are rich in two of the main types of omega-3 fatty acid—eicosapentaenoic acid (EPA) and docosahexaenoic acid (DHA). As such, these species are cultivated for their high oil content [12,14]. It is easy to increase the amount of lipids in algae by manipulating their temperature, exposure to light, and access to nitrogen. This then ensures the high lipid concentrations that give optimum production of biodiesel [15], bioethanol, pyrolysis oil, refuse-derived fuel (RDF), biogas [16], and other advanced biofuels, all of which are mainly used in transportation [17], electricity, and heat [18]. These contribute to a sustainable quality of life which is well-reported as the basic driver for providing clean, safe, reliable, and secure energy supplies around the globe, as advanced biofuels are a fast-track solution for reducing transport emissions. Microalgae-sourced renewable biofuels such as biodiesel, biogas, bioethanol, and biohydrogen offer solutions for green transport without requiring costly investment in new cars or fuel distribution systems, especially for biodiesel. The combination of CO_2 fixation [19,20], biofuel production, and wastewater treatment, as well as the production of high value end products such as human and bull meal, pharmaceuticals, and cosmetics makes microalgae very promising for use in industrial processes [21].

Techniques for biofuel productions from microalgae are microalgae oil to biodiesel [15], anaerobic digestion to biomethane [22], and photobiological conversion to biohydrogen. Despite the promise of fuel production from microalgae, there are limitations in terms of their cultivation which will prevent them from completely replacing fossil fuels for the time being [23]. For some species such as *Scenedesmus vacuolatus*, the optimum growing conditions were achieved at a light intensity and CO_2 concentration below the natural levels; given artificial systems such as a twin layer porous substrate photobioreactor are a significant advantage [24].

Light intensity is the most important parameter that affects microalgae growth and biomass productivity. The term light intensity is used to describe the strength or amount of light produced by a specific light source. Hence, an important consideration to enhance the feasibility of scaling microalgae production from laboratory to pilot and then to industry scale would be adequate light distribution and control that translates into high growth rates and enhanced biomass productivity. This study addresses the effect of different light wavelengths and photoperiods (light/dark cycles) on the growth rate and biomass productivity of two microalgae species: *C. vulgaris* and *S. quadricauda*. *C. vulgaris* is a unicellular microalga with spherical shape and 1 to 6 μm diameter; it is of interest as it is reported to have short doubling time and high biomass productivity. It can produce high amounts of lipids which are converted to biodiesel. *S. quadricauda* is also a good candidate for biodiesel production as it contains a high percentage of lipids; mainly fatty acids of carbon chain length C14–C18 [25].

2. Materials and Methods

2.1. Microalgae Source and Cultivation Medium

In this study, experiments were performed with microalgae *C. vulgaris* and *S. quadricauda*. These microalgae species were obtained from University of Texas (UT) at Austin and Sciento Manchester, respectively. They were cultivated for 16 days. Biomass was cultivated in a sterilised BG 11 medium with the following composition: sodium nitrate $NaNO_3$ 17.6 mM; dipotassium phosphate K_2HPO_4 0.23 mM; magnesium sulphate $MgSO_4{\cdot}7H_2O$ 0.3 mM; calcium chloride, dihydrate $CaCl_2{\cdot}2H_2O$ 0.24 mM; citric acid·H_2O 0.031 mm; ferric ammonium citrate 0.021 mM; EDTA disodium dihydrate $Na_2EDTA{\cdot}2H_2O$ 0.0027 mM; sodium carbonate, Na_2CO_3 0.19 mM. The cultures were cultivated at pH 7.0 (cultivation took place at pH 7.0 because algae strains thrive at neutral pH) and at 25 °C (ambient temperature was used as it is the normal laboratory/room temperature for experiment).

2.2. Photobioreactor Setup and Operation

The two microalgae samples were cultivated in laboratory scale PBR as shown in Figure 1 with a 2 litres maximum capacity and 137 mm diameter. The PBR is designed to ensure efficient light penetration to produce rapid and robust growth of algae cultures.

This photobioreactor configuration has superior performance, for example, when compared to flasks, traditional light sources, and open ponds with respect to controlling culture environment, preventing contamination, good mixing, and increased light utilisation. Better values for all of these parameters are achievable for closed reactors, which translates to higher algae productivity.

Figure 1. Closed PBR used in this research. (1) Tube head; (2) CO_2 tube; (3) LED lighting remote; (4) PBR outer cover; (5) pump for mixing; (6) handheld photometer.

The LED lighting platform consists of LEDs with different light wavelengths. For this study, red (620 to 700 nm), white (400 to 700 nm), and green (494 to 577 nm) were chosen to optimise the light source for the microalgae. The gas exchange agitator further enhances growth and is likewise customisable and testable. The cultivation was carried out and maintained in different patterns over a 24 h cycle, each shown as the number of hours of light (L) followed by number of hours of darkness (D). The three different photoperiods of 24L:0D, 19L:5D, and 12L:12D, each carried out with white, red, and green LED wavelengths, gave a total of nine patterns.

2.3. Cell Concentration and Productivity

Growth was monitored every 24 h (Figure 2) by checking absorbance at 680 nm using an Evolution 220 Thermo Scientific USA visible density spectrophotometer. Cultures were taken in three replicates and the average value was recorded.

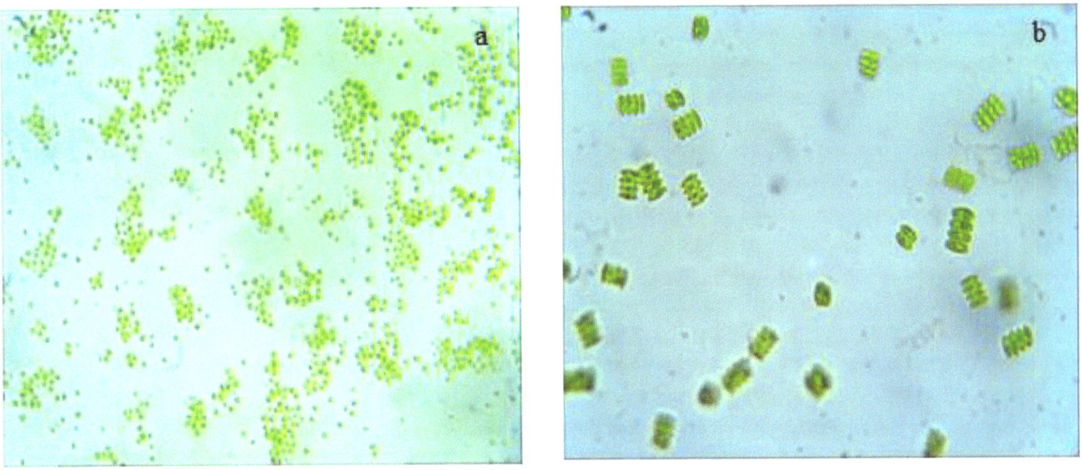

Figure 2. Microalgae species: (**a**) *C. vulgaris*; (**b**) *S. quadricauda*.

Three 10 mL samples were taken every day at the same time (circa 10 a.m.) for 16 days where optical density (OD) readings were taken. Insight 2 software was used to perform the absorbance analysis. The obtained values were used to construct growth curves.

Specific growth rate and biomass productivity were calculated using:

$$\mu = \frac{\ln X - \ln X_0}{t} \tag{1}$$

$$Q_x = \frac{X - X_0}{t} \tag{2}$$

where μ is the specific growth rate (d^{-1}), Q_x is the biomass productivity ($g\ L^{-1}\ d^{-1}$), X_0 is the initial biomass concentration, X is the final biomass concentration, t is time (d).

Cell concentration was measured by haemocytometry. The haemocytometer by Fisher Scientific, U.S. is composed of nine equally sized large squares. The central one is divided into 25 small squares, each one with a volume of 4×10^{-6} mL. *C. vulgaris* and *S. quadricauda* were counted in the small squares since their effective diameters are less than 10 µm. The cell concentration was calculated. The concentrations after 13 days of cultivation for *C. vulgaris* and *S. quadricauda* suspensions were $7.6 \pm 1.8 \times 10^6$ cells mL^{-1} and $15.8 \pm 0.9 \times 10^6$ cells mL^{-1}, respectively.

3. Results and Discussion

3.1. Effects of White, Red, and Green Wavelengths and Duration Results on Growth Rate

The growth rate for cultures was calculated under white, red, and green light exposures. It was observed that *C. vulgaris* achieved the highest growth rate under 24L:0D photoperiod with white light. *C. vulgaris* growth curves remained almost in the linear range for the whole 24L:0D experimentations under white and especially red light. Under various light sources at constant light (24L:0D) the variation trends of the microalgae growth rate values were similar throughout the experiments, as shown in Figure 3.

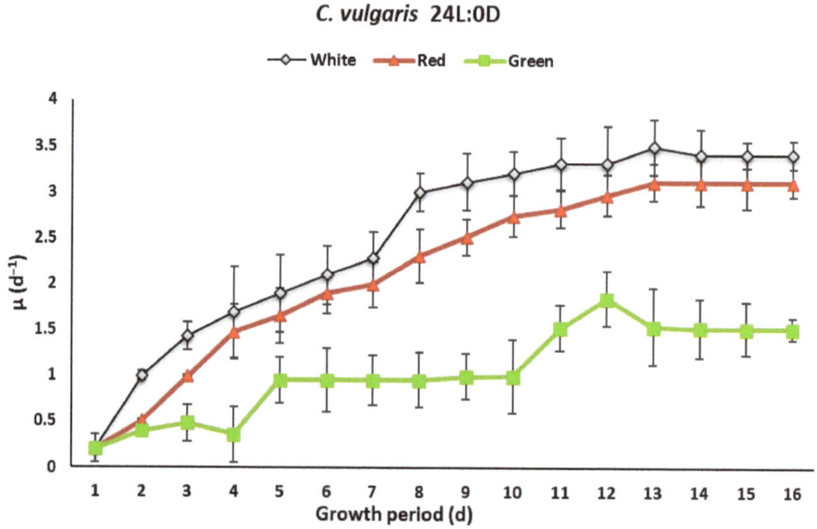

Figure 3. *C. vulgaris* growth curves under different lights at 24L:0D.

The present results are 41% higher than those reported by Maroneze et al. [26] for the highest growth rate of *Tetraselmis* sp. under blue light. Comparative analysis is shown in Table 1. The same pattern was observed by Maroneze et al. [26] when they cultivated *Scenedesmus obliquus* under white light; it was reported that higher growth rate of 2.51 μd^{-1} was achieved and Chen et al. [27] reported 1.26 μd^{-1} for *C. vulgaris* under blue light wavelength.

Table 1. Comparison of specific growth rate with other species.

Algal Species	Specific Growth Rate, μ (day^{-1})	Photoperiod (L:D)	LED and Intensity (μmol Photon m^{-2} s^{-1})	References
C. vulgaris	3.49	24:0	White (200)	Current work
S. quadricauda	2.84	24:0	White (200)	Current work
S. obliquus	2.51	24:0	White (30)	[26]
Tetraselmis sp.	1.47	24:0	Blue (100)	[26]
Nannochloropsis sp.	1.64	24:0	Blue (100)	[26]
C.vulgaris	1.26	12:12	Blue (70)	[26]

Additionally, the green light resulted in lowest growth rate for both species in all photoperiods compared to white and red lights as shown in Figures 4 and 5 respectively for *C. vulgaris* and *S. quadricauda*. In all lighting periods under 24 h *C. vulgaris* had the highest growth rate.

Regarding growth under white light, the average growth rates were respectively 2.390 d^{-1} and 1.409 d^{-1} for *C. vulgaris* and *S. quadricauda* under 24L:0D. For red light it was 2.023 μd^{-1} and 1.148 d^{-1} C. *vulgaris* and *S. quadricauda* under 24L:0D. For green light, the average growth rates were 0.456 d^{-1} and 0.725 d^{-1} photoperiod. Results proved that both species maintained a steady growth rate under white and red light compared to much slower growth under green light.

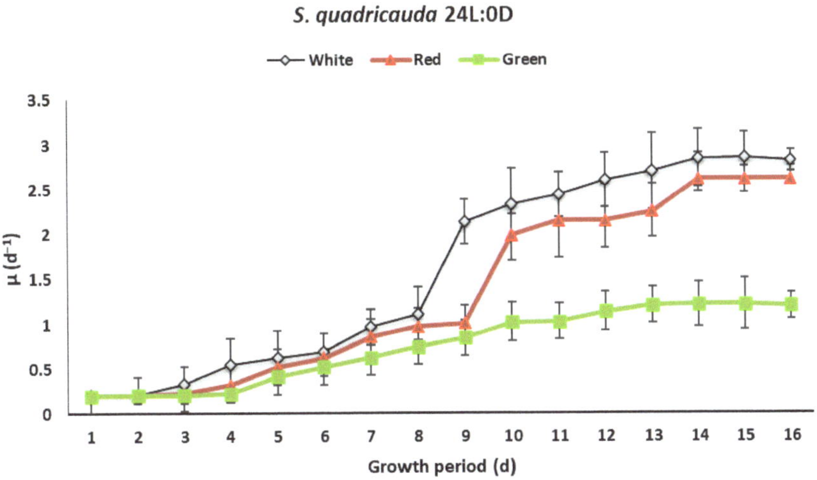

Figure 4. *S. quadricauda* growth curves under 24L:0D.

In addition, it was observed that *S. quadricauda* recorded slightly higher growth under green light when compared to *C. vulgaris* under green light. This was perhaps due to the ability of *S. quadricauda* to grow with stored energy.

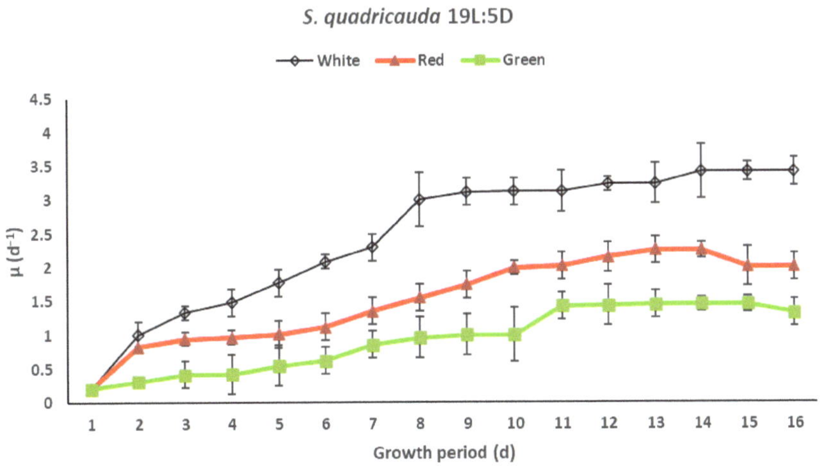

Figure 5. *S. quadricauda* growth curves under 19L:5D.

During 19L:5D, it was however observed that *S. quadricauda* recorded the highest growth rate of 3.41 d^{-1} on day 14 (Figure 5) compared to a growth rate at 24L:0D light period with value of 2.84 d^{-1} on day 14 as shown in Figure 5 above. Both curves illustrate the specific growth rate for *S. quadricauda* under all lights for 24L:0D and 19L:5D light duration (Figure 4). Furthermore, it was observed that growth rate remained constant on days 15 and 16 as represented in the curves in Figure 5.

In 19D:5D, *C. vulgaris'* highest growth was observed on days 14 and 15 (Figure 6). Furthermore, it was observed that growth rate remained constant on days 14, 15, and 16 as represented in the plots in Figure 3. However, it was found that the difference on growth rate between both species under red light wavelength was not statistically significant.

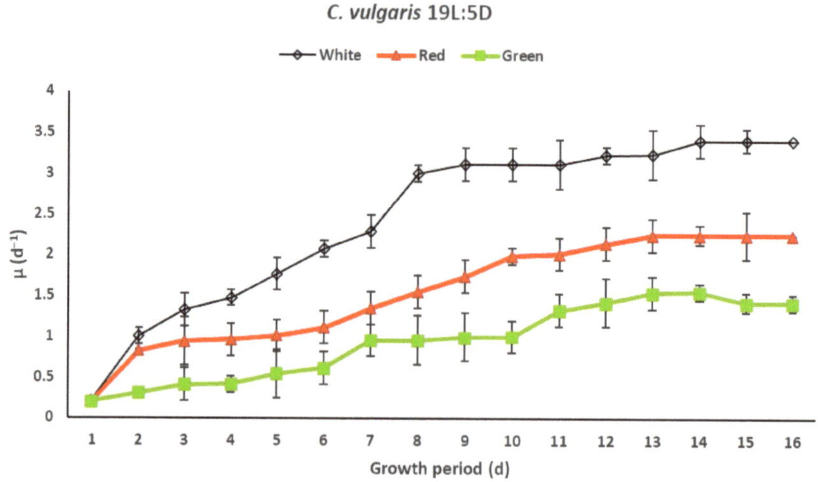

Figure 6. *C. vulgaris* growth curves under 19L:5D.

During a 12L:12D photoperiod, under white light, *C. vulgaris* growth rate increased. Growth rate increased from day 2 to day 11. The highest specific growth rate was observed on day 11 with a value of 1.99 d^{-1} as shown in Figure 7. From day 12 to day 16 growth rate declined as the species reached death phase. *C. vulgaris* cultures subjected to 12L:12D under red light had a growth rate of 1.354 d^{-1} on day 9.

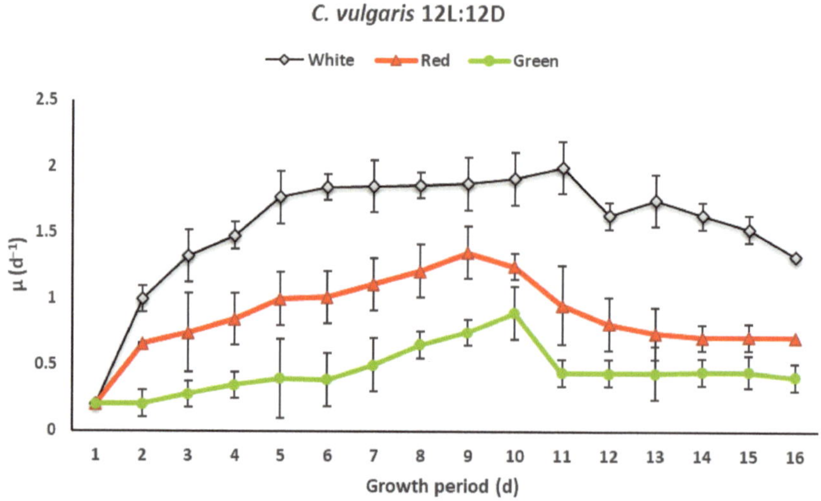

Figure 7. *C. vulgaris* growth curves under 12L:12D.

The growth rates under red light for *S. quadricauda* were 1.305, 1.452, and 0.521 d^{-1}, while growth rates for *C. vulgaris* were recorded as 2.023, 1.451, and 0.900 d^{-1} (Figure 8), under 24:0, 19:5, and 12L:12D photoperiods, respectively. Additionally, it was observed that the average growth rate of *S. quadricauda* under 19L:5D 1.452 d^{-1} was slightly higher compared to 24L:0D 1.305 d^{-1}. *S. quadricauda* also recorded slightly higher growth rate under 19L:5D photoperiod compared to *C. vulgaris*. In the light limiting pattern, cell growth rate decreased with the biomass productivity.

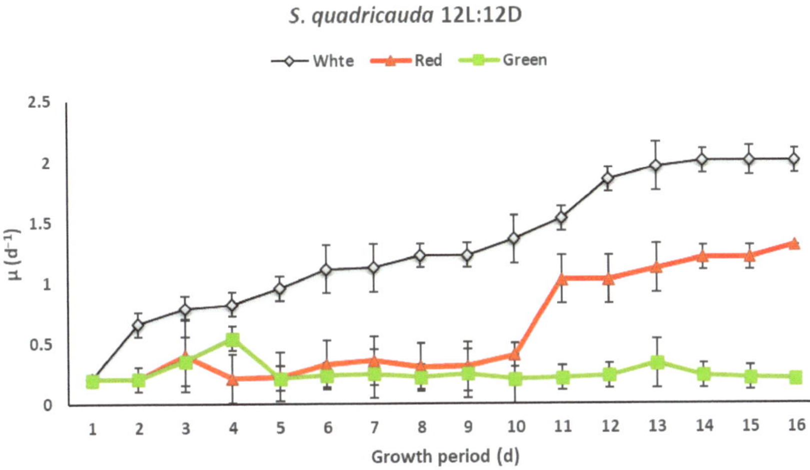

Figure 8. *S. quadricauda* growth curves under 12L:12D.

A different scenario was observed under green light as the average specific growth was the lowest for both species in all light durations compared with white and red lights. Both species recorded 0.978 d^{-1} and 0.685 d^{-1} for *C. vulgaris* and *S. quadricauda* under a 24L:0D photoperiod. Green light was not absorbed but reflected by the cell pigments. A similar pattern was reported by Cetin et al. [28], who determined the growth of *C. vulgaris* under green light and found that algal growth rate under green light was low.

Additionally, it was reported by Niizawa et al. [29] that the emission profile of green LEDs largely coincides with the "green window" of the visible radiation spectrum explaining the low value observed in all cultures. From these results, consequently, it can be concluded that the microalgae species had a low average growth rate under green light during all photoperiods. This could be that microalgae do not require green light as it has green chlorophyll that reflects it, unlike colour attracts effective growth rate. Therefore, the best suitable light for effective cultivation and biomass productivity is white light.

3.2. Light Wavelength and Duration Results on Biomass Productivity

The biomass productivities were divided into nine patterns, the same as the specific growth rate. In pattern 1, under white light (24L:0D) *C. vulgaris* and *S. quadricauda* obtained the biomass productivity rate corresponding to 2.438 g $L^{-1}d^{-1}$ and 1.559 g $L^{-1}d^{-1}$, since constant light supply had deeper light penetration thus higher photosynthetic activity in the microalgae culture; this is in line with Yan et al. [30]. In pattern 2, under white light (19L:5D) high biomass productivity was also observed. The average biomass productivity of *C. vulgaris* and *S. quadricauda* respectively correspond to 2.421 g $L^{-1}d^{-1}$ and 2.369 g $L^{-1}d^{-1}$. In pattern 3, (12L:12D) 1.482 g L^{-1} and 1.242 g L^{-1} were achieved at 16 days of cultivation.

In pattern 4, red light was used at 24L:0D, giving 2.112 g $L^{-1}d^{-1}$ and 1.305 g $L^{-1}d^{-1}$ for *C. vulgaris* and *S. quadricauda*. In pattern 5, under 19L:5D, the key finding was that *S. quadricauda* had the same biomass productivity as *C. vulgaris* as shown in Figure 9 where results at different conditions (i.e., light wavelength and durations) are presented. In pattern 6, 12L:12D was used giving 0.820 g $L^{-1}d^{-1}$ and 0.521 g $L^{-1}d^{-1}$ for *C. vulgaris* and *S. quadricauda*. It has been reported that red light allows higher biomass production by increasing the tolerance to high radiation in cultures. Different microalgae species respond differently to light conditions perhaps due to the presence and activity of different types of photoreceptors [31].

In pattern 7, green light was used, and biomass productivity was observed as 1.007 g $L^{-1}d^{-1}$ and 0.725 g $L^{-1}d^{-1}$ for *C. vulgaris* and *S. quadricauda* during 24L:0D light period. During pattern 8 experiments, 0.909 g $L^{-1}d^{-1}$ and 0.892 g $L^{-1}d^{-1}$ for *C. vulgaris* and *S. quadricauda* was recorded. In pattern 9, biomass productivity yield of 0.427 g $L^{-1}d^{-1}$ and 0.228 g $L^{-1}d^{-1}$ was recorded. Overall, dwindling biomass productivity was observed for cultivation under green light. This is correlated with culture growth rate; as already mentioned, green LEDs emit three times fewer photons than other lights. Furthermore, it was observed that in all 12L:12D photoperiods that light limitation resulted in decreased biomass productivity in addition to growth rate. From this study, it was observed that white and red light stimulates higher growth rate and biomass productivity for both *C. vulgaris* and *S. quadricauda*. The highest biomass productivity was observed for white LED light. Additionally, it was found that *S. quadricauda* had slightly higher biomass productivity under 19L:5D compared to *C. vulgaris* because *S. quadricauda* can thrive under short periods of darkness. Some species benefited from darkness periods given their photosynthetic cycles, where very fast photic reactions are followed by slower thermochemical reactions.

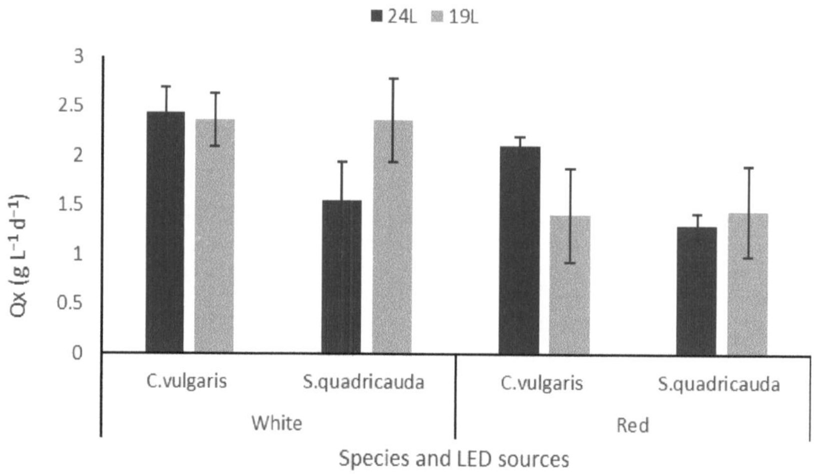

Figure 9. Biomass productivity.

Biomass productivity of *C. vulgaris* cultivated under white light in all photoperiods was higher than *S. quadricauda*. *C. vulgaris* and *S. quadricauda* were cultivated under three different photoperiods (24L:0, 19L:5D, 12L;12D) and under three different LED lights (white, red, and green) to determine the conditions that gave the highest species growth rate and biomass productivity. Under various light sources at constant light 24 h the variation trends of the biomass productivity values were similar throughout the experiments. They were both maintained at an almost consistent rate. When the biomass productivity increases, duplication time reduces; this was observed for both species, but the reverse was observed for low biomass productivity—with low biomass productivity duplication time increases. From this study, it can be concluded and further supported that light is essential for biomass productivity and is directly related to microalgae growth rate [32].

C. vulgaris cultures under a photoperiod of 24L:0D with white, red, and green lights obtained the highest biomass productivity. The reverse was observed during the dark cycle cultivation. A consequence of long dark periods is that basic carbon compounds, which have been synthesised in the light, will be respired. This results in reduction of the overall growth rate and biomass yield [26].

In addition, it is evident that optimum light wavelengths and photoperiods for maximised biomass productivity vary for different microalgae. Biomass productivity can be controlled by different light sources. *S. quadricauda* achieved growth rate and biomass productivity under 19L:5D for red LED as 1.451 d^{-1} and 1.452 gL^{-1} d^{-1}, respectively; this is slightly higher for white under same experimental conditions. Consequently, it can be concluded that *S. quadricauda* thrived the most for growing patterns including short periods of darkness, while *C. vulgaris* achieved both highest growing rates and biomass productivities for growth under continuous light.

3.3. *Implications of LED on Cost, Biomass Productivity on Transport Biofuels, and Way Forward*

LEDs are great for their energy efficiency, robustness, long lifetime, and environmental benefits. The ability to adjust LED brightness and colour wavelength in this study was not for cosmetic reasons but, as shown from our results, they reflected how each species responded to each light wavelength. Although LEDs have high initial cost, they have a longer life span than filament or other traditional light sources, thus making the cost over time for LEDs lower than for traditional lighting. For example, LEDs consume less power than a traditional light source to do the same job. This translates to real energy savings, as well as reduced carbon emissions and a positive impact on the environment. Including periods of darkness then further enhances energy saving.

Cultivation efficiency will play a significant role in the steady growth for biodiesel production as algal biodiesel use can improve air quality, reduce atmospheric carbon dioxide concentrations, and decrease engine maintenance [33]. Biodiesel fuel is easily blended with petroleum diesel to make a premium fuel with improved performance [34].

Furthermore, mandatory blending requirements is an essential factor that would lead to the increased consumption and production of biofuels now and in the future. Similarly, a preferential tax system which subsidises biofuel prices from the contribution of state oil fund from conventional fuels to lower the consumer price of biofuel will also increase customer preference [35]. Improved regulation is required to prevent spread of price spikes [36]; this could be done by increasing the price of conventional fuels while reducing biodiesel fuel prices to encourage usage.

Overall, while considering the implications for biofuel production in OECD countries, and increasingly in non-OECD countries, it is evident that biofuels are more environmentally friendly. As a result, adequate cultivation that improves lipid abundance and lowers production cost, along with efficient dewatering techniques with higher drying efficiency, are keys to enhanced commercial production of biofuels for transport applications.

4. Conclusions

This experimental study assessed the effect of using laboratory-scale closed PBRs to efficiently cultivate *C. vulgaris* and *S. quadricauda.* Both species were cultivated at three different light periods (24L:0D, 19L:5D, 12L;12D) using three different LED lights sources (white, red, and green) to determine the conditions resulting in the highest growth rate and biomass productivity. *S. quadricauda* and *C. vulgaris* responded differently to lighting conditions. It was found that, depending on LED source and light period, different growth rates and biomass productivity were obtained.

Additionally, the key findings were under light intensity of 200 μmol m^{-2} s^{-1}, high growth rate and biomass productivity for *C. vulgaris* under white light at 24L:0D was observed, due to a faster growth rate induced by a continuous light intensity. Regarding, *S. quadricauda*, slightly higher growth rate and biomass productivity was observed under 19L:5D for red light, hence, *S. quadricauda* thrived the most during dark cycles. Growth rate for both species under green light wavelength for all durations was generally low.

Based on the results obtained, the following conclusions can be drawn:

- White LED light resulted in the highest growth rate and highest biomass productivity. The best light colours for the species studied by order of enhanced productivity are white and red lights. For photoperiod the best order is 24L:0D and 19L:5D for *C. vulgaris* and *S. quadricauda*, respectively,
- Green LED light and a long dark period (i.e., 12L:12D) did not show an impressive positive trend in either growth rate or biomass productivity for both microalgae species compared to white and red LEDs and 24L:0D and 19L:5D light durations. It can be concluded that longer dark periods were detrimental to the species mainly under green LED light. Therefore, it is concluded that microalgae have a very short life cycle in the dark under green LED light, and exposure to green wavelength apparently led to slow growth, which would have led stagnation if continued.
- *C. vulgaris* had shorter duplication time compared to *S. quadricauda*.
- *S. quadricauda* thrived the most with a 5 h dark cycle.
- Short periods of darkness (5 h) could be a new strategy to improve biomass productivity under specific light (red). Therefore, under the experimental conditions, the best approach to cultivate both species for enhanced biomass production with short period of darkness was 19L:5D. For commercial process, 5 h of darkness would translate to energy saving over time, when compared to running a 24 h light source. Furthermore, there is not much difference comparing white and red light.

Author Contributions: Conceptualization, R.C.A. and A.G.O.; methodology, R.C.A., C.R. and A.G.O.; software, R.C.A.; validation, R.C.A., C.R., A.D. and A.G.O.; investigation, R.C.A.; writing—original draft preparation, R.C.A., C.R. and A.D. writing—review and editing, R.C.A., C.R., A.D. and A.G.O. All authors have read and agreed to the published version of the manuscript.

Funding: This research received no external funding.

Institutional Review Board Statement: Not applicable.

Informed Consent Statement: Not applicable.

Data Availability Statement: Not applicable.

Conflicts of Interest: The authors declare no conflict of interest.

References

1. Ohunakin, O.S.; Adaramola, M.S.; Oyewola, O.M. Wind Energy Evaluation for Electricity Generation Using WECS in Seven Selected Locations in Nigeria. *Appl. Energy* **2011**, *88*, 3197–3206. [CrossRef]
2. Ohunakin, O.S.; Adaramola, M.S.; Oyewola, O.M.; Fagbenle, R.O. Solar Energy Applications and Development in Nigeria: Drivers and Barriers. *Renew. Sustain. Energy Rev.* **2014**, *32*, 294–301. [CrossRef]
3. Rodriguez, C.; Alaswad, A.; El-Hassan, Z.; Olabi, A.G. Improvement of Methane Production from *P. canaliculata* through Mechanical Pretreatment. *Renew. Energy* **2018**, *119*, 73–78. [CrossRef]
4. Rodriguez, C.; Alaswad, A.; El-Hassan, Z.; Olabi, A.G. Mechanical Pretreatment of Waste Paper for Biogas Production. *Waste Manag.* **2017**, *68*, 157–164. [CrossRef] [PubMed]
5. Huang, Y.-T.; Lai, C.-W.; Wu, B.-W.; Lin, K.-S.; Wu, J.C.S.; Hossain, M.S.A.; Yamauchi, Y.; Wu, K.C.-W. Advances in Bioconversion of Microalgae with High Biomass and Lipid Productivity. *J. Taiwan Inst. Chem. Eng.* **2017**, *79*, 37–42. [CrossRef]

6. Russell, C.; Rodriguez, C.; Yaseen, M. High-Value Biochemical Products & Applications of Freshwater Eukaryotic Microalgae. *Sci. Total Environ.* **2022**, *809*, 151111. [CrossRef]
7. Olabi, A.G. 100% Sustainable Energy. *Energy* **2014**, *77*, 1–5. [CrossRef]
8. Anyanwu, R.C.; Rodriguez, C.; Durrant, A.; Olabi, A.G. Optimisation of Tray Drier Microalgae Dewatering Techniques Using Response Surface Methodology. *Energies* **2018**, *11*, 2327. [CrossRef]
9. Shahi, T.; Beheshti, B.; Zenouzi, A.; Almasi, M. Bio-Oil Production from Residual Biomass of Microalgae after Lipid Extraction: The Case of *Dunaliella* sp. *Biocatal. Agric. Biotechnol.* **2020**, *23*, 101494. [CrossRef]
10. Collotta, M.; Champagne, P.; Mabee, W.; Tomasoni, G.; Leite, G.B.; Busi, L.; Alberti, M. Comparative LCA of Flocculation for the Harvesting of Microalgae for Biofuels Production. *Procedia CIRP* **2017**, *61*, 756–760. [CrossRef]
11. Zhu, L.D.; Li, Z.H.; Hiltunen, E. Strategies for Lipid Production Improvement in Microalgae as a Biodiesel Feedstock. *BioMed Res. Int.* **2016**, *2016*, 7–9. [CrossRef] [PubMed]
12. Anyanwu, R.C.; Rodriguez, C.; Durrant, A.; Olabi, A.G. Microalgae Cultivation Technologies. *Encycl. Smart Mater.* **2018**, *2*, 709–731. [CrossRef]
13. Hadley, K.B.; Bauer, J.; Milgram, N.W. The Oil-Rich Alga *Schizochytrium* sp. as a Dietary Source of Docosahexaenoic Acid Improves Shape Discrimination Learning Associated with Visual Processing in a Canine Model of Senescence. *Prostaglandins Leukot. Essent. Fat. Acids* **2017**, *118*, 10–18. [CrossRef] [PubMed]
14. Makkar, H.P.S.; Tran, G.; Heuzé, V.; Giger-Reverdin, S.; Lessire, M.; Lebas, F.; Ankers, P. Seaweeds for Livestock Diets: A Review. *Anim. Feed. Sci. Technol.* **2016**, *212*, 1–17. [CrossRef]
15. Onumaegbu, C.; Alaswad, A.; Rodriguez, C.; Olabi, A. Optimization of Pre-Treatment Process Parameters to Generate Biodiesel from Microalga. *Energies* **2018**, *11*, 806. [CrossRef]
16. Rodriguez, C.; Alaswad, A.; Benyounis, K.Y.; Olabi, A.G. Pretreatment Techniques Used in Biogas Production from Grass. *Renew. Sustain. Energy Rev.* **2017**, *68*, 1193–1204. [CrossRef]
17. Wilberforce, T.; El-Hassan, Z.; Khatib, F.N.; Al Makky, A.; Baroutaji, A.; Carton, J.G.; Olabi, A.G. Developments of Electric Cars and Fuel Cell Hydrogen Electric Cars. *Int. J. Hydrog. Energy* **2017**, *42*, 25695–25734. [CrossRef]
18. OECD; IEA; FAO. *Technology Roadmap: How2Guide for Bioenergy Roadmap—Development and Implementation*; OECD: Paris, France; IEA: Paris, France; FAO: Rome, Italy, 2017.
19. Kaewkannetra, P.; Enmak, P.; Chiu, T. The Effect of CO_2 and Salinity on the Cultivation of Scenedesmus Obliquus for Biodiesel Production. *Biotechnol. Bioprocess Eng.* **2012**, *17*, 591–597. [CrossRef]
20. Tang, D.; Han, W.; Li, P.; Miao, X.; Zhong, J. CO_2 Biofixation and Fatty Acid Composition of Scenedesmus Obliquus and Chlorella Pyrenoidosa in Response to Different CO_2 Levels. *Bioresour. Technol.* **2011**, *102*, 3071–3076. [CrossRef]
21. Anyanwu, R.C.; Rodriguez, C.; Durrant, A.; Olabi, A.G. Micro-Macroalgae Properties and Applications. *Ref. Modul. Mater. Sci. Mater. Eng.* **2018**. [CrossRef]
22. Rodriguez, C.; Alaswad, A.; El-Hassan, Z.; Olabi, A.G. Waste Paper and Macroalgae Co-Digestion Effect on Methane Production. *Energy* **2018**, *154*, 119–125. [CrossRef]
23. Dutta, K.; Daverey, A.; Lin, J.G. Evolution Retrospective for Alternative Fuels: First to Fourth Generation. *Renew. Energy* **2014**, *69*, 114–122. [CrossRef]
24. Allegra Carbone, D.; Olivieri, G.; Pollio, A.; Melkonian, M. Growth and Biomass Productivity of Scenedesmus Vacuolatus on a Twin Layer System and a Comparison with Other Types of Cultivations. *Appl. Microbiol. Biotechnol.* **2017**, *101*, 8321–8329. [CrossRef] [PubMed]
25. Onumaegbu, C.; Alaswad, A.; Rodriguez, C.; Olabi, A. Modelling and Optimization of Wet Microalgae Scenedesmus Quadricauda Lipid Extraction Using Microwave Pre-Treatment Method and Response Surface Methodology. *Renew. Energy* **2019**, *132*, 1323–1331. [CrossRef]
26. Maroneze, M.M.; Siqueira, S.F.; Vendruscolo, R.G.; Wagner, R.; de Menezes, C.R.; Zepka, L.Q.; Jacob-Lopes, E. The Role of Photoperiods on Photobioreactors—A Potential Strategy to Reduce Costs. *Bioresour. Technol.* **2016**, *219*, 493–499. [CrossRef]
27. Chen, C.Y.; Yeh, K.L.; Chang, H.Y.; Chang, J.S. Strategies to Improve Oil/Lipid Production of Microalgae in Outdoor Cultivation Using Vertical Tubular-T Ype Photobioreactors. *Energy Procedia* **2014**, *61*, 2755–2758. [CrossRef]
28. Baidya, A.; Akter, T.; Islam, M.R.; Shah, A.; Hossain, M.A.; Salam, M.A.; Paul, S.I. Effect of different wavelengths of LED light on the growth, chlorophyll, β-carotene content and proximate composition of *Chlorella ellipsoidea*. *Heliyon* **2021**, *7*, e08525. [CrossRef] [PubMed]
29. Niizawa, I.; Leonardi, R.J.; Irazoqui, H.A.; Heinrich, J.M. Light Wavelength Distribution Effects on the Growth Rate of Scenedesmus Quadricauda. *Biochem. Eng. J.* **2017**, *126*, 126–134. [CrossRef]
30. Yan, C.; Muñoz, R.; Zhu, L.; Wang, Y. The Effects of Various LED (Light Emitting Diode) Lighting Strategies on Simultaneous Biogas Upgrading and Biogas Slurry Nutrient Reduction by Using of Microalgae *Chlorella* sp. *Energy* **2016**, *106*, 554–561. [CrossRef]
31. Chiarini, A.; Quadrio, M. The Light/Dark Cycle of Microalgae in a Thin-Layer Photobioreactor. *J. Appl. Phycol.* **2021**, *33*, 183–195. [CrossRef]
32. Santos, K.R.; Mariano, A.B. Determination of Optimal Algae Concentration for Continuous Growth of *Scenedesmus* sp. in CHU and Modified CHU Media. *Rev. Eng. Térmica* **2014**, *13*, 52–58. [CrossRef]

33. Charles, M.B.; Ryan, R.; Ryan, N.; Oloruntoba, R. Public Policy and Biofuels: The Way Forward? *Energy Policy* **2007**, *35*, 5737–5746. [CrossRef]
34. Gallagher, B.J. The Economics of Producing Biodiesel from Algae. *Renew. Energy* **2011**, *36*, 158–162. [CrossRef]
35. Rawat, I.; Ranjith Kumar, R.; Mutanda, T.; Bux, F. Biodiesel from Microalgae: A Critical Evaluation from Laboratory to Large Scale Production. *Appl. Energy* **2013**, *103*, 444–467. [CrossRef]
36. Foley, A.; Olabi, A.G. Renewable Energy Technology Developments, Trends and Policy Implications That Can Underpin the Drive for Global Climate Change. *Renew. Sustain. Energy Rev.* **2017**, *68*, 1112–1114. [CrossRef]

sustainability

MDPI

Article

A New Fractional-Order Load Frequency Control for Multi-Renewable Energy Interconnected Plants Using Skill Optimization Algorithm

Ahmed Fathy [1,2,*], Hegazy Rezk [3], Seydali Ferahtia [4], Rania M. Ghoniem [5], Reem Alkanhel [5] and Mohamed M. Ghoniem [6]

1 Electrical Engineering Department, Faculty of Engineering, Jouf University, Sakaka 72388, Saudi Arabia
2 Electrical Engineering Department, Faculty of Engineering, Zagazig University, Zagazig 44519, Egypt
3 Department of Electrical Engineering, College of Engineering in Wadi Alddawasir, Prince Sattam Bin Abdulaziz University, Al-Kharj 11942, Saudi Arabia
4 Laboratoire de Genie Electrique, Department of Electrical Engineering, University of M'Sila, M'Sila 28000, Algeria
5 Department of Information Technology, College of Computer and Information Sciences, Princess Nourah bint Abdulrahman University, Riyadh 11671, Saudi Arabia
6 Department of Computer, Mansoura University, Mansoura 35516, Egypt
* Correspondence: afali@ju.edu.sa

Abstract: Connection between electric power networks is essential to cover any deficit in the generation of power from any of them. The exchange powers of the plants during load disturbance should not be violated beyond their specified values. This can be achieved by installing load frequency control (LFC); therefore, this paper proposes a new metaheuristic-based approach using a skill optimization algorithm (SOA) to design a fractional-order proportional integral derivative (FOPID)-LFC approach with multi-interconnected systems. The target is minimizing the integral time absolute error (ITAE) of frequency and exchange power violations. Two power systems are investigated. The first one has two connected plants of photovoltaic (PV) and thermal units. The second system contains four plants, namely, PV, wind turbine, and two thermal plants, with governor dead-band (GDB) and generation rate constraints (GRC). Different load disturbances are analyzed in both considered systems. Extensive comparisons to the use of chef-based optimization algorithm (CBOA), jumping spider optimization algorithm (JSOA), Bonobo optimization (BO), Tasmanian devil optimization (TDO), and Atomic orbital search (AOS) are conducted. Moreover, statistical tests of Friedman ANOVA table, Wilcoxon rank test, Friedman rank test, and Kruskal Wallis test are implemented. Regarding the two interconnected areas, the proposed SOA achieved the minimum fitness value of 1.8779 pu during 10% disturbance on thermal plant. In addition, it outperformed all other approaches in the case of 1% disturbance on the first area as it achieved ITAE of 0.0327 pu. The obtained results proved the competence and reliability of the proposed SOA in designing an efficient FOPID-LFC in multi-interconnected power systems with multiple sources.

Keywords: LFC; PV plant; wind energy; multi-interconnected system; renewable energy; skill optimization algorithm

1. Introduction

Achieving stability of power system operation is essential to guarantee the continuity of customer services, especially during load disturbances. The frequency of the power system is greatly affected by the load violation. This problem arises in multi-interconnected systems, as the load violation causes a change in both frequency and exchange of power between the connected plants. Therefore, load frequency control (LFC) is essential to banish the violations in both frequency and tie-line power [1–3]. Many researchers have dealt with designing the LFC in multi-area multi-sources, which may be conventional or renewable energy. Yousri et al. [4] introduced Harris Hawks optimization (HHO)-based methodology to obtain the optimal gains of LFC-proportional-integral (PI) inserted in interconnected systems with renewable energy sources (RESs). The authors considered the integral time absolute error (ITAE) as the target to be minimized. Ali et al. [5] recommended multi-verse optimizer (MVO) as an efficient tool to design the model predictive control (MPC) LFC inserted in six-interconnected systems with renewables-based plants with storage systems. Moreover, the effect of generation rate constraint (GRC) and governor dead band (GDB) zones for thermal plant have been considered. Fathy et al. [6] designed the fractional-order proportional integral derivative (FOPID) LFC via movable damped wave algorithm (MDVA) to banish the variations of frequency deviations and tie-line powers of an interconnected system with RESs during load disturbance. Moreover, ITAE was the target in the considered optimization problem. In [7], the optimal gains of PI and PID-LFC have been determined via particle swarm optimizer (PSO); the designed controller has been used with hybrid solar-wind-micro-hydro interconnected systems. An extensive review of many approaches employed in designing LFC with solar-wind interconnected systems has been conducted in [8]. Additionally, the authors designed the FOPID-controller via the flower pollination algorithm (FPA). Fathy et al. [9] presented LFC simulated via an adaptive neuro fuzzy inference system (ANFIS) trained by antlion optimizer (ALO) installed in two and four-interconnected power plants with renewable energy sources. The authors considered the ITAE of frequency and tie-line power deviations as the fitness function to be minimized. Many reported techniques assigned to design LFC insertion in classical and modern power systems have been reviewed in [10], including nonlinear models, controller parameter identification, soft computing approaches, integration of renewable energy sources, future trends, and challenges. In [11], LFC designed via decentralized MPC has been installed within the Egyptian power system with traditional and renewable-based plants. The considered renewable-based plants were the wind farms of Zafarana and Gabel El-Zeit, and Benban solar plant. Additionally, both GRC and GDB of thermal plants have been considered. In [12], the authors identified the optimal parameters of proportional-derivative LFC with a cascaded filter via coyote optimization algorithm (COA); the controller has been used with renewable energy-based plants. Arora et al. [13] introduced an approach-based on moth flame optimizer (MFO) to minimize the frequency constraints of renewable energy generation systems via installing LFC. A comprehensive review of different LFC structures in both single and multi-interconnected systems has been reported in [14–17]. Takayama et al. [18] introduced a coordination method between the LFC and economic dispatching control (EDC) for large-scale renewable energy generation plants. Additionally, the optimal size of a battery storage system was analyzed based on the presented LFC and EDC. An improved twin delayed deep deterministic policy gradient deep reinforcement learning-based LFC has been introduced and incorporated in RESs with variable loads and electric vehicles [19]. The integral absolute error (IAE) of the frequency and exchange power deviations was selected as the target to be minimized. In [20], the marine predators algorithm (MPA) has been presented to find the parameters of LFC-PID inserted in interconnected systems with RESs and storage systems. An enhanced COA has been employed to design PI-PI and PD with filter-based LFC installed with RESs-based interconnected system [21]. The Ziegler–Nichols method has been used by Subham et al. [22] to adjust the parameters of automatic LFC-PID inserted in hybrid power systems. Moreover, hardware-in-loop based OP4510 has been simulated to assess the presented controller. PI, PID, and fuzzy-based LFC have been presented to mitigate the frequency and tie-line power violations of interconnected systems with renewable energy sources [23]. In [24], a hybrid approach comprising anopheles search algorithm and artificial intelligence techniques has been introduced to design PID-LFC incorporated in a triple interconnected hybrid system including solar, biomass, and fuel cell-based plants. Masuta et al. [25] presented a coordinated LFC for conventional power plants, battery energy storage system, heat pump water heater, and electric vehicles in addition to renewable energy sources.

Dutta et al. [26] introduced an emotional controller for LFC inserted in a two area hybrid power system with solar and biomass generating sources in addition to electric vehicles. Salp swarm algorithm (SSA) has been used to tune the parameters of LFC-PID inserted in a multi-area hybrid power system with renewable energy sources [27]. A sliding mode-based LFC has been incorporated in a multi-area interconnected system with integrated RESs [28]; moreover, the authors investigated the system asymptotic stability via Lyapunov theory on the basis of a linear matrix inequality technique. In [29], the authors identified the optimal gains of FOPID-LFC using an improved chaotic atom search optimizer (IASO); the controller was inserted in a multi-area system with multi-hybrid sources such that the ITAE of the frequency and exchange power violations is minimized. A multi-area hybrid source interconnected system including photovoltaics (PV), diesel engines, micro hydro generating units, and fuel cells as storage system has been established and controlled via fuzzy logic-based LFC [30]. Three types of LFC controllers were introduced in [31]: cascaded fractional order controller, three degrees of freedom PID, and tilt integral derivative one. The presented controllers have been installed in an interconnected power system with wind generation systems. Moreover, equilibrium optimizer (EO) has been employed to tune the gains of different controllers. A fractional order two degrees of freedom-based LFC has been designed using Quasi-oppositional HHO to achieve stable frequency for two identical area power systems including PV, biogas unit, wind turbine, and thermal power plant [32]. Xu et al. [33] presented an artificial sheep algorithm-based LFC installed in an interconnected-area with RESs. In [34], the author used teaching learning-based optimizer (TLBO) to get the optimal parameters of automatic LFC incorporated in a multi-source system of thermal, hydro, and gas plants. The authors used different fitness functions such as integral squared error (ISE), IAE, integral time squared error (ITSE), and ITAE to assess the presented approach. A reference offset governor approach has been used by Tedesco et al. [35] to simulate the LFC in a multi-area microgrid with renewable energy sources. A power control approach of hybrid renewable energy systems has been introduced in [36]. A parallel buck-boost converter controlled via fuzzy logic control has been constructed and installed in a hybrid renewable energy-based system [37].

The reported methods employed many metaheuristic optimizers and other artificial intelligence techniques, such as fuzzy logic and ANFIS, in designing the LFC. Many of the used optimizers are unable to obtain the desired results due to getting stuck in local optima. Moreover, the others need several parameters that should be defined by the user. Furthermore, the artificial intelligence-based approaches are not accurate due to the imprecise defined data, and also, they require excessive data for training. All these shortcomings are taken in consideration when conducting the present analysis.

The aim of this work is to design a FOIPD controller based LFC via a recent approach using the skill optimization algorithm (SOA). The algorithm is responsible for identifying the unknown parameters of the considered controller such that the integral time absolute error of the frequency and exchange power deviations is minimized.

The work contributions can be listed as follows:

- A new skill optimization algorithm (SOA)-based methodology is proposed to design FOPID-LFC installed with interconnected systems with RESs.
- Two power systems are investigated, PV/thermal and thermal/wind turbine/thermal/PV, at different load disturbances.
- An extensive comparison of CBOA, JSOA, BO, TDO, and AOS is conducted.
- Statistical tests of Friedman ANOVA table, Wilcoxon rank test, Friedman rank test, and Kruskal Wallis test are implemented.
- The competence and reliability of the proposed SOA are confirmed via the obtained results.

The paper is organized as follows: Section 2 presents the mathematical model of the interconnected system. Section 3 explains the main principle of a fractional-order PID controller (FOPID). The proposed skill optimization algorithm is presented in Section 4, while Section 5 introduces a formulation using the proposed optimization problem. The numerical analysis is given in Section 6, while Section 7 handles the conclusions.

2. Model of Interconnected Systems

Two multi-interconnected multi-sources systems are constructed and analyzed in this work. The first one covers two areas with photovoltaic (PV) with maximum power point tracker (PV) and thermal plants. The second system has four-interconnected plants comprising PV, wind turbine (WT), and two thermal units with GDB and GRC. Many reported works have been conducted to model PV and WT-based generating systems [38–40]. The considered systems are investigated under different load disturbances, and the mathematical model of each plant is presented in this section.

2.1. PV Plant Model

The PV generation system comprises solar cells that may be configured in series and/or parallel to generate the required power and cover the customer's needs. The generated power from the PV system is affected by the variation in weather conditions such as irradiance and temperature. Hence, the PV panel output voltage can be written as follows [41]:

$$V_{pv} = \left(\frac{n_s \varepsilon k T}{Q}\right) ln\left(\frac{n_p G I_{ph} - I + n_p I_o}{n_p I_o}\right) - \left(\frac{n_s I R_s}{n_p}\right) \quad (1)$$

where I and V_{pv} denote the PV panel output current and voltage, n_s and n_p are the number of series and parallel cells, ε is the factor of completion, k represents the Boltzmann constant, T is the PV panel temperature, Q represents the electron charge, G is the irradiance in W/m^2, I_{ph} and I_o are the photo and saturation currents, and R_s denotes the cell series resistance. The characteristics of the PV panel are nonlinear. The PV panel power-voltage (P-V) curve has a unique global maximum power (GMP) as shown in Figure 1, and it is essential to monitor this point to enhance the PV panel performance and maximize its efficiency. This target is achieved via MPPT which tunes the duty cycle of the DC-DC converter at the PV panel terminals. Many approaches have been used in simulating the MPPT; one of the most popular is called incremental conductance (INC) [42], which is used in this work.

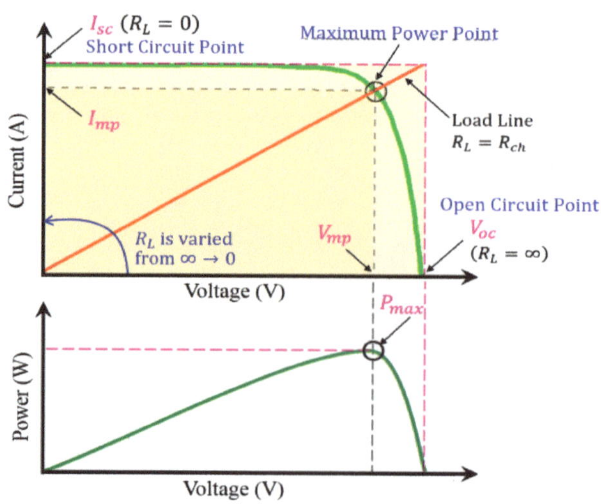

Figure 1. I-V and P-V characteristics of PV panel.

This approach depends on the derivative of the PV power to the voltage in three considered regions which are GMP right, GMP, and GMP left as follows:

$$\begin{cases} \frac{dP_{pv}}{dV_{pv}} > 0 & At\ right\ of\ GMP \\ \frac{dP_{pv}}{dV_{pv}} = 0 & At\ GMP \\ \frac{dP_{pv}}{dV_{pv}} > 0 & At\ left\ of\ GMP \end{cases} \tag{2}$$

where P_{pv} represents the PV module output power. Finally, the PV generating unit including PV panel, DC-DC converter, MPPT, and inverter can be represented as two cascaded blocks with two gains as follows [43]:

$$G_1 = \left(\frac{S^2}{S^2 + \omega^2}\right)\left(\frac{V_{pv}(S^2 + \omega^2)(S^2 + 2\omega^2)}{kS^2(S^2 + 4\omega^2)}\right)\left(\frac{1 - e^{-ST_s}}{ST_s}\right) \tag{3}$$

$$G_2 = \left(\frac{\frac{M_1}{LC}}{S^2 + \left(\frac{1}{RC}\right)S + \frac{1}{LC}}\right)\left(\frac{1 - e^{\frac{-ST_s}{2}}}{1 + e^{\frac{-ST_s}{2}}}\right)\left(\frac{M_2}{1 + ST_s}\right) \tag{4}$$

where ω represents the angular frequency of the grid, R, C, and L denote the output resistance, capacitance, and inductance of the converter, T_s represents the simulation time, M_1 and M_2 represent the voltage gains of the buck converter and inverter, respectively.

2.2. Model of WT

The wind turbine (WT) operation is characterized by the coefficient of power (C_p); this depends on important parameters known as the tip ratio (λ) and the pitch angle of blade (β). The WT output power can be maximized by keeping the value of the tip speed ratio at its optimum value. The value of λ can be calculated as [44],

$$\lambda = \frac{\omega_t r}{V_w} \tag{5}$$

where ω_t is the mechanical angular speed of the turbine, r denotes the radius of the turbine, and V_w is the wind speed. The wind power can be calculated as [45],

$$P_W = \frac{1}{2}\rho A C_p(\lambda, \beta) V_W{}^3 \tag{6}$$

where ρ is the air density and A denotes the swept area of turbine blades. The value of the power coefficient can be expressed as follows:

$$C_p = (0.44 - 0.0167\beta)\sin\left(\frac{\pi(\lambda - 2)}{15 - 0.3\beta}\right) - 0.00184(\lambda - 3)\beta \tag{7}$$

The wind turbine output power can be calculated as [46],

$$P_m = C_p(\lambda, \beta) P_W \tag{8}$$

The WT output power variations versus the rotor speed are shown in Figure 2.
The wind plant transfer function can be expressed as [47],

$$G_{WT}(s) = \left(\frac{K_{pw1}(1 + sT_{pw1})}{1 + s}\right)\left(\frac{K_{pw2}}{1 + sT_{pw2}}\right)\left(\frac{K_{pw3}}{1 + s}\right) \tag{9}$$

where K_{pw1}, K_{pw2}, and K_{pw3} denote the wind plant gains while T_{pw1}, T_{pw2}, and T_{pw3} are the wind plant time constants.

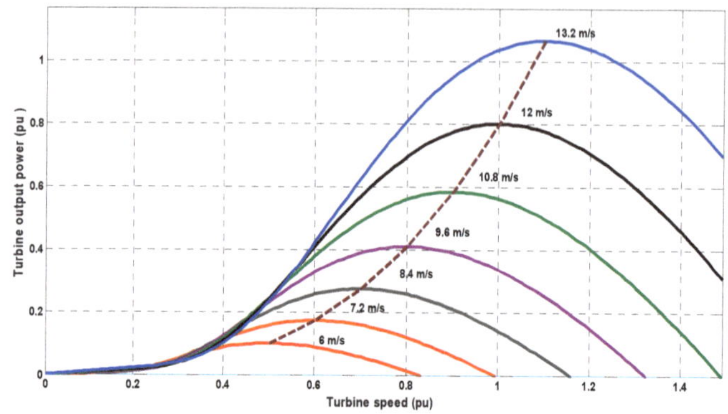

Figure 2. WT output power versus the rotor speed.

2.3. *Thermal Plant Model*

A thermal generating unit contains a steam turbine as prime mover, speed governor, reheater, and generating unit. The transfer functions of all these components can be written as follows [48]:

$$G_t = \frac{K_t}{1 + T_t S} \tag{10}$$

$$G_g = \frac{K_g}{1 + T_g S} \tag{11}$$

$$G_r = \frac{1 + K_r T_r S}{1 + T_r S} \tag{12}$$

$$G_{gen} = \frac{K_p}{1 + T_p S} \tag{13}$$

where G_t, G_g, G_r, and G_{gen} denote the turbine, governor, reheater, and generator transfer functions, respectively, K_t, K_g, K_r, and K_p denote the gains of turbine, governor, reheater, and generator, respectively, T_t, T_g, T_r, and T_p are the time constants of the stated components, respectively. As stated before, the authors considered two interconnected multi-source power systems. The first one is a PV/thermal system, and Figure 3 shows the block diagram of the connected system. The second system is thermal/WT/thermal/PV system; both GDB and GRC of thermal units are considered; the architecture of such a system is shown in Figure 4.

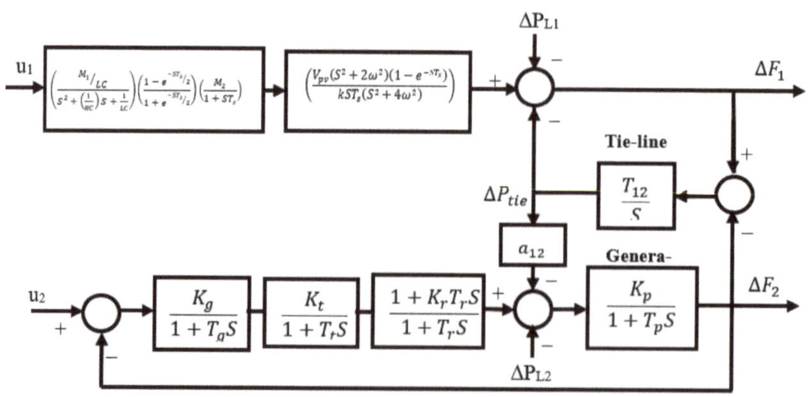

Figure 3. The considered PV/thermal connected power system.

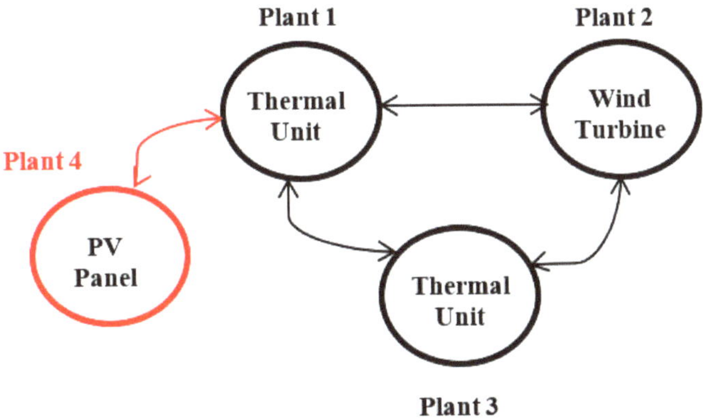

Figure 4. The architecture of thermal/WT/thermal/PV system.

3. Fractional-Order PID Controller (FOPID)

A fractional-order PID controller (FOPID) was presented in [49]; it has superior performance compared to the conventional PID for closed-loop systems. FOPID is different from the conventional PID as the order of its integral and derivative is not an integer. This gives the controller more freedom in the controller tuning, resulting in a better dynamic performance of FOPID than the conventional PID. Five parameters are used to identify such a controller, k_p, k_i, k_d, λ_d, and μ, and the controller transfer function can be expressed as,

$$G_c = k_p + \frac{k_i}{S^{\lambda_d}} + k_d S^{\mu}, \lambda, \mu > 0 \tag{14}$$

By solving Equation (12), one can get the controller output as follows:

$$u(t) = k_p e(t) + k_i D^{-\lambda_d} e(t) + k_d D^{-\mu} e(t) \tag{15}$$

where $e(t)$ denotes the input of FOPID; the controller is inserted before the plant to feed it with the reference input such that the error between the reference input and the plant actual output is minimized. The FOPID performance can be enhanced via tuning the parameters λ_d and μ. which is This is the approach proposed here.

The FOPID controller is simulated in Simulink via FOMCON Toolbox. It depends on fractional-order calculus to model, design, and control the system. There are block sets provided by this toolbox from which $PI^{\lambda}D^{\mu}$ (FOPID) controller can operate.

4. The Proposed Skill Optimization Algorithm

The skill optimization algorithm (SOA) was introduced by Givi1 et al. [50]. It is inspired by human efforts to develop enhanced skills. In life, the people (members of SAO) strive to improve their skills via learning. The approach begins by initializing the members randomly. The population matrix of SOA can be formulated as follows:

$$X = \begin{bmatrix} X_1 \\ \vdots \\ X_i \\ \vdots \\ X_N \end{bmatrix} = \begin{bmatrix} x_{1,1} & \cdots & x_{1,d} & \cdots & x_{1,m} \\ \vdots & \vdots & \vdots & \vdots & \vdots \\ x_{i,1} & \cdots & x_{i,d} & \cdots & x_{i,m} \\ \vdots & \vdots & \vdots & \vdots & \vdots \\ x_{N,1} & \cdots & x_{N,d} & \cdots & x_{N,m} \end{bmatrix} \tag{16}$$

where X_i is the ith candidate, $x_{i,d}$ denotes the dth variable value proposed via the ith member of the population, N represents the number of members, and m denotes the number of considered variables. Each row in the population matrix represents the candidate solution; the fitness function of each one is calculated and expressed as follows:

$$F = \begin{bmatrix} F_1 \\ \vdots \\ F_i \\ \vdots \\ F_N \end{bmatrix} = \begin{bmatrix} F(X_1) \\ \vdots \\ F(X_i) \\ \vdots \\ F(X_N) \end{bmatrix} \quad (17)$$

where F_i denotes the fitness value of the ith candidate solution; the best fitness value recognizes the best member and vice versa. During the iterative process followed in SOA, the fitness values are updated in addition to the worst and best members. Two phases are followed in SOA to update the population members, which are exploration and exploitation. In the first stage, the skill learning process is conducted via experts, while the second stage depends on the activities and individual efforts. The main objective of the exploration phase is to search for the global solution in the search space via moving the algorithm members under the guidance of other members. The original optimal area is identified properly when the exploration power is increased. On the other hand, the exploitation phase aims at local search, which helps in converging better solutions.

4.1. Exploration Phase

In this phase, the population member strives to enhance his skill via an expert member who has a good condition based on fitness value. The expert member is selected randomly from the members with better fitness values than the considered member. It guides the population member to different locations in the search space via learning the skill to do this. The new location is accepted when its fitness value is improved, this can be modeled as follows:

$$x_{i,d}^{P1} = x_{i,d} + rand \times (E_{i,d} - I \times x_{i,d}), E_i = X_k if\ F_k < F_i \quad (18)$$

$$x_{i,d} = \begin{cases} x_{i,d}^{P1} & if\ F_i^{P1} < F_i \\ x_{i,\,d} & if\ F_i^{P1} \geq F_i \end{cases} \quad (19)$$

where $x_{i,d}^{P1}$ denotes the updated position of member i, d in the first phase (P1), $F_{i,d}^{P1}$ is the fitness value of $x_{i,d}^{P1}$, E_i represents the selected expert member, $E_{i,d}$ is the dth dimension of the expert member, $rand$ is a random number in range [0, 1], and I represents a random number that has a value of either 1 or 2.

4.2. Exploitation Phase

In this phase, each member strives to enhance the skill gained in the exploration phase via individual activity and practice. It is modeled as a local search to increase the exploitation such that the member tries to enhance his fitness value. This can be conducted as,

$$x_{i,d}^{P2} = \begin{cases} x_{i,d} + \frac{1-2\times rand\times x_{i,d}}{t} & if\ rand < 0.5 \\ x_{i,d} + \frac{LB_j + rand(UB_j - LB_j)}{t} & if\ rand < 0.5 \end{cases} \quad (20)$$

$$x_{i,d} = \begin{cases} x_{i,d}^{P2} & if\ F_i^{P2} < F_i \\ x_{i,\,d} & if\ F_i^{P2} \geq F_i \end{cases} \quad (21)$$

where $x_{i,d}^{P2}$ is the updated position of member i, d in the second phase (P1), $F_{i,d}^{P2}$ represents the fitness value of $x_{i,d}^{P2}$, t is the number of iterations, UB_j and LB_j are the upper and lower limits of the jth variable. The steps followed in SOA are given in Figure 5.

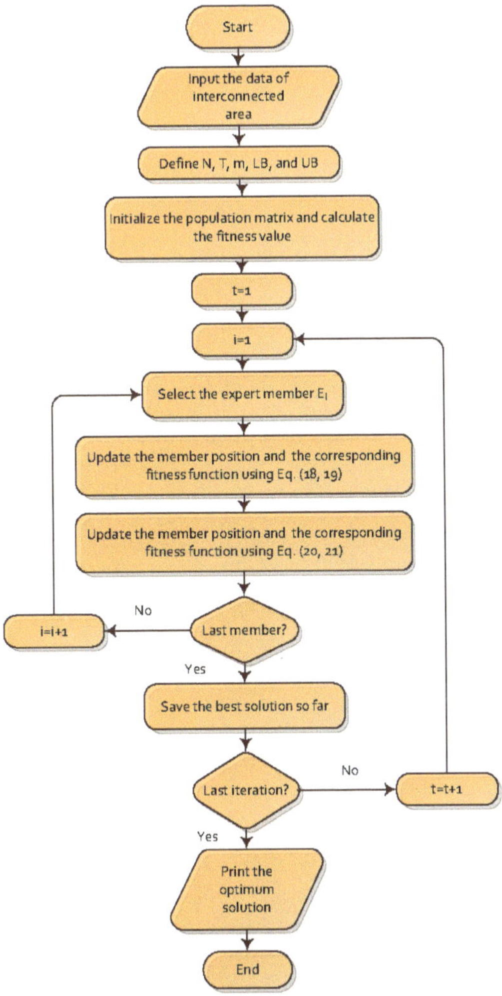

Figure 5. SOA flowchart.

5. The Proposed Optimization Problem

The problem of designing an LFC inserted in interconnected RESs is formulated as an optimization problem with a constrained fitness function, the formulation of the proposed optimization problem and the solution methodology is presented in this section.

5.1. The Fitness Function

In this work, the authors considered the integral time absolute error (ITAE) of frequency and exchange power violations as the target to be minimized. ITAE integrates the time multiplied by absolute error over a specified time; the ITAE tuning crops systems that settle abundantly more rapidly than the other tuning methods that use IAE and ISE. The considered variables to be identified are the FOPID parameters, k_p, k_i, k_d, λ_d, and μ.

The fitness function can be written as follows [6]:

$$Minimize\ ITAE = \int_0^t \left((\textstyle\sum_{i=1}^{n_a} |\Delta F_i + \Delta P_{tie,i}|).t \right) dt \quad (22)$$

where ΔF_i and $P_{tie,i}$ are the violations in ith area frequency and exchange power, respectively, t denotes the specified time, and n_a is the number of interconnected plants. The constraints accompanying the optimization problem can be written as,

$$\begin{array}{c} k_p{}^{min} \leq k_p < k_p{}^{max} \\ k_i{}^{min} \leq k_i < k_i{}^{max} \\ k_d{}^{min} \leq k_d < k_d{}^{max} \\ \lambda_d{}^{min} \leq \lambda_d < \lambda_d{}^{max} \\ \mu^{min} \leq \mu < \mu^{max} \end{array} \quad (23)$$

where min and max denote the minimum and maximum limits of the scaling variable, and have been set in the range 0.1–2 [51].

5.2. The SOA-Based Solution Methodology

The SOA-based methodology is responsible for identifying the optimal parameters of FOPID, resulting in minimum ITAE. The controller with the best parameters has a small input signal ($e(t)$) which is the difference between the plant's actual output (y(t)) and reference input (w(t)). The controller feeds the power plant with the required signal that helps in mitigating the violations in both frequency and exchange power. The configuration of a power plant with the proposed SOA-FOPID is shown in Figure 6. The adapted parameters of the FOPID controller are k_p, k_i, k_d, λ_d, and μ. They are identified via the proposed SOA to minimize the error signal ($e(t)$).

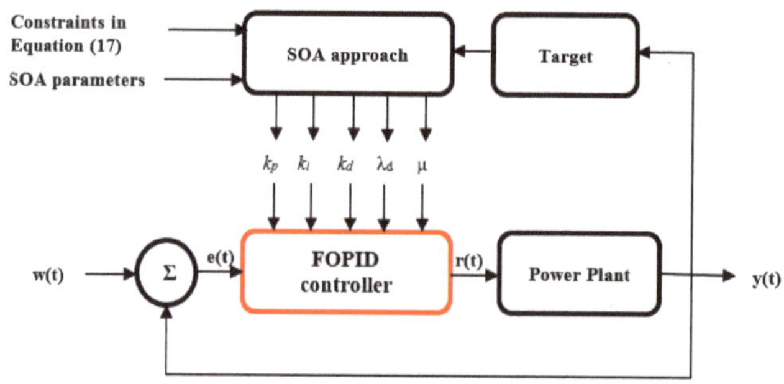

Figure 6. The configuration of a power plant with the proposed SOA-FOPID.

6. Numerical Analysis

In this work, the designed FOPID via SOA is investigated via two interconnected multi-source power systems. The first system is a PV plant connected to a thermal generating unit. The second system contains four power systems, namely, two thermal generation plants, a wind turbine, and PV. Additionally, the effect of GDB and GRC are considered in the thermal plants. Furthermore, different load disturbances have been analyzed in both considered systems. The proposed SOA is compared to other approaches using the chef-based optimization algorithm (CBOA), jumping spider optimization algorithm (JSOA), Bonobo optimization (BO), Tasmanian devil optimization (TDO), and Atomic orbital search (AOS). All approaches are implemented for 50 population size, 100 iterations, and 10 independent runs.

6.1. Two-Interconnected Power System

The two-part interconnected power system with the proposed FOPID is modeled in Simulink/Matlab. The system configuration is shown in Figure 7. The system parameters given in [9] are used in this work. The PV plant and thermal generator capacities are 500 MW and 2000 MW, respectively. The adapted parameters of the FOPID controller installed with the PV plant are k_{p1}, k_{i1}, k_{d1}, λ_{d1}, and μ_1, while those of the controller of the thermal plant are k_{p2}, k_{i2}, k_{d2}, λ_{d2}, and μ_2. The source of contingency is a sudden change in the load in one area, which leads to high oscillations in the power system frequency and exchange power. This oscillation must be damped via LFC.

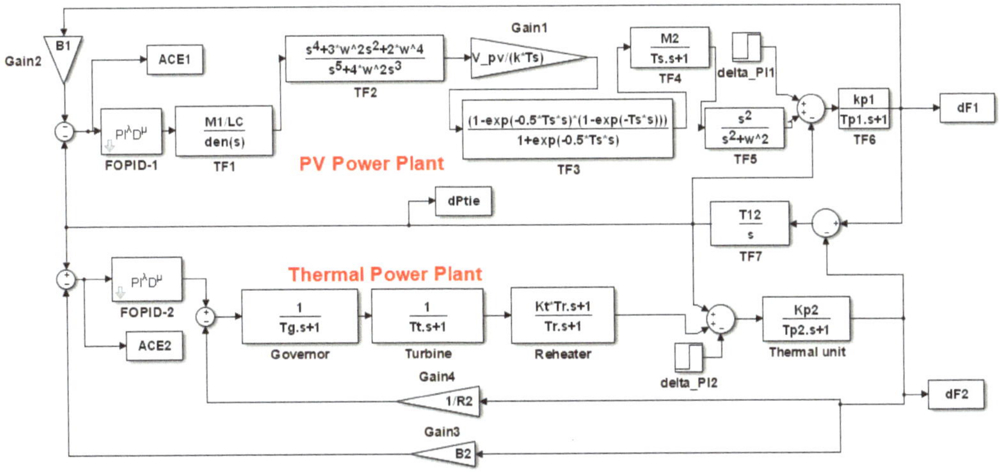

Figure 7. PV/thermal system in Simulink.

The first considered load disturbance is 10% on the PV plant. The optimal gains of the FOPID and the value of the best fitness function obtained via the proposed SOA and the others during ΔP_{L1} = 10% are tabulated in Table 1. The results reveal that the best fitness value is 6.7506 obtained via the proposed SOA; the TDO comes in the second rank achieving ITAE of 6.7558. On the other hand, the worst optimizer is AOS, with a fitness value of 8.0539. Moreover, the elapsed time of each optimizer is measured. The proposed SOA is not the best in terms of computational time; however, it achieved the best fitness value which is the target of solving the considered problem. The variations of fitness value versus the iteration number for all considered approaches are shown in Figure 8. Moreover, the statistical parameters that assess the optimizer performance during the iteration process are given in Table 2. As the reader can see, the proposed SOA achieved the best variance and standard deviation (Std. dev.) with values of 1.1083 and 0.5277, respectively, outperforming all other optimizers. Furthermore, the time responses of ΔF_1, ΔF_2, and ΔP_{tie} for the PV/thermal system at ΔP_{L1} = 10% are shown in Figure 9. The curves confirm the preference for the FOPID designed via the proposed SOA in such a case.

Table 1. FOPID parameters at ΔP_{L1} = 10% in PV/thermal system.

	CBOA	JSOA	BO	TDO	AOS	SOA
k_{p1}	1.0000	1.0000	1.0000	0.99998	1.0000	1.0000
k_{i1}	1.0000	1.0000	1.0000	0.48677	0.97191	0.9971
k_{d1}	1.0000	1.0000	1.0000	1.0000	1.0000	1.0000
λ_{d1}	0.7725	0.23273	0.64308	0.33975	0.98977	0.35946
μ_1	0.65477	1.0000	0.69726	0.83521	0.65565	0.87552
k_{p2}	1.0000	1.0000	1.0000	0.26872	0.99882	0.99059
k_{i2}	1.0000	1.0000	1.0000	0.99909	1.0000	0.8911
k_{d2}	0.22547	1.0000	0.086092	0.99938	1.0000	0.13447
λ_{d2}	0.53196	1.0000	0.00000	0.20817	1.0000	0.87058
μ_2	0.83065	1.0000	0.53543	0.77218	1.0000	0.977
Elapsed time (Sec.)	8924.533	3868.488	4265.285	6622.671	4707.0259	8151.158
Fitness value	7.1173	8.0539	6.9705	6.7558	8.0943	6.7506

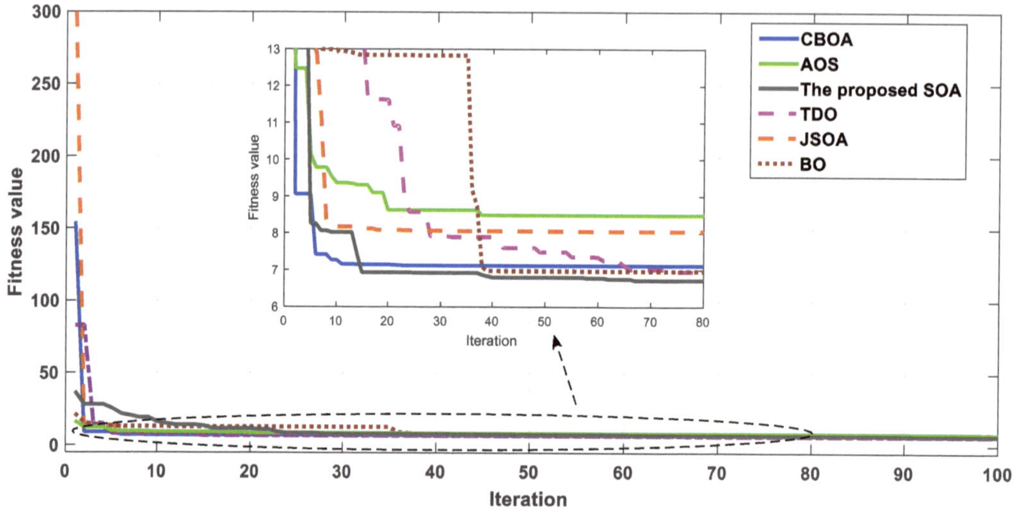

Figure 8. The fitness value versus the iteration number for PV/thermal system at ΔP_{L1} = 10%.

Table 2. Statistical parameters of the considered optimizers for PV/thermal system at ΔP_{L1} = 10%.

	CBOA	JSOA	BO	TDO	AOS	SOA
Best	7.1173	8.0539	6.9705	6.6985	8.0943	6.7506
Worst	9.8586	16.552	98.501	9.2745	11.947	40.980
Mean	7.9378	12.403	26.497	7.4199	9.8271	11.064
Median	7.6988	12.107	10.046	7.3483	9.3298	7.7636
Variance	0.6422	16.461	994.99	0.5719	1.8280	1.1083
Std. dev.	0.8013	4.0572	31.544	0.7562	1.3520	0.5277

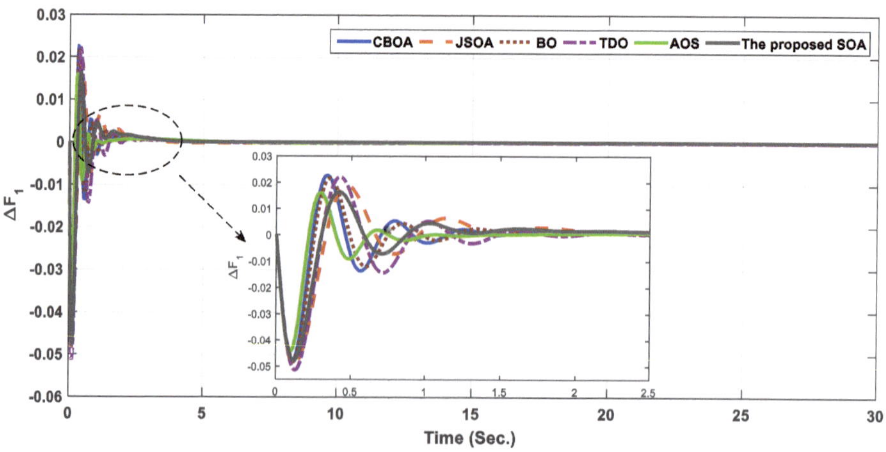

(**a**)

Figure 9. *Cont.*

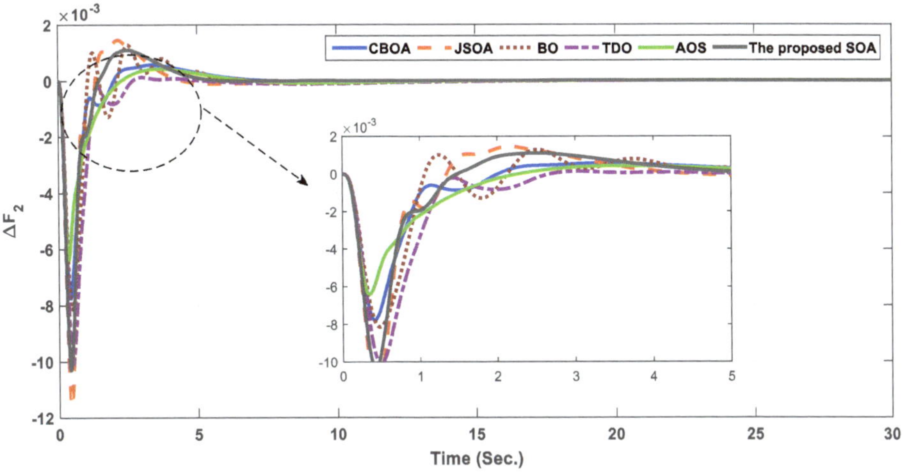

(**b**)

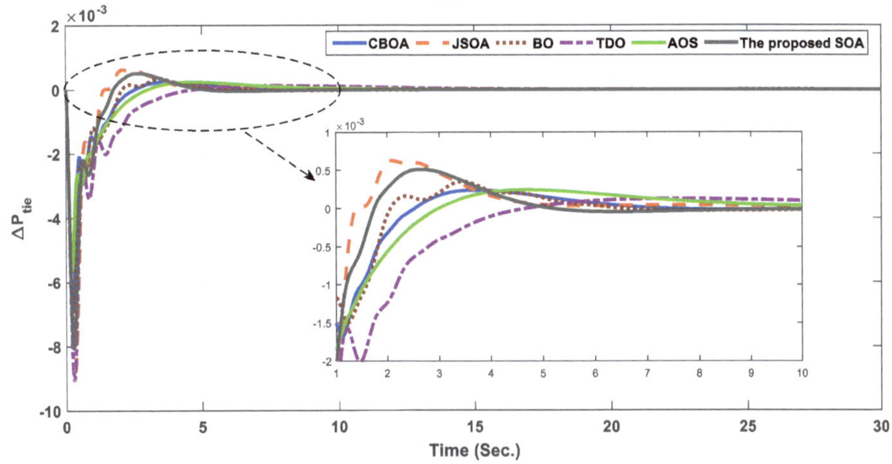

(**c**)

Figure 9. The time responses of (**a**) ΔF_1, (**b**) ΔF_2, and (**c**) ΔP_{tie} for PV/thermal system at ΔP_{L1} = 10%.

To confirm the proposed SOA validity, statistical tests were conducted using the Friedman ANOVA table, Wilcoxon rank test, Friedman rank test, and Kruskal Wallis test. The results of these tests are tabulated in Table 3. The Wilcoxon test results revealed that, JSOA and AOS reject the null equal medians while CBOA, BO, and TDO accept the null equal medians. Therefore, the SOA has a significant difference from the other approaches. Regarding the Friedman rank test, the SOA came in the first rank with a value of 3.2, then CBOA, TDO, BO, AOS, and JSOA came next to the proposed approach. The *p*-values for the Friedman, ANOVA, and Kruskal Wallis tests were 6.8140e-05, 0.0308, and 1.3583e-04, respectively. The Friedman test *p*-value confirmed the significant difference in column means, while Kruskal Wallis *p*-value revealed the rejection of the null hypothesis of data with the same distribution. The fitness functions during separate trials via ANOVA Wallis for the PV/thermal system at ΔP_{L1} = 10% are shown in Figure 10. Regarding the boxplots given in Figure 10, the penguins obtained via the proposed SOA, TDO, and CBOA have the smallest flippers, while the others have larger flippers. This confirms that CBOA, TDO, and the proposed SOA reject the null hypothesis.

Table 3. Statistical tests of Friedman ANOVA, Wilcoxon rank, Friedman rank, and Kruskal Wallis for PV/thermal system at ΔP_{L1} = 10%.

		CBOA	JSOA	BO	TDO	AOS	SOA
Wilcoxon rank test	*p*-value	0.7913	0.0090	0.1405	0.1212	0.0073	-
	h-value	0	1	0	0	1	-
Null hypothesis rejection		×	√	×	×	√	-
Friedman rank		4.4	9.6	8.1	5.0	8.7	3.2
Friedman test *p*-value		6.8140e-05					
p-value based on ANOVA		0.0308					
p-value based on Kruskal Wallis test		1.3583e-04					

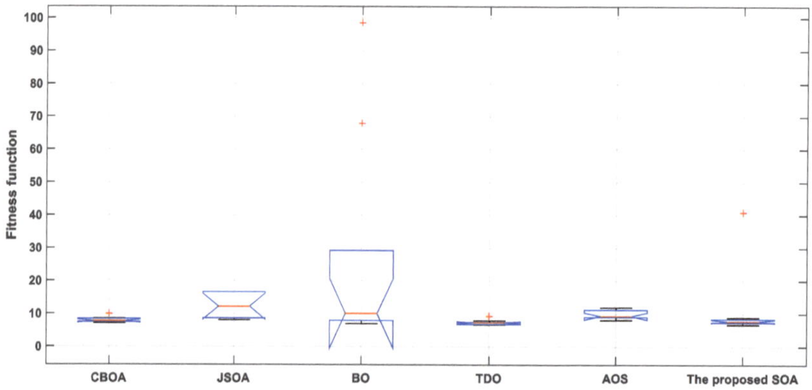

Figure 10. The fitness functions during separate trials via ANOVA Wallis for PV/thermal system at ΔP_{L1} = 10%.

A second load disturbance of 10% is assumed at the thermal plant: the optimal parameters of the controllers via SOA and others are given in Table 4 in addition to the other approaches. The proposed SOA succeeded in achieving the best ITAE of 1.8779, outperforming all considered approaches. The worst approach is JSOA, with a fitness value of 3.8959.

Table 4. FOPID parameters at ΔP_{L2} = 10% in PV/thermal system.

	CBOA	JSOA	BO	TDO	AOS	SOA
k_{p1}	0.36569	0.1000	0.1000	0.10003	0.11668	0.1001
k_{i1}	0.1	1.0000	0.1000	0.10485	0.10334	0.1
k_{d1}	0.3708	0.1000	0.71061	0.99984	0.11471	0.14113
λ_{d1}	0.8262	1.0000	0.61687	0.67797	0.72813	0.20952
μ_1	0.92171	0.90165	0.99761	0.94303	0.96683	0.78574
k_{p2}	0.87133	1.0000	0.99612	0.20311	0.11802	0.39052
k_{i2}	1.0000	1.0000	1.0000	0.99984	0.99783	1.0000
k_{d2}	1.0000	1.0000	1.0000	0.99577	0.96641	1.0000
λ_{d2}	0.28196	1.0000	0.2517	0.10308	0.96431	0.28383
μ_2	0.95274	0.1000	0.99088	0.87324	0.11028	0.73404
Fitness value	2.9092	3.8959	2.0181	2.0489	2.4266	1.8779

The performances of the approaches during the iterative process are shown in Figure 11; the curves confirmed the preference for the proposed SOA. The time responses of frequency and power exchange are given in Figure 12. The proposed SOA-based methodology achieved excellent performance for the PV/thermal interconnected power system subjected to different load disturbances.

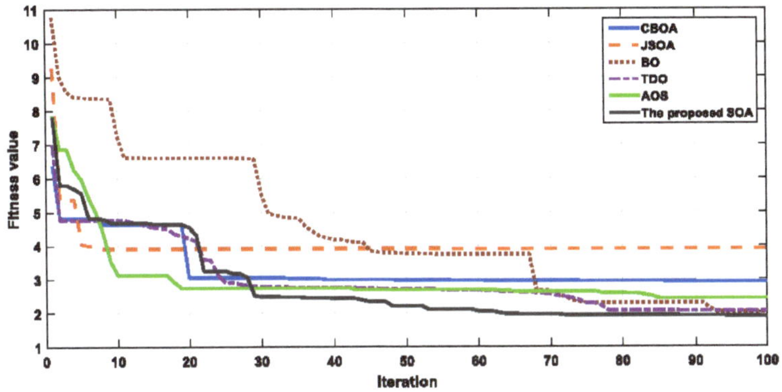

Figure 11. The fitness value versus the iteration number for PV/thermal system at ΔP_{L2} = 10%.

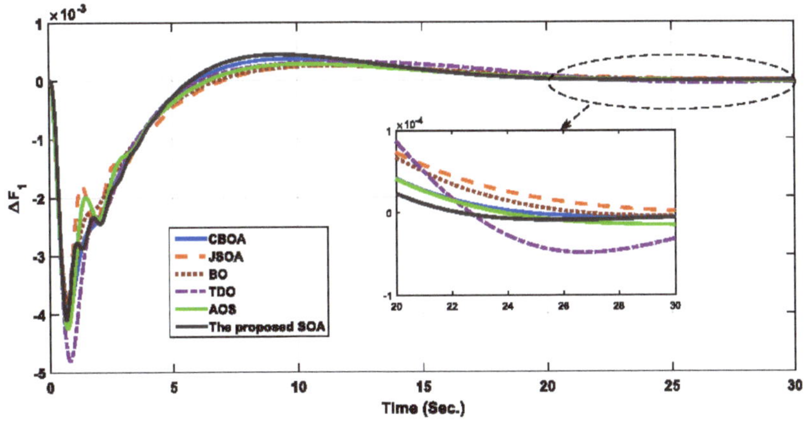

(a)

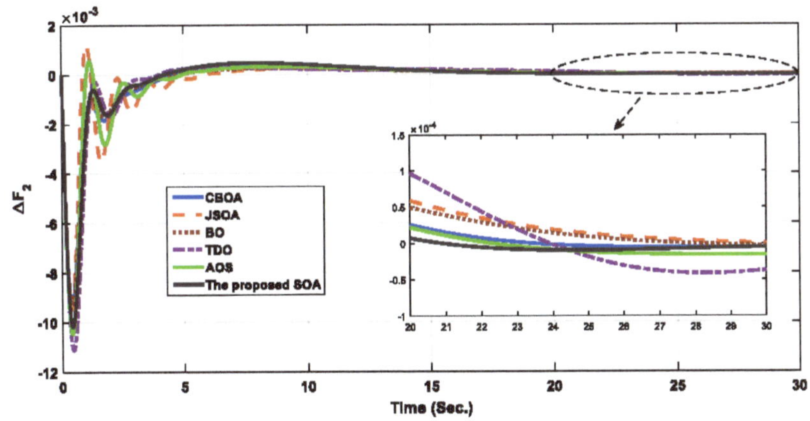

(b)

Figure 12. *Cont.*

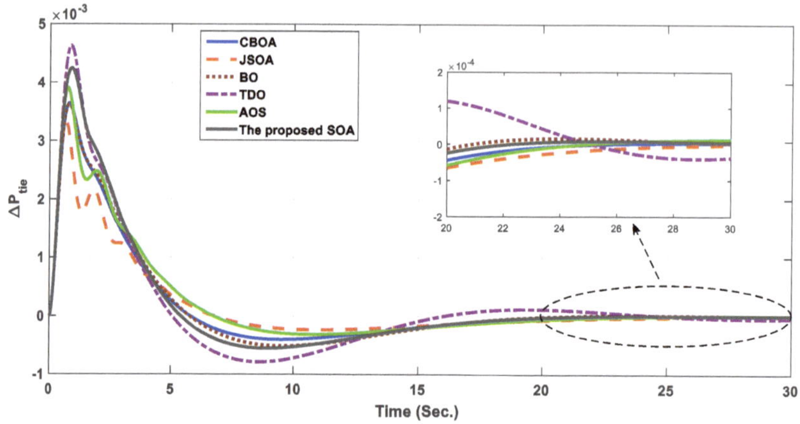

(c)

Figure 12. The time responses of (**a**) ΔF_1, (**b**) ΔF_2, and (**c**) ΔP_{tie} for PV/thermal system at ΔP_{L2} = 10%.

6.2. Four Interconnected Power System

Most power systems have many generating units, which may be conventional and/or renewable energy. In this section, the authors analyze the proposed LFC-FOPID with a thermal/WT/thermal/PV system; the GDB and GRC of thermal units are considered (see Figure 4). This configuration is constructed in Simulink/Matlab, as shown in Figure 13. In such a system, four FOPID controllers are installed with the four considered plants. The controller with the first thermal plant has adapted parameters of k_{p1}, k_{i1}, k_{d1}, λ_{d1}, and μ_1, while the second with the wind energy plant (Area 2) has k_{p2}, k_{i2}, k_{d2}, λ_{d2}, and μ_2. The third controller is installed with the second thermal plant (Area 3), it has adapted parameters of k_{p3}, k_{i3}, k_{d3}, λ_{d3}, and μ_3. The last controller is installed with the PV plant (Area 4) with parameters of k_{p4}, k_{i4}, k_{d4}, λ_{d4}, and μ_4. These parameters are identified via the proposed SOA. The first considered disturbance is 1% on the first thermal plant; the results obtained in such a case are tabulated in Table 5. The obtained results revealed that, the proposed SOA achieved the first rank in term of fitness value with a value of 0.0327, TDO comes in the second rank with a fitness value of 0.0368, while the worst optimizer is BO with ITAE of 0.3565. Moreover, the time responses of frequency and exchange power violations obtained via all considered approaches are shown in Figure 14; the zoomed curves show the better performance of SOA compared to the others. The time response performance specifications of frequency and tie-line power violations, including rise time (t_r), settling time (t_s), minimum settling time ($t_{s,min}$), maximum settling time ($t_{s,max}$), overshoot (OS), undershoot (Us), and peak time (t_p) are calculated and given in Table 6. These parameters are helpful in clarifying the preference for the proposed SOA in achieving system stability after an acceptable time.

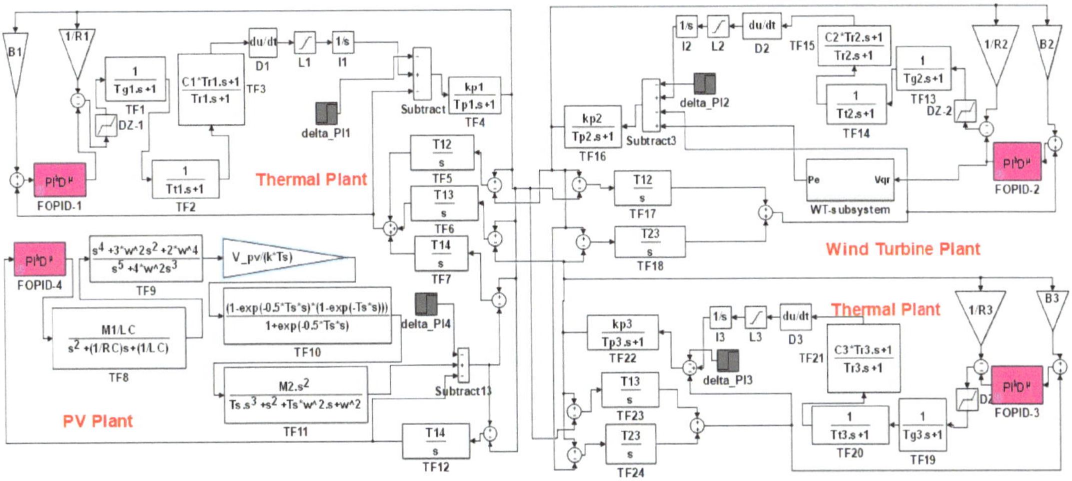

Figure 13. Simulink model of thermal/WT/thermal/PV interconnected system.

Table 5. FOPID parameters at ΔP_{L1} = 1% in thermal/WT/thermal/PV system.

	CBOA	JSOA	BO	TDO	AOS	SOA
k_{p1}	0.030335	0.015797	1.0000	0.01	0.013113	0.027242
k_{i1}	0.013487	0.020344	0.010001	0.102469	0.150434	0.014646
k_{d1}	0.01	0.01	0.0100	0.01	0.033762	0.032416
λ_{d1}	0.016799	0.01	0.999969	0.010007	0.0100	0.031366
μ_1	0.033807	0.01	0.010001	0.01	0.0100	0.03194
k_{p2}	0.054376	0.01	0.278846	0.010621	0.059971	0.024249
k_{i2}	0.014237	0.01	0.253138	0.029362	0.0100	0.032453
k_{d2}	0.010332	0.064301	0.066672	0.01001	0.0100	0.023304
λ_{d2}	0.014425	0.01	0.482488	0.493617	0.025474	0.012931
μ_2	0.032054	0.01	0.0100	0.119511	0.02015	0.030971
k_{p3}	0.023032	0.01	0.834948	0.018221	0.05298	0.012431
k_{i3}	0.025541	0.016533	0.858446	0.010075	0.085947	0.025044
k_{d3}	0.088584	0.065911	1.0000	0.170223	0.034121	0.031075
λ_{d3}	0.01	0.106448	0.642914	0.167097	0.0100	0.030966
μ_3	0.042053	0.012318	0.444737	0.01	0.32664	0.012818
k_{p4}	0.01	0.02	0.0100	0.036939	0.012773	0.018763
k_{i4}	0.01134	0.015331	0.020992	0.010394	0.032627	0.019244
k_{d4}	0.028713	0.01	0.47806	0.01	0.011926	0.015534
λ_{d4}	0.076208	0.028389	0.0100	0.012869	0.010865	0.0312
μ_4	0.01	0.018096	1.000	0.098181	0.253268	0.032036
Elapsed time (Sec.)	10,889.171	6675.141	6287.736	10,042.1075	6630.9976	9471.469
Fitness value	0.0511	0.06301	0.3565	0.0368	0.0571	0.0327

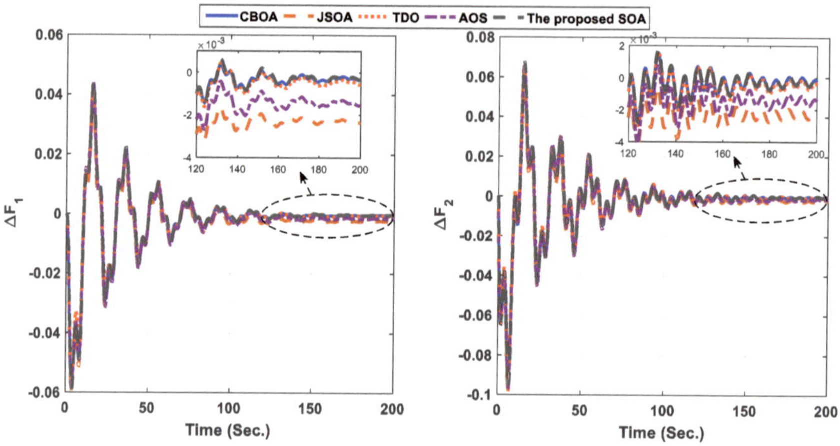

(a)

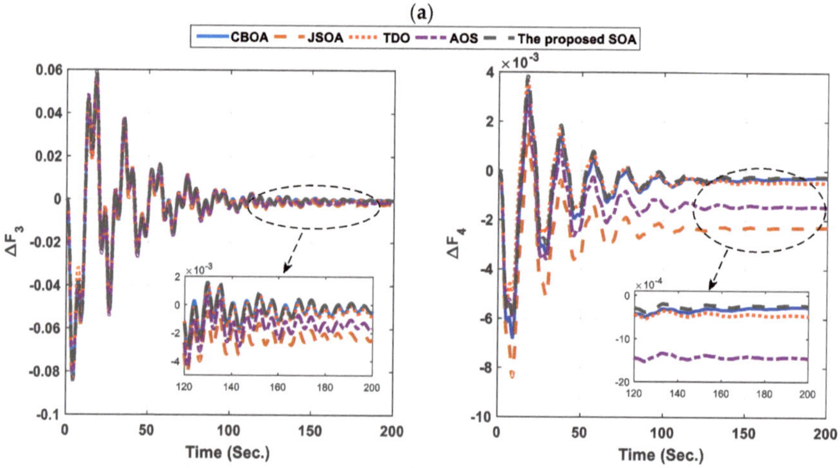

(b)

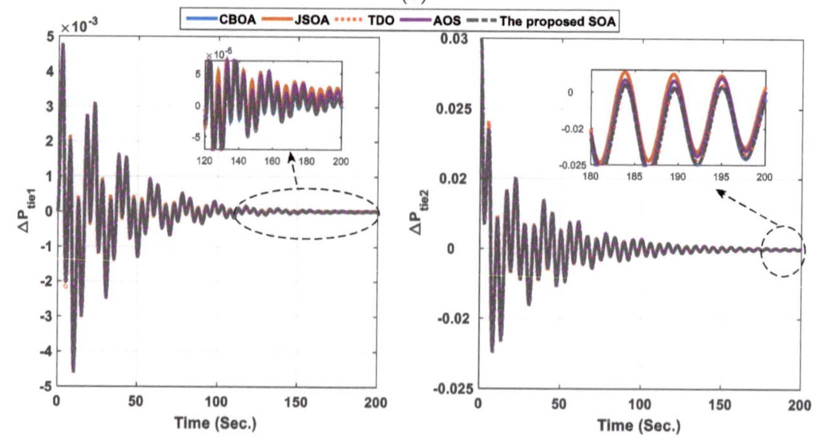

(c)

Figure 14. *Cont.*

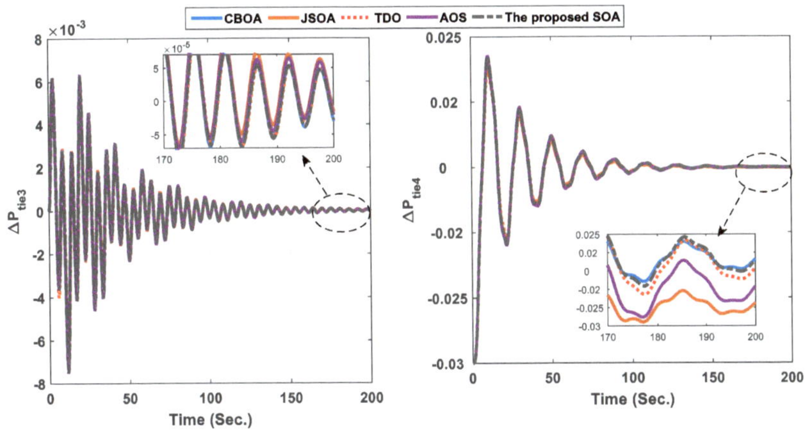

(**d**)

Figure 14. The time responses of (**a**) ΔF_1, ΔF_2 (**b**) ΔF_3, ΔF_4, (**c**) ΔP_{tie1}, $\Delta P_{tie2,}$ and (**d**) ΔP_{tie3}, ΔP_{tie4} of thermal/WT/thermal/PV system at ΔP_{L1} = 1%.

Table 6. Performance specifications of ΔF_1, ΔF_2, ΔF_3, ΔF_4, ΔP_{tie1}, $\Delta P_{tie2,}$ ΔP_{tie3}, and ΔP_{tie4} of four interconnected system at ΔP_{L1} = 1%.

		t_r (Sec.)	t_s (Sec.)	$t_{s,min}$ (Sec.)	$t_{s,max}$ (Sec.)	Os (pu)	Us (pu)	t_p (Sec.)
ΔF_1	CBOA	0.218126	112.4181	−0.05914	0.042865	17530.68	12779.44	3.967595
	JSOA	0.42883	112.3608	−0.05925	0.041337	2430.683	1765.517	3.976048
	BO	5.394888	99.65318	−0.04436	0.064188	533.3823	782.7876	6.314557
	TDO	0.259345	116.7282	−0.05873	0.040585	10456.03	7294.803	3.943959
	AOS	0.369491	116.4693	−0.05909	0.043968	3735.791	2854.158	3.967459
	SOA	0.213651	116.1438	−0.05913	0.044413	18632.35	14071.23	3.966584
		t_r (Sec.)	t_s (Sec.)	$t_{s,min}$ (Sec.)	$t_{s,max}$ (Sec.)	Os (pu)	Us (pu)	t_p (Sec.)
ΔF_2	CBOA	0.002726	124.7015	−0.09662	0.065934	47160.75	32251.03	6.758115
	JSOA	0.029639	124.4023	−0.09886	0.064269	4352.023	2894.324	6.707654
	BO	0.004961	99.71404	−0.04461	0.077011	37488.43	37156.86	73.48295
	TDO	0.005394	133.0857	−0.08998	0.063528	22146.13	15705.53	6.754989
	AOS	0.018928	132.897	−0.09696	0.067587	6734.498	4764.106	6.739581
	SOA	0.002528	132.5271	−0.09692	0.067685	51019.06	35699.57	6.745544
		t_r (Sec.)	t_s (Sec.)	$t_{s,min}$ (Sec.)	$t_{s,max}$ (Sec.)	Os (pu)	Us (pu)	t_p (Sec.)
ΔF_3	CBOA	0.243729	135.2427	−0.08438	0.057852	17936.97	12365.7	4.385474
	JSOA	0.432962	134.8931	−0.08456	0.05576	3334.456	2264.804	4.394015
	BO	0.661445	99.91756	−0.06307	0.087805	744.1215	1175.131	76.12008
	TDO	0.281637	135.799	−0.08369	0.056156	11618.06	7862.521	4.354946
	AOS	0.378296	135.5671	−0.0844	0.058481	4932.842	3487.269	4.386527
	SOA	0.240396	135.5081	−0.08441	0.059358	18688.96	13212.79	4.385127
		t_r (Sec.)	t_s (Sec.)	$t_{s,min}$ (Sec.)	$t_{s,max}$ (Sec.)	Os (pu)	Us (pu)	t_p (Sec.)
ΔF_4	CBOA	0.646564	129.2411	−0.00681	0.003278	2336.508	1173.331	9.174419
	JSOA	1.339719	125.0202	−0.00841	0.001541	265.737	67.00468	9.349339
	BO	77.84681	99.29976	0.013333	0.019746	21.57953	28.14631	84.9515
	TDO	0.772649	137.297	−0.00538	0.003661	1025.934	766.6746	5.254582
	AOS	1.150486	136.6	−0.00596	0.002898	308.7192	198.8507	9.114361
	SOA	0.625838	126.1221	−0.00607	0.00385	2357.405	1559.14	9.069103

Table 6. *Cont.*

		t_r (Sec.)	t_s (Sec.)	$t_{s,min}$ (Sec.)	$t_{s,max}$ (Sec.)	Os (pu)	Us (pu)	t_p (Sec.)
$\Delta P_{tie\text{-}1}$	CBOA	0.000637	118.5854	−0.00452	0.003022	161006.8	170204.2	2.364085
	JSOA	0.049687	118.455	−0.0046	0.004776	36540.76	35302.54	2.36612
	BO	0.882234	99.87703	−0.01834	−0.00146	104.4635	0	7.694692
	TDO	0.013741	118.7534	−0.00409	0.004767	505700.3	433561.9	2.360746
	AOS	0.035581	125.6115	−0.00461	0.004773	71280.16	68873.31	2.36558
	SOA	0.000582	118.6747	−0.00459	0.003083	178819.4	186219	2.366509
		t_r (Sec.)	t_s (Sec.)	$t_{s,min}$ (Sec.)	$t_{s,max}$ (Sec.)	Os (pu)	Us (pu)	t_p (Sec.)
$\Delta P_{tie\text{-}2}$	CBOA	6.057111	125.7488	−0.02237	−0.01003	49.54956	0	8.337896
	JSOA	6.007973	125.6896	−0.0224	−0.01006	49.87361	0	8.2875
	BO	1.163071	99.60019	−0.01051	0.011322	179.7168	259.602	80.0753
	TDO	6.183642	125.8925	−0.02206	−0.01013	47.45189	0	8.469989
	AOS	6.040323	125.8636	−0.02244	−0.00994	50.11094	0	8.328108
	SOA	6.046111	125.7889	−0.02242	−0.00996	49.86594	0	8.331606
		t_r (Sec.)	t_s (Sec.)	$t_{s,min}$ (Sec.)	$t_{s,max}$ (Sec.)	Os (pu)	Us (pu)	t_p (Sec.)
$\Delta P_{tie\text{-}3}$	CBOA	0.004796	159.2212	−0.00745	0.00622	26006.25	21791.22	11.3151
	JSOA	0.002346	159.1255	−0.00744	0.00625	53840.32	45316.67	11.271
	BO	1.105488	99.86355	−0.00686	0.005781	50.67886	178.6869	9.799218
	TDO	0.003439	159.3877	−0.00722	0.006008	33852.81	29085.48	11.43047
	AOS	0.003184	159.2393	−0.00752	0.00634	39593.22	33481.55	11.30361
	SOA	0.004757	159.2366	−0.00751	0.006278	26456.8	22202.1	11.30935
		t_r (Sec.)	t_s (Sec.)	$t_{s,min}$ (Sec.)	$t_{s,max}$ (Sec.)	Os (pu)	Us (pu)	t_p (Sec.)
$\Delta P_{tie\text{-}4}$	CBOA	3.811909	99.87232	0.00929	0.02338	55.94675	0	10.6354
	JSOA	3.795319	99.11462	0.00931	0.023423	56.73186	0	10.57403
	BO	0.472669	98.00973	−0.00196	0.024023	2113.074	180.1916	12.07693
	TDO	3.871557	110.1162	0.009571	0.022714	51.59852	0	10.77417
	AOS	3.765408	109.9693	0.009032	0.023574	57.54973	0	10.62046
	SOA	3.789229	100.6543	0.009114	0.023548	57.07933	0	10.63256

It is essential to investigate the designed FOPID controller via the proposed SOA during variable disturbances. Therefore, the second disturbance considered in the four interconnected power systems is variable step one, as shown in Figure 15. The time responses of the frequency and tie-line power violations for this case are shown in Figure 16. The proposed FOPID controllers succeeded in achieving a stable system by banishing the changes in frequencies and exchange powers occurring during that disturbance.

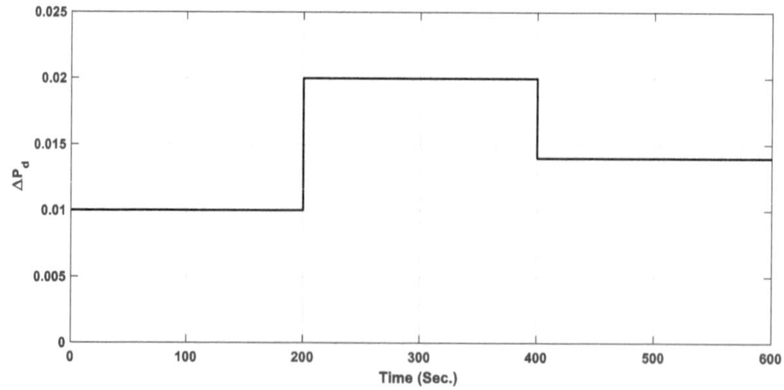

Figure 15. Pattern of variable disturbance.

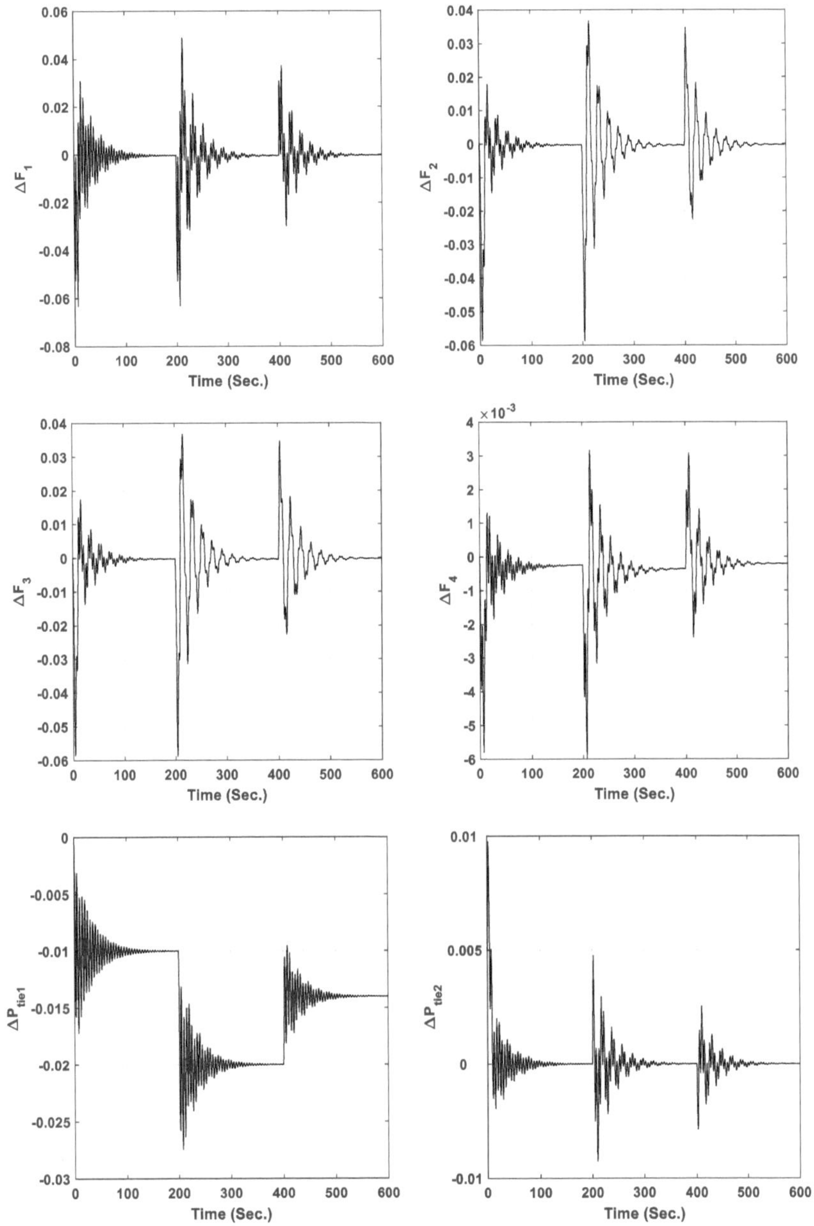

Figure 16. *Cont.*

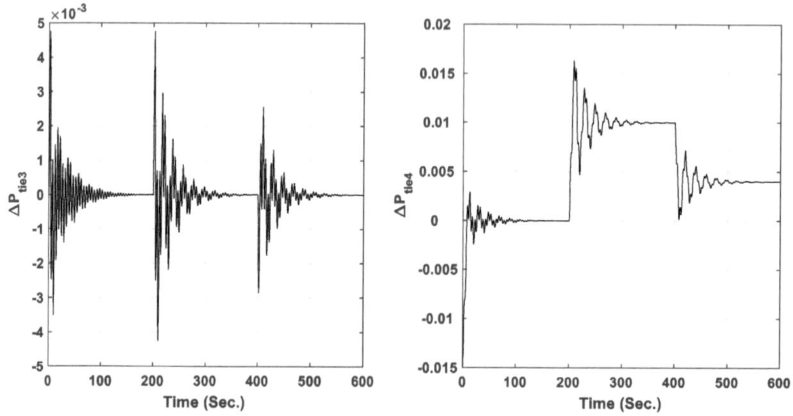

Figure 16. The time responses of frequency and exchange power violations for thermal/WT/thermal/PV system with variable disturbance in area 1.

Finally, the authors recommend the SOA as an efficient tool to design FOPID-LFC installed in multi-interconnected multi-sources power systems, as the obtained results confirmed its preferability in all studied load disturbances.

7. Conclusions

This work introduces a recent skill optimization algorithm to identify the optimal parameters of fractional-order proportional integral derivative load frequency control installed with an interconnected system, including renewable energy sources. The target is to mitigate the integral time absolute error of frequency and exchange power violations. Two-area and four-area power systems are analyzed. The first system has photovoltaic and thermal plants, while the second one comprises photovoltaic, wind turbine, and two thermal plants with governor dead-band and generation rate constraints. In the photovoltaic/thermal system, two disturbances are analyzed, which are 10% on the first plant and the same on the thermal plant. In the thermal/wind turbine/thermal/photovoltaic system, 1% and variable load disturbances on the first thermal area are simulated. The proposed algorithm is compared to chef-based optimization algorithm, jumping spider optimization algorithm, Bonobo optimization, Tasmanian devil optimization, and Atomic orbital search. The authors also conducted statistical tests (Friedman ANOVA table, Wilcoxon rank test, Friedman rank test, and Kruskal Wallis test) to assess the proposed approach. The best fitness values are obtained via the proposed approach with values of 1.8779 pu and 0.0327 pu for the two and four areas models, respectively. The capability and robustness of the proposed controller designed via the proposed approach are confirmed. Accelerating the response of the designed controllers in a multi-interconnected area via a hybrid optimization approach is recommended for future work. Moreover, investigating the stability of the system will be considered in the next paper.

Author Contributions: Data curation, A.F. and H.R.; Formal analysis, S.F.; Methodology, R.M.G. and R.A.; Software, M.M.G. All authors have read and agreed to the published version of the manuscript.

Funding: This work was funded by the Deanship of Scientific Research at Princess Nourah bint Abdulrahman University, through the Research Groups Program Grant no. (RGP-1443-0046).

Institutional Review Board Statement: Not applicable.

Informed Consent Statement: Not applicable.

Data Availability Statement: Not applicable.

Conflicts of Interest: The authors declare no conflict of interest.

Nomenclature

I	PV panel output current	r	Radius of turbine
V_{pv}	PV panel output voltage	V_w	Wind speed
n_s	Number of series cells	ρ	Air density
n_p	Number of parallel cells	A	Swept area of turbine blades
ε	Factor of completion	G_t	Steam turbine transfer function
k	Boltzmann constant	G_g	Governor transfer function
T	PV panel temperature	G_r	Reheater transfer function
Q	Electron charge	G_{gen}	Generator transfer function
G	Irradiance in W/m^2	K_t	Steam turbine gain
I_{ph}	Photo current	K_g	Governor gain
I_o	Saturation current	K_r	Reheater gain
R_s	Cell series resistance	K_p	Generator gain
P_{pv}	PV module output power	T_t	Steam turbine time constant
ω	Angular frequency of grid	T_g	Governor time constant
R	Converter output resistance	T_r	Reheater time constant
C	Converter output capacitance	T_p	Generator time constant
L	Converter output inductance	k_p, k_i, k_d, λ_d, and μ	Parameters of FOPID controller
T_s	Simulation time	K_{pw1}, K_{pw2}, and K_{pw3}	Wind plant gains
M_1	Buck converter voltage gain	T_{pw1}, T_{pw2}, and T_{pw3}	Wind plant time constants
M_2	Inverter converter voltage gain	$e(t)$	FOPID controller input
C_p	Power coefficient	ΔF_i	Violations in *i*th area frequency
λ	Tip ratio	$P_{tie,i}$	Violations in *i*th area exchange power
β	Pitch angle of blade	t	Specified time
ω_t	Mechanical angular speed of turbine	n_a	Number of interconnected plants

References

1. Shayeghi, H.; Shayanfar, H.; Jalili, A. Load frequency control strategies: A state-of-the-art survey for the researcher. *Energy Convers. Manag.* **2009**, *50*, 344–353. [CrossRef]
2. Xu, D.; Liu, J.; Yan, X.-G.; Yan, W. A Novel Adaptive Neural Network Constrained Control for a Multi-Area Interconnected Power System With Hybrid Energy Storage. *IEEE Trans. Ind. Electron.* **2017**, *65*, 6625–6634. [CrossRef]
3. Su, X.; Liu, X.; Song, Y.-D. Fault-Tolerant Control of Multiarea Power Systems via a Sliding-Mode Observer Technique. *IEEE/ASME Trans. Mechatron.* **2017**, *23*, 38–47. [CrossRef]
4. Yousri, D.; Babu, T.S.; Fathy, A. Recent methodology based Harris Hawks optimizer for designing load frequency control incorporated in multi-interconnected renewable energy plants. *Sustain. Energy Grids Netw.* **2020**, *22*, 100352. [CrossRef]
5. Ali, H.H.; Kassem, A.M.; Al-Dhaifallah, M.; Fathy, A. Multi-Verse Optimizer for Model Predictive Load Frequency Control of Hybrid Multi-Interconnected Plants Comprising Renewable Energy. *IEEE Access* **2020**, *8*, 114623–114642. [CrossRef]
6. Fathy, A.; Alharbi, A.G. Recent Approach Based Movable Damped Wave Algorithm for Designing Fractional-Order PID Load Frequency Control Installed in Multi-Interconnected Plants With Renewable Energy. *IEEE Access* **2021**, *9*, 71072–71089. [CrossRef]
7. Datta, A.; Bhattacharjee, K.; Debbarma, S.; Kar, B. Load frequency control of a renewable energy sources based hybrid system. In Proceedings of the 2015 IEEE Conference on Systems, Process and Control (ICSPC), Bandar Sunway, Malaysia, 18–20 December 2015; pp. 34–38. [CrossRef]
8. Ramesh, M.; Yadav, A.K.; Pathak, P.K. An extensive review on load frequency control of solar-wind based hybrid renewable energy systems. *Energy Sources Part A Recover. Util. Environ. Eff.* **2021**, 1–25. [CrossRef]
9. Fathy, A.; Kassem, A.M. Antlion optimizer-ANFIS load frequency control for multi-interconnected plants comprising photovoltaic and wind turbine. *ISA Trans.* **2019**, *87*, 282–296. [CrossRef]
10. Ranjan, M.; Shankar, R. A literature survey on load frequency control considering renewable energy integration in power system: Recent trends and future prospects. *J. Energy Storage* **2022**, *45*, 103717. [CrossRef]
11. Magdy, G.; Bakeer, A.; Alhasheem, M. Robust decentralized model predictive load-frequency control design for time-delay renewable power systems. *Int. J. Emerg. Electr. Power Syst.* **2021**, *22*, 617–628. [CrossRef]
12. El-Ela, A.A.A.; El-Sehiemy, R.A.; Shaheen, A.M.; Diab, A.E.-G. Design of cascaded controller based on coyote optimizer for load frequency control in multi-area power systems with renewable sources. *Control Eng. Pract.* **2022**, *121*, 105058. [CrossRef]
13. Arora, K.; Kumar, A.; Kamboj, V.K.; Prashar, D.; Shrestha, B.; Joshi, G.P. Impact of Renewable Energy Sources into Multi Area Multi-Source Load Frequency Control of Interrelated Power System. *Mathematics* **2021**, *9*, 186. [CrossRef]
14. Gulzar, M.M.; Iqbal, M.; Shahzad, S.; Muqeet, H.A.; Shahzad, M.; Hussain, M.M. Load Frequency Control (LFC) Strategies in Renewable Energy-Based Hybrid Power Systems: A Review. *Energies* **2022**, *15*, 3488. [CrossRef]

15. Latif, A.; Suhail Hussain, S.M.; Das, D.C.; Ustun, T.S. State-of-the-art of controllers and soft computing techniques for regulated load frequency management of single/multi-area traditional and renewable energy based power systems. *Appl. Energy* **2020**, *266*, 114858. [CrossRef]
16. Tungadio, D.H.; Sun, Y. Load frequency controllers considering renewable energy integration in power system. *Energy Rep.* **2019**, *5*, 436–453. [CrossRef]
17. Pandey, S.K.; Mohanty, S.R.; Kishor, N. A literature survey on load–frequency control for conventional and distribution generation power systems. *Renew. Sustain. Energy Rev.* **2013**, *25*, 318–334. [CrossRef]
18. Takayama, S.; Matsuhashi, R. Development of model for load frequency control in power system with large-scale integration of renewable energy. In Proceedings of the 2016 IEEE Power and Energy Conference at Illinois (PECI), Urbana, IL, USA, 19–20 February 2016; pp. 1–8. [CrossRef]
19. Khalid, J.; Ramli, M.A.; Khan, M.S.; Hidayat, T. Efficient Load Frequency Control of Renewable Integrated Power System: A Twin Delayed DDPG-Based Deep Reinforcement Learning Approach. *IEEE Access* **2022**, *10*, 51561–51574. [CrossRef]
20. Sobhy, M.A.; Abdelaziz, A.Y.; Hasanien, H.M.; Ezzat, M. Marine predators algorithm for load frequency control of modern interconnected power systems including renewable energy sources and energy storage units. *Ain Shams Eng. J.* **2021**, *12*, 3843–3857. [CrossRef]
21. El-Ela, A.A.A.; El-Sehiemy, R.A.; Shaheen, A.M.; Diab, A.E.-G. Enhanced coyote optimizer-based cascaded load frequency controllers in multi-area power systems with renewable. *Neural Comput. Appl.* **2021**, *33*, 8459–8477. [CrossRef]
22. Subham, G.T.; Ramachandran, R.; Kandasamy, J.; Muralidharan, R. Automatic Load Frequency Control of Renewable Energy Integrated Hybrid Power System. *J. Trends Comput. Sci. Smart Technol.* **2022**, *4*, 10–16. [CrossRef]
23. Abhilash, T.; Pavani, A.P. Multi area load frequency control of power system involving renewable and non-renewable energy sources. In Proceedings of the 2017 Innovations in Power and Advanced Computing Technologies (i-PACT), Vellore, India, 21–22 April 2017; pp. 1–5. [CrossRef]
24. Munisamy, V.; Sundarajan, R.S. Hybrid technique for load frequency control of renewable energy sources with unified power flow controller and energy storage integration. *Int. J. Energy Res.* **2021**, *45*, 17834–17857. [CrossRef]
25. Masuta, T.; Shimizu, K.; Yokoyama, A. Load Frequency Control by use of a Number of Both Heat Pump Water Heaters and Electric Vehicles in Power System with a Large Integration of Renewable Energy Sources. *IEEJ Trans. Power Energy* **2012**, *132*, 23–33. [CrossRef]
26. Dutta, A.; Prakash, S. Utilizing Electric Vehicles and Renewable Energy Sources for Load Frequency Control in Deregulated Power System Using Emotional Controller. *IETE J. Res.* **2022**, *68*, 1500–1511. [CrossRef]
27. Hasanien, H.; El-Fergany, A. Salp swarm algorithm-based optimal load frequency control of hybrid renewable power systems with communication delay and excitation cross-coupling effect. *Electr. Power Syst. Res.* **2019**, *176*, 105938. [CrossRef]
28. Huynh, V.; Minh, B.; Amaefule, E.; Tran, A.-T.; Tran, P. Highly Robust Observer Sliding Mode Based Frequency Control for Multi Area Power Systems with Renewable Power Plants. *Electronics* **2021**, *10*, 274. [CrossRef]
29. Irudayaraj, A.X.R.; Wahab, N.I.A.; Premkumar, M.; Radzi, M.A.M.; Bin Sulaiman, N.; Veerasamy, V.; Farade, R.A.; Islam, M.Z. Renewable sources-based automatic load frequency control of interconnected systems using chaotic atom search optimization. *Appl. Soft Comput.* **2022**, *119*, 108574. [CrossRef]
30. Rawat, S.; Jha, B.; Panda, M.K.; Rath, B.B. Load frequency control of a renewable hybrid power system with simple fuzzy logic controller. In Proceedings of the 2016 International Conference on Computing, Communication and Automation (ICCCA), Greater Noida, India, 29–30 April 2016; pp. 918–923. [CrossRef]
31. Guha, D.; Roy, P.K.; Banerjee, S. Equilibrium optimizer-tuned cascade fractional-order 3DOF-PID controller in load frequency control of power system having renewable energy resource integrated. *Int. Trans. Electr. Energy Syst.* **2021**, *31*, e12702. [CrossRef]
32. Saxena, A.; Shankar, R. Improved load frequency control considering dynamic demand regulated power system integrating renewable sources and hybrid energy storage system. *Sustain. Energy Technol. Assess.* **2022**, *52*, 102245. [CrossRef]
33. Xu, Y.; Li, C.; Wang, Z.; Zhang, N.; Peng, B. Load Frequency Control of a Novel Renewable Energy Integrated Micro-Grid Containing Pumped Hydropower Energy Storage. *IEEE Access* **2018**, *6*, 29067–29077. [CrossRef]
34. Barisal, A. Comparative performance analysis of teaching learning based optimization for automatic load frequency control of multi-source power systems. *Int. J. Electr. Power Energy Syst.* **2015**, *66*, 67–77. [CrossRef]
35. Tedesco, F.; Casavola, A. Load/Frequency Control in the presence of Renewable Energy Systems: A Reference-Offset Governor approach. *IFAC-PapersOnLine* **2020**, *53*, 12548–12553. [CrossRef]
36. Ahin, M.E.; Blaabjerg, F. A hybrid PV-battery/supercapacitor system and a basic active power control proposal in MATLAB/simulink. *Electronics* **2020**, *9*, 129.
37. Ahin, M.E.; Okumuş, H.İ. Parallel-Connected Buck–Boost Converter with FLC for Hybrid Energy System. *Electr. Power Compon. Syst.* **2021**, *48*, 2117–2129.
38. Walker, G. Evaluating MPPT converter topologies using a MATLAB PV model. *J. Electr. Electron. Eng. Aust.* **2001**, *21*, 49–55.
39. Ahin, M.E.; Okumuş, H.İ. Physical structure, electrical design, mathematical modeling and simulation of solar cells and modules. *Turk. J. Electromech. Energy* **2016**, *1*. [CrossRef]
40. Mustafa, E.Ž.; Adel, M.S.; Halil, O. A novel filter compensation scheme for single phase-self-excited induction generator micro wind generation system. *Sci. Res. Essays* **2012**, *7*, 3058–3072.

41. Renaudineau, H.; Donatantonio, F.; Fontchastagner, J.; Petrone, G.; Spagnuolo, G.; Martin, J.-P.; Pierfederici, S. A PSO-Based Global MPPT Technique for Distributed PV Power Generation. *IEEE Trans. Ind. Electron.* **2014**, *62*, 1047–1058. [CrossRef]
42. Adly, M.; Besheer, A. A meta-heuristics search algorithm as a solution for energy transfer maximization in stand-alone photovoltaic systems. *Int. J. Electr. Power Energy Syst.* **2013**, *51*, 243–254. [CrossRef]
43. Alam, M.K.; Khan, F.H. Transfer function mapping for a grid connected PV system using reverse synthesis technique. In Proceedings of the 2013 IEEE 14th Workshop on Control and Modeling for Power Electronics (COMPEL), Salt Lake City, UT, USA, 23–26 June 2013; pp. 1–5. [CrossRef]
44. Kassem, A.M.; Abdelaziz, A.Y. Reactive power control for voltage stability of standalone hybrid wind–diesel power system based on functional model predictive control. *IET Renew. Power Gener.* **2014**, *8*, 887–899. [CrossRef]
45. Kassem, A.M.; Yousef, A.M. Voltage and frequency control of an autonomous hybrid generation system based on linear model predictive control. *Sustain. Energy Technol. Assess.* **2013**, *4*, 52–61. [CrossRef]
46. Chekkal, S.; Lahaçani, N.A.; Aouzellag, D.; Ghedamsi, K. Fuzzy logic control strategy of wind generator based on the dual-stator induction generator. *Int. J. Electr. Power Energy Syst.* **2014**, *59*, 166–175. [CrossRef]
47. Sahu, R.K.; Gorripotu, T.S.; Panda, S. A hybrid DE–PS algorithm for load frequency control under deregulated power system with UPFC and RFB. *Ain Shams Eng. J.* **2015**, *6*, 893–911. [CrossRef]
48. Santy, T.; Natesan, R. Load frequency control of a two area system consisting of a grid connected PV system and diesel generator. *Int. J. Emerg. Technol. Comput. Electron.* **2015**, *13*, 456–461.
49. Podlubny, I. Fractional-order systems and fractional-order controllers. *Inst. Exp. Phys. Slovak Acad. Sci. Kosice* **1994**, *12*, 1–18.
50. Givi, H.; Marie, H. Skill optimization algorithm: A new human-based metaheuristic technique. *Comput. Mater. Contin.* **2022**, *74*, 179–202. [CrossRef]
51. Sahu, B.K.; Pati, S.; Panda, S. Hybrid differential evolution particle swarm optimisation optimised fuzzy proportional–integral derivative controller for automatic generation control of interconnected power system. *IET Gener. Transm. Distrib.* **2014**, *8*, 1789–1800. [CrossRef]

sustainability

MDPI

Article

System Design, Optimization and 2nd Law Analysis of a 100 MWe Double Reheat s-CO_2 Power Plant at Full Load and Part Loads

Sreekanth Manavalla [1,2], Feroskhan M. [2,*], Joseph Daniel [2], Sivakumar Ramasamy [2], T. M. Yunus Khan [3], Rahmath Ulla Baig [4], Naif Almakayeel [4] and Bhanu Kiran Voddin Tirumalapur [2]

1 Electric Vehicle Incubation, Testing and Research Centre, Vellore Institute of Technology Chennai, Chennai 600127, India; manavalla.sreekanth@vit.ac.in
2 School of Mechanical Engineering, Vellore Institute of Technology Chennai, Chennai 600127, India; joseph.daniel@vit.ac.in (J.D.); sivakumar.r@vit.ac.in (S.R.); vtbk1971@gmail.com (B.K.V.T.)
3 Department of Mechanical Engineering, College of Engineering, King Khalid University, Abha 61421, Saudi Arabia; yunus.tatagar@gmail.com
4 Department of Industrial Engineering, College of Engineering, King Khalid University, Abha 61421, Saudi Arabia; rub786@gmail.com (R.U.B.); halmakaeel@kku.edu.sa (N.A.)
* Correspondence: feroskhan.m@vit.ac.in

Abstract: Super-critical Carbon dioxide (s-CO_2) power plants are considered to be efficient and environmentally friendly compared to the traditional Rankine cycle-based steam power plants and Brayton cycle-based gas turbine power plants. In this work, the system design of a coal-fired 100 MWe double reheat s-CO_2 power plant is presented. The system is also optimized for efficiency with turbine inlet pressures and the recompression ratio as the variables. The components needed, mass flow rates of various streams and their pressures at various locations in the system have been established. The plant has been studied based on 1st and 2nd laws at full load and at part loads of 80%, 60% and 40%. Operating parameters such as mass flow rate, pressure and temperature have considerably changed in comparison to full load operation. It was also observed that the 1st law efficiency is 53.96%, 53.93%, 52.63% and 50% while the 2nd law efficiency is 51.88%, 51.86%, 50.61% and 48.1% at 100%, 80%, 60% and 40% loads, respectively. The power plant demonstrated good performance even at part loads, especially at 80% load, while the performance deteriorated at lower loads. At full load, the highest amount of exergy destruction is found in the main heater (36.6%) and re-heaters (23.2% and 19.6%) followed by the high-temperature recuperator (5.7%) and cooler (4.1%). Similar trends were observed for the part load operation. It has been found that the recompression ratio should be kept high (>0.5) at lower loads in order to match the performance at higher loads. Combustion and heat exchange due to finite temperature differences are the main causes of exergy destruction, followed by pressure drop.

Keywords: supercritical carbon dioxide; power plant; double reheat; optimization; exergy analysis; part load

1. Introduction

The modern world is undergoing rapid development which is reflected in the increased per capita energy consumption. Developing countries such as India are experiencing exponential growth and, at the same time, awareness of climate change is catching up and governments worldwide are looking at environmentally friendly power generation technologies that are also efficient and sustainable. Renewable energy sources are known to be environmentally friendly and even though power from solar and wind energy is becoming competitive [1], they are not reliable and sustainable without the support of bulky energy storage devices such as batteries. Conventional power generation technologies involving fossil fuels have the advantage of being mature and well understood while they suffer from an environmental regulations viewpoint [2]. Several innovative ways of utilizing fossil fuels and reducing their impact on the environment have been explored and are still being investigated [3]. Some of the methods are fluidized bed combustion [4], oxy-fuel combustion [5], chemical looping combustion [6], etc., which aim at reducing the pollutants at the combustion stage itself. In addition, the application of fossil fuels is being investigated in high-end thermal power plants such as ultra super-critical power plants [7]. However, they need expensive materials due to high pressure and temperatures. Some technologies are designed to handle the pollutants after the combustion process, and carbon capture and storage (CCS) is one among them [8]. It would be useful if the captured and stored carbon dioxide could be put to further use. One such application is to use carbon dioxide as a refrigerant while another application is to use it in power generation [9], such as in s-CO_2 power plants.

The primary benefit of using CO_2 as a working fluid in power generation is that it can replace water as a working fluid, thereby preserving and conserving a more important natural resource. Moreover, large amounts of CO_2 generated by burning fossil fuels (~963 tons/day by a 100 MWe plant) [10] can be utilized and thereby decrease its contribution to global warming. Other benefits of CO_2 as a working fluid are the fact that it is inert, non-toxic, non-corrosive, has low ozone depletion potential, is accessible and affordable, has good thermal characteristics for heat exchange, a low compression work around critical point, higher thermal efficiency and the equipment is compact due to high density [11]. In addition to fossil fuels, the necessary heat for a s-CO_2 power plant can also be sourced from geothermal, solar and/or waste heat energies [12].

Despite having several benefits mentioned above and having been introduced several decades ago in the 1960s by Angelino [13] and Feher [14], the power generation technology using s-CO_2 is still at the testing stage with test facilities in the 1–250 kW range and the largest pilot plant being of 10 MWe capacity built by the Gas Turbine Institute, Southwest Research Institute and General Electric Global Research together in San Antonio, Texas, USA [15]. Several review articles were recently published highlighting the benefits, drawbacks, state-of-the-art and future work to be taken up to make s-CO_2 power a reality at a higher scale [11,12,16–20]. Sarkar [16] emphasized the need for more experimental research on transcritical Rankine cycles to carry out a techno-economic comparison of CO_2 with other working fluids. Crespi et al. [17] have stated that high efficiencies in the range of 50–60% can be achieved by using combined cycle layouts involving s-CO_2. Ahn et al. [12] suggested that the focus should be on developing pilot power plants bigger than 10 MWe so as to explore the commercial benefits due to scale-up. Yin et al. [18] carried out a review on the application of s-CO_2 with concentrated solar power and suggested that future research should focus on the materials used, thermodynamic analysis at the design point and off-design point as well as transient modeling. White et al. [19] highlighted areas of research in the turbomachinery, heat exchangers, materials and control systems. Yu et al. [20] found that identifying an efficient cycle for a given heat source is an unanswered question and needs to be studied for various heat sources. They also suggested that CO_2-based gas mixtures also need to be evaluated for usage in these power plants.

Based on the reported research and review articles, it can be found that several thermodynamic configurations involving s-CO_2 have been studied. All these studies aimed at maximizing the thermal efficiency either by using a single cycle or a combined cycle. Sun et al. [21] proposed a bottoming cycle for a coal-fired s-CO_2 power plant to effectively utilize the heat in the exhaust gases. They reported an efficiency of 47% with a double reheat recompression cycle with CO_2 parameters as 620 °C temperature and 30 MPa pressure. Chen et al. [22] compared two recompression cycles, one with a single reheat and another with a double reheat. They found that the double reheat cycle is more efficient than the single reheat cycle. Moreover, they have evaluated the exergy destruction in the combustion, heat transfer and pressure drop and found that most of the exergy is destructed during combustion followed by heat transfer. Bai et al. [23] proposed an improved recompression s-CO_2 coal-fired power plant. Rogalev et al. [24] explored the possibility of replacing steam with s-CO_2 in a nuclear power plant and found it to be an attractive option for a working fluid temperature above 455 °C.

Some studies on the part load performance of s-CO_2 plants have been reported in the literature. Sanchez et al. [25] analyzed a hybrid system comprising a molten carbonate fuel cell and s-CO_2 cycle to evaluate the control strategies at part loads and concluded that mixed strategies give better results compared to individual control strategies. Fan et al. [26] considered simple supercritical and trans-critical plants to conduct thermodynamic analysis and determine the control strategy. Yang et al. [27] studied four simple and separate layouts powered by solar energy and compared different cycles at full and part loads. Tong et al. [28] studied a coal-powered single reheat recompression cycle and evaluated the boiler's part load performance. Lee et al. [29] developed a CFD model to study the part load performance of radial flow turbines used in s-CO_2 plants with a maximum deviation of 10%. Alfani et al. [30] considered four different system layouts powered by waste heat and carried out a 1st law and techno-economic analysis and optimization. They also compared the performance with organic Rankine cycle technology and concluded that s-CO_2 has a better performance even at part loads. Fan and Dai [31] combined a simple s-CO_2 with the Kalina cycle and a recompression s-CO_2 with the Kalina cycle and carried out thermo-economic analysis and optimization of the system. They found the recompression-Kalina cycle to be a better performer both thermodynamically and economically. Fan et al. [32] studied a nuclear energy-powered s-CO_2 plant combined with an organic Rankine cycle to carry out 1st law and economic analysis. Wang et al. [33] considered a fuel oil-powered recompression cycle and compared different control strategies at part loads. They concluded that inventory and anti-surge control is better than valve controls. Xingyan et al. [34] studied four different cycles and evaluated their performance based on layout and control strategies. Gini et al. [35] evaluated a simple recuperated cycle powered by molten salt for control strategy development at part loads. It is evident that some studies on the part load performance of s-CO_2 plants have been reported, and a few studied the specific components [28,29] in the plant. There is no study that reported a part load study on a double reheat recompression cycle powered by coal. Table 1 summarizes the reported literature on part load studies of s-CO_2 power plants.

Table 1. Part load or off-design studies reported in the literature.

S. No	Reference	Source of Energy	Type of System	Objective
1	Sanchez et al. [25]	Natural gas	MCFC and s-CO_2 hybrid system	1st law analysis, inventory control strategy
2	Fan et al. [26]	Nuclear energy	Combined super critical and trans-critical	1st and 2nd law analysis, identify a control strategy
3	Yang et al. [27]	Solar energy	Simple recuperative, reheat, recompression and intercooling cycles studied separately	1st law analysis, performance comparison between different cycles
4	Tong et al. [28]	Coal	Single reheat recompression	Focusses on boiler furnace part load performance
5	Lee et al. [29]	-	-	Focusses on turbine part load performance
6	Alfani et al. [30]	Waste heat	Simple recuperative, simple recuperative cycle with bypass and turbine split flow configurations	1st law and Techno-economic analysis and optimization
7	Fan and Dai [31]	Nuclear energy	Combined s-CO_2 and Kalina cycle	1st and 2nd law analysis, Thermo-economic optimization and evaluate different control strategies

Table 1. *Cont.*

S. No	Reference	Source of Energy	Type of System	Objective
8	Fan et al. [32]	Nuclear energy	Combined s-CO_2 with Organic Rankine Cycle	1st law and economic analysis
9	Wang et al. [33]	Fuel oil	Recompression s-CO_2 Brayton cycle	Comparison of control strategies
10	Xingyan et al. [34]	-	Simple recuperated, recompression, single reheat, intercooling cycles considered separately	Performance evaluation based on layout and control strategies
11	Gini et al. [35]	Molten salt	Simple recuperated s-CO_2 cycle	Control strategy development
	Present work	Coal	Double reheat recompression	System design, 1st and 2nd law analysis and parameter optimization for better efficiency

From the summary of the literature presented above, it can be seen that several thermodynamic cycle layouts have been investigated and reported in the open literature. However, to the best of the authors' knowledge, there is no study which considered the system design and optimization of a 100 MWe double reheat s-CO_2 power plant and thermodynamic analysis at part loads. Since the existing systems would eventually be scaled up and part load operation would be inevitable and important, as emphasized by Yin et al. [18], the present study aims to predict the performance at full load as well as part loads of an s-CO_2 power plant. Since it is generally understood that a double reheat recompression cycle has a high efficiency in comparison to other configurations [17,22], the same has been considered in the present study. The objectives of the present study are listed below:

i. To perform a system design of a 100 MWe coal-fired double reheat recompression s-CO_2 system based on the parameters and configurations reported by Chen et al. [22].
ii. To evaluate its performance at full load as well as at part loads of 80%, 60% and 40% based on 1st and 2nd laws of thermodynamics.
iii. To optimize the performance based on thermal efficiency (1st law efficiency) at full load.
iv. Identify the crucial operational parameters that influence thermal efficiency.
v. Identify the system components that contribute to high exergy destruction.

2. System Layout Description

The s-CO_2 power plant system layout considered for the present study is a double reheat recompression cycle, as described by Chen et al. [22]. The schematic of the layout is shown in Figure 1. Carbon dioxide is heated to the rated supercritical condition in the heater using the heat generated by coal firing. This s-CO_2 expands in the high-pressure turbine, then enters the reheater-1 to undergo a rise in temperature, further expands in the intermediate-pressure turbine followed by a temperature rise in reheater-2 and, finally, expands in the low-pressure turbine. The low-pressure turbine exhaust is still at moderate pressure and temperature and its energy is partly recovered in the high and low-temperature recuperators in which heat is transferred from the low-pressure turbine exhaust to the CO_2 emerging from the compressors. This is meant to save input heat energy in the form of fuel. After the low-temperature recuperator, the s-CO_2 is split into two streams, one of which is cooled in the cooler using ambient air before it enters the main compressor while the other stream is directed into the re-compressor. The main compressor handles cooler s-CO_2 and hence saves energy. After compression and heat exchange in the low-temperature recuperator, both the compressor exit streams are mixed in a direct contact heat exchanger named a 'junction' in Figure 1. Subsequently, heat exchange takes place in the high-temperature recuperator and the CO_2 enters the heater to receive heat from coal combustion. In the figure, the coal flow stream and flue gas stream have not been shown for the sake of simplicity. Part of the power produced by the turbines is consumed by the compressors as indicated by the shaft connecting the turbines and compressors.

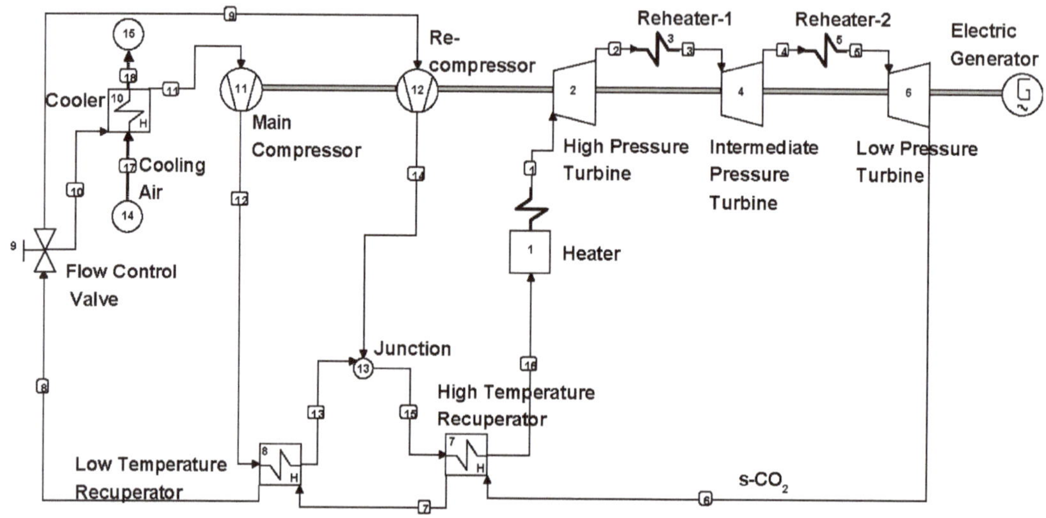

Figure 1. Schematic layout of the double reheat recompression s-CO_2 power generation system.

The corresponding Temperature vs. Entropy diagram is shown in Figure 2. Important processes have been marked in the T-S diagram.

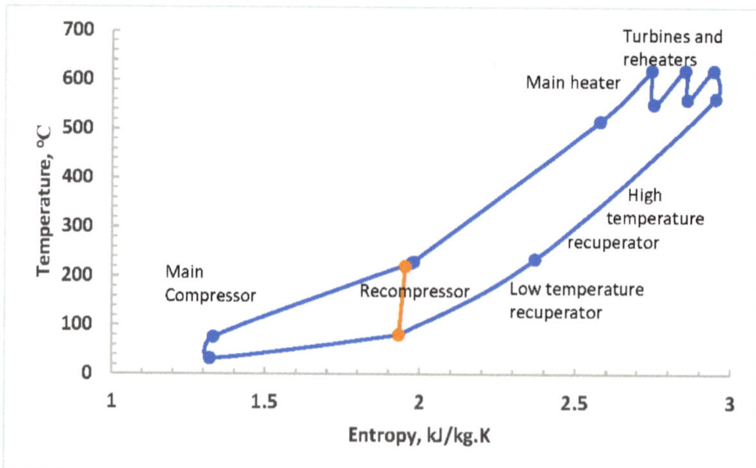

Figure 2. Temperature–Entropy diagram of the double reheat recompression s-CO_2 plant at full load.

The fuel properties and important operating parameters of the plant are shown in Table 2. Operating parameters are chosen in such a way that the CO_2 remains in a super critical state through the entire operation.

Table 2. Fuel and operational parameters used in the simulation at full load [22].

Parameter	Value
Lower heating value of coal	26.51 MJ/kg
Exergy of coal	27.57 MJ/kg
Pressure/Temperature of CO_2 at HP turbine inlet	320 bar/620 °C
Temperature of CO_2 at IP and LP turbine inlet	620 °C
Pressure of CO_2 at IP turbine inlet	192.3 bar
Pressure of CO_2 at LP turbine inlet	123.3 bar
Pinch points in heat exchangers	5 °C
Turbine isentropic efficiency	93%
Compressor isentropic efficiency	90%
Pressure drop in heater	1 bar
Pressure drop in recuperators	0.5 bar
Recompression ratio	0.683

Model Formulation:

The following assumptions were made in the model formulation and analysis:

i. All components operate at a steady state.
ii. Changes in kinetic and potential energies are neglected for all fluid streams.
iii. All components are perfectly insulated and there is no heat exchange with the surroundings.
iv. The heat exchangers (recuperators and cooler) are the shell and tube type.
v. The dead state conditions are taken to be 1 bar and 30 °C.

The following equations are framed and solved for all the components in the steady flow system.

$$\textbf{Mass Balance}: \quad \sum_{in} \dot{m} = \sum_{out} \dot{m} \tag{1}$$

$$\textbf{Energy Balance}: \quad \sum_{in} \dot{E} = \sum_{out} \dot{E} \tag{2}$$

$$\text{For multiple streams,} \quad \dot{Q} - \dot{W} = \sum_{out} \dot{m}h - \sum_{in} \dot{m}h \tag{3}$$

$$\text{For multiple streams,} \quad \dot{Q} - \dot{W} = \dot{m}[h_2 - h_1] \tag{4}$$

$$\textbf{Energy Balance}: \quad \underbrace{\dot{X}_{in} - \dot{X}_{out}}_{\substack{\text{Rate of net exergy} \\ \text{transfer by heat,} \\ \text{work, and mass}}} - \dot{X}_{destroyed} = 0 \tag{5}$$

$$\dot{X}_{heat} = \left(1 - \frac{T_0}{T}\right)\dot{Q} \tag{6}$$

$$\dot{X}_{work} = \dot{W}_{useful} \tag{7}$$

$$\dot{X}_{mass} = \dot{m}\psi \tag{8}$$

$$\psi = (h - h_0) - T_0(s - s_0) \tag{9}$$

$$\dot{X}_{destroyed} = T_0\dot{S}_{gen} \tag{10}$$

$$\eta_I = \frac{\text{Net Power}}{\dot{m}_{\text{fuel}} \times \text{LCV of fuel}} \tag{11}$$

$$\eta_{II} = \frac{\text{Net Power}}{\dot{m}_{\text{fuel}} \times \text{Exergy of fuel}} \tag{12}$$

$$\text{Fuel exergy} = 1.04 \times \text{LCV of fuel} \tag{13}$$

$$\text{Recompression ratio, } \Phi = \frac{\text{Main compressor flow rate}}{\text{Total flow rate}} \tag{14}$$

The chemical exergy factorized in Equation (13), 1.04, is estimated using the method proposed by Kotas [36]. The individual equations for mass, energy and exergy balance for each component are shown in Table 3. Heat and power terms shown in Table 2 are positive quantities. The subscripts in the equations correspond to the pipe numbers shown in Figure 1. The exergy destroyed in the equations corresponds to the respective component. Exergy associated with heat transfer is computed using Equation (6) while the flow exergy of each component is evaluated using Equation (9).

Table 3. Equations of mass, energy and exergy balance of individual components.

Component	Mass Balance	Energy Balance	Exergy Balance
Heater	$\dot{m}_{16} = \dot{m}_1$	$\dot{Q} = \dot{m}_1(h_1 - h_{16})$	$\dot{X}_{16} + \dot{X}_Q - \dot{X}_1 - \dot{X}_{des} = 0$
High Pressure Turbine	$\dot{m}_1 = \dot{m}_2$	$\dot{W} = \dot{m}_1(h_1 - h_2)$	$\dot{X}_1 - \dot{W} - \dot{X}_2 - \dot{X}_{des} = 0$
Reheater-1	$\dot{m}_2 = \dot{m}_3$	$\dot{Q} = \dot{m}_2(h_3 - h_2)$	$\dot{X}_2 + \dot{X}_Q - \dot{X}_3 - \dot{X}_{des} = 0$
Intermediate Pressure Turbine	$\dot{m}_3 = \dot{m}_4$	$\dot{W} = \dot{m}_3(h_3 - h_4)$	$\dot{X}_3 - \dot{W} - \dot{X}_4 - \dot{X}_{des} = 0$
Reheater-2	$\dot{m}_4 = \dot{m}_5$	$\dot{Q} = \dot{m}_4(h_5 - h_4)$	$\dot{X}_4 + \dot{X}_Q - \dot{X}_5 - \dot{X}_{des} = 0$
Low Pressure Turbine	$\dot{m}_5 = \dot{m}_6$	$\dot{W} = \dot{m}_5(h_5 - h_6)$	$\dot{X}_5 - \dot{W} - \dot{X}_6 - \dot{X}_{des} = 0$
High Temperature Recuperator	$\dot{m}_6 = \dot{m}_7$ $\dot{m}_{15} = \dot{m}_{16}$	$\dot{m}_6(h_6 - h_7) = \dot{m}_{15}(h_{16} - h_{15})$	$\dot{X}_6 + \dot{X}_{15} - \dot{X}_7 - \dot{X}_{16} - \dot{X}_{des} = 0$
Low Temperature Recuperator	$\dot{m}_7 = \dot{m}_8$ $\dot{m}_{12} = \dot{m}_{13}$	$\dot{m}_7(h_7 - h_8) = \dot{m}_{12}(h_{13} - h_{12})$	$\dot{X}_7 + \dot{X}_{12} - \dot{X}_8 - \dot{X}_{13} - \dot{X}_{des} = 0$
Flow Control valve	$\dot{m}_8 = \dot{m}_9 + \dot{m}_{10}$	$\dot{m}_8h_8 = \dot{m}_9h_9 + \dot{m}_{10}h_{10}$	$\dot{X}_8 - \dot{X}_9 - \dot{X}_{10} - \dot{X}_{des} = 0$
Cooler	$\dot{m}_{10} = \dot{m}_{11}$ $\dot{m}_{17} = \dot{m}_{18}$	$\dot{m}_{10}(h_{10} - h_{11}) = \dot{m}_{17}(h_{18} - h_{17})$	$\dot{X}_{10} + \dot{X}_{17} - \dot{X}_{11} - \dot{X}_{18} - \dot{X}_{des} = 0$
Main Compressor	$\dot{m}_{11} = \dot{m}_{12}$	$\dot{W} = \dot{m}_{11}(h_{12} - h_{11})$	$\dot{X}_{11} + \dot{W} - \dot{X}_{12} - \dot{X}_{des} = 0$
Re-Compressor	$\dot{m}_9 = \dot{m}_{14}$	$\dot{W} = \dot{m}_9(h_{14} - h_9)$	$\dot{X}_9 + \dot{W} - \dot{X}_{14} - \dot{X}_{des} = 0$
Junction	$\dot{m}_{13} + \dot{m}_{14} = \dot{m}_{15}$	$\dot{m}_{13}h_{13} + \dot{m}_{14}h_{14} = \dot{m}_{15}h_{15}$	$\dot{X}_{13} + \dot{X}_{14} - \dot{X}_{15} - \dot{X}_{des} = 0$

The equations shown in Table 3 are framed and solved by the flow-sheeting software Cycle-Tempo 5.1 based on the inputs given using the graphical user interface. The property values of carbon dioxide are computed using RefProp 10 add-in software.

The system shown in Figure 1 and the inputs shown in Table 3 are used to solve the system and the output is validated by comparing the present results with those of Chen et al. [22] after adjusting some inputs so as to make the model relevant to theirs. The results have been compared to validate the present model and are shown in Table 4.

Table 4. Model validation by comparing with the results of Chen et al. [22].

	Temperature, °C			Pressure, bar			Mass Flow Rate, kg/s		
S. No.	Chen et al. [22]	This Work	% Error	Chen et al. [22]	This work	% Error	Chen et al. [22]	This Work	% Error
1	620	620	0	320	320	0	39.565	40.982	3.58
2	550.92	550.76	0.03	192.3	192.3	0	39.565	40.982	3.58
3	620	620	0	192.3	192.3	0	39.565	40.982	3.58
4	560.73	560.46	0.05	123.3	123.3	0	39.565	40.982	3.58
5	620	620	0	123.3	123.3	0	39.565	40.982	3.58
6	561.9	562.88	0.17	79	80	1.27	39.565	40.982	3.58
7	235	235	0	79	79.5	0.63	39.565	40.982	3.58
8	86.73	86.73	0	79	79	0	39.565	40.982	3.58
9	86.73	86.73	0	79	79	0	27.023	27.991	3.58
10	32.65	32.44	0.64	79	79	0	27.023	27.991	3.58
11	80.89	77.7	3.94	334.5	334.5	0	27.023	27.991	3.58
12	230	225.42	1.99	334	334	0	27.023	27.991	3.58
13	229.7	229.57	0.06	334	334	0	12.542	12.991	3.58
14	515.63	516.13	0.1	333.5	333.5	0	39.565	40.982	3.58
15	543.63	545.5	0.34	1	1	0	5.965	6.011	0.77
16	506.7	512.72	1.19	1	1	0	5.534	5.532	0.04

The maximum error is found to be less than 4%. This could be due to the different software used by Chen et al. [22] and in the present work to compute fluid properties. Moreover, the convergence criteria of the two software could be different. Moreover, some commonly used parameters (such as the recompression ratio, chemical exergy factor) which are not provided by Chen et al. [22] have been assumed or indirectly estimated in the present work. Considering all these reasons, a maximum error of less than 4% is concluded to be reasonable and hence the present model is concluded to be satisfactory.

3. Results and Discussion

The results of mass and energy balance calculations are shown in the Figure 3 at full load (100 MWe) conditions. The compressors, turbines and generator are shown mounted on the same shaft indicating that part of the power generated in the turbines is consumed by the compressor while the rest is converted to electric energy in the generator. Due to the heat recovery processes, the amount of heat supplied in the main heater is lower and hence results in better efficiency. It can also be seen that the HP turbine, IP turbine and LP turbines produce 51 MW, 43.6 MW and 42.47 MW, respectively, while the main compressor and re-compressor consume 16.75 MW and 20.4 MW, respectively, resulting in a net power output of 100 MW. The exergy flow is shown in Figure 4. The total exergy supplied to the main heater and two re-heaters is 192.79 MW, out of which 100 MW is converted to net power output while 92.87 MW is destroyed in various components, with the major culprit being the main heater. This results in a 2nd law efficiency of 51.87%.

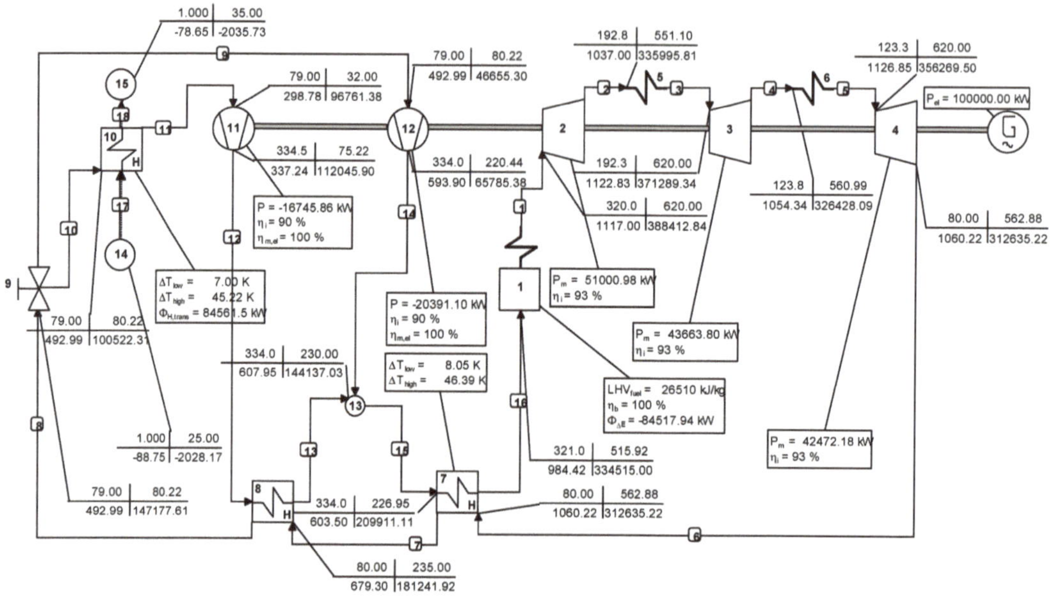

Figure 3. Values of pressure, temperature, enthalpy and mass flow rates at various locations in the double reheat 100 MW_e s-CO_2 power plant. The major legend is to be read as shown here p | T / h | Φ_{ex}.

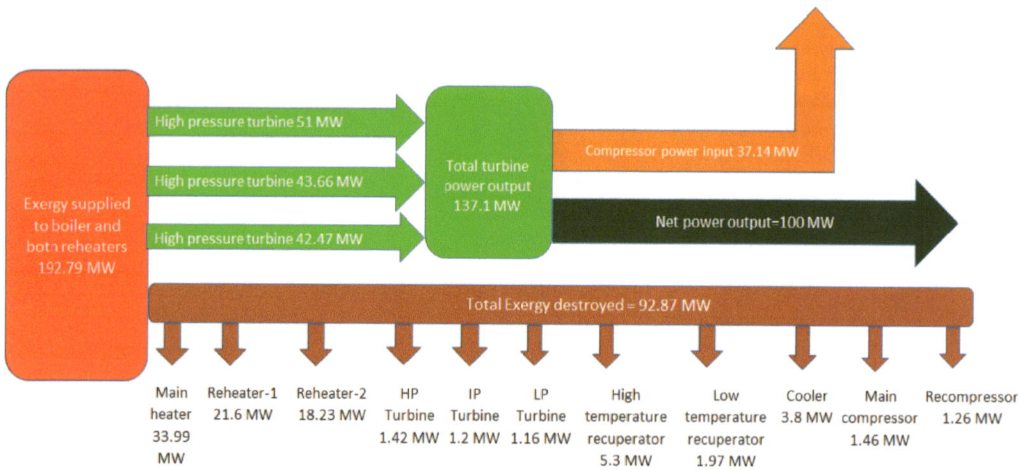

Figure 4. Exergy flow in the full load (100 MWe) plant.

3.1. Optimization

Recompression ratio, ϕ, is an important operational parameter of choice that can decide the performance of the plant [24,37]. Hence, we studied the influence of it on the 1st law efficiency of the power plant. The recompression ratio has been varied from 0.1 to 0.75 at full load and the efficiency is estimated and plotted in Figure 5 for two different pressure ratios of the turbines. At this point, it should be mentioned that as per the design of Chen et al. [22], the high-pressure turbine pressure ratio is 1.66, the intermediate-pressure turbine pressure ratio is 1.56 and that of the low-pressure turbine is 1.54. For studying the influence of the recompression ratio in the present study, this has been kept the same for all turbines at 1.5 for the 1st and 2 for the second case.

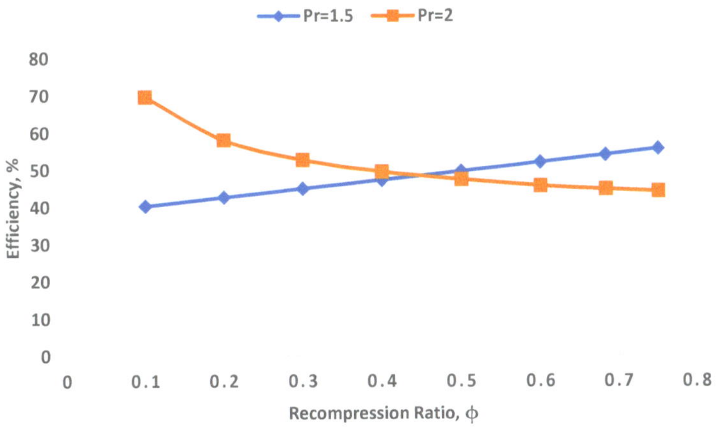

Figure 5. Influence of recompression ratio on 1st law efficiency.

Opposite trends can be noticed. At a pressure ratio of 1.5, there is a monotonous rise in efficiency while there is a drop for a pressure ratio of 2. Rogalev et al. [24] observed a maximum efficiency at a recompression ratio of around 0.35 for a fixed pressure ratio. Hence, it can be understood that there is a possibility of optimization with the recompression ratio and turbine pressure ratio as parameters. However, there are other parameters to be considered for optimization. The base case of full load has been considered for optimization with respect to the thermal efficiency. For this purpose, the present work has used the Optimization tool in Cycle-Tempo to maximize the efficiency of the plant. The turbine inlet parameters and recompression ratio have been chosen as the variables and 1st law efficiency has been chosen as the target parameter to be optimized. At the same time, 2nd law efficiency has also been evaluated with the dead state taken as 1 bar and 30 °C. The results of the optimization study have been tabulated in Table 5.

Table 5. Design and optimized parameters at different recompression ratios at full load.

	Design	Optimized at Different Recompression Ratio				
Recompression ratio, φ	0.683	0.683	0.75	0.8	0.9	1
High pressure turbine inlet pressure, bar	320	320	320	320	320	320
Intermediate pressure turbine inlet pressure, bar	192.3	206.9	206.9	206.9	206.9	206.9
Low pressure turbine inlet pressure, bar	123.3	130.5	130.5	130.5	130.5	130.5
1st law efficiency, η_{I}, %	53.92	53.94	55.57	56.77	59.2	61.7
2nd law efficiency, η_{II}, %	51.85	51.87	53.42	54.59	56.94	59.3

After optimization, few changes in the parameters have taken place. The inlet pressure of the high-pressure turbine remains the same while that of the intermediate and low-pressure turbine marginally increases to 206.9 and 130.5 bar from 192.3 and 123.3 bar, respectively. In the range of study, it can be seen that the 1st law efficiency reached a maximum of 61.7% while the 2nd law efficiency reached 59.3%. There was a very small rise in the 1st law efficiency when the recompression ratio was kept the same. However, when the recompression ratio was increased, there was a good amount of rise in efficiency. The maximum efficiencies were seen at a recompression ratio of 1 which means all the working fluid is compressed in the main compressor, thereby minimizing the power consumed. There is no change in the turbine inlet pressures at different recompression ratios. The pressure ratios of the optimized system turbines are 1.55, 1.58 and 1.63 for the high-pressure turbine, intermediate-pressure turbine and low-pressure turbine, respectively. The same for the base case are 1.66, 1.55 and 1.54, respectively. The optimization process is aimed at increasing the 1st law efficiency and hence the pressures and temperatures are altered by the software algorithm to provide heat addition at a higher possible temperature.

3.2. Part Load Performance

Part load performance is inevitable for all power plants and pressure control plays a crucial role in operating power plants at off-design conditions [38]. The part load calculations have been carried out using the part load option in Cycle-Tempo which is based on Traupel's formula. The base case of full load operation at 100 MW has been used as the basis for the parameters. In this procedure, the data from the full load simulation are input into the part load simulations along with the heat exchanger surfaces, turbine operational parameters as well as compressor parameters. Since the available sizes of heat exchangers, turbine and compressor sizes do not change even at part load, the turbine inlet pressure, temperature, fluid flow rate and a few other parameters should be altered at part load. Table 6 shows the power produced in the turbines and that consumed in the compressors at various loads. The negative sign for compressors indicates power consumption. As per intuition, the power produced and consumed decreases at part load compared to full load. However, the powers shown in Table 5 at part load are not simply the fraction of it at full load. In plain terms, at 80% load, the high-pressure turbine's output is not 80% of 51 MW but 81.7% of 51 MW. Likewise, at 60% load and 40% load, the power output is 64.3% and 45.7%, respectively.

Table 6. Power production/consumption details at various loads.

Net Power (Turbine/Compressor)	Full Load, 100 MW	80% Load, 80 MW	60% Load, 60 MW	40% Load, 40 MW
High Pressure Turbine	51	41.7	32.8	23.34
Intermediate Pressure Turbine	43.6	34.1	25.2	16.2
Low Pressure Turbine	42.47	29.9	19.66	10.98
Main Compressor	−16.75	−11.6	−7.92	−4.74
Re-compressor	−20.4	−14.17	−9.7	−5.8

The fluid mass flow rates are supposed to decrease at part loads. This is investigated and shown below. Table 7 compares the mass flow rates of the fluids (CO_2 and air in pipes 17 and 18) in various pipelines at different loads. The pipe numbers can be referred to in Figure 1. It must be noted that the flow rates at different loads in pipes 1 to 8 are the same because these pipes carry the s-CO_2 through the main heater, turbines, reheaters and recuperators before bifurcating at the cooler. These flows merge downstream in pipe 15. Intuitively, the flow rates should decrease to 80%, 60% and 40% of that at full load. However, the flow rate of CO_2 has reduced to 86.1%, 72.4% and 57.7% of the full load for part load operation at 80%, 60% and 40%, respectively. This is because the pressures and temperatures at a few salient points have changed at part loads when compared to full loads. These have been shown in Tables 8 and 9.

Table 7. Comparison of fluid mass flow rates (in kg/s) at full and part loads in different pipelines (refer to Figure 1).

Pipe	100 MW	80 MW	60 MW	40 MW
1	637.5	548.69	461.9	367.97
2	637.5	548.69	461.9	367.97
3	637.5	548.69	461.9	367.97
4	637.5	548.69	461.9	367.97
5	637.5	548.69	461.9	367.97
6	637.5	548.69	461.9	367.97
7	637.5	548.69	461.9	367.97
8	637.5	548.69	461.9	367.97
9	202.1	173.94	146.4	116.65
10	435.4	374.76	315.48	251.33
11	435.4	374.76	315.48	251.33
12	435.4	374.76	315.48	251.33
13	435.4	374.76	315.48	251.33
14	202.1	173.94	146.4	116.65
15	637.5	548.69	461.9	367.97
16	637.5	548.69	461.9	367.97
17	8370.5	6836.36	5481	4093.88
18	8370.5	6836.36	5481	4093.88

Table 8. Comparison of fluid pressures (in bar) at full and part loads in different pipelines (refer to Figure 1).

Pipe	100 MW	80 MW	60 MW	40 MW
1	320	277.3	236.7	194.3
2	192.8	170.9	150.3	129.4
3	192.3	170.4	149.8	128.9
4	123.8	114.2	105.6	97.27
5	123.3	113.7	105.1	96.77
6	80	80	80	80
7	80	80	80	80
8	79	79	79	79
9	79	79	79	79
10	79	79	79	79
11	79	79	79	79
12	334.5	280.8	240.2	197.8
13	332.5	278.8	238.2	195.8
14	334	278.8	238.2	195.8
15	334	278.8	238.2	195.8
16	333.5	278.3	237.7	195.3
17	1	1	1	1
18	1	1	1	1

Table 9. Comparison of temperatures (in °C) at full and part loads in different pipelines (refer to Figure 1).

Pipe	100 MW	80 MW	60 MW	40 MW
1	620	620	620	620
2	551.1	554.4	558.75	565.4
3	620	620	620	620
4	560.99	566.53	573.26	582.33
5	620	620	620	620
6	562.88	573.39	583.71	594.62
7	235	235	235	235
8	80.22	73.73	68.32	62.03
9	80.22	73.73	68.32	62.03
10	80.22	73.73	68.32	62.03
11	32	32	32	32
12	75.22	68.73	63.32	57.03
13	230	230	230	230
14	220.44	192.93	169.95	142.82
15	227.08	217.93	210.03	200.17
16	516.64	522.43	529.32	536.52
17	25	25	25	25
18	35	35	35	35

Some of the pressures at various locations in the power plant are also supposed to change at part loads. This has been investigated and compared in Table 8. These pressure changes would be brought about by pressure control methods such as sliding pressure control, etc. Unlike the mass flow rate, the pressures in the cycle keep reducing from pipe 1 to pipe 11 as the expansion takes place and before being compressed. This can also be observed in the part load operation. The general trend is that the pressures have decreased at part loads while they remained the same at a few points. This was observed by Adibhatla and Kaushik [38] in their part load performance study on a super-critical steam thermal power plant. They also noted that sliding pressure control is better than constant pressure control. The high-pressure turbine inlet pressure decreased to 86.6%, 73.9% and 60.7% of the full load pressure of 320 bar at 80%, 60% and 40% loads, respectively. At the exit of the high-pressure turbine, the pressures are 88.6%, 78.1% and 67.2% of the full load value of 192.8 bar at 80%, 60% and 40% loads, respectively. Pressure is an important parameter from an operation and safety viewpoint. Since the pressures at the part load did not increase, the design of the pressure vessels and pipelines for pressure can be based on the full load value without worrying about pressure spikes at part loads.

Temperatures at various locations also change at part loads and are compared in Table 9. Similar to the pressures, the temperatures drop as the cycle proceeds from pipe 1 to 11. The lowest temperature and pressure in the cycle are kept above the critical point of CO_2. It can be seen that the turbine inlet temperatures have remained the same as that in the full load (620 °C) while the exit temperatures have increased. This is in contrast to the behavior of power, mass flow rate, and pressure, which reduced at part loads considered in the present study. This is to increase the temperature of heat addition in the re-heaters so as to keep the overall temperature of heat addition high and thereby operate at the best possible thermal efficiency at that particular part load. It can be observed that the exit temperature of the high-pressure turbine has increased by 0.6%, 1.4% and 2.6% for 80%, 60% and 40% loads, respectively. This implies that the temperatures at the inlet and outlet of the high-pressure turbine are practically remaining constant. A similar observation can be made for the remaining salient points.

From the comparison of mass flow rates, pressures and temperatures, it can be concluded that there is considerable change in the first two parameters while there is only a minute change in the last one when compared to the full load values.

Figure 6 shows the percentage exergy loss in various components compared to the total exergy loss. Most of the exergy is lost in the main and re-heaters and this can be attributed to the heat transfer by finite temperature difference. In the study carried out by Adibhatla and Kaushik [38], it was found that the boiler experiences the highest amount of exergy destruction followed by the turbines, which is similar to what can be noticed in Figure 6. Moreover, these components operate at high temperatures which contributes more to the lost exergy. The same can be observed to a certain extent in the high-temperature recuperator and cooler. Even though the temperature difference between the hot and cold fluids in the cooler is high, the exergy lost is low due to both fluids being at a relatively low temperature.

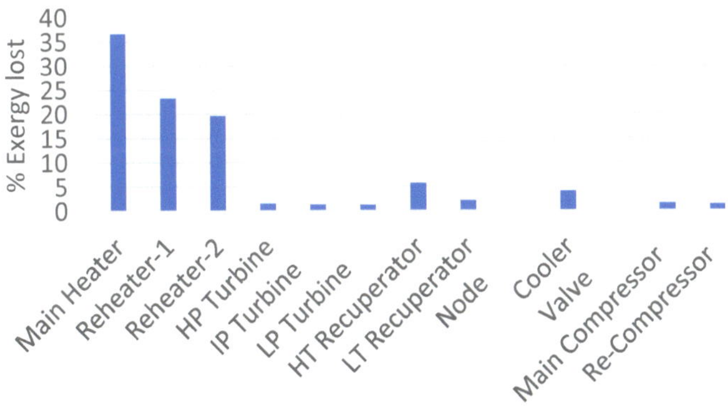

Figure 6. Percentage of exergy lost in various components.

Figure 7 shows the value diagrams of the heat-exchanging equipment in the plant, namely, the high-temperature recuperator (HTR), low-temperature recuperator (LTR) and Cooler. The value diagram gives an understanding of the amount of useful work that can be produced if a heat engine is operated between the ambient and the heat exchanger. This can be seen on the y-axis showing the Carnot efficiency and also as the shaded region between the two curves. This value is greatest for the HTR and least for the LTR. The cooler also has some exergy lost on par with the two recuperators because of the high-temperature difference between the two fluids at the hot end. This is difficult to avoid as the ambient air is used for cooling the CO_2 in the cooler. However, the exergy lost in the cooler is lower because it operates in the low-temperature region compared to the recuperators. To decrease this exergy loss, the temperature of carbon dioxide in the cooler needs to be reduced further, which implies greater heat transfer in the HTR and LTR, which in turn can increase the exergy lost in them. So, a balance needs to be struck between the exergy lost in the cooler and that in the recuperators.

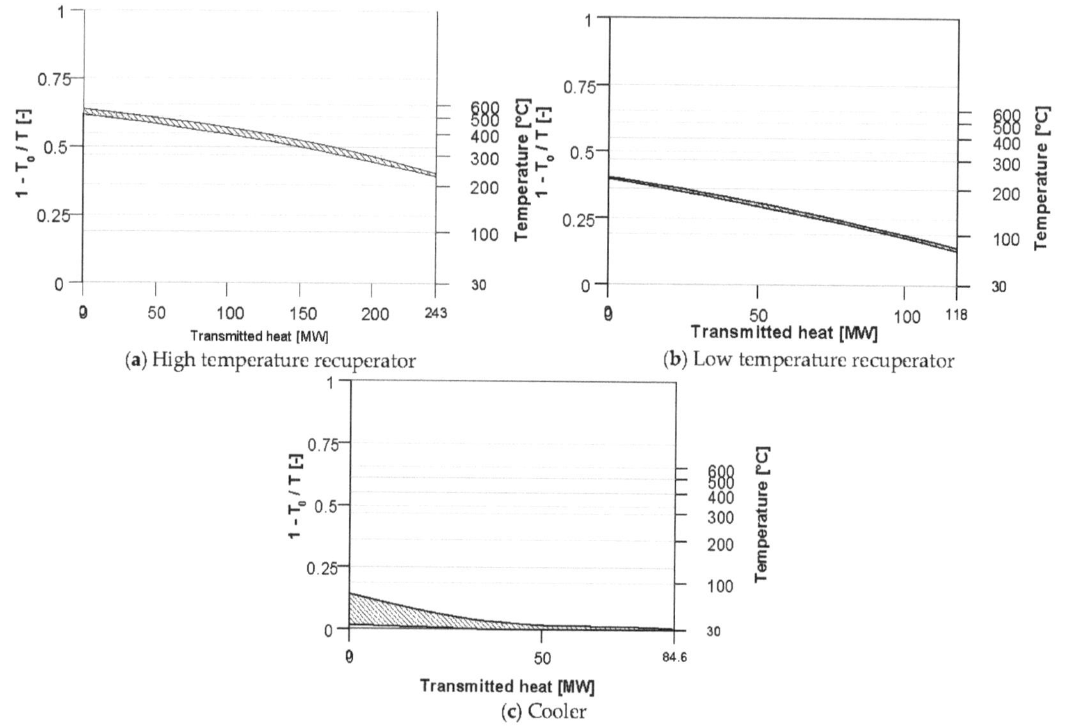

Figure 7. Value diagram of (**a**) HT recuperator, (**b**) LT recuperator and (**c**) Cooler.

Specific work is an indicator of the amount of working fluid necessary to generate a given quantity of work. Higher magnitude indicates better utilization of the working fluid by the choice of its operational parameters such as pressure and temperature [26]. Figure 8 shows the specific work as a function of the recompression ratio at different loads. As anticipated, the full-load plant has the highest amount of specific work while the 40% load has the lowest amount. The specific work also increases as the recompression ratio increases, which is in agreement with the results of the optimization study. This is because at a low recompression ratio between 0.1 and 0.2, 80–90% of the flow takes place in the re-compressor, which operates at higher temperatures, resulting in greater compression work and hence lower specific work. Moreover, the specific work at 80% load is almost equal to that at full load, especially at a low recompression ratio. An interesting observation for the recompression ratio between 0.1 and 0.2 is that at full load the specific work is slightly lower than that at 80% load. This could be due to the magnitude of differences in the mass flow rates, pressures and temperatures in the main and re-compressor at 100% and 80% loads. At the full load and 80% load, the mass flow rates are 435.4 and 374.76 kg/s, the main and re-compressor exit pressures are 334.5/280.8 bar and 334/278 bars while the respective temperatures are 75.22/68.3 and 220.44/192.93 °C, respectively, for the main and re-compressors. It can be seen that the mass flow rates and pressures have changed greatly during part load operation while the temperatures changed only by a small fraction. This along with the fact that about 80–90% of the flow takes place in the re-compressor for the recompression ratio 0.1–0.2 could have resulted in a slight reduction in specific work at full load. At lower loads, the drastic drop in the mass flow rate could have resulted in consistent behavior in the entire range of the recompression ratio considered.

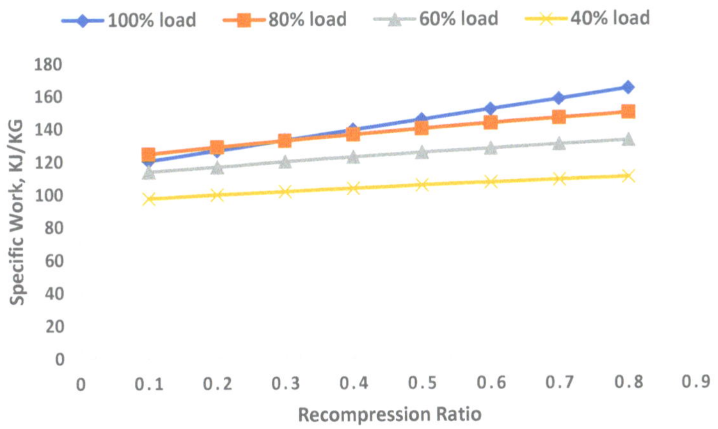

Figure 8. Specific work as a function of recompression ratio at various loads.

The 1st law and 2nd law efficiencies at various loads are compared in Figure 9. The 2nd law efficiency is understandably always lower than the 1st law efficiency. In addition, both the efficiencies remain almost the same at full and 80% loads while there is a drastic drop at 60% and 40% loads. A similar observation was made by Yang et al. [27] during their study on the 1st law-based performance of an s-CO_2 power plant at full and part load even though their system is not a double reheat recompression one like in the present study. This drop in efficiencies is due to the off-design operation where the components are designed for full load. This results in unnecessary losses in the fluid flow as well as increased irreversibilities. Hence, it is not advisable to operate this power plant below 60% load. Moreover, exergo-economic studies need to be conducted at part loads to see if it is worthwhile to operate even at 80% or not.

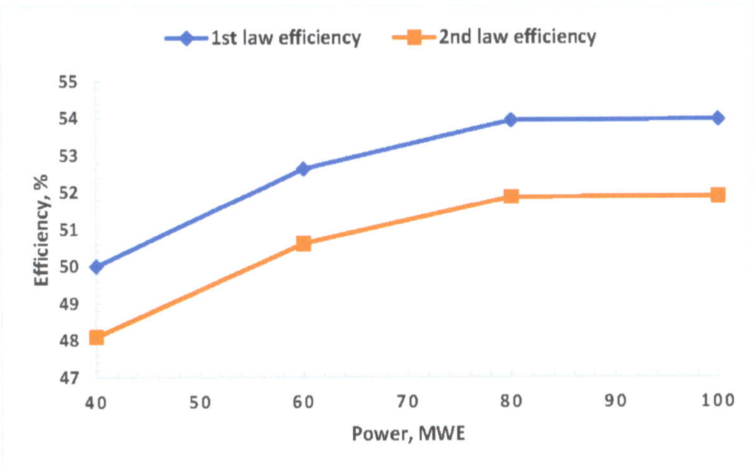

Figure 9. Comparison of 1st and 2nd law efficiencies at various loads.

Sanchez et al. [25] reported about 34% 1st law efficiency at full load and about 32% at 40% load. Fan et al. [26] reported a thermal efficiency of 35 to 38% from 35% load to 100% load, respectively, under both an inventory control strategy as well as a turbine throttling strategy. In comparison to the present study, their efficiencies have a drop of 3 percentage points while in the present study the drop is by 4 percentage points. This could be due to the combined cycles of Sanchez et al. [25] and Fan et al. [26], which evened out the variation to a greater extent. This is supported by the results reported by Yang et al. [27] where their thermal efficiencies varied between 37% and 43% for a load range of 40% to 100%. These efficiencies are lower than those reported in the present work. This could be due to the simple single reheat cycle they implemented while the present work implemented a double reheat recompression system. Wang et al. [33] studied the influence of different control strategies on thermal efficiency and found that inventory control and coupled control are the best choices for part load operation. The present work implemented the inventory control method for part load operation. They also found the efficiency to be ranging between 22% and 30% for a load ranging between 40% and 100% and the lower efficiencies can be attributed to the simplicity of the system. Gini et al. [35], who considered a simple recuperated s-CO_2 system, also showed that the thermal efficiency dropped to 80% of the full load value at 50% load. From this comparison with the studies reported in the literature, it can be understood that a combined or hybrid system involving s-CO_2 yields more or less constant thermal efficiency at loads even down to 40% while simple systems perform poorly at low loads.

4. Conclusions

In the present study, the system design of a 100 MWe double reheat recompression s-CO_2 power plant was carried out, various operational parameters were estimated and the system was optimized at full load. Part load performance at 80% down to 40% was evaluated. The recompression ratio was found to be a crucial parameter that influences the performance both at full and part loads. Based on the 2nd law analysis, the components that are responsible for high exergy destruction were also identified. Key observations are objectively listed below:

i. Recompression ratio and turbine inlet pressures as well as pressure ratios were identified as crucial for improving the performance of the power plant. The power plant at full load was optimized and an estimated 1st law efficiency of 61.7% and 2nd law efficiency of 59.3% were attained.
ii. The specific work was highest (~160 kJ/kg) at full load and almost the same at 80% load, especially at low recompression ratios. It dropped rapidly at lower loads, indicating that the recompression ratio should be kept high at lower loads.
iii. The power plant operated almost equally well at full load and 80% load while a drastic drop in efficiency was noticed at 60% and 40% loads compared to those at 100% load.
iv. Operational pressures decreased at part loads while few temperatures, especially at the turbine exit, increased marginally. The mass flow rate reduced at part load when compared to full load.
v. Most of the exergy is destroyed in the main heater (>35%), re-heater (>20%), followed by the high-temperature recuperator and cooler.

Author Contributions: Conceptualization, S.M. and F.M.; methodology, J.D. and S.R.; software, S.M., T.M.Y.K., R.U.B. and N.A.; validation, S.M., T.M.Y.K., N.A. and J.D.; formal analysis, B.K.V.T., N.A. and S.M.; investigation, S.M., F.M. and J.D.; data curation, S.M.; writing—original draft, S.M., F.M. and J.D. All authors have read and agreed to the published version of the manuscript.

Funding: This work was funded by King Khalid University under grant number R.G.P. 2/118/44.

Institutional Review Board Statement: Not Applicable.

Informed Consent Statement: No humans were involved in the research.

Data Availability Statement: Data not available.

Acknowledgments: The authors extend their appreciation to the Deanship of Scientific Research at King Khalid University for funding this work through research groups program under grant number R.G.P. 2/118/44.

Conflicts of Interest: The authors declare no conflict of interest.

Nomenclature

Symbol	Name
$\dot{E}$	Rate of energy transfer, kW
h	Specific enthalpy, kJ/kg
h_0	Dead state specific enthalpy, kJ/kg
$\dot{m}$	Mass flow rate, kg/s
$\dot{Q}$	Rate of heat transfer, kW
S	Specific entropy, kJ/kg-K
S_0	Dead state specific entropy, kJ/kg-K
S_{gen}	Entropy generated, kJ/K
T	Absolute temperature, K
T_0	Dead state temperature, K
$\dot{W}$	Rate of work, kW
$\dot{X}$	Rate of Exergy Transfer, kW
$\dot{X}_{des}$	Exergy destroyed, kW
ϕ	Recompression ratio, (-)

References

1. Furlan, C.; Mortarino, C. Forecasting the impact of renewable energies in competition with non-renewable sources. *Renew. Sustain. Energy Rev.* **2018**, *81*, 1879–1886. [CrossRef]
2. Javidkia, F.; Hashemi-Tilehnoee, M.; Zabihi, V. A comparison between fossil and nuclear power plants pollutions and their environmental effects. *J. Energy Power Eng.* **2011**, *5*, 811–820.
3. Ryzhkov, A.F.; Bogatova, T.F.; Maslennikov, G.E.; Osipov, P.V.; Nizov, V.A. Creation of energy-efficient and environmentally friendly energy sources on fossil fuels to address global climate issues. *J. Physics: Conf. Ser.* **2020**, *1677*, 012115. [CrossRef]
4. Sreekanth, M.; Kumar, R.R.; Kolar, A.K.; Leckner, B. Estimation of wood char size at the end of devolatilization in a bubbling fluidized bed combustor. *Fuel* **2008**, *87*, 3393–3402. [CrossRef]
5. Yin, C.; Yan, J. Oxy-fuel combustion of pulverized fuels: Combustion fundamentals and modeling. *Appl. Energy* **2016**, *162*, 742–762. [CrossRef]
6. Pragadeesh, K.S.; Regupathi, I.; Sudhakar, D.R. Insitu gasification—Chemical looping combustion of large coal and biomass particles: Char conversion and comminution. *Fuel* **2021**, *292*, 120201. [CrossRef]
7. Yang, M.; Duan, L.; Tong, Y. Design and Performance Analysis of New Ultra-Supercritical Double Reheat Coal-Fired Power Generation Systems. *Energies* **2021**, *14*, 238. [CrossRef]
8. Osman, A.I.; Hefny, M.; Maksoud, M.I.A.A.; Elgarahy, A.M.; Rooney, D.W. Recent advances in carbon capture storage and utilisation technologies: A review. *Environ. Chem. Lett.* **2021**, *19*, 797–849. [CrossRef]
9. Pipitone, G.; Bolland, O. Power generation with CO_2 capture: Technology for CO_2 purification. *Int. J. Greenh. Gas Control.* **2009**, *3*, 528–534. [CrossRef]
10. Available online: https://www.epa.gov/energy/greenhouse-gases-equivalencies-calculator-calculations-and-references#:~:text=Home%20electricity%20use&text=The%20national%20average%20carbon%20dioxide,EIA%202020b%3B%20EPA%202021 (accessed on 9 February 2023).
11. Liao, G.; Liu, L.; Jiaqiang, E.; Zhang, F.; Chen, J.; Deng, Y.; Zhu, H. Effects of technical progress on performance and application of supercritical carbon dioxide power cycle: A review. *Energy Convers. Manag.* **2019**, *199*, 111986. [CrossRef]
12. Ahn, Y.; Bae, S.J.; Kim, M.; Cho, S.K.; Baik, S.; Lee, J.I.; Cha, J.E. Review of supercritical CO_2 power cycle technology and current status of research and development. *Nucl. Eng. Technol.* **2015**, *47*, 647–661. [CrossRef]
13. Angelino, G. Carbon Dioxide Condensation Cycles for Power Production. *J. Eng. Power* **1968**, *90*, 287–295. [CrossRef]
14. Feher, E.G. The supercritical thermodynamic power cycle. *Energy Convers.* **1968**, *8*, 85–90. [CrossRef]
15. Marion, J. Mwe Supercritical Carbon Dioxide (sCO_2) Pilot Power Plant. December. Available online: https://www.gti.energy/wp-content/uploads/2019/01/STEP-Project-Detailed-Description-Dec2018.pdf (accessed on 9 February 2023).
16. Sarkar, J. Review and future trends of supercritical CO_2 Rankine cycle for low-grade heat conversion. *Renew. Sustain. Energy Rev.* **2015**, *48*, 434–451. [CrossRef]
17. Crespi, F.; Gavagnin, G.; Sánchez, D.; Martínez, G.S. Supercritical carbon dioxide cycles for power generation: A review. *Appl. Energy* **2017**, *195*, 152–183. [CrossRef]

18. Yin, J.; Zheng, Q.; Peng, Z.; Zhang, X. Review of supercritical CO_2 power cycles integrated with CSP. *Int. J. Energy Res.* **2019**, *44*, 1337–1369. [CrossRef]
19. White, M.T.; Bianchi, G.; Chai, L.; Tassou, S.A.; Sayma, A.I. Review of supercritical CO_2 technologies and systems for power generation. *Appl. Therm. Eng.* **2021**, *185*, 116447. [CrossRef]
20. Yu, A.; Su, W.; Lin, X.; Zhou, N. Recent trends of supercritical CO_2 Brayton cycle: Bibliometric analysis and research review. *Nucl. Eng. Technol.* **2020**, *53*, 699–714. [CrossRef]
21. Sun, E.; Hu, H.; Li, H.; Liu, C.; Xu, J. How to Construct a Combined S-CO_2 Cycle for Coal Fired Power Plant? *Entropy* **2018**, *21*, 19. [CrossRef]
22. Chen, Z.; Wang, Y.; Zhang, X. Energy and exergy analyses of S–CO_2 coal-fired power plant with reheating processes. *Energy* **2020**, *211*, 118651. [CrossRef]
23. Bai, W.; Li, H.; Zhang, L.; Zhang, Y.; Yang, Y.; Zhang, C.; Yao, M. Energy and exergy analyses of an improved recompression supercritical CO_2 cycle for coal-fired power plant. *Energy* **2021**, *222*, 119976. [CrossRef]
24. Rogalev, N.; Rogalev, A.; Kindra, V.; Komarov, I.; Zlyvko, O. Structural and Parametric Optimization of S–CO_2 Nuclear Power Plants. *Entropy* **2021**, *23*, 1079. [CrossRef] [PubMed]
25. Sánchez, D.; Chacartegui, R.; de Escalona, J.M.; Muñoz, A.; Sánchez, T. Performance analysis of a MCFC & supercritical carbon dioxide hybrid cycle under part load operation. *Int. J. Hydrogen Energy* **2011**, *36*, 10327–10336.
26. Fan, G.; Li, H.; Du, Y.; Chen, K.; Zheng, S.; Dai, Y. Preliminary design and part-load performance analysis of a recompression supercritical carbon dioxide cycle combined with a transcritical carbon dioxide cycle. *Energy Convers. Manag.* **2020**, *212*, 112758. [CrossRef]
27. Yang, J.; Yang, Z.; Duan, Y. Part-load performance analysis and comparison of supercritical CO_2 Brayton cycles. *Energy Convers. Manag.* **2020**, *214*, 112832. [CrossRef]
28. Tong, Y.; Duan, L.; Pang, L. Off-design performance analysis of a new 300 MW supercritical CO_2 coal-fired boiler. *Energy* **2020**, *216*, 119306. [CrossRef]
29. Lee, S.; Yaganegi, G.; Mee, D.J.; Guan, Z.; Gurgenci, H. Part-load performance prediction model for supercritical CO_2 radial inflow turbines. *Energy Convers. Manag.* **2021**, *235*, 113964. [CrossRef]
30. Alfani, D.; Binotti, M.; Macchi, E.; Silva, P.; Astolfi, M. sCO_2 power plants for waste heat recovery: Design optimization and part-load operation strategies. *Appl. Therm. Eng.* **2021**, *195*, 117013. [CrossRef]
31. Fan, G.; Dai, Y. Thermo-economic optimization and part-load analysis of the combined supercritical CO_2 and Kalina cycle. *Energy Convers. Manag.* **2021**, *245*, 114572. [CrossRef]
32. Fan, G.; Du, Y.; Li, H.; Dai, Y. Off-design behavior investigation of the combined supercritical CO_2 and organic Rankine cycle. *Energy* **2021**, *237*, 121529. [CrossRef]
33. Wang, R.; Wang, X.; Shu, G.; Tian, H.; Cai, J.; Bian, X.; Li, X.; Qin, Z.; Shi, L. Comparison of different load-following control strategies of a sCO_2 Brayton cycle under full load range. *Energy* **2022**, *246*, 123378. [CrossRef]
34. Xingyan, B.; Wang, X.; Wang, R.; Cai, J.; Tian, H.; Shu, G. Optimal selection of supercritical CO_2 Brayton cycle layouts based on part-load performance. *Energy* **2022**, *256*, 124691. [CrossRef]
35. Gini, L.; Maccarini, S.; Traverso, A.; Barberis, S.; Guedez, R.; Pesatori, E.; Bisio, V. A prototype recuperated supercritical CO_2 cycle: Part-load and dynamic assessment. *Appl. Therm. Eng.* **2023**, *225*, 120152. [CrossRef]
36. Kotas, T.J. *The Exergy Method of Thermal Plant Analysis*; Krieger Publishing, Co.: Malabar, FL, USA, 1995.
37. Sathish, S.; Kumar, P.; Nassar, A. Analysis of a 10 MW recompression supercritical carbon dioxide cycle for tropical climatic conditions. *Appl. Therm. Eng.* **2020**, *186*, 116499. [CrossRef]
38. Adibhatla, S.; Kaushik, S. Energy and exergy analysis of a super critical thermal power plant at various load conditions under constant and pure sliding pressure operation. *Appl. Therm. Eng.* **2014**, *73*, 51–65. [CrossRef]

sustainability

MDPI

Article

A Comparison of Different Renewable-Based DC Microgrid Energy Management Strategies for Commercial Buildings Applications

Hegazy Rezk [1,*], Rania M. Ghoniem [2], Seydali Ferahtia [3], Ahmed Fathy [4,5], Mohamed M. Ghoniem [6] and Reem Alkanhel [2]

[1] Department of Electrical Engineering, College of Engineering in Wadi Alddawasir, Prince Sattam bin Abdulaziz University, Wadi Alddawasir 11991, Saudi Arabia
[2] Department of Information Technology, College of Computer and Information Sciences, Princess Nourah bint Abdulrahman University, Riyadh 11671, Saudi Arabia
[3] Laboratoire de Génie Electrique, Department of Electrical Engineering, University of M'sila, M'Sila 28000, Algeria
[4] Electrical Engineering Department, Faculty of Engineering, Jouf University, Sakaka 72388, Saudi Arabia
[5] Electrical Engineering Department, Faculty of Engineering, Zagazig University, Zagazig 44519, Egypt
[6] Department of Computer, Mansoura University, Mansoura 35516, Egypt
* Correspondence: hr.hussien@psau.edu.sa

Abstract: DC microgrid systems allow commercial buildings to use locally generated energy and achieve an optimal economy efficiently. Economical and eco-friendly energy can be achieved by employing renewable energy sources. However, additional controllable sources, such as fuel cells, are required because of their reduced efficiency and fluctuated nature. This microgrid can use energy storage systems to supply transient power and enhance stability. The functioning of the microgrid and its efficiency are related to the implemented energy management strategy. In this paper, a comparison of several reported energy management strategies is fulfilled. The considered EMSs include the fuzzy logic control (FLC) strategy, the state machine control (SMC) strategy, the equivalent consumption minimization strategy (ECMS), and external energy maximization strategy (EEMS). These strategies are compared in terms of power-saving, system efficiency, and power quality specifications. The overall results confirm the ability of EEMS (high efficiency of 84.91% and economic power-saving 6.11%) and SMC (efficiency of 84.18% with high power-saving 5.07%) for stationary applications, such as building commercial applications. These strategies provide other advantages, which are discussed in detail in this paper.

Keywords: energy management strategies; DC microgrids; energy efficiency; photovoltaic; fuel cells

1. Introduction

In recent years, numerous factors have caused an increase in energy consumption, motivating many countries to create adaptable, sustainable, affordable energy solutions [1]. In addition, the transition to clean alternative energy is required to avoid impending environmental and economic catastrophes [2]. The new energy paradigm is predicated on shifting the present power system's dependence on traditional fossil-based energy resources to an energy mix, mainly comprised of renewable energy resources (RESs) [3]. However, incorporating new RESs can bring technical and economic challenges, such as stability and financial profitability [4]. The microgrid concept based on advanced control and management technologies is the most promising solution to these issues [5]. Microgrids are a set of power sources (also called distributed generators (DGs)), including renewables, storage systems, and loads connected with control and management systems. These microgrids can be connected to the utility grid at various power and voltage levels using electronic converters [6]. DGs may often deliver reduced generating costs, improved dependability, and increased security that traditional generators cannot. This technology expands the possibilities for efficient electric power generation, transportation, and distribution [7]. Microgrids based on RESs are located near final users to reduce transportation losses. In addition, they help reduce greenhouse gas emissions and ensure power supply during the utility grid outage [8]. Moreover, this type of power system provides an uninterrupted power supply (UPS) and is appropriate for rural and isolated arias, islands, and military applications [9]. Optimal EMS, based on the salp swarm algorithm (SSA), has been proposed in [10]. A techno-economic multi-level EMS for commercial buildings has been suggested in [11]. In this paper, the bald eagle search algorithm (BES) has been used to address operating cost minimization. An adaptive fuzzy logic controller (AFLC) has been implemented as an EMS for a DC microgrid [12], where the adaption mechanism is based on metaheuristic optimization algorithms.

Most of the previously mentioned EMSs have been designed for electric vehicles (EVs) and off-grid applications. Updated versions of these strategies are utilized in this paper. The novelty is to modify these strategies to respond to the energy requirements of a commercial building under the grid-connected mode. The building's power system, represented in Figure 1, is based on a DC microgrid, including two sources: a photovoltaic generation unit (PV) and a fuel cell system (FC), a battery storage system, and an AC/DC grid-connection converter. Then, a comparative study between the pre-mentioned EMSs is carried out to evaluate the performance of each one in terms of efficiency, energy consumption, final storage system SoC and power quality. The main contributions of this paper can be presented as follows:

- Design-updated EMSs based on the pre-mentioned conventional EMS: FLC, SMC, ECMS, and EEMS for stationary applications such as commercial building power systems.
- Evaluate the performance of each one of the proposed EMS.
- Realization of comparative study in terms of the power-saving, system efficiency, and power quality specifications.

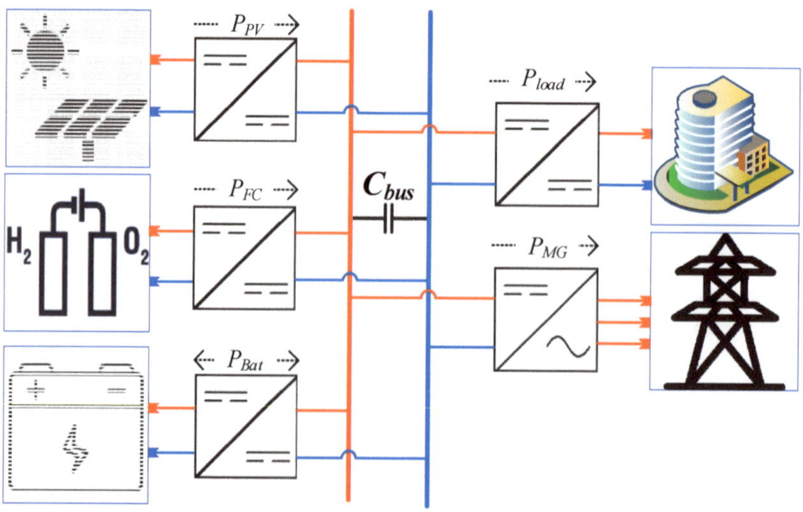

Figure 1. The architecture of the proposed DC microgrid, where all the power sources are connected to the DC bus using their converters.

2. Building Power System Modeling

2.1. Power System Description

In contrast to the alternating current (AC) building power systems that incorporate photovoltaics and DC loads, direct current (DC) power systems with modern power electronics can reduce much of the power conversion loss. According to the literature, DC microgrids can save up to 15% of electricity in buildings [13], where 308 V infrastructure microgrid voltage is the standard value [14]. The structure of the studied building's power system is illustrated in Figure 1. A fuel cell system, a photovoltaic array, and a Lithium-ion battery storage system mainly electrify the building. The microgrid can be connected to the utility grid through a bidirectional DC/AC converter. These sources are connected to the 380 V bus using their static converters.

2.2. Modeling of PV Array

Installing solar PV on commercial roofs would make good economic sense, minimizing carbon emissions and allowing unused rooftop space to be used. A single-diode equivalent circuit model is presented in Figure 2. According to [15], using the Kirchhoff Law, the governing equation of this equivalent circuit is:

$$\begin{aligned} I &= I_{ph} - I_d - I_r \\ &= I_{ph} - \underbrace{I_0\left[e^{\frac{V+I\cdot R_s}{nV_{th}}} - 1\right]}_{I_d} - \underbrace{\frac{V+I\cdot R_s}{R_{sh}}}_{I_r} \end{aligned} \quad (1)$$

where I_{ph}, I_d, and I_r are the photocurrent, the diode, and the shunt resistance currents, respectively; n represents the ideality factor; I_0 is the saturation current; R_s and R_{sh} are the series and parallel equivalent resistant; and V_{th} donates the thermal voltage provided, as follows:

$$V_{th} = \frac{k \cdot T_c}{q} \quad (2)$$

where q is the elementary charge (1.602×10^{-13} C); T_c represents the cell temperature in K; and k is the Boltzmann constant (1.381×10^{-23} J/K).

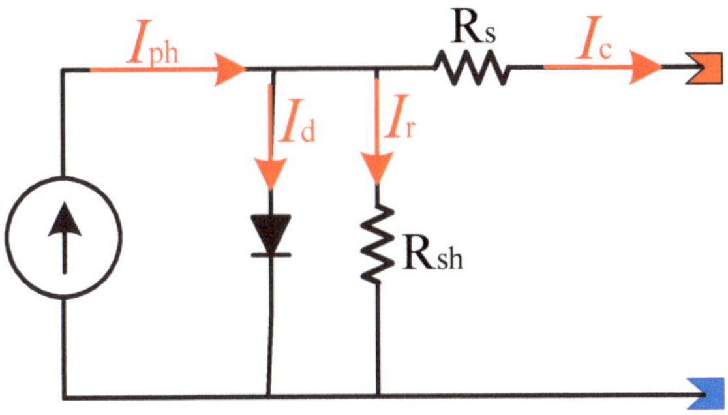

Figure 2. PV equivalent circuit.

The photocurrent is expressed as:

$$I_{ph} = \left[I_{SC} - \alpha\left(T_c - T_{ref}\right)\right]\lambda \quad (3)$$

where I_{SC} is the short circuit current in A; T_{ref} represents the measured and reference temperature in K; α is the temperature coefficient; and λ is the radiance value in kW/m^2. Therefore, the equation of current delivered by the PV array given in function of the number of cells in series N_S and parallel N_P can be written as:

$$I = I_{ph} - I_0 \left[e^{\frac{V_{out} + I_{out} \cdot N_S \cdot R_S}{n \cdot N_S \cdot V_{th}}} - 1 \right] - \frac{V_{out} + I_{out} \cdot N_S \cdot R_S}{R_{sh} \cdot N_S} \quad (4)$$

where V_{out} and I_{out} are the output voltage and current, respectively.

2.3. Battery Description and Modeling

Electrochemistry is employed to address the issue of intermittent renewable energy generation, as well as the sluggish dynamics of the fuel cell. The battery storage device is deployed to reduce the peak load power of the commercial building. At present, there are several types of batteries commercialized. Table 1 includes the technical information for the most popular varieties [16].

Table 1. Batteries' characteristics.

	Unit	Lead-Acid	Nickel-Metal	Nickel-Cadmium	Lithium-Ion
Cell voltage	V	2	1.25	1.25	3.6
Energy density	Wh/Kg	30–50	60–120	45–80	110–160
Power density	W/Kg	180	250–1000	150	1800
Overcharge/deep-discharge tolerance		Not tolerant	Good	Moderate	Excellent
Self-discharge rate		Low	High	Moderate	Neglected
Number of cycles		200–300	300–500	1500	500–1000

The model used in this study is a lithium-ion based on the Thevenin model [17], illustrated in Figure 3. This model is one of the most common battery models. It is based on an internal resistance model used in Matlab with a parallel RC network. It consists mainly of three components: the open-circuit voltage U_{oc}, the internal resistances, and equivalent capacitances. The internal resistances combine the ohmic resistance R_0 and the polarization resistance R_{Th}. The equivalent capacitance C_{Th} defines the battery dynamics during charging and discharging. U_{Th} is the voltage across C_{Th}.

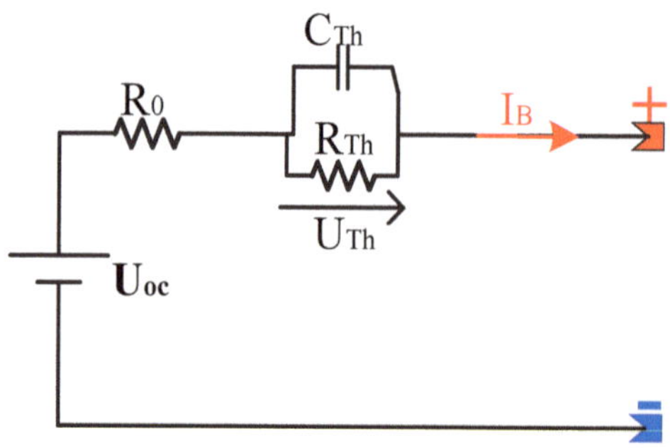

Figure 3. Li-ion battery equivalent circuit.

The source of the voltage is defined as:

$$\begin{aligned} U_B &= U_{oc} - R_0 \cdot I_B - U_{Th} \\ \dot{U}_{Th} &= -\frac{U_{Th}}{R_{Th} \cdot C_{Th}} + \frac{I_B}{C_{Th}} \end{aligned} \quad (5)$$

2.4. SOFC Description and Modeling

As reported in [18], Solid Oxide Fuel Cells (SOFCs) have excellent characteristics, deeming them the most convenient choice for stationary applications. In a SOFC, the electrolyte separates the two electrodes, creating two boundary layers. The polarization effect can charge these layers, identified as the electrochemical double-layer charging effect [19]. According to [20], the SOFC single-cell model, based on its V-I and P-I characteristics curves (Figure 4), is given as:

$$U_{FC,cell} = U_{oc} - U_{act} - U_{cons} - U_{ohm} \tag{6}$$

where U_{oc} presents the open-circuit voltage; U_{act} is the activation voltage loss; U_{cons} is the concentration voltage loss; and U_{ohm} is the ohmic voltage loss.

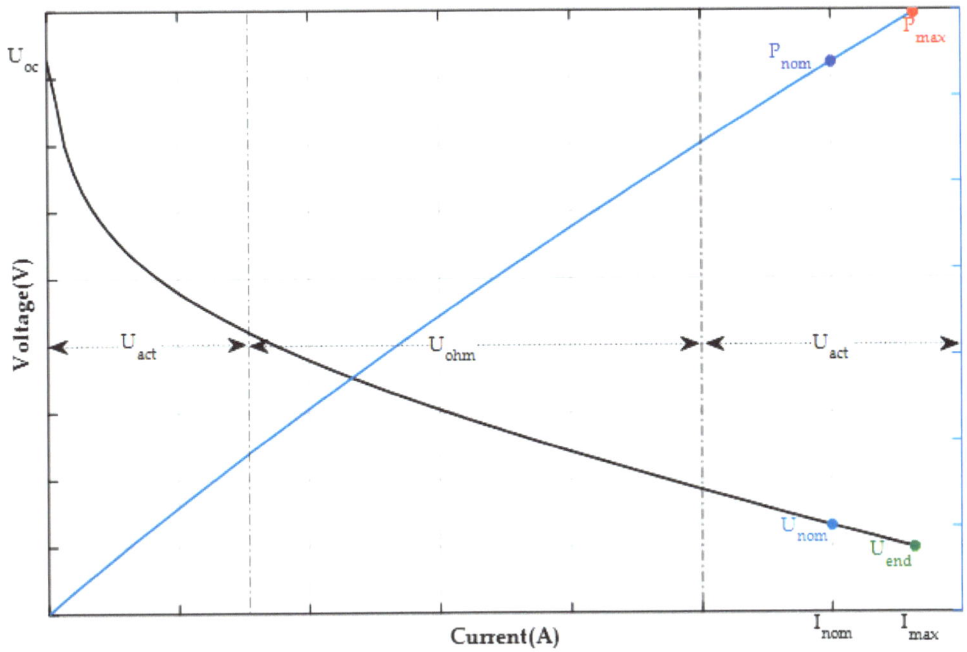

Figure 4. V-I and P-I characteristics curves.

According to [21], the open-circuit voltage and the losses voltage are given as:

$$U_{oc} = U_0 + \frac{RT}{2F} \ln\left[\frac{P_{H_2} \cdot P_{O_2}^{0.5}}{P_{H_2O}}\right] \tag{7}$$

where R and F express the universal gas constant (8:314 KJ (mol·K)$^{-1}$) and Faraday constant (96,486 \$C·mol^{-1}), respectively. P_{H2}, P_{O2}, and P_{H2O} are the partial pressures of hydrogen, oxygen, and water. T is the cell's temperature. E_0 represents the standard potential can be calculated as follows [22]:

$$U_0 = E_{std} + \frac{\Delta s}{nF}(T - 298) \tag{8}$$

where E_{std} is the voltage value under the standard conditions; n represents the number of electrons transported for each fuel molecule involved in the reaction; and Δs denotes entropy change.

The partial pressures of hydrogen, oxygen, and water formulas are reported as follows:

$$P_{H_2} = \frac{{k_{H_2}}^{-1}}{1 + s \cdot \tau_{H_2}} (q_{H_2}^{in} - 2 \cdot k_r \cdot I) \tag{9}$$

$$P_{O_2} = \frac{{k_{O_2}}^{-1}}{1 + s \cdot \tau_{O_2}} (q_{O_2}^{in} - k_r \cdot I) \tag{10}$$

$$P_{H_2O} = \frac{{k_{H_2O}}^{-1}}{1 + s \cdot \tau_{H_2O}} (2 \cdot k_r \cdot I) \tag{11}$$

where τ_{H2}, τ_{O2}, and τ_{H2O} are the hydrogen, oxygen, and water dynamic constants; k_r is a constant that represents the reaction molar flow rate at the anode.

The activation losses can be represented using the Butler-Volmer equation [23], as follows:

$$\begin{aligned} U_{act} &= \tfrac{2 \cdot R \cdot T}{z \cdot F} \sinh^{-1}\left(\tfrac{I}{2 \cdot I_0}\right) \\ &= A \cdot \sinh^{-1}\left(\tfrac{I}{2 \cdot I_0}\right) \end{aligned} \tag{12}$$

where z presents the number of moles of electrons and I_0 denotes the density of the exchange current, which can be expressed as:

$$I_0 = k_1 \cdot T \cdot e^{\frac{-k_2}{T}} \tag{13}$$

where k_1 and k_2 are empirical constants that express the apparent exchange current.

The concentration voltage loss can be expressed as follows:

$$\begin{aligned} U_{cons} &= \tfrac{R \cdot T}{z \cdot F} \ln\left(1 - \tfrac{I}{I_{\max}}\right) \\ &= B \cdot \ln\left(1 - \tfrac{I}{I_{\max}}\right) \end{aligned} \tag{14}$$

$$I_{\max} = k_3 \frac{\ln(1 - C_{re})}{T} \tag{15}$$

where C_{re} expresses the reactant concentration and k_3 is a constant indicating the limiting current density factor.

The ohmic voltage formula is a function of its internal resistance (R_{int}), which can be presented as follows:

$$U_{ohm} = R_{\text{int}} \cdot I = \gamma e^{\delta(298^{-1} - T^{-1})} \tag{16}$$

where γ and δ are constants.

For a fuel cell stack with multiple cells, the output voltage can be expressed as follows:

$$U_{FC} = N_{cells} \cdot U_{FC,cell} = N_{cells}(U_{oc} - U_{act} - U_{cons} - U_{ohm}) \tag{17}$$

where N_{cells} express the number of cells; the operating diagram of the SOFC can be presented in Figure 5.

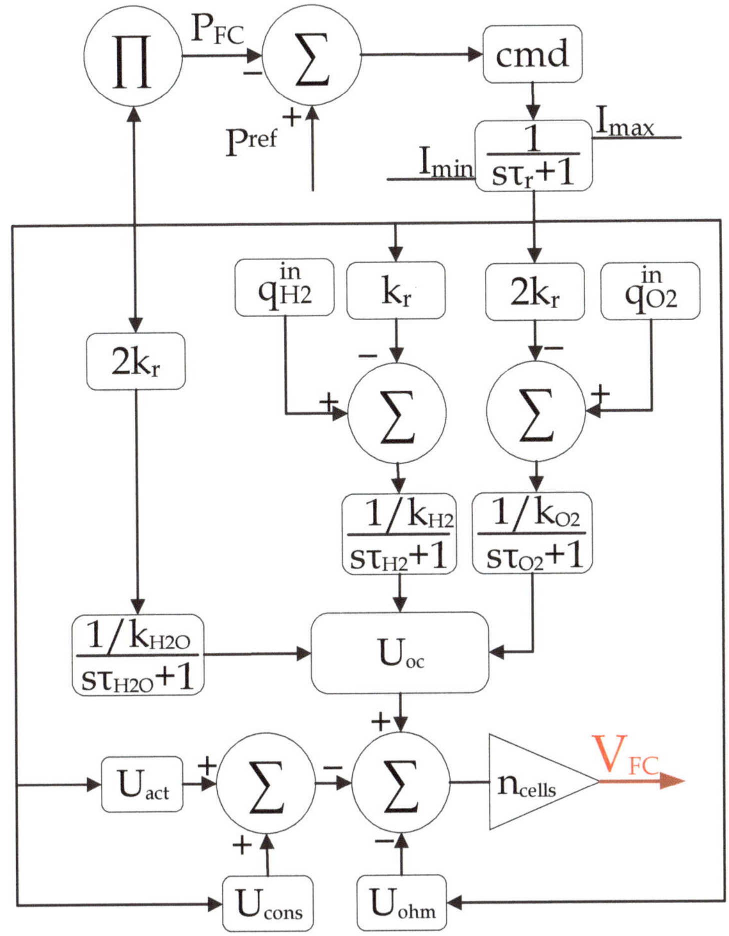

Figure 5. SOFC equivalent circuit.

3. Energy Management Strategy

Electricity efficiency is critical for environmental, economic, and for technological applications. As a result, developing MG energy management is critical, according to [24]. Over the last two decades, the design of energy management systems (EMSs) has piqued the interest of the energy research community. This interest arises from the ongoing and improved development of renewable energy systems and the optimization of EMS technology employing sophisticated energy storage systems, according to [25].

Commercial buildings consume a large quantity of energy; thus, a robust energy management system (EMS) is essential. The EMS has a considerable impact on the system's overall performance. For the investigated system, the net power (ΔP) is the demanded power by the load that is not provided by solar power. The fuel cell and batteries will be operated under the EMS command to provide net power. The EMS must run the battery within the authorized SoC range to prevent overcharge and deep discharge.

$$\Delta P = P_{Load} - P_{PV} \tag{18}$$

On the other hand, the FC dynamics must be restricted to prevent reactant starvation. Figure 6 depicts the global EMS system where the EMS receives system state data from the acquisition system. The control signals are generated based on these data.

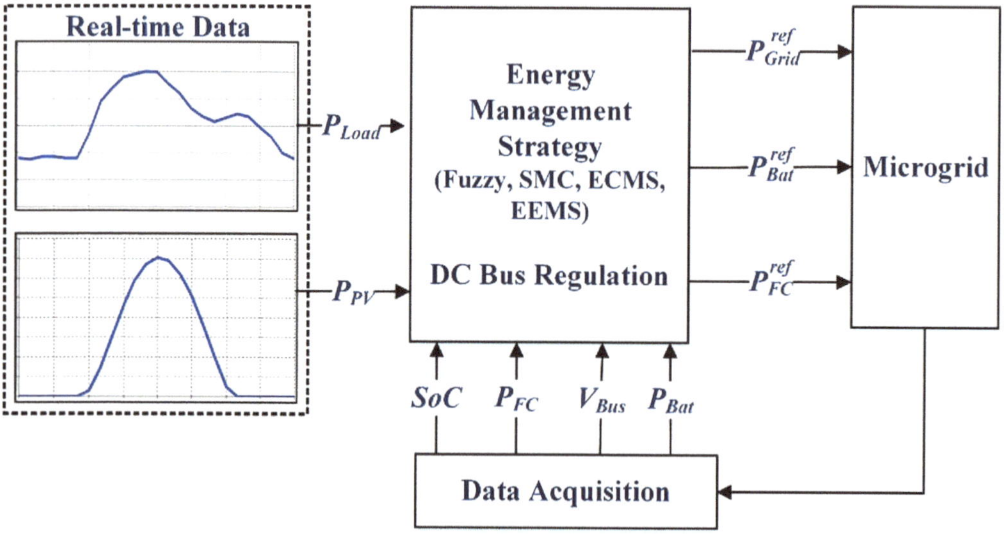

Figure 6. An illustration of the global EMS scheme.

3.1. Fuzzy-Based EMS

The first investigated EMS relies on the fuzzy logic approach to obtain the desired results. The PV generator operates in MPPT mode in this study, and the FC system and battery power are regulated by the generated references, using the fuzzy inference system (FIS)-based EMS. The load power, PV power, and battery SoC are the control inputs, while the outputs are the FC power reference, battery power reference, and utility grid power reference.

Sugeno-type FIS is utilized. The membership functions are shown in Figures 7 and 8, the output surface for the fuzzy inference system is shown in Figure 9, and the fuzzy roles are shown in Table 2.

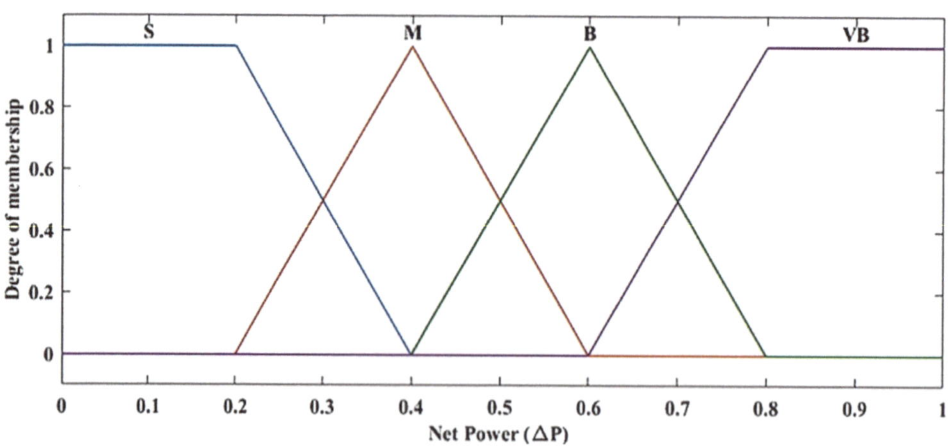

Figure 7. Membership functions for the first input (net power).

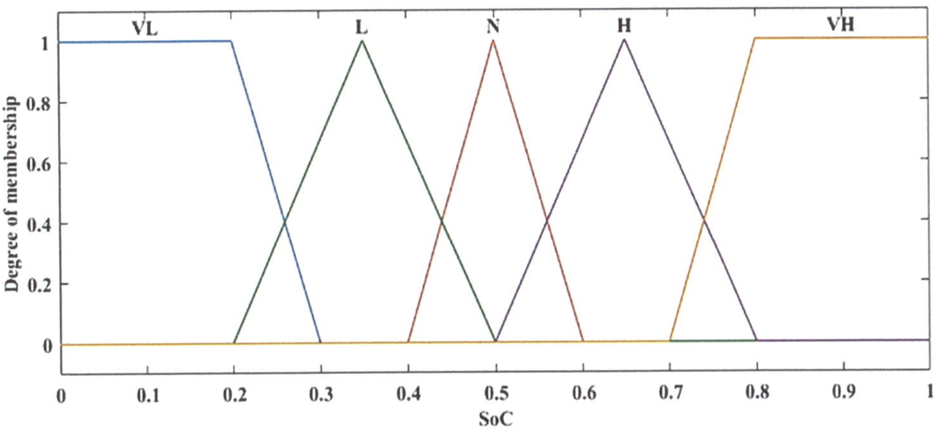

Figure 8. Membership functions for the second input (SoC).

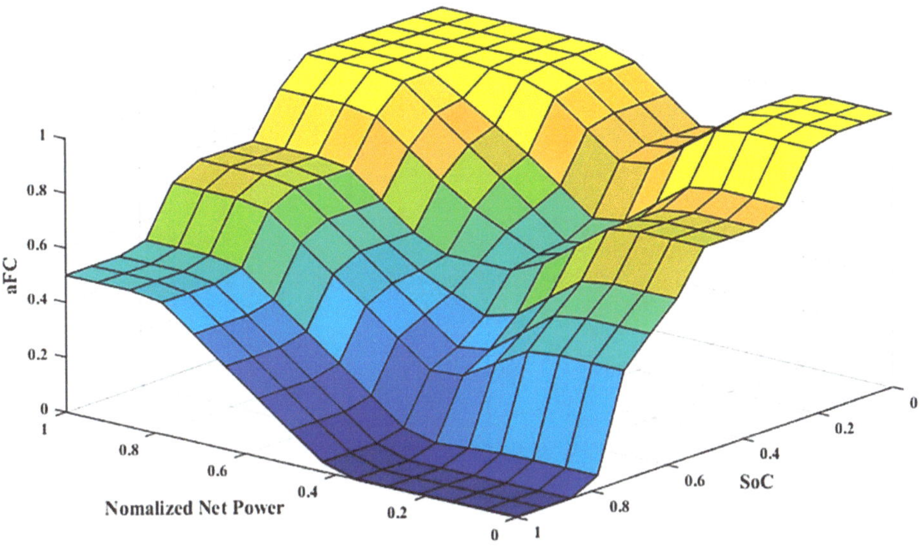

Figure 9. The output surface for the fuzzy inference system.

Table 2. Fuzzy engine roles.

		SoC					
		Very low (VL)	Low (L)	Nature (N)	High (H)	Very high (VH)	Off = 0
ΔP	Small (V)	Max	High	Medium	Low	Off	Low = 0.25
	Medium (M)	Max	High	Medium	Low	Off	Medium = 0.5
	Big (B)	Max	High	Medium	Medium	Low	High = 0.75
	Very big (VB)	Max	Max	High	High	Medium	Max = 1

The current dynamics are limited by using a first-order filter as:

$$I_{FC}^{ref} = \frac{1}{\tau_{FC} \cdot s + 1} \frac{P_{FC}^{ref}}{V_{FC}} \text{ where } I_{FC}^{ref} \in [0, I_{FC}^{\max}] \tag{19}$$

where τ_{FC} has to be equivalent to or greater than the FC response constant time.

The battery power is generated using a PI controller to stabilize the DC bus voltage, whereas the grid power is generated based on Equation (28).

3.2. Equivalent Consumption Minimization Strategy

The ECMS seeks to reduce fuel use while keeping the battery SOC within allowable limits. The performance of the ECMS is dependent on the accuracy of the empirical computation of the related fuel consumption, as reported in [26]. The associated goal function is to reduce:

$$P_{Batt}^{opt} = \min(C_{FC} + \beta \cdot C_{Batt}) \tag{20}$$

$$\beta = 1 - 2 \cdot \mu \frac{SoC - 0.5(SoC_{\min} + SoC_{\max})}{SoC_{\min} + SoC_{\max}} \tag{21}$$

where C_{Bat} expresses the battery fuel consumption according to the provided energy; μ is a constant (0.6). This equation is subjected to:

$$\begin{array}{c} SoC_{\min} \leq SoC \leq SoC_{\max} \\ V_{bus}^{\min} \leq V_{bus} \leq V_{bus}^{\max} \\ P_{FC}^{\min} \leq P_{FC} \leq P_{FC}^{\max} \end{array} \tag{22}$$

The FC output power can be calculated as follows:

$$P_{FC}^{ref} = \Delta P - P_{Grid}^{ref} - P_{Batt}^{opt} \tag{23}$$

The ECMS scheme is illustrated in Figure 10.

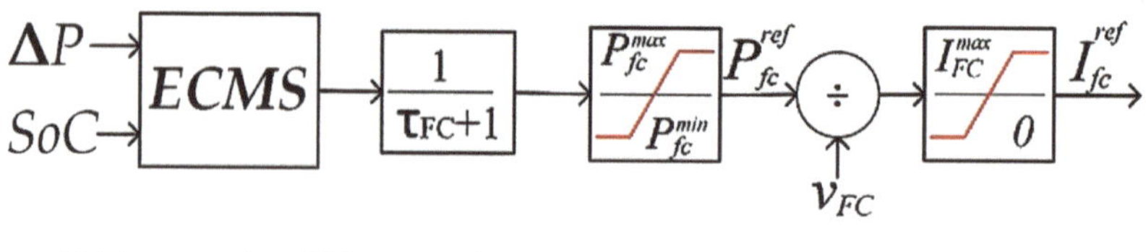

Figure 10. The generation of FC power reference using ECMS.

3.3. External Energy Maximization Strategy

The primary goal of this strategy is to supply the least amount of fuel possible while respecting the battery and DC bus capacitor power limits [27]. Its objective function is to maximize the energy provided by the battery and the bus capacitor, as follows:

$$P_{Batt}^{opt} = \min(\Delta T \cdot P_{Batt} + 0.5 \cdot C_{bus} \cdot (\Delta v)^2) \tag{24}$$

where ΔT represents the sampling time, and Δv represents the charge/discharge voltage. This equation is subjected to:

$$\begin{array}{c} P_{Batt} \leq (SoC - SoC_{\min}) V_{Batt} \cdot Q_{Batt} \\ V_{bus}^{\min} - V_{bus} \leq \Delta v \leq V_{bus}^{\max} - V_{bus} \\ P_{Batt}^{charge} \leq P_{Batt} \leq P_{Batt}^{discharge} \end{array} \tag{25}$$

where V_{batt} and Q_{batt} are the battery nominal voltage and capacity; the operating scheme of the EEMS is given in Figure 11. In contrast to Figure 10, the inputs in Figure 11 are the net power and the bus voltage, where the used algorithm is EEMS.

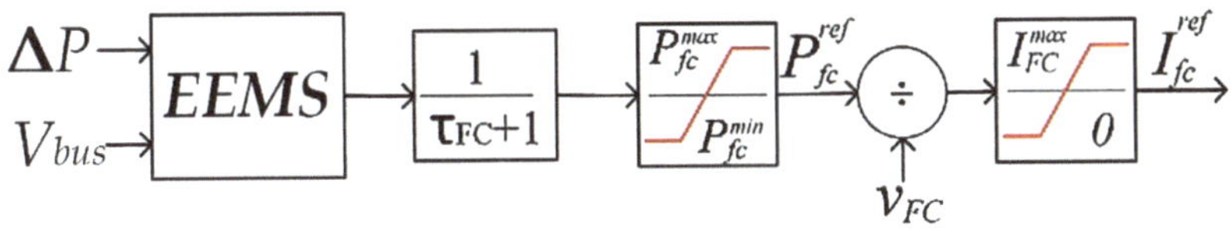

Figure 11. The generation of FC power reference using EEMS.

3.4. State Machine Control

State Machine Control is a strategy based on switching rules [28]. Its principle is to select the operating state according to the inputs: ΔP and battery SOC. According to the state of its inputs, it charges or discharges the battery, as explained in Figure 12.

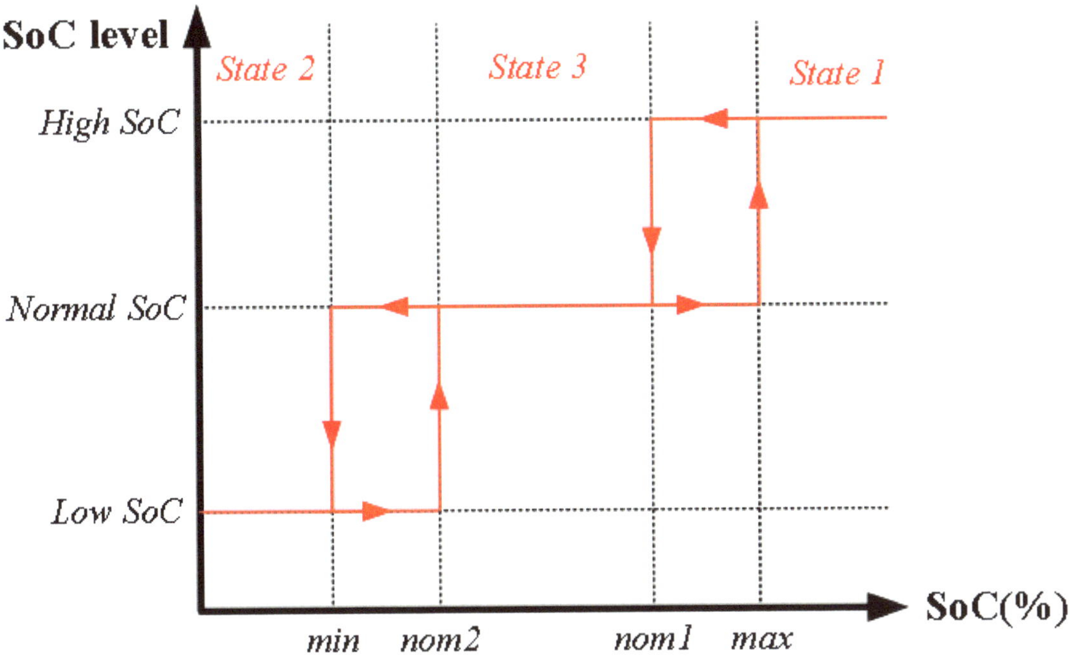

Figure 12. Illustration of the SMC control scheme states.

3.5. Grid Power Reference Generation

The microgrid will turn into (on-grid) mode if the battery SoC exceeds its allowable limits by injecting or absorbing the power from the common line. Its power reference is given as:

$$P_{Grid}^{ref} = \Delta P - P_{Batt}^{req} - P_{FC} \quad (26)$$

where P_{Batt}^{req} is the required power by the battery to maintain its SoC, which can be calculated as follows:

$$P_{Batt}^{req} = k_{grid} \cdot V_{Batt} \cdot Q_{Batt} \quad (27)$$

where k_{grid} is again generated by a state machine controller. Its value depends on the batter's SoC, and its base roles are given as Algorithm 1:

Algorithm 1 Utility grid connection roles

if $SoC < SoC_{\min}$ *do*
if $\Delta P - P_{FC}^{\max}$ *do*
$P_{Grid}^{ref} = \Delta P - P_{FC} + (SoC_{\min} - SoC)Q_{Batt} \cdot V_{Batt}$
else
$P_{Grid}^{ref} = (SoC_{\min} - SoC)Q_{Batt} \cdot V_{Batt}$
end
elseif $SoC > SoC_{\max}$ *and* $\Delta P < 0$ *do*
$P_{Grid}^{ref} = \Delta P$
elseif $SoC < SoC_{\max}$ *and* $SoC > SoC_{\min}$ *do*
$P_{Grid}^{ref} = 0$
end

The inverter dynamics have to be limited to protect the battery from the overcurrent charge and to restrict the voltage from the high overshoot as follows:

$$I_{Gid}^{ref} = \frac{1}{\tau_{grid} \cdot s + 1} \frac{P_{Grid}^{ref}}{V_{Bus}} \tag{28}$$

where τ_{grid} is the constant time of the inverter, the operation of the grid inverter is illustrated in Figure 13.

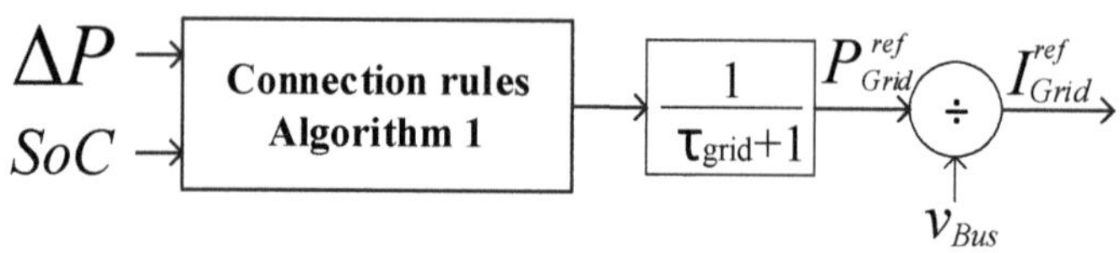

Figure 13. Grid connector block scheme.

3.6. DC Bus Stabilization

The principal role of the battery is to stabilize the DC bus voltage by regulating the power in the commune line. In this study, a PI controller is employed to generate the reference DC bus power, which ensures a stable voltage and satisfies the load power. The dc bus energy E_{bus} can be expressed as [29]:

$$\dot{E}_{Bus} = -\Delta P + P_{FC} + P_{Batt} + P_{Grid} \tag{29}$$

Therefore, the battery power reference can be obtained as a function of the dc bus energy, as:

$$P_{Batt}^{ref} = \dot{E}_{Bus} + \Delta P - P_{FC} - P_{Grid} \tag{30}$$

The bus energy is zero at the steady state. However, its value will typically charge depending on the difference between the reference and the measured voltage. Its value will be generated using a PI controller, which regulates the bus energy as follow:

$$\dot{E}_{Bus} = (E_{Bus}^{ref} - E_{Bus})\left(k_p + \frac{k_i}{s}\right) \tag{31}$$

$$\begin{aligned} E_{Bus}^{ref} &= 0.5 \cdot V_{Bus}^{ref} \cdot C_{Bus} \\ E_{Bus} &= 0.5 \cdot V_{Bus} \cdot C_{Bus} \end{aligned} \tag{32}$$

where V_{Bus}^{ref} is the bus voltage reference and C_{Bus} is the bus capacitance.

The exchanged power must be limited to protect the battery. The maximum amount of the delivered electrical energy should be between 10 and 20% of its total capacity. Therefore, the reference power can be written as:

$$\left|P_{Batt}^{ref}\right| < V_{Batt} \cdot Q_{Batt} \cdot (0.1 \sim 0.2) \tag{33}$$

The reference current can be written as:

$$I_{Batt}^{ref} = \frac{P_{Batt}^{ref}}{V_{Batt}} \tag{34}$$

The maximum discharge current must be lighter than 10% of the capacity, and the maximum charge current must be smaller than 2%.

$$-0.02 \cdot Q_{Batt} < I_{Batt}^{ref} < 0.1 \cdot Q_{Batt} \tag{35}$$

The bus stabilization control low is illustrated in Figure 14.

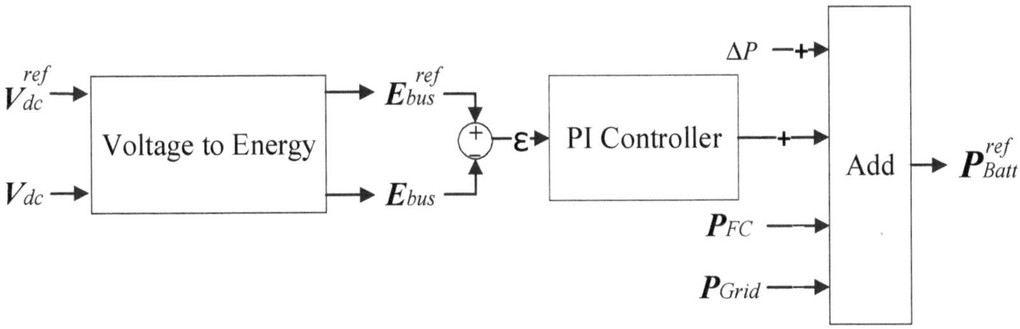

Figure 14. The proposed DC Bus stabilization control scheme.

4. Results and Discussion

The power system parameters are based on those reported in [11]. In the considered building, the DC bus voltage is set at 380 V; the fuel cell maximum power was set at 250 kW. The used battery of 220 V 1500 Ah. Table 3 presents the power system parameters. In this study, the min SoC is supposed to be 30%, whereas the remaining 30% is reserved for the crucial loads in case of a blackout.

Table 3. Power System Parameters.

Parameters	Value	Unit
Bus voltage reference	380	V
Battery capacity	1500	Ah
Battery voltage	220	V
SoC max	90	%
SoC min	30	%
FC max output power	250	kW

To verify the performance of each EMS, a simulation was carried out in the MATLAB/Simulink framework for five days (120 h). In many papers dealing with RTO strategies, the studied load and solar power profiles are 24 h profiles, such as [30,31]. A five day profile is chosen to validate the system performance in real time, where the chosen profile proposes different solar and load behaviors and scenarios to emulate the behavior of the physical system. The solar irradiance is provided in Figure 15, and the power profiles are provided in Figure 16. In this study, two profiles for the load and the solar power are provided by the House Load toolbox [32]. Generally, these papers take different power profiles to demonstrate the performance of the proposed EMS under different operating conditions. In this paper, a single profile of 120 h has different operating conditions, such as cloudy, partially cloudy, and sunny days; the load behavior also changes to emulate the working and the weekends.

A cloudy profile is operated, so the solar power is very fluctuating. This will affect the battery output power responsible for stabilizing the DC bus voltage. The EMS has to supply the net power, represented in Figure 17.

The delivered power from each source with EEMS-, ECMS-, SMC-, and FLC-based EMS are illustrated in Figures 18–21. The recorded SoC for each strategy is illustrated in Figure 22.

The SoC decreases much faster using ECMS and Fuzzy strategies, meaning the battery shares more power when employing these strategies. The grid power will prevent the battery from overcharging or deep-discharge; however, charging/discharging cycles will reduce the battery lifespan. Furthermore, it raises grid dependency by increasing the demand for utility grid power. It is observed that the EEMS and SMC strategies utilize the FC power more than the battery. The charging/discharging cycles are limited in utilizing the strategies.

As a consequence, an extended battery lifespan is achieved. Although the EEMS are based on the DC bus capacitance, it offers extraordinary efficiency (84.91%). The appropriate choice of the SMC rules will increase its efficiency (84.18%) compared with ECMS (76.27%) and FLC (76.30%).

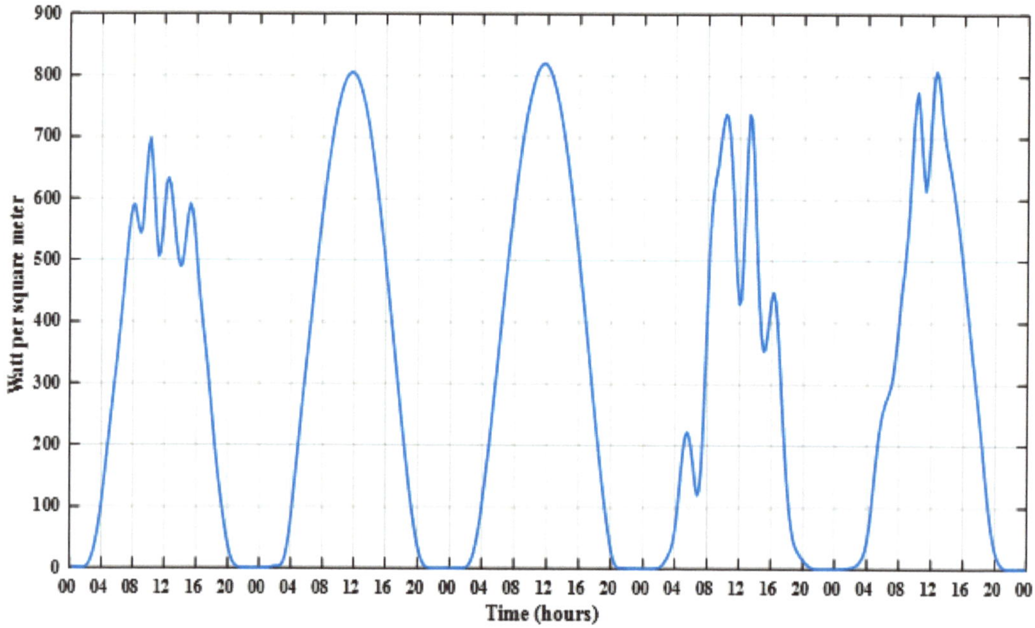

Figure 15. Solar irradiance profile for five days (W/m^2) [32].

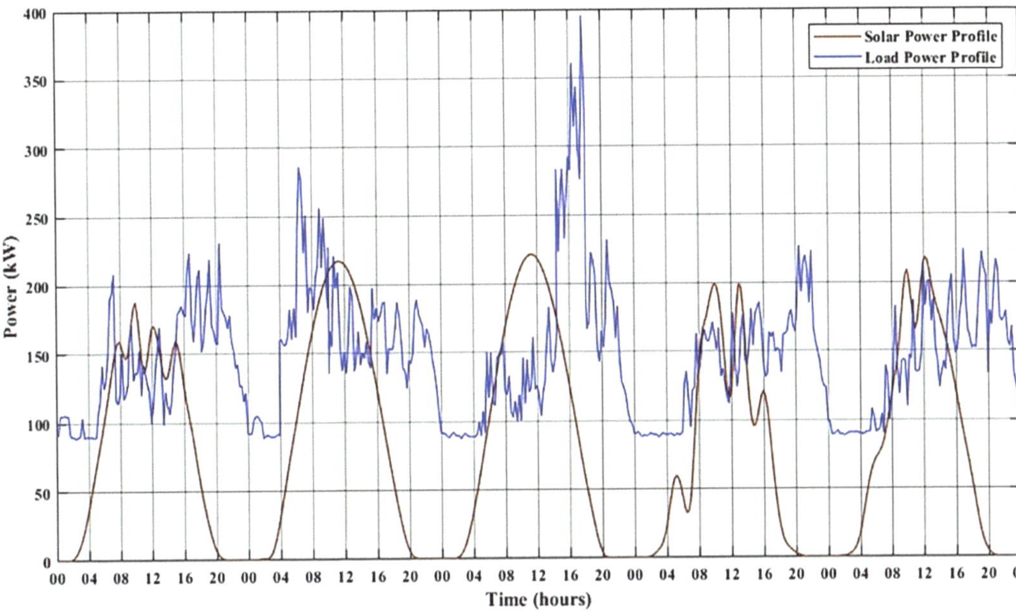

Figure 16. Used load and solar power profiles for five days (kW) [32].

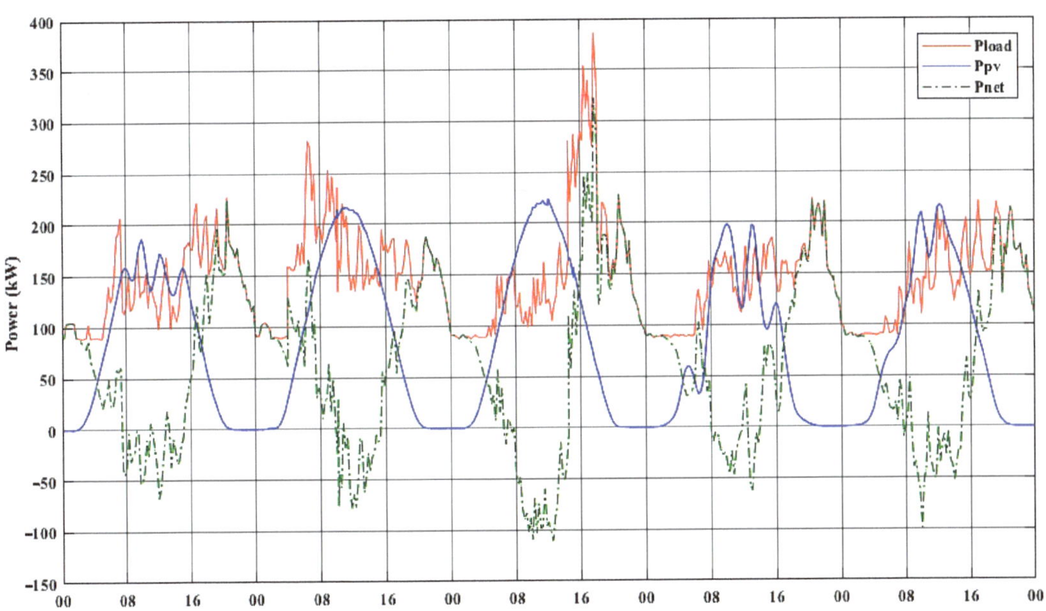

Figure 17. The measured load, the renewable, and the net power (kW).

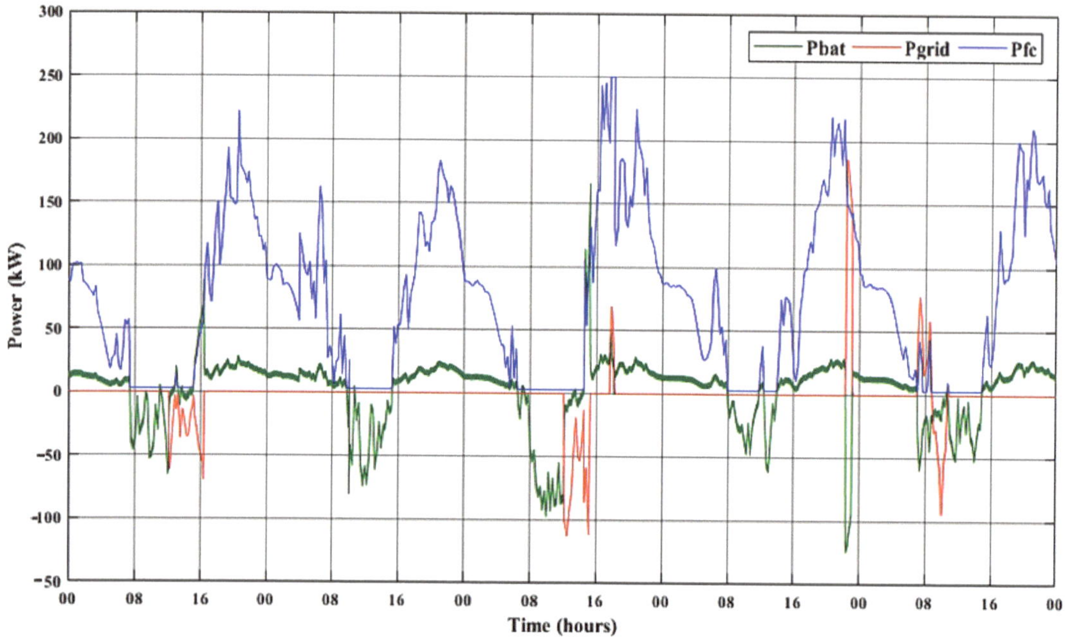

Figure 18. FC, battery, and grid power using EEMS (kW).

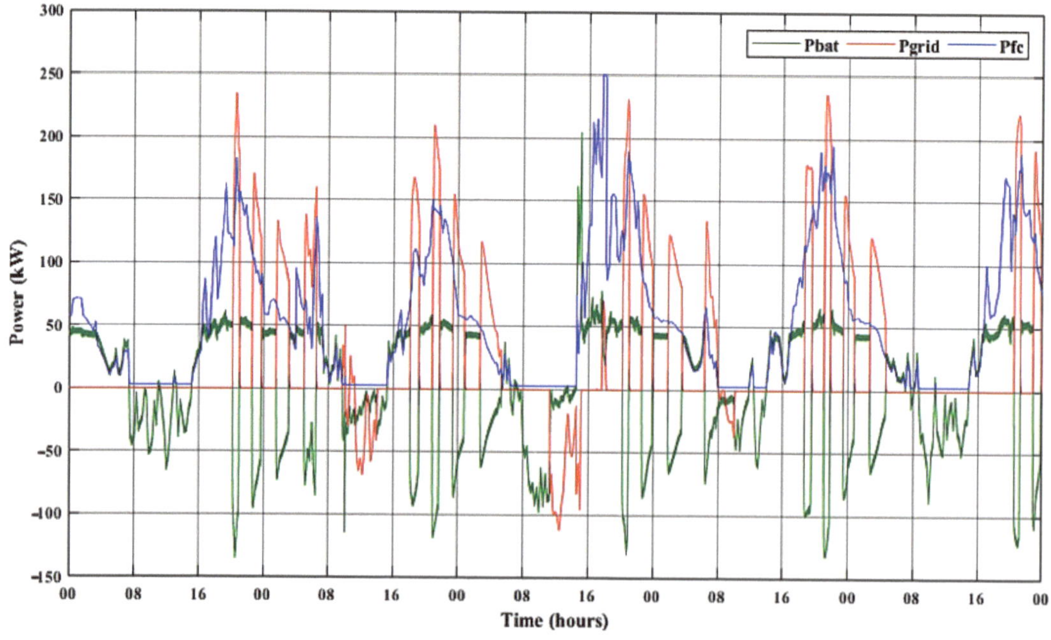

Figure 19. FC, battery, and grid power using ECMS (kW).

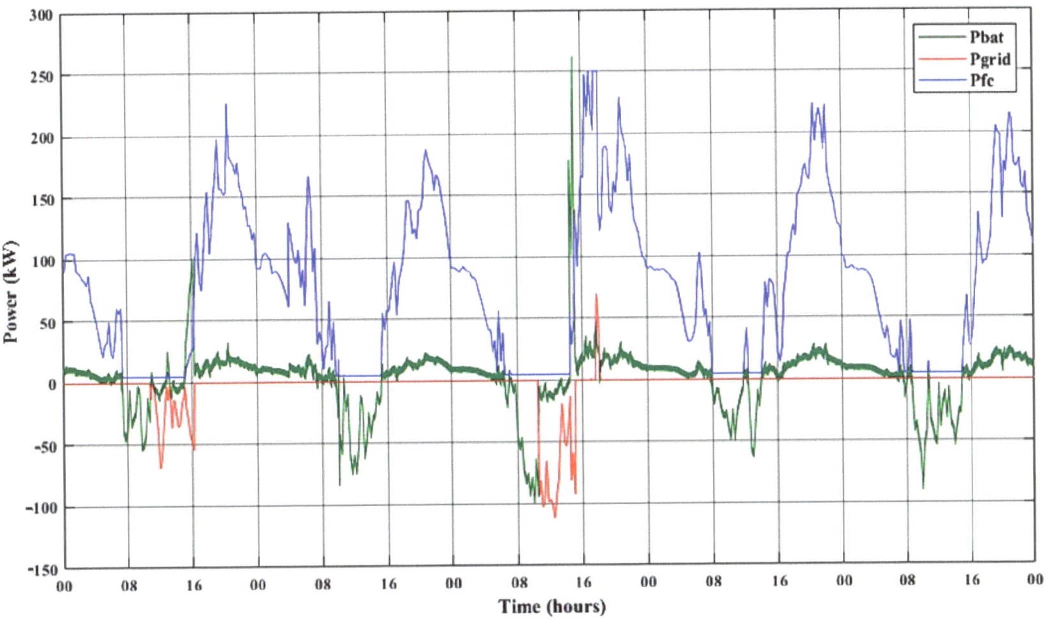

Figure 20. FC, battery, and grid power using SMC (kW).

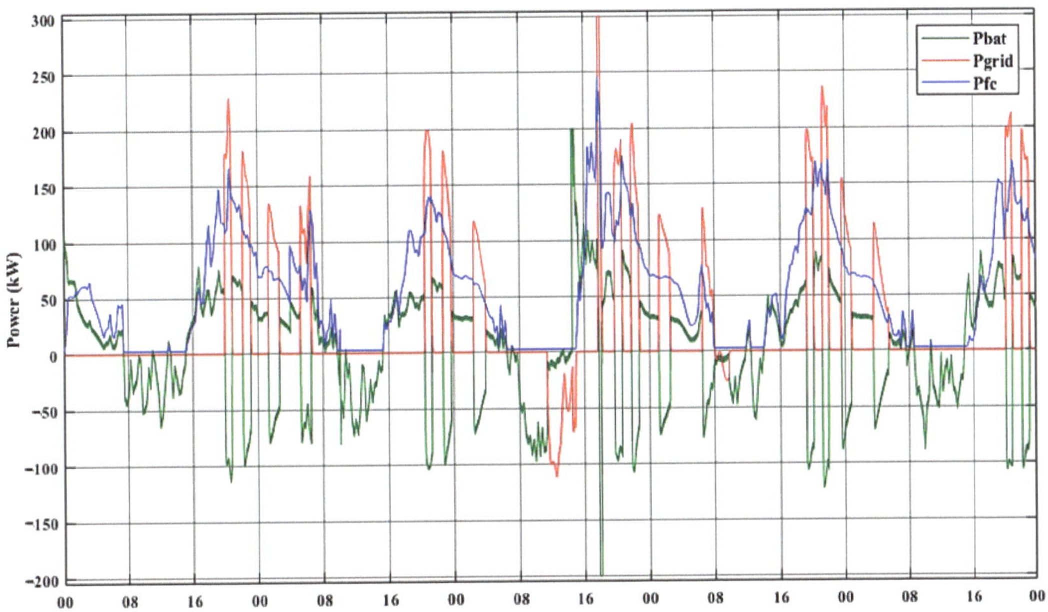

Figure 21. FC, battery, and grid power using FLC-based EMS (kW).

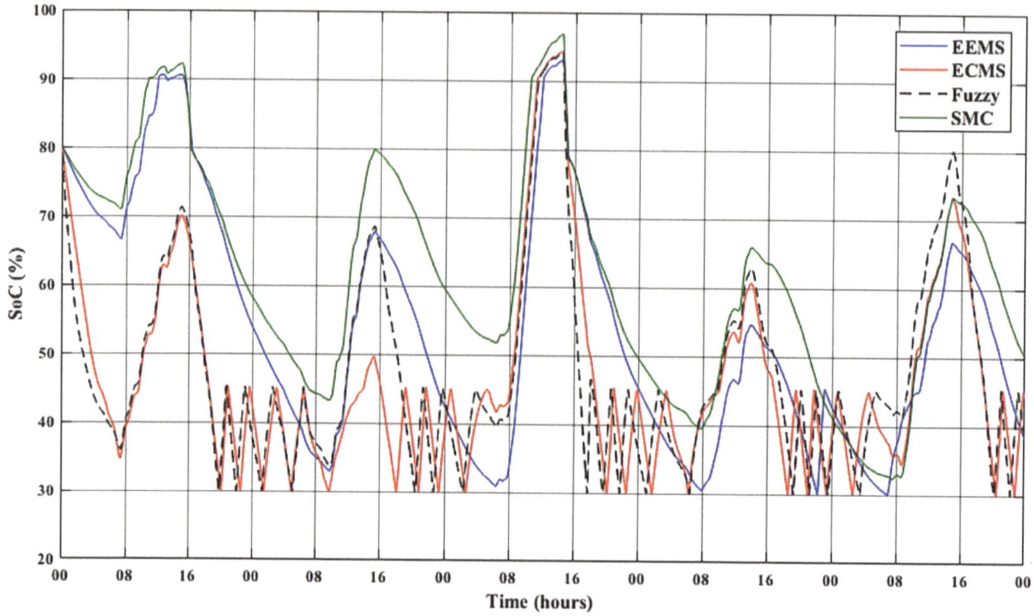

Figure 22. Battery State of Charge (%).

The simulation result statistics are given in Table 4 and Figure 23.

Table 4. Simulation results.

	Efficiency (%)	Final SoC (%)	FC Energy (KWh)	Grid Energy (KWh)	Paid Energy (KWh)	Power Saving (%)
EEMS	**84.91**	38.46	83,712	−1826	**81,886**	**6.11**
ECMS	76.27	44.15	62,070	24,199	86,270	1.83
SMC	80.18	**50.20**	87,175	−4380	82,794	5.07
Fuzzy	76.30	38.55	62,865	24,349	87,215	NA

The results illustrated above prove the superiority of the EEMS as a suitable strategy for commercial buildings in terms of electrical efficiency (84.91%), power saving (6.11%), and paid energy (81,886).

As mentioned before, each strategy controls the battery power. Accordingly, the strategy performance will directly affect the DC bus voltage. Figure 24 shows the obtained DC bus voltage of each strategy. The measurement statistics are given in Table 5 and Figure 25. Although these results show a relatively small static error (2 V), the EEMS ensures excellent power quality. As a result of the low standard deviation (std) provided by the EEMS, the latter will prevent the FC system and the battery from premature aging.

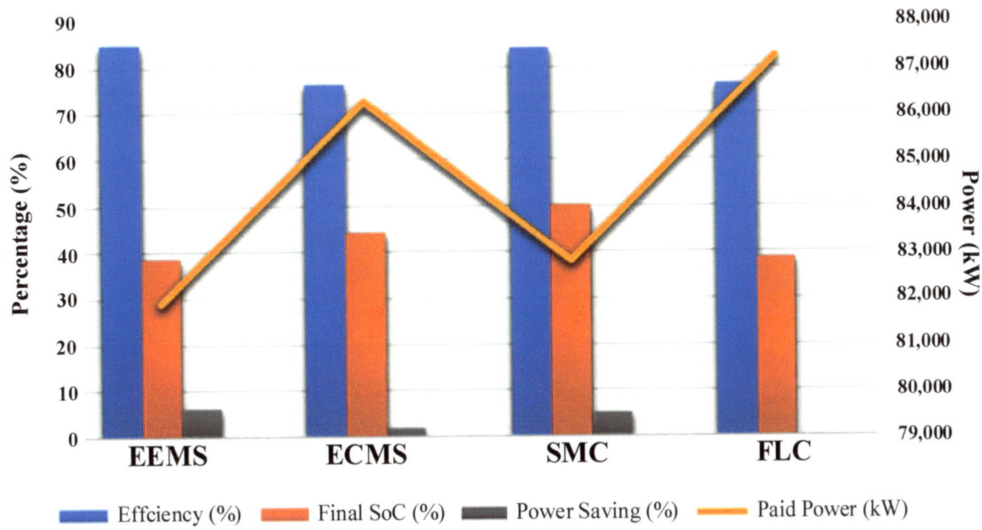

Figure 23. Simulation Statistics.

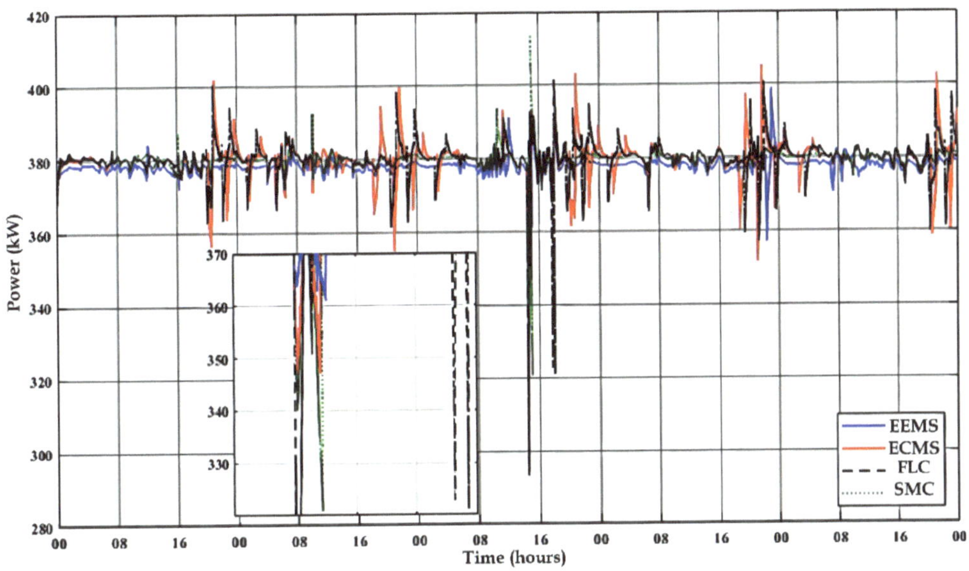

Figure 24. DC bus voltage (V).

Table 5. Bus voltage statistics.

	Min	Max	Mean	Median	Range	StD
EEMS	**357**	**398.9**	378	378.2	**41.95**	**2.07**
ECMS	346.9	405	380	380.3	58.12	4.59
SMC	320	413	380	**380.2**	93.16	2.54
Fuzzy	293.3	401	380	380.3	108.1	4.88

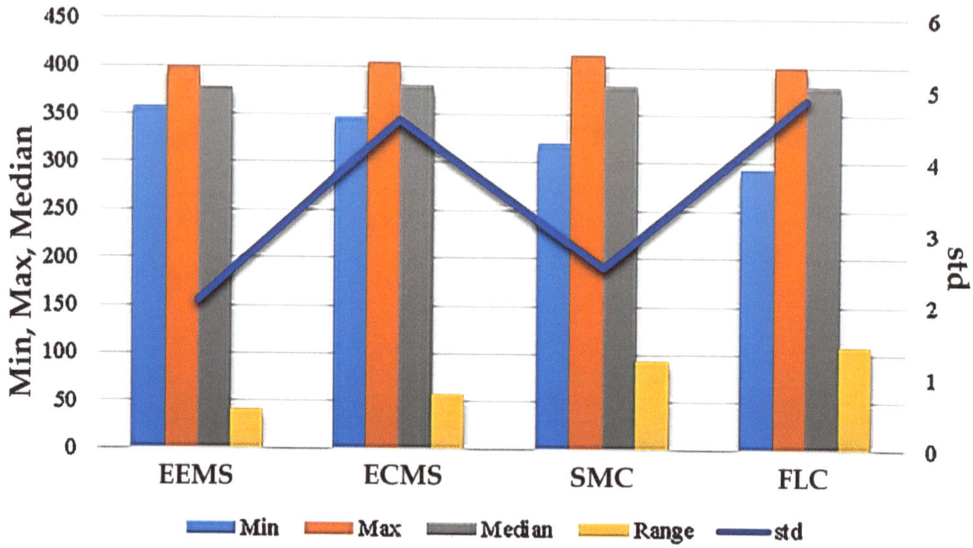

Figure 25. Bus voltage graphical statistics.

5. Conclusions

A comprehensive comparison of four energy management strategies is accomplished. These strategies include fuzzy logic control (FLC), equivalent consumption minimization strategy (ECMS), external energy maximization strategy (EEMS), and state machine control (SMC). These strategies are adopted to achieve optimal performance against severe load and solar power fluctuations. The fundamental comparison indicators are overall system efficiency, power-saving, and the final SoC. The obtained results prove that EEMS provides optimal performance. It operates the system with an efficiency of 84.91% and a power-saving of 6.11%. This strategy can also ensure the power quality for the building and improve the lifespan of the power sources. The system can utilize battery safety with the most extended possible lifespan using the EEMS and SMC (low charging/discharging cycles). In addition, the limited voltage provided by EEMS (41.95 V) gives the battery more safety. The reduced bus voltage (StD 2.068) ensures safe operation for the FC. Eventually, and from the previous analysis, the EEMS provides the most satisfactory performance compared to the other strategies. The results obtained showed that the EEMS performed better than the other studied strategies. However, optimized versions of these strategies may provide a superior performance, such as adaptive optimal fuzzy or salp swarm algorithm (SSA)-based EEMS, and so on. The performance of these strategies will be investigated in future works.

Author Contributions: Conceptualization, H.R., S.F. and A.F.; Methodology, S.F. and A.F.; Software, S.F.; Formal analysis, H.R. and S.F.; Writing—original draft, H.R., R.M.G., S.F., A.F., M.M.G. and R.A.; Writing—review & editing, H.R., R.M.G., S.F., A.F., M.M.G. and R.A.; Project administration, H.R. and R.M.G. All authors have read and agreed to the published version of the manuscript.

Funding: This work was funded by the Deanship of Scientific Research at Princess Nourah bint Abdulrahman University, through the Research Groups Program Grant no. (RGP-1443-0046).

Institutional Review Board Statement: Not applicable.

Informed Consent Statement: Not applicable.

Data Availability Statement: Not applicable.

Conflicts of Interest: The authors declare no conflict of interest.

References

1. Wang, F.; Xie, J.; Wu, S.; Li, J.; Barbieri, D.M.; Zhang, L. Life cycle energy consumption by roads and associated interpretative analysis of sustainable policies. *Renew. Sustain. Energy Rev.* **2021**, *141*, 110823. [CrossRef]
2. Gui, E.M.; Diesendorf, M.; MacGill, I. Distributed energy infrastructure paradigm: Community microgrids in a new institutional economics context. *Renew. Sustain. Energy Rev.* **2017**, *72*, 1355–1365. [CrossRef]
3. Mohabat Doost, D.; Buffa, A.; Brunetta, G.; Salata, S.; Mutani, G. Mainstreaming Energetic Resilience by Morphological Assessment in Ordinary Land Use Planning. The Case Study of Moncalieri, Turin (Italy). *Sustainability* **2020**, *12*, 4443. [CrossRef]
4. Ehigiamusoe, K.U.; Dogan, E. The role of interaction effect between renewable energy consumption and real income in carbon emissions: Evidence from low-income countries. *Renew. Sustain. Energy Rev.* **2022**, *154*, 111883. [CrossRef]
5. Srivastava, C.; Tripathy, M. DC microgrid protection issues and schemes: A critical review. *Renew. Sustain. Energy Rev.* **2021**, *151*, 111546. [CrossRef]
6. Mirsaeidi, S.; Dong, X.; Said, D.M. Towards hybrid AC/DC microgrids: Critical analysis and classification of protection strategies. *Renew. Sustain. Energy Rev.* **2018**, *90*, 97–103. [CrossRef]
7. Majeed Butt, O.; Zulqarnain, M.; Majeed Butt, T. Recent advancement in smart grid technology: Future prospects in the electrical power network. *Ain. Shams. Eng. J.* **2021**, *12*, 687–695. [CrossRef]
8. Dagar, A.; Gupta, P.; Niranjan, V. Microgrid protection: A comprehensive review. *Renew. Sustain. Energy Rev.* **2021**, *149*, 111401. [CrossRef]
9. Sandelic, M.; Peyghami, S.; Sangwongwanich, A.; Blaabjerg, F. Reliability aspects in microgrid design and planning: Status and power electronics-induced challenges. *Renew. Sustain. Energy Rev.* **2022**, *159*, 112127. [CrossRef]
10. Ferahtia, S.; Djeroui, A.; Rezk, H.; Houari, A.; Zeghlache, S.; Machmoum, M. Optimal control and implementation of energy management strategy for a DC microgrid. *Energy* **2022**, *238*, 121777. [CrossRef]
11. Ferahtia, S.; Rezk, H.; Abdelkareem, M.A.; Olabi, A.G. Optimal techno-economic energy management strategy for building's microgrids based bald eagle search optimization algorithm. *Appl. Energy* **2022**, *306*, 118069. [CrossRef]
12. Fathy, A.; Ferahtia, S.; Rezk, H.; Yousri, D.; Abdelkareem, M.A.; Olabi, A.G. Optimal adaptive fuzzy management strategy for fuel cell-based DC microgrid. *Energy* **2022**, *247*, 123447. [CrossRef]
13. Vossos, V.; Gerber, D.; Bennani, Y.; Brown, R.; Marnay, C. Techno-economic analysis of DC power distribution in commercial buildings. *Appl. Energy* **2018**, *230*, 663–678. [CrossRef]
14. AlLee, G.; Tschudi, W. Edison Redux: 380 Vdc Brings Reliability and Efficiency to Sustainable Data Centers. *IEEE Power Energy Mag.* **2012**, *10*, 50–59. [CrossRef]
15. Huang, J.; Fan, J.; Furbo, S.; Chen, D.; Dai, Y.; Kong, W. Economic analysis and optimization of household solar heating technologies and systems. *Sustain. Energy Technol Assess.* **2019**, *36*, 100532. [CrossRef]
16. Hidalgo León, R.; Almeida Pazmiño, G.; Jácome Ruiz, P.; Soriano Idrovo, G.; Urquizo Guevara, J. A Survey on Technologies to Implement Battery Emulators based on DC/DC Power Converters. In Proceedings of the 14th Latin American and Caribbean Conference for Engineering and Technology, San Jose, Costa Rica, 20–22 July 2016. [CrossRef]
17. He, H.; Xiong, R.; Fan, J. Evaluation of Lithium-Ion Battery Equivalent Circuit Models for State of Charge Estimation by an Experimental Approach. *Energies* **2011**, *4*, 582–598. [CrossRef]
18. US Department of Energy. *Comparison of Fuel Cell Technologies*; Energy Efficiency & Renewable Energy: Washington, DC, USA, 2016.
19. Wang, C.; Nehrir, M.H.H. A Physically Based Dynamic Model for Solid Oxide Fuel Cells. *IEEE Trans. Energy Convers.* **2007**, *22*, 887–897. [CrossRef]
20. Belila, A.; Amirat, Y.; Benbouzid, M.; Berkouk, E.M.; Yao, G. Virtual synchronous generators for voltage synchronization of a hybrid PV-diesel power system. *Int. J. Electr. Power Energy Syst.* **2020**, *117*, 105677. [CrossRef]
21. Wang, N.; Wang, D.; Xing, Y.; Shao, L.; Afzal, S. Application of co-evolution RNA genetic algorithm for obtaining optimal parameters of SOFC model. *Renew. Energy* **2020**, *150*, 221–233. [CrossRef]
22. Chakraborty, U. Reversible and Irreversible Potentials and an Inaccuracy in Popular Models in the Fuel Cell Literature. *Energies* **2018**, *11*, 1851. [CrossRef]
23. Chakraborty, S.; Simões, M.G.; Kramer, W.E. (Eds.) *Power Electronics for Renewable and Distributed Energy Systems*; Springer: London, UK, 2013. [CrossRef]
24. Jiang, Q.; Xue, M.; Geng, G. Energy management of microgrid in grid-connected and stand-alone modes. *IEEE Trans. Power Syst.* **2013**, *28*, 3380–3389. [CrossRef]
25. Vásquez, L.O.P.; Ramírez, V.M.; Thanapalan, K. A Comparison of Energy Management System for a DC Microgrid. *Appl. Sci.* **2020**, *10*, 1071. [CrossRef]
26. Motapon, S.N.; Dessaint, L.-A.; Al-Haddad, K. A Robust H2-Consumption-Minimization-Based Energy Management Strategy for a Fuel Cell Hybrid Emergency Power System of More Electric Aircraft. *IEEE Trans. Ind. Electron.* **2014**, *61*, 6148–6156. [CrossRef]
27. Ahmadi, S.; Bathaee, S.M.T.; Hosseinpour, A.H. Improving fuel economy and performance of a fuel-cell hybrid electric vehicle (fuel-cell, battery, and ultra-capacitor) using optimized energy management strategy. *Energy Convers. Manag.* **2018**, *160*, 74–84. [CrossRef]

28. Han, Y.; Yang, H.; Li, Q.; Chen, W.; Zare, F.; Guerrero, J.M. Mode-triggered droop method for the decentralized energy management of an islanded hybrid PV/hydrogen/battery DC microgrid. *Energy* **2020**, *199*, 117441. [CrossRef]
29. Ferahtia, S.; Djerioui, A.; Zeghlache, S.; Houari, A. A hybrid power system based on fuel cell, photovoltaic source and supercapacitor. *SN Appl. Sci.* **2020**, *2*, 940. [CrossRef]
30. Liu, N.; Chen, Q.; Liu, J.; Lu, X.; Li, P.; Lei, J.; Zhang, J. A Heuristic Operation Strategy for Commercial Building Microgrids Containing EVs and PV System. *IEEE Trans. Ind. Electron.* **2015**, *62*, 2560–2570. [CrossRef]
31. Al-Sakkaf, S.; Kassas, M.; Khalid, M.; Abido, M.A. An energy management system for residential autonomous DC microgrid using optimized fuzzy logic controller considering economic dispatch. *Energies* **2019**, *12*, 1457. [CrossRef]
32. Sandels, C.; Widén, J.; Nordström, L. Forecasting household consumer electricity load profiles with a combined physical and behavioral approach. *Appl. Energy* **2014**, *131*, 267–278. [CrossRef]

sustainability

MDPI

Article

Performance Investigation of Renewable Energy Integration in Energy Management Systems with Quantum-Inspired Multiverse Optimization

Dilip Kumar [1,*], Yogesh Kumar Chauhan [2] and Ajay Shekhar Pandey [2]

[1] Department of Electrical Engineering, Institute of Engineering and Technology, Dr. Rammanohar Lohia Avadh University, Ayodhya, Faizabad 224001, India
[2] Department of Electrical Engineering, Kamla Nehru Institute of Technology, Sultanpur 228118, India
* Correspondence: dilipyadav@rmlau.ac.in

Abstract: The study introduces a novel standalone hybrid Energy Management System that combines solar PV, wind energy conversion systems, battery storage, and microturbines in order to provide reliable and efficient power under various operating conditions. The developed Quantum-Inspired Multiverse Optimization (QI-MVO) algorithm has thus far allowed for a remarkable efficiency of 99.9% and a 40% reduction in power losses when compared to conventional approaches. A rather speedy convergence to best solutions is exhibited by the methods, which take about 0.07 s for calculation, hence ensuring accurate optimization in complex energy systems. The QI-MVO-based EMS brings in improved reliability and optimal utilization of the system through balanced energy distribution and by maintaining system operational stability. In conclusion, the present work showcases QI-MVO as a sustainable and scalable energy management solution, which sets the stage for optimization strategies wherein hybrid energy management assumes a very important role.

Keywords: solar photovoltaic (PV); wind turbine (WT); battery system (BS); microturbine (MT); MPPT; energy management system (EMS); hybrid isolated system

1. Introduction

The rapid increase in global energy demand, coupled with the urgent need to reduce carbon emissions, has led to a growing reliance on renewable energy sources such as solar and wind power. Hybrid energy systems, which integrate multiple renewable energy sources with backup and storage technologies, have emerged as a sustainable solution to address the intermittency and variability associated with renewable energy generation [1,2]. However, the effective management of such systems remains a significant challenge, particularly under dynamic load and environmental conditions. This study proposes a novel Quantum-Inspired Multiverse Optimization (QI-MVO) algorithm to address these challenges, offering a computationally efficient and reliable solution for energy management in hybrid systems.

1.1. Problem Statement

Hybrid energy systems face several critical challenges, including long convergence times, high computational demands, and suboptimal performance in complex systems with multiple interconnected components [3,4]. These limitations are particularly pronounced in dynamic environments where load demands and renewable energy availability fluctuate frequently. Existing optimization techniques, such as Genetic Algorithms (GA) and Particle

https://doi.org/10.3390/su17083734

Swarm Optimization (PSO), often struggle to achieve efficient energy distribution and system stability under such conditions [5,6]. This highlights the need for more advanced optimization algorithms that can deliver faster convergence, lower computational complexity, and improved performance in real-time energy management.

1.2. Current Solutions and Their Limitations

Several optimization algorithms have been developed to improve the performance of Energy Management Systems (EMS) in hybrid energy systems. Table 1 provides an overview of recent research in this area, summarizing key methodologies, findings, and research gaps.

Table 1. Overview of recent research publications.

Reference No.	Sources								Techniques	Experimental Setup	Key Findings	Research Gap
	PV	WECS	BESS	Super Capacitor	Hydrogen	Electrolyser	Fuel Cell	Fly Wheel				
[7]	✓	✓	✓						Non-optimized control	Simulation study of hybrid renewable energy system	Demonstrated high efficiency and reliability	Integration challenges with the grid
[8]	✓	✓	✓	✓	✓	✓	✓		Honey Badger Algorithm (HBA)	Multi-objective optimization for HRES with PV, WT, and ESS	Effective energy flow management, minimized LPSP, and cost efficiency	Further validation in diverse geographical and climatic conditions
[9]	✓	✓	✓						Particle Swarm Optimization (PSO)	Hybrid microgrid system under uncertainty	Improved system robustness and performance	Scalability to larger systems
[10]	✓	✓	✓						Fuzzy Logic Control	Hybrid systems in microgrids	Enhanced control efficiency and stability	Implementation complexity in real-world scenarios
[11]	✓	✓	✓						Model Predictive Control	Consensus-based EMS for DC microgrid	Improved energy management and reduced losses	Integration with existing infrastructures
[12]	✓	✓	✓						Adaptive Neuro-Fuzzy Interface System (ANFIS)	Grid-connected hybrid system	Superior performance compared to classical EMS	Real-time control under dynamic conditions
[13]		✓	✓						Reinforcement Learning	Microgrid energy management	Effective learning-based optimization	Data requirements for training models
[14]	✓		✓			✓	✓		Hybrid Intelligent Control	Adaptive microgrid optimization	Integration of rule-based control and deep learning	Adaptability to changing environmental conditions
[15]	✓	✓							Adaptive Fuzzy Logic Control	Microgrid-connected hybrid systems	Improved system adaptability and efficiency	Testing under diverse climatic conditions
[16]	✓	✓	✓					✓	Hybrid micro-grid system	Innovative strategy for energy management	Empirical validation needed	Hybrid micro-grid system

Despite these advancements, existing algorithms such as GA and PSO often exhibit slow convergence and high computational overhead, particularly in scenarios with dynamic energy demands [7–9]. Fuzzy Logic Control and Model Predictive Control have shown promise but face challenges in implementation complexity and real-time adaptability [10–12]. These limitations underscore the need for a more efficient optimization approach that can address the specific challenges of hybrid energy systems.

This is a bibliographic analysis of the categorization and visualization of 41 research papers on Energy Management Systems, Renewable Energy Integration, optimization techniques, battery storage systems, MPPT and control systems, and microgrids and grid integration. Accordingly, the papers are sorted in six primary groups: Energy Management Systems (EMS) [1–4,6,12,13,17–19]; Renewable Energy Integration [5,7–9,15,16,20–22]; optimization techniques [10,14,23–28]; battery storage systems [29–34]; MPPT and control systems [35–39], and microgrids and grid two visualizations are offered to help understand the distribution of these papers as shown in Figure 1a a pie chart pie graph with the percentage distribution of papers among the categories and Figure 1b a bar graph bar chart with the number of papers cited in a year across the categories. Energy Management Systems (EMS) have the biggest share (24%), Renewable Energy Integration has 22%, and optimization techniques have 20%. With peaks for specific years, including 2020 and 2023, it clearly demarcates that areas such as optimization techniques and Renewable Energy Integration have been showing increased research activity. They provide an overall overview of the research landscape around the matters of energy management and renewable energy systems, highlighting the major role they play in coping with modern energy problems.

1.3. Knowledge Gap and Novelty of the Study

The current literature lacks a comprehensive optimization algorithm that can efficiently balance computational complexity with rapid convergence, particularly in dynamic environments with fluctuating energy demands and renewable energy availability. This study addresses this gap by introducing the Quantum-Inspired Multiverse Optimization (QI-MVO) algorithm, which is specifically designed to overcome the limitations of existing methods. The QI-MVO algorithm improves upon traditional optimization techniques by offering faster convergence, lower computational complexity, and enhanced performance in real-time energy management.

The novelty of this work lies in the integration of quantum-inspired principles with the Multiverse Optimizer (MVO) framework, resulting in an algorithm that is both computationally efficient and highly effective in optimizing hybrid energy systems. Unlike existing methods, QI-MVO is specifically tailored to address the challenges of dynamic energy management, ensuring continuous energy supply while minimizing power losses and maximizing operational efficiency.

1.4. Contribution of the Study

The key contribution of the proposed studies is as follows:

- QI-MVO is a novel approach to optimize hybrid energy systems, which is better than current GA and PSO.
- The proposed studies minimize power loss, ensure balanced power distribution, and enhance system reliability under different loads and environmental conditions.
- It presents the study of a dynamic model of a hybrid energy system incorporating solar PV, WECS, battery storage, and microturbine backup to guarantee a continuous energy supply.

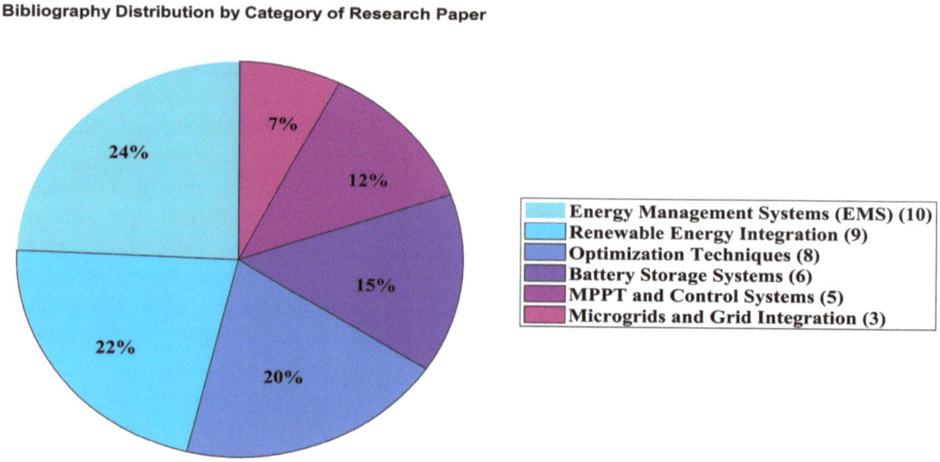

(a)

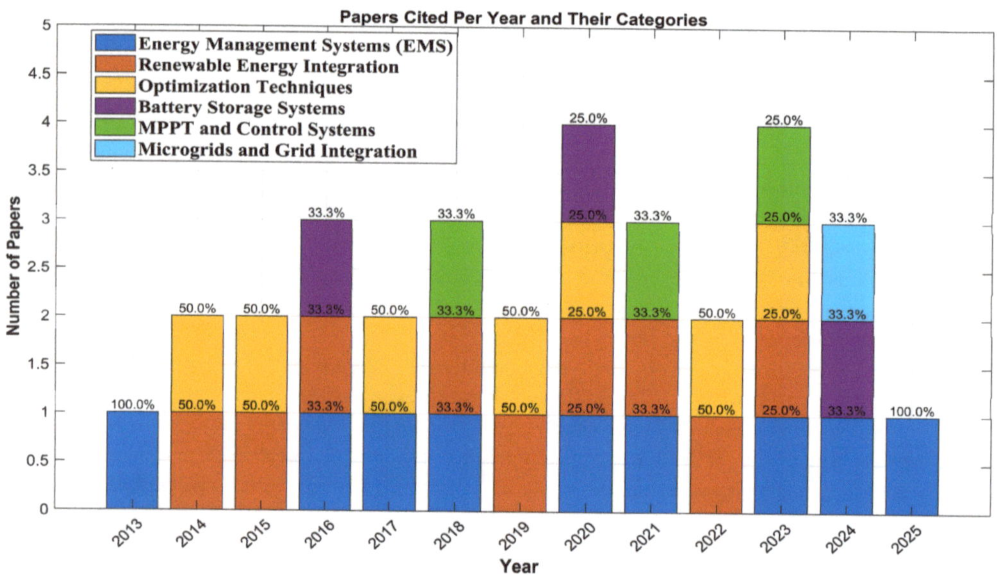

(b)

Figure 1. (**a**) The pie chart suggests the plot of research papers by category. (**b**) The bar graph represents the variation in the number of papers cited per year in terms of category.

1.5. Objectives of the Research

The goals of this study consist of the following:

- Maximize power extraction from solar photovoltaic (PV) and wind energy conversion systems (WECS).
- Control of backup power is improved through microturbine and battery storage.

- Implement a hybrid power system topology that integrates power electronics to AC and DC loads and test the model under different operational modes.
- Design enhanced Energy Management System (EMS) with optimal control strategy for standalone hybrid systems.

1.6. Structure of the Paper

The remainder of this paper is organized as follows: Section 2 elaborates on the design and modeling of the hybrid EMS system. Section 3 provides a detailed description of the proposed QI-MVO algorithm. Section 4 presents the simulation results and performance comparison. Finally, Section 5 summarizes the findings and suggests future research directions.

By introducing the QI-MVO algorithm, this study aims to contribute to the broader goal of sustainable energy development by enhancing the performance and reliability of hybrid energy systems. The proposed approach offers a computationally efficient solution to the challenges of energy management in dynamic environments, ensuring continuous energy supply while minimizing power losses and maximizing operational efficiency.

2. System Configuration and Design

The hybrid energy system comprises four basic components: solar PV systems, wind energy conversion systems (WECSs), battery storage systems (BSSs), and microturbine (MT). This section relates to the complete design of the system, its connection, and mathematical modeling in conjunction with how effectively it is able to control energy in different operational conditions. This hybrid energy system conducts nonstop power generation and delivery by incorporating different types of technologies. It has a 15 kw solar PV system that has been designed to be reliable enough and efficient under different solar irradiance conditions. It consists of an additional part, which is a 15-kilowatt wind energy conversion system designed to efficiently harness wind energy for different wind speed runs. The energy-storing component of this system is the 100 amp hour battery, which is meant to stabilize the load. Power outages are avoided through the backup supplied by the 30 kw microturbine, which is operational when the load is high or renewable energy production is inadequate. An intricate Energy Management System (EMS) manages the entire centralized DC bus within those components. Benefits extend to improving the overall efficiency and reliability of the system, with reduced energy losses, while ensuring the effective control of loads. The overall basic structure of the hybrid energy management standalone system is shown in Figure 2.

2.1. Solar PV System

The solar PV system converts solar irradiance into electrical energy [35,36]. The output power of a PV module is governed by its characteristic equation, as given in Equation (1):

$$I = I_{ph} - I_{sat}\left(e^{\frac{q(V+IR_s)}{nkT}} - 1\right) - \frac{V + IR_s}{R_{sh}} \quad (1)$$

where:

I_{ph}: photo-generated current (A).
I_{sat}: saturation current of the diode (A).
q: electron charge (1.6×10^{-19} C).
V: output voltage of the module (V).
R_s: series resistance (Ω).
R_{sh}: shunt resistance (Ω).
n: diode ideality factor.

k: Boltzmann constant (1.38×10^{-23} J/K).
T: temperature of the module (K).

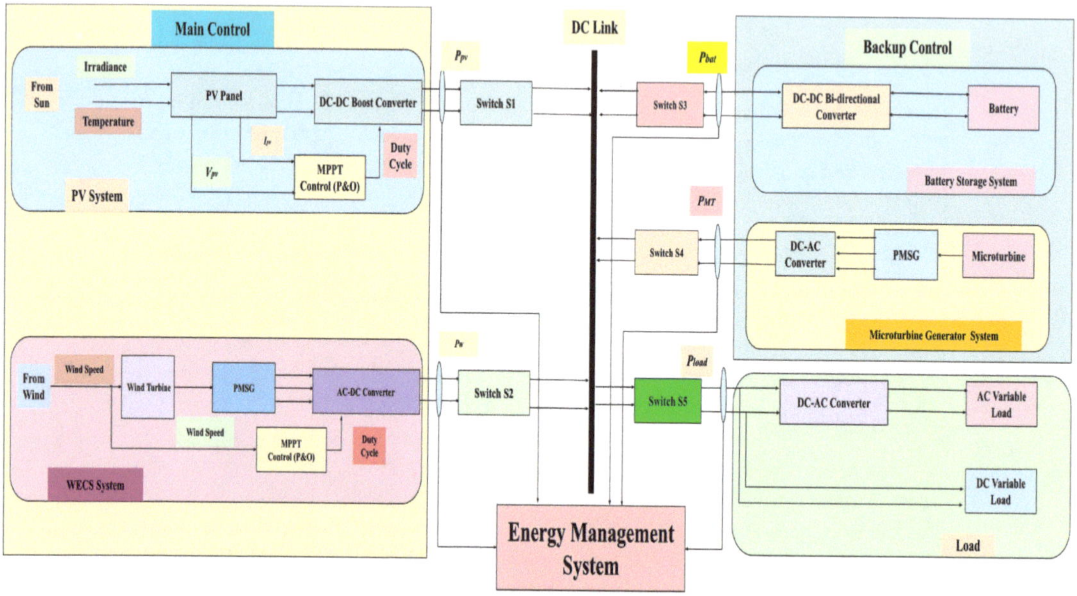

Figure 2. Block diagram of the hybrid energy management standalone system.

Information about the solar PV system is provided in Table 2. The maximum power point is tracked using MPPT techniques like Perturb and Observe (P&O) and Incremental Conductance (INC) to optimize efficiency under varying irradiance conditions. The total power output of the PV array is given by Equation (2):

$$P_{pv} = V_{pv} \times I_{pv} \tag{2}$$

Table 2. Specification of solar PV system.

Parameter	Value
Rated Power	15,000 W
DC Link Voltage	400 V
Short Circuit Current (Isc)	8.21 A
Open Circuit Voltage (Voc)	37.6 V
Number of Series Modules (Ns)	10
Number of Parallel Modules (Np)	5

Implementation of Maximum Power Point Tracking (MPPT) for Solar PV

The maximum power point tracking (MPPT) technique is implemented in the solar PV system to ensure that the system operates at its maximum power point (MPP) under varying solar irradiance and temperature conditions. The MPPT algorithm continuously adjusts the operating voltage of the PV array to maximize power output.

The Perturb and Observe (P&O) method is used for MPPT in this study. The P&O algorithm works by periodically perturbing (increasing or decreasing) the PV array voltage and observing the resulting change in power output. If the power output increases, the algorithm continues to perturb the voltage in the same direction. If the power output decreases, the algorithm reverses the direction of the perturbation. This process continues until the MPP is reached.

The mathematical representation of the P&O algorithm is as follows:

1. Measure the current voltage $V(k)$ and current $I(k)$ of the PV array.
2. Calculate the power $P(k) = V(k) \times I(k)$
3. Compare $P(k)$ with the previous power $P(k-1)$.
4. If $P(k) > P(k-1)$, continue perturbing the voltage in the same direction.
5. If $P(k) < P(k-1)$, reverse the direction of the voltage perturbation.

This method ensures that the PV system operates at its maximum efficiency, even under rapidly changing environmental conditions.

2.2. Wind Energy Conversion System (WECS)

The WECS captures wind energy and converts it into electrical energy [37,38]. The mechanical power output of the wind turbine is expressed in Equation (3):

$$P_w = \frac{1}{2}\rho A v^3 C_P(\lambda, \beta) \quad (3)$$

where:

ρ: air density (1.225 kg/m^3).
A: rotor swept area (m^2).
C_P: power coefficient, function of tip-speed ratio λ and pitch angle β.
v: wind speed (m/s).

Implementation of Tip-Speed Ratio (λ) Optimization for WECS

The tip-speed ratio (λ) optimization is implemented in the wind energy conversion system (WECS) to maximize the power output of the wind turbine. The tip-speed ratio is defined as the ratio of the speed of the rotor blades to the speed of the wind. It is a critical parameter that determines the efficiency of the wind turbine.

The tip-speed ratio (λ) is calculated as in Equation (4):

$$\lambda = \frac{\omega_r R}{V_\omega} \quad (4)$$

where:

- ω_r: angular velocity of the rotor (rad/sec).
- R: radius of the turbine blades (m).
- V_ω: wind speed (m/sec).

To optimize λ, the WECS uses a Perturb and Observe (P&O) method similar to the one used in the solar PV system. The algorithm adjusts the rotor speed to maintain the optimal tip-speed ratio, ensuring maximum power extraction from the wind.

The steps for λ optimization are as follows:

1. Measure the current wind speed V_ω and rotor speed ω_r.
2. Calculate the current tip-speed ratio λ.
3. Compare λ with the optimal tip-speed ratio λ_{opt}.
4. If $< \lambda_{opt}$, increase the rotor speed.
5. If $\lambda > \lambda_{opt}$, decrease the rotor speed.

This optimization ensures that the wind turbine operates at its maximum efficiency, regardless of variations in wind speed. Table 3 presents the specification of WECS.

Table 3. Specification of wind energy conversion system.

Parameter	Value
Rated Power	15,000 W
DC Link Voltage	400 V
Power Coefficient (Cp)	0.4
Air Density (ρ)	1.225
Blade Radius (R)	10 m
Swept Area (A)	$\pi \times (10)^2 = 314.16\ m^2$

2.3. Battery Storage System (BSS)

The BSS smooths power fluctuations and stores surplus energy [20,29]. The state of charge (SOC) of the battery is modeled as in Equation (5):

$$SOC(t) = SOC(t-1) + \frac{\eta_{ch} P_{ch} \Delta t}{E_{bat}} - \frac{P_{dis} \Delta t}{\eta_{dis} E_{bat}} \quad (5)$$

where:

$SOC(t)$: battery state of charge at time t.

η_{ch}, η_{dis}: charging and discharging efficiency.

P_{ch}, P_{dis}: charging and discharging power (W).

E_{bat}: battery energy capacity (Wh).

Δt: time interval (h).

The battery operates between SOC_{min} and SOC_{max} to avoid overcharging or deep discharge. Table 4 provides information on specifications used in the battery storage system.

Table 4. Specification of battery storage system.

Parameter	Value
Nominal Voltage	48 V
Capacity	100 Ah
Maximum Power	5000 W
Initial SOC	50%
Minimum SOC	20%
Maximum SOC	90%

2.4. Microturbine

The MT serves as a backup energy provider whenever renewable power generation cannot meet the demand [21,22]. Table 5 shows a detailed specification list for the microturbine system. The microturbine output power depends on its input fuel energy and operational efficiency according to Equation (6):

$$P_{MT} = \eta_{MT} \times P_{fuel} \quad (6)$$

where η_{MT} is the efficiency of the microturbine and P_{fuel} is the fuel power input.

Table 5. Specification of microturbine system.

Parameter	Value
Rated Power	30,000 W
Efficiency	85%

2.5. Boost Converter Configuration

The boost converter plays a crucial role in the hybrid energy system by stepping up the voltage from lower levels generated by the renewable energy sources (Solar PV and WECS) to the desired DC bus voltage [30,31]. This ensures efficient integration and compatibility of the sources within the hybrid system. The output voltage of the boost converter is given by Equation (7):

$$V_o = \frac{V_{in}}{1 - D} \tag{7}$$

where:

V_o: output voltage.

V_{in}: input voltage from the PV or WECS.

D: duty cycle of the switching signal ($0 < D < 1$).

2.6. Bidirectional Converter

The bidirectional converter is integral to the battery storage system (BSS), enabling energy flow in both directions to charge and discharge the battery depending on the system requirements [32,33]. It facilitates energy management and stabilizes the DC bus voltage during dynamic load and generation conditions.

Boost Mode (Discharging): The converter steps up the battery voltage to match the DC bus voltage during discharging, as given in Equation (8):

$$V_{DC} = \frac{V_{bat}}{1 - D} \tag{8}$$

Buck Mode (Charging): The converter steps down the DC bus voltage to charge the battery efficiently, as given in Equation (9):

$$V_{bat} = V_{DC}(1 + D) \tag{9}$$

where:

V_{DC}: DC bus voltage.

V_{bat}: battery voltage.

2.7. Energy Management System (EMS)

The hybrid energy system integrates solar PV, wind energy conversion system (WECS), battery storage (BSS), and microturbine (MT) through a centralized Energy Management System (EMS). The EMS prioritizes renewable energy sources (solar and wind) to meet load demand, stores surplus energy in the battery, and activates the microturbine only when renewable sources and battery storage are insufficient. This integration ensures balanced power distribution, minimizes power losses, and maintains system reliability under varying environmental and load conditions.

The objective of the Energy Management System is to minimize power losses while ensuring system dependability by ensuring appropriate power distribution [17,34]. The power balance equation is given by Equation (10):

$$P_{load} = P_{pv} + P_w + P_{bat} + P_{MT} - P_{loss} \tag{10}$$

where:

P_{load}: total load power (W).

P_{loss}: power losses in the system (W).

2.8. Simulation Parameter

For precise modeling and performance evaluation under actual operating conditions, the Energy Management System (EMS) simulation depends on accurately specified parameters. A summary of the main simulation parameters used in this study is given in Table 6.

Table 6. Summary of the primary simulation parameters.

Parameter	Value
Sampling Time (T_s)	0.001 s
Simulation Duration (T_{end})	24 h
DC Link Voltage	400 V
DC Link Capacitance	1000×10^{-6} F

The system follows priority rules for energy dispatch: solar PV and wind energy are prioritized first to meet load demand, utilizing renewable sources. Surplus energy is stored in the battery, which is discharged during low renewable generation. The microturbine is activated only when renewable sources and battery storage are insufficient, ensuring minimal reliance on non-renewable energy and optimal resource utilization.

2.9. Load Parameter

By establishing a balance between the energy supply and the fluctuating needs of dynamic operating situations, the Energy Management System (EMS) efficiently controls both AC and DC load profiles [40,41]. The sample load parameters used to evaluate the system's performance are listed in Table 7.

Table 7. Sample load parameters utilized for evaluating the system's performance.

Parameter	Value
AC Load Profile (Sample)	1500–30,000 W
DC Load Profile (Sample)	700–40,000 W

The efficiency of different systems in the hybrid energy setup is as follows: the solar PV system operates at 22% efficiency, converting solar irradiance to electrical energy, while the wind energy conversion system (WECS) achieves 45% efficiency in harnessing wind power. The battery storage system (BSS) has a round-trip efficiency of 95% for charging and discharging, and the microturbine (MT) operates at 85% efficiency in converting fuel to electrical energy. The inverter efficiency ranges from 95% for DC to AC conversion, contributing to an overall system efficiency of 90%, accounting for all losses. These metrics ensure optimal performance and reliability of the hybrid energy system.

3. Methodology

Quantum-Inspired Multiverse Optimisation (QI-MVO) is a potent and sophisticated optimization tool based on unified physics and multiverse theory. This method has shown commendable results in the optimality of Energy Management Systems (EMS) in hybrid energy systems comprising renewable energy sources, energy storage devices, and turbine-based backup generators. The hybrid energy system under consideration for this study comprises a microturbine (MT), battery storage system (BSS), wind energy conversion system (WECS), and solar photovoltaic (PV) system. The objectives include minimizing power losses, achieving balanced power distribution and improving reliability, and optimizing the interaction of all these elements.

The Energy Management System (EMS) employs a rule-based control strategy to prioritize renewable energy sources (solar PV and wind) over non-renewable sources (microturbine) and utilizes battery storage to smooth power fluctuations. For scenarios with high uncertainty, the EMS incorporates fuzzy logic to make adaptive decisions based on real-time data, ensuring optimal energy dispatch and system stability under varying conditions. The system follows priority rules for energy dispatch: solar PV and wind energy are prioritized first to meet load demand, utilizing renewable sources. Surplus energy is stored in the battery, which is discharged during low renewable generation. The microturbine is activated only when renewable sources and battery storage are insufficient, ensuring minimal reliance on non-renewable energy and optimal resource utilization.

During peak demand, the battery storage system (BSS) and microturbine (MT) work together to ensure continuous power supply. The BSS discharges stored energy to meet the load demand, while the microturbine is activated if the battery's state of charge (SOC) falls below a predefined threshold (i.e., 20%). This ensures that the system can handle high demand without interruptions.

3.1. QI-MVO Optimization Process

3.1.1. Initialization

The initial phase of the QI-MVO algorithm is to generate a population of potential solutions, or "universes". Every universe represents a possible arrangement for the distribution of electricity among the four elements: microturbine, solar PV, wind, and battery storage. These universes' starting locations are dispersed at random throughout the specified search space to make sure that the solutions are diverse. The initialization equation is given by Equation (11):

$$X_{i,j} = LB_j + \left(UB_j - LB_j\right) \text{rand}(0,1) \quad (11)$$

where:

$X_{i,j}$: the position of the i-th universe in the j-th dimension.

LB_j, UB_j: lower and upper bounds for the j-th dimension (representing power values from solar, wind, battery, or microturbine).

$\text{rand}(0,1)$: random number between 0 and 1.

3.1.2. Fitness Function

The fitness function evaluates the quality of each solution based on power loss and efficiency, as shown in Equation (12):

$$\text{Minimize } f(x) = \sum_{t=1}^{T} \left| P_{pv,t} + P_{w,t} + P_{bat,t} + P_{MT,t} - P_{load,t} \right| \quad (12)$$

where:

$P_{pv,t}, P_{w,t}, P_{bat,t}, P_{MT,t}$ are the power outputs from the solar, wind, battery, and microturbine at time t.

$P_{load,t}$ is the total load demand at time t.

3.1.3. Multiverse Exploration

The core of the QI-MVO algorithm involves the multiverse exploration principle, which combines quantum and classical strategies for solution refinement [23,24]. Figure 3 illustrates the concept map of the QI-MVO algorithm phases:

White Hole Transport: Universes exchange solutions based on their fitness. Universes with better fitness (lower power loss) are used to update worse solutions, as shown in Equation (13):

$$X_{i,j} = X_{donor,j}, \text{ if } \text{rand}(0,1) < WH_{prob} \quad (13)$$

where:

$X_{i,j}$: the current position of the i-th universe in the j-th dimension.

$X_{donor,j}$: the position of a randomly chosen donor universe in the j-th dimension.

$rand(0,1)$: random number between 0 and 1.

WH_{prob}: the probability of a white hole event, controlling the likelihood of replacing $X_{i,j}$ with $X_{donor,j}$.

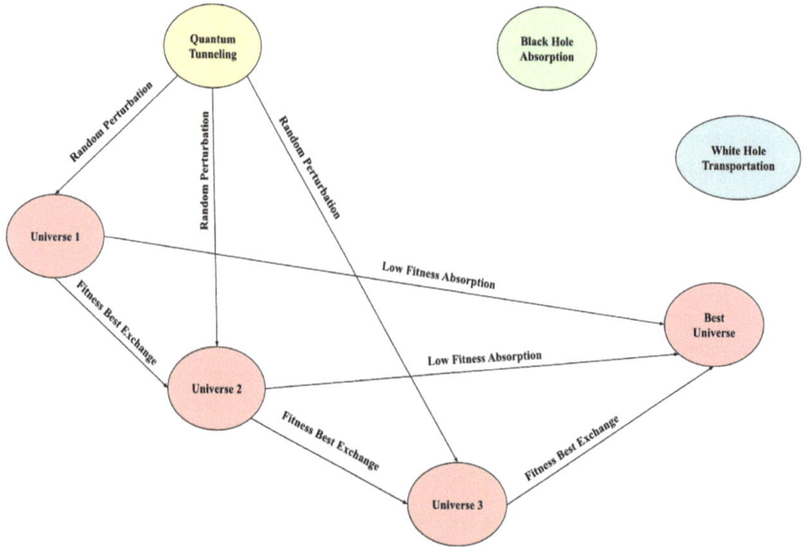

Figure 3. Concept map of QI-MVO algorithm phases.

Black Hole Absorption: Universes with poor solutions are absorbed by universes with better solutions, as given by Equation (14):

$$X_{i,j} = X_{best,j}, \text{ if fitness}(X_{i,j}) > \text{if fitness}\left(X_{best,j}\right) \tag{14}$$

where:

$X_{i,j}$: the current position of the i-th universe in the j-th dimension.

$X_{best,j}$: the best-known position in the j-th dimension (from the universe with the best fitness value).

$\text{fitness}(X_{i,j})$: the fitness value of the i-th universe, determined by the objective function.

This condition updates the position of $X_{i,j}$ with the position of $X_{best,j}$ when the fitness of $X_{i,j}$ is worse than the fitness of the current best solution.

Quantum Tunneling: To enhance diversity and exploration, Quantum Tunneling is applied by introducing small random perturbations to the solution positions, as described in Equation (15):

$$X_{i,j} = X_{i,j} + A_q \cdot rand(0,1) \tag{15}$$

where:

A_q is the quantum amplitude and $rand(0,1)$ generates random perturbations.

These perturbations help escape local optima and ensure better exploration of the solution space [25,26].

3.1.4. Optimization Process

At each iteration, the positions of all universes are updated based on the multiverse exploration principles. The fitness values are recalculated, and the best solution is tracked. The algorithm stops when the maximum number of iterations is reached or when no significant improvement is observed. A flowchart of the QI-MVO optimization process is shown in Figure 4.

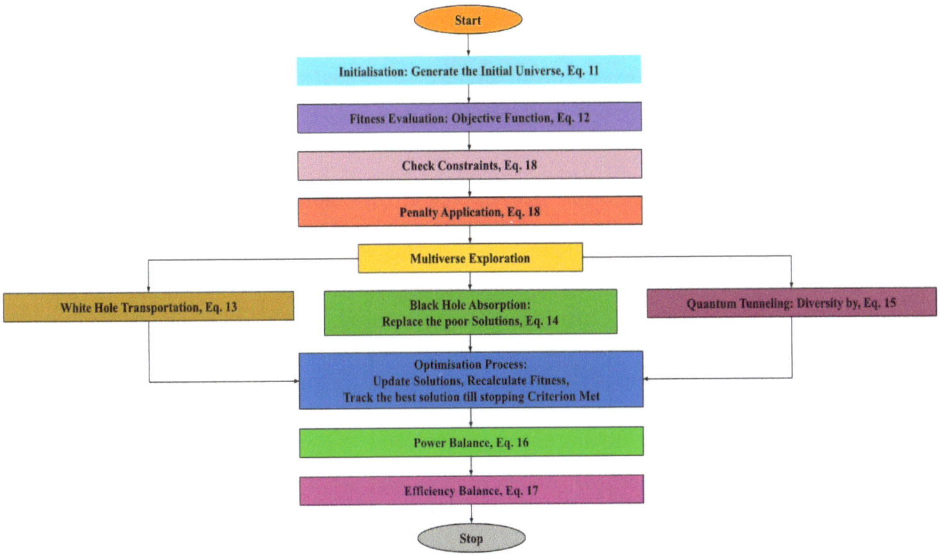

Figure 4. Flowchart of QI-MVO algorithm.

3.1.5. Power Balance and Efficiency

At each time step, the total power supplied by all energy sources must meet the load demand while minimizing losses, as expressed in Equation (16):

$$P_{load,t} = P_{pv,t} + P_{w,t} + P_{bat,t} + P_{MT,t} - P_{loss,t} \tag{16}$$

The system efficiency at each time step is determined using Equation (17):

$$\eta_t = \frac{P_{load,t}}{P_{pv,t} + P_{w,t} + P_{bat,t} + P_{MT,t}} \cdot 100 \tag{17}$$

where $P_{loss,t}$ is the system power loss at time t.

3.1.6. Penalty for Constraints Violation

When the optimum total power surpasses the predetermined operational limits, the Energy Management System (EMS) penalizes the fitness function. For dependable and effective energy distribution, this guarantees adherence to system restrictions [18,19]. If the total optimized power falls outside the specified bounds, a penalty is applied to the fitness function, as defined in Equation (18):

$$\text{Penalty} = \begin{cases} 0,\text{if } P_{min} \leq P_{total,optim} \leq P_{max} \\ 10^6,\text{Otherwise} \end{cases} \tag{18}$$

4. Results and Discussion

This study investigates the combinations along with the operating performance of a hybrid energy system that incorporates the microturbine system (MT), battery storage system (BSS), wind energy conversion system (WECS), and solar photovoltaic (PV) system. The hybrid system is supposed to ensure a reliable and efficient power supply against fluctuating climatic conditions and varying load demands.

The theoretical data and recent literature support the findings. This can be proved as the efficiency improvement attained by QI-MVO (up to 99.9%) is consistent with the work of [7], which exhibits the effectiveness of quantum-inspired algorithms in solving the optimization problem of hybrid energy systems. The results reported in [38] have shown that advanced optimization techniques are essential for minimizing energy waste, and this is consistent with the reduction in power losses (up to 40%). This is also validated by [29], which also shows the need for dynamic energy management strategies for hybrid systems. Being a series of references, these serve as a strong theoretical basis for the results and control the potential bias in the interpretation.

The execution and integration evaluations of hybrid energy systems occurred through assessment simulations and optimized methodologies. Simulation with optimization of the system utilized MATLAB 2018a alongside Quantum-Inspired Multiverse Optimization (QI-MVO). QI-MVO represented the most appropriate advanced algorithm, which combined multiverse theory with quantum mechanics to handle the dynamic nonlinear energy flows observed in hybrid systems. The QI-MVO algorithm achieved optimal power loss reduction by precisely controlling energy generation and energy storage against energy consumption. The system accomplished its performance-based requirements by satisfying remarkable improvements in efficiency and reliability along with versatility enhancement.

4.1. Simulation Results

The investigation seeks the performance of a hybrid energy system over a 24 h operation study on 1 May 2024, at coordinates 11.8474° N, 92.6614° E, Havelock Island, India, using NASA/POWER CERES/MERRA2 Native Resolution Hourly Data. The multi-panel graph delineates significant aspects of energy consumption, meteorological conditions, and renewable generation.

The solar irradiance demonstrates a profound hourly cycle, with 0 W/m^2 earlier in the dawn period, a high point of approximately 728.18 W/m^2 about 10 a.m., and decreasing steadily to 0 W/m^2 after sunset. This trend points out the stochasticity engendered in solar energy. Regarding attentive control of environmental temperature, 30.06 °C in the morning—with the highest afternoon temperature of 31.83 °C—is essential for the optimal functioning of photovoltaic modules. Finally, a moderate wind energy potential is illustrated by wind speeds throughout the day, peaking at 4.70 m/s at 4 p.m., decreasing to 0.95 m/s at midnight, and starting at 2.36 m/s at 1 a.m.

AC loads exhibit distinct patterns driven by cooling and residential demands, with a noon high of 30,000 W and a lower nighttime peak of 25,000 W. On the other hand, DC loads show significant variability, indicating specific equipment utilization spikes, with a range of 700 W to a maximum of 40,000 W at 3 PM. The solar PV system's output, which peaks at 11,239 W at noon and closely resembles the solar irradiance curve, demonstrates how dependent it is on sunshine levels.

From 65.99 W at midnight to a maximum of 7298.11 W in the afternoon, wind energy maintains a steady contribution throughout the day, balancing the variation in solar energy output. The balance between solar and wind energy demonstrates the hybrid system's ability to effectively manage the intermittent nature of individual renewable sources.

The analysis highlights the interdependence between renewable energy sources and load requirements. While solar energy provides a regular 24 h cycle, wind energy ensures a steady output, preventing changes in solar generation. Since the consumption patterns of AC and DC loads differ, customized energy management strategies are needed. The research highlights the necessity of integrating energy storage and backup systems to manage the sporadic nature of renewable energy sources and offers a reliable power source for both AC and DC applications.

The battery system likewise maintains a more stable and even SOC and SOD from approximately 50% to 49.9998% throughout the day. This averts needless variations in its input and output, which defines its ability to allow changing energy demands while ensuring a relatively stable output of power must be demonstrated. Coupled with energy production peaks, such as when the battery exports 5000 W between hours 15 and 20, the system demonstrates the ability to deal with energy shortages occurring as demands peak. In those cases, the microturbine effectively plays a backup role by generating 19,259.15 W at hour 15 and 16,197.96 W at hour 20. While it demonstrates how the microturbine fulfills these outputs capably and acts to make up shortfalls of energy when only renewable generation is inadequate for energy balance, this, once again, underpins the important role of the microturbine.

Only small power surpluses or deficits indicate the system's ability to balance generation and load needs throughout most of the day. Nonetheless, heavy reliance on the microturbine at times of high demand, for example, in hour 15, warrants a more efficient contingency plan and careful energy storage management of the latter. The three major components, battery, microturbine, and renewable sources, act cooperatively in this case to balance their surpluses and shortages and provide a steady supply of electricity.

Though the SOC and SOD characters remain uniform all through, the system considers the efficiency of storage and backup to enhance flexibility toward meeting energy demands. This ensures longer operational life for the system, maximizes the efficiency of operations, reduces dependency on external power, and ensures great reliability of the system. The system is well-optimized and contains the ability to maintain energy balance along a vast realm of functionalities, as expressed by the stable SOC and SOD values.

Figures 5–8 show the performance of the system. Figure 5 illustrates input parameters with time, including wind speed (m/s), ambient temperature (°C), and solar irradiation (W/m^2), which demonstrates that the trend shows solar irradiance peaking at noon, temperature rising in the afternoon, and wind speed fluctuating moderately. These trends reflect typical daily environmental variations, validating the system's ability to handle real-world conditions. Figure 6 shows how load varies throughout the day by showing AC and DC load powers (Watt), which represents AC load peaks during the day, while DC load shows variability, reflecting typical energy consumption patterns. This supports the system's adaptability to varying load demands. Figure 7 presents the power generation trends of solar PV and wind energy conversion systems (Watt) as time progresses, alongside the outputs of the major control system. The solar PV output follows irradiance levels, while wind energy provides steady output, which demonstrates the system's ability to balance intermittent renewable sources. Figure 8 focuses on the outputs of backup control and presents the trends in battery power (Watt), microturbine power (Watt), SOC and SOD percent, and Watt trends in surpluses or deficits with time. The battery SOC remains stable, and the microturbine activates during peak demand, showing effective backup management. These results validate the EMS's ability to maintain power balance and ensure system reliability. Together, these figures provide the reader with a complete view of the hybrid energy system's operational conditions and capture interconnections among all its various components.

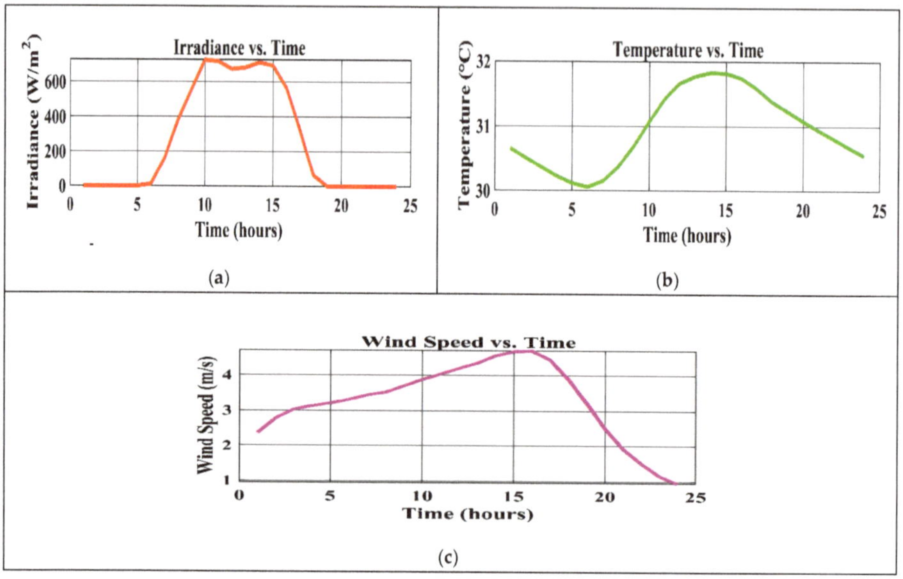

Figure 5. Input parameters of Energy Management System: (**a**) irradiance (W/m^2) vs. time (hours); (**b**) temperature (°C) vs. time (hours); (**c**) wind (m/s) vs. time (hours).

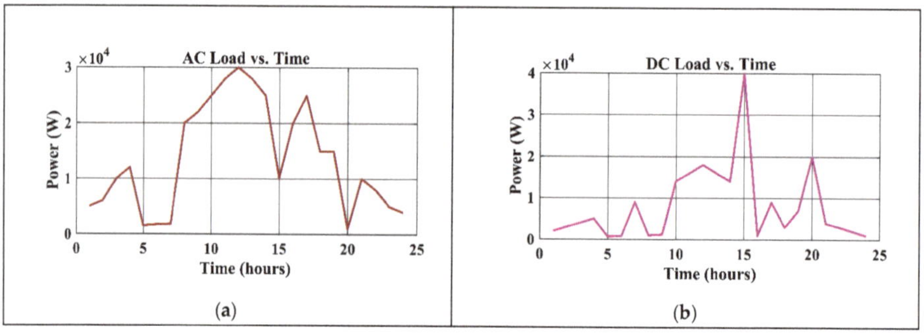

Figure 6. Load samples for Energy Management System: (**a**) AC load power (Watt) vs. time (hours); (**b**) DC load power (Watt) vs. time (hours).

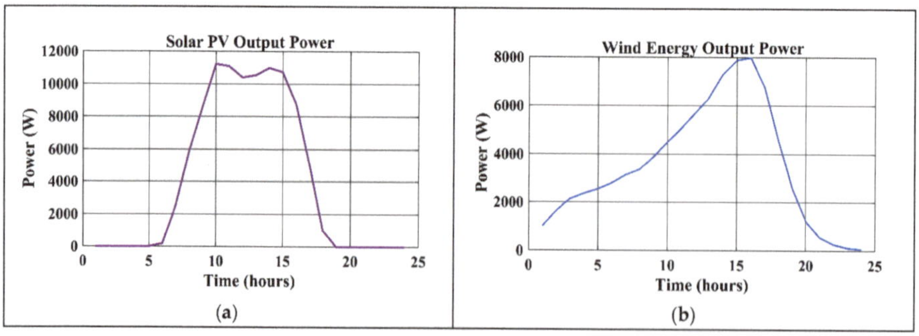

Figure 7. Main control outputs: (**a**) solar PV power versus time (hours); (**b**) wind energy conversion system power versus time (hours).

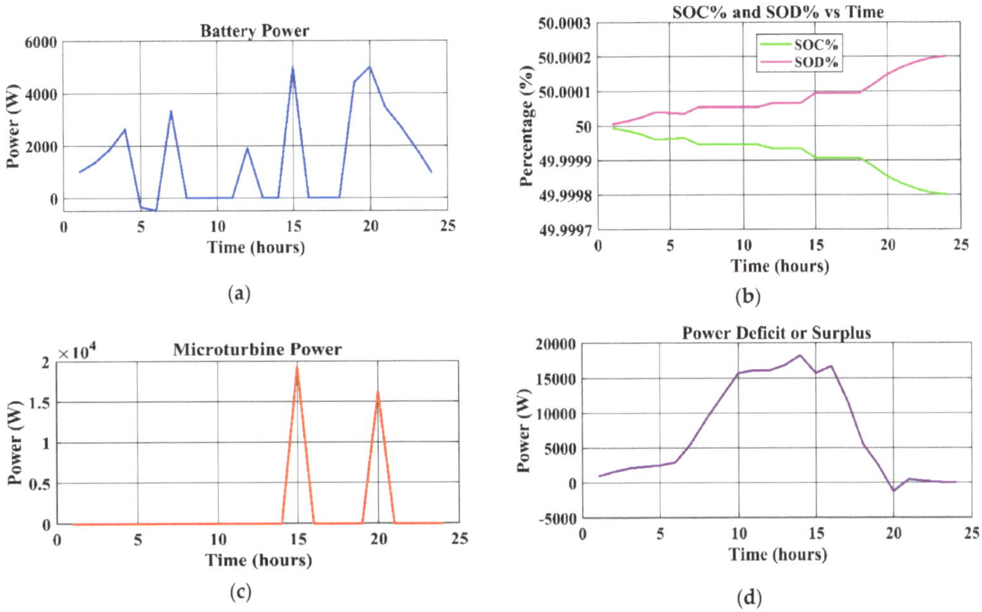

Figure 8. Backup control outputs: (**a**) battery time (hours) versus battery power (Watt); (**b**) SOC and SOD (%) versus time (hours); (**c**) time (hours) versus microturbine power (watts); (**d**) Watts of power deficit or surplus versus time (hours).

4.2. Comparative Model: Non-Optimized vs. GA vs. PSO vs. QI-MVO-Based Energy Management System

In a comparative evaluation of Energy Management Systems (EMS) applying Quantum-Inspired Multiverse Optimisation (QI-MVO), Particle Swarm Optimisation (PSO), Genetic Algorithm (GA), and non-optimized methods, some marked optimization-driven advantages concerning stability, energy management, and efficiency should be realized. In this section, the different techniques are examined for their effectiveness based on crucial features of power output, system efficiency, evolution of fitness, and load-handling capacities.

Figures 9–11 display the performance results for each of GA, PSO, and QI-MVO. GA gives good power management, a higher efficiency feature, and convergence of the fitness property. The subfigures show the improvements seen by the technique concerning power management, efficiency, and individual performance indices. However, PSO shows that with rapid convergence, constant efficiency, and good power control, it depicts the potential of energy optimization. The most compelling power stability, impressive efficiency improvements, and blemish-free optimization, even when tackling complex and difficult problems, are presented by QI-MVO, which absolutely pulverizes both methods.

A numerical representation of comparative simulation results of the non-optimized, GA, PSO, and QI-MVO techniques is given in Table 8. Based on its superior convergence, high efficiency, and stability, QI-MVO gives the most consistent results of the four. In many instances, GA is seen to perform a good optimization, whereas PSO is seen to show better results than the non-optimized method in terms of speed of convergence.

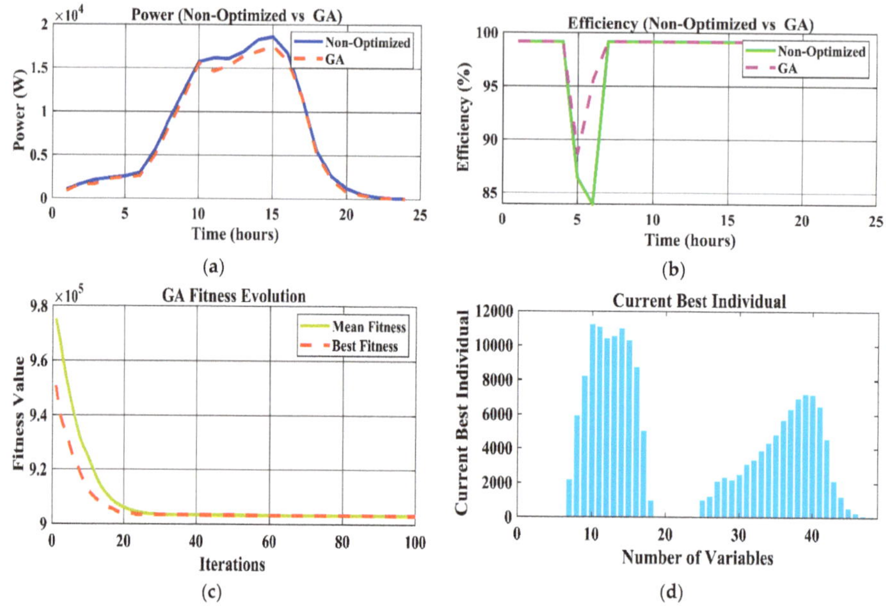

Figure 9. GA optimization graph: (**a**) time (hours) vs. power (watt) (non-optimized vs. GA); (**b**) efficiency (%) vs. time (hours) (comparison graph non-optimized vs. GA); (**c**) iteration comparative graph vs. GA fitness value (mean and best fitness); (**d**) number of variables compared to current best individual.

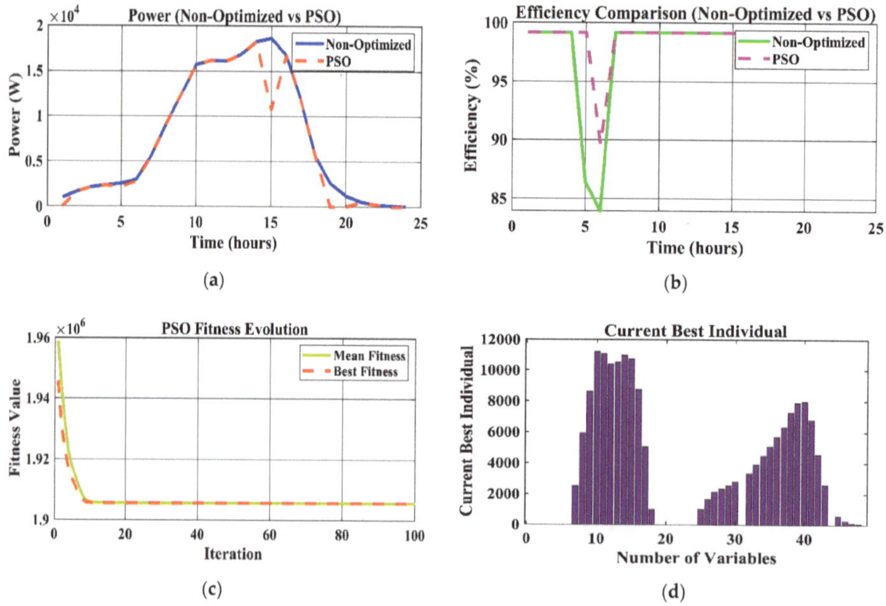

Figure 10. PSO optimization graph: (**a**) time (hours) vs. power (watts) (non-optimized vs. PSO); (**b**) efficiency (%) vs. time (hours) (non-optimized vs. PSO); (**c**) comparison graph PSO fitness value (mean and best fitness) vs. iteration (non-optimized vs. PSO); (**d**) current best individual vs. number of variables.

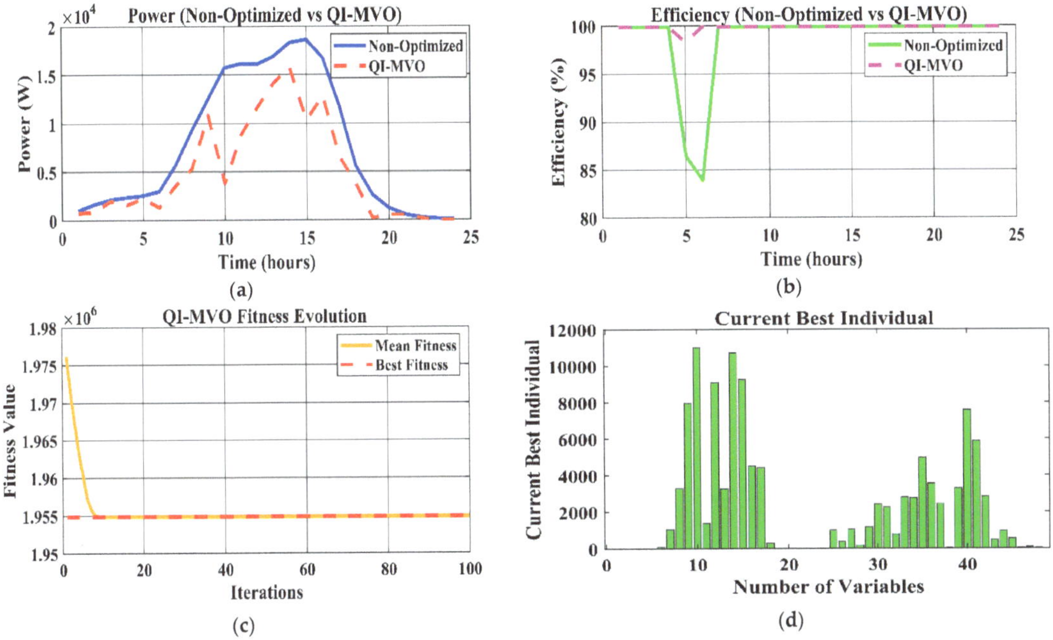

Figure 11. QI-MVO optimization graph: (**a**) power (Watt) vs. time (hours); (**b**) efficiency (%) vs. time (hours) comparison graph (non-optimized vs. QI-MVO comparison graph) (**c**) iteration vs. QI-MVO fitness value (mean and best fitness); (**d**) current best individual vs. variable number in QI-MVO.

Table 8. Evaluation of non-optimized, GA PSO, and QI-MVO simulation results.

S.NO	Metric	Non-Optimized	GA [27,28,39]	PSO [28,39]	QI-MVO
1.	Efficiency (%) (Minimum)	83.4	88	90	99.2
2.	Efficiency (%) (Maximum)	95.7	99.2	99.4	99.9
3.	Number of Iterations	100	100	100	100
4.	Convergence Time (seconds)	12	3.67	0.18	0.07
5.	Fitness Value (Best)	-	903,227.62	1,904,026.98	1,954,885.85
6.	Fitness Value (Mean)	-	908,734.32	1,906,129.46	1,963,291.94

The EMS efficiency study shows significant differences across the methods. The non-optimized EMS achieves efficiency levels between 83.4% and 95.7%, resulting in large energy losses because static control systems cannot adapt to changing loads and environmental conditions. GA can boost efficiency by 88% to 99.2% because of its capacity to dynamically adjust system settings and cut down on energy waste. With a range of 90% to 99.4%, PSO's real-time optimization capabilities, which ensure effective resource utilization, greatly boost efficiency. With values between 99.2% and 99.9%, QI-MVO provides the highest efficiency due to its precise optimization, which minimizes energy losses and maximizes utilization.

The results of all optimization approaches, including GA, PSO, and QI-MVO, are obtained using 100 iterations in order to fairly compare convergence characteristics and computation efficiency. It takes 12 s for the EMS that is not optimized to converge. GA reduces this time to 3.67 s, but it is still slower than other methods. PSO drastically cuts the convergence time to 0.18 s by utilizing its efficient swarm-based search techniques and streamlined computational environment. Using quantum-inspired principles that enable efficient solution space search and exploitation, QI-MVO outperforms all other methods with a convergence time of only 0.07 s.

Fitness values offer more proof that optimization techniques work. Fitness measures have no value because the non-optimized EMS does not apply any algorithms. GA shows its ability to enhance performance with a mean of 908,734.32 and a highest fitness value of 903,227.62. The best fitness value of 1,904,026.98 and mean of 1,906,129.46 show that PSO performs better than GA, proving that it can identify better solutions with less variability. QI-MVO achieves the highest fitness with a mean of 1,963,291.94 and a best value of 1,954,885.85, demonstrating its ability to consistently discover optimal solutions with little deviations.

The parameter settings for GA, PSO, and QI-MVO are now clearly detailed to ensure reproducibility and facilitate comparison. For GA, the population size is set to 50, with 100 iterations, a crossover rate of 0.8, and a mutation rate of 0.01. For PSO, the population size is also 50, with 100 iterations, an inertia weight of 0.7, and cognitive and social coefficients of 1.5. For QI-MVO, the population size is 50, with 100 iterations, a white hole probability (WH_{prob}) of 0.2, and a quantum amplitude (A_q) of 0.1. These standardized settings allow readers to replicate the experiments and validate the results effectively.

The non-optimized EMS has trouble managing peak loads and reducing power loss, which leads to frequent decreases in efficiency and significant power losses. GA has greater power losses than PSO and QI-MVO, but it enhances load control through adaptive changes. PSO successfully balances load demands by reducing power losses by 25% as compared to the non-optimized system. The most efficient technology is QI-MVO, which reduces power losses by 40% and has excellent load-matching capabilities that guarantee stable and dependable operation in a variety of scenarios.

A number of significant findings are highlighted in this investigation. QI-MVO performs close to optimally and outperforms non-optimized and general analysis frameworks by a large margin. QI-MVO has unparalleled computing power, alongside the rapid convergence era and the highest fitness beliefs. It is scheduled to be the superior option due to its exceptional optimization capacity and flexibility in managing the complex development of hybrid energy structures. QI-MVO consistently exceeds the unoptimized structure in all prosody, while GA and PSO provide significant additions.

The conclusions above show how, through optimization, hybrid renewable energy systems could revolutionize. QI-MVO consistently outperforms them, thanks to its excellent performance, rapid convergence, and high efficiency despite the significant developments in GA and PSO. The above findings show QI-MVO's commitment to a new approach to improving EMS performance and provide a solid basis for future studies that aim to expand their use to more and more sophisticated energy relationships.

5. Conclusions

This examination addresses the impediment to delivering reliable and productive control sources by carefully inspecting a hybrid energy control system integrating solar PV, wind turbine, battery storage, and microturbine. Therefore, in order to improve organizational performance, we shall use a powerful optimization method to improve organizational performance. Quantum-Inspirated Multiverse Optimisation (QI-MVO), Particle Swarm Optimisation (PSO), and Heritable Algorithm (GA) show that QI-MVO is better suited to requirements concerning authority loss, convergence speed, and organizational effectiveness.

The conclusions show the QI-MVO algorithm's ability to maximize energy use within the hybrid energy structure, achieving a maximum efficiency of 99.9%. Moreover, compared with GA and PSO, QI-MVO achieves a remarkable 40% decrease in influence losses, demonstrating its efficiency in improving overall structural performance. One's extraordinary computational productivity, which converges in just 0.07 s with the same number of iterations, demonstrates the resilience and speed of the algorithm. Moreover, the improved hybrid energy standalone governance framework successfully responds to changing load conditions and exhibits outstanding stability and reliability.

The results indicate the importance of advanced optimization techniques for the establishment of hybrid energy structures, together with QI-MVO, which promises promising future functions. This aims to promote the development of efficient and green energy structures by opening the way for the expansion of the hybrid energy administration framework capabilities to larger relationships. In order to extend QI-MVO's use in multidisciplinary environments, upcoming investigations may focus on the real-time implementation of QI-MVO in the smart grid and the use of adaptive control strategies.

Author Contributions: Conceptualization, D.K., Y.K.C. and A.S.P.; methodology, D.K., Y.K.C. and A.S.P.; software, D.K.; validation, D.K., Y.K.C. and A.S.P.; formal analysis, D.K.; investigation, D.K., Y.K.C. and A.S.P.; writing—original draft preparation, D.K.; writing—review and editing, D.K., Y.K.C. and A.S.P. All authors have read and agreed to the published version of the manuscript.

Funding: This research received no external funding.

Institutional Review Board Statement: Not applicable.

Informed Consent Statement: Not applicable.

Data Availability Statement: Data is contained within the article. The authors acknowledge that data were taken from NASA/POWER (https://power.larc.nasa.gov) access date (19 December 2024) used in this study. The wind speed, temperature and solar irradiance data at Havelock Island, India (Latitude: 11.8474, Longitude: 92.6614) on 1 May 2024 has enabled detailed analysis and validation of the proposed models and methodology.

Conflicts of Interest: The authors declare that there are no conflicts of interest.

Abbreviations

List of Symbols

Symbol	**Description**
I_{ph}	Photo-generated current (A).
I_{sat}	Saturation current of the diode (A).
q	Electron charge (1.6×10^{-19} C).
V	Output voltage of the module (V).
R_s	Series resistance (Ω).
R_{sh}	Shunt resistance (Ω).
n	Diode ideality factor.
k	Boltzmann constant (1.38×10^{-23} J/K).
T	Temperature of the module (K).
P_{pv}	Power output from solar PV (W)
V_{pv}	Voltage output from solar PV (W)
I_{pv}	Current output from solar PV (W)
ρ	Air density (1.225 kg/m^3).
A	Rotor swept area (m^2).
C_P	Power coefficient, function of tip-speed ratio λ and pitch angle β.
v	Wind speed (m/s).
ω	The angular velocity of the rotor (rad/s)
R	The radius of the turbine blades (m)
P_w	Power output from wind (W)

$SOC(t)$	Battery state of charge at time t
η_{ch}, η_{dis}	Charging and discharging efficiency
P_{ch}, P_{dis}	Charging and discharging power (W).
E_{bat}	Battery energy capacity (Wh).
Δt	Time interval (h).
P_{bat}	Power output from battery (W)
η_{MT}	The efficiency of the microturbine
P_{fuel}	The fuel power input.
P_{MT}	Power output from microturbine (W)
V_o	Output voltage.
V_{in}	Input voltage from the PV or WECS.
D	Duty cycle of the switching signal (0 < D< 1).
V_{DC}	DC bus voltage.
V_{bat}	Battery voltage.
P_{load}	Total load power (W).
P_{loss}	Power losses in the system (W).
$X_{i,j}$	is the position of the i-th universe in the j-th dimension.
LB_j, UB_j	Lower and upper bounds for the j-th dimension (representing power values from solar, wind, battery, or microturbine).
$rand(0, 1)$	Random number between 0 and 1.
$X_{donor,j}$	The position of a randomly chosen donor universe in the j-th dimension.
P_{min}	Minimum power limit
$P_{total,optim}$	Optimized total power
P_{max}	Maximum power limit
WH_{prob}	White hole probability
A_q	Quantum amplitude
Abbreviation	**Full Form**
SOC	State of charge
SOD	State of discharge
PV	Photovoltaic
WECS	Wind energy conversion system
BSS	Battery storage system
MT	Microturbine
EMS	Energy Management System
GA	Genetic Algorithm
PSO	Particle Swarm Optimization
QI-MVO	Quantum-Inspired Multiverse Optimization
DC	Direct current
AC	Alternating current
MPPT	Maximum power point tracking
INC	Incremental Conductance
P&O	Perturb and Observe
LB	Lower Bound
UB	Upper bound
WH	White hole
BH	Black hole
QT	Quantum Tunneling

References

1. Hemmati, R. Technical and economic analysis of home energy management system incorporating small-scale wind turbine and battery energy storage system. *J. Clean. Prod.* **2017**, *159*, 106–118. [CrossRef]
2. Tawfik, T.M.; Badr, M.A.; El-Kady, E.Y.; Abdellatif, O.E. Optimization and energy management of hybrid standalone energy system: A case study. *Renew. Energy Focus* **2018**, *25*, 48–56. [CrossRef]

3. Kumar, R.; Gupta, R.A.; Bansal, A.K. Economic analysis and power management of a stand-alone wind/photovoltaic hybrid energy system using biogeography based optimization algorithm. *Swarm Evol. Comput.* **2013**, *8*, 33–43. [CrossRef]
4. Chauhan, A.; Saini, R.P. Techno-economic optimization based approach for energy management of a stand-alone integrated renewable energy system for remote areas of India. *Energy* **2016**, *94*, 138–156. [CrossRef]
5. Behzadi, M.S.; Niasati, M. Comparative performance analysis of a hybrid PV/FC/battery stand-alone system using different power management strategies and sizing approaches. *Int. J. Hydrogen Energy* **2015**, *40*, 538–548. [CrossRef]
6. Murty, V.V.S.N.; Kumar, A. Optimal Energy Management and Techno-economic Analysis in Microgrid with Hybrid Renewable Energy Sources. *J. Mod. Power Syst. Clean Energy* **2020**, *8*, 929–940. [CrossRef]
7. Hossain, M.J.; Arafath, S.M. Performance evaluation of PV-Wind-BESS hybrid renewable energy systems: A comprehensive study. *Renew. Energy* **2023**, *200*, 123–135. [CrossRef]
8. Shaier, A.A.; Elymany, M.M.; Enany, M.A.; Elsonbaty, N.A. Multi-objective optimization and algorithmic evaluation for EMS in a HRES integrating PV, wind, and backup storage. *Sci. Rep.* **2025**, *15*, 1147. [CrossRef]
9. Mquqwana, M.A.; Krishnamurthy, S. Particle Swarm Optimization for an Optimal Hybrid Renewable Energy Microgrid System under Uncertainty. *Energies* **2024**, *17*, 422. [CrossRef]
10. Feddaoui, O.; Toufouti, R.; Jamel, L.; Meziane, S. Fuzzy logic control of hybrid systems including renewable energy in microgrids. *Int. J. Electr. Comput. Eng. (IJECE)* **2020**, *10*, 5559–5569. [CrossRef]
11. Ali, S.U.; Waqar, A.; Aamir, M.; Qaisar, S.M.; Iqbal, J. Model predictive control of consensus-based energy management system for DC microgrid. *PLoS ONE* **2023**, *18*, e0278110. [CrossRef]
12. García, P.; García, C.A.; Fernández, L.M.; Llorens, F.; Jurado, F. ANFIS-Based Control of a Grid-Connected Hybrid System Integrating Renewable Energies, Hydrogen and Batteries. *IEEE Trans. Ind. Inform.* **2014**, *10*, 1107–1117. [CrossRef]
13. Kuznetsova, E.; Li, Y.-F.; Ruiz, C.; Zio, E.; Bell, K.; Ault, G. Reinforcement learning for microgrid energy management. *Energy* **2013**, *59*, 133–146. [CrossRef]
14. Akbulut, O.; Cavus, M.; Cengiz, M.; Allahham, A.; Giaouris, D.; Forshaw, M. Hybrid Intelligent Control System for Adaptive Microgrid Optimization: Integration of Rule-Based Control and Deep Learning Techniques. *Energies* **2024**, *17*, 2260. [CrossRef]
15. Khallouf, K.; Zarour, L.; Benbouhenni, H.; Debdouche, N.; Elbarbary, Z.; El-Barbary, Z. Adaptive fuzzy logic control for microgrid-connected hybrid photovoltaic/wind generation systems. *Energy Rep.* **2024**, *12*, 4741–4756. [CrossRef]
16. Ndeke, C.B.; Adonis, M.; Almaktoof, A. Energy management strategy for a hybrid micro-grid system using renewable energy. *Discov. Energy* **2024**, *4*, 1. [CrossRef]
17. Zhang, N.; Yan, J.; Hu, C.; Sun, Q.; Yang, L.; Gao, D.W.; Guerrero, J.M.; Li, Y. Price-Matching-based regional energy market with hierarchical reinforcement learning algorithm. *IEEE Trans. Ind. Inform.* **2024**, *20*, 11103–11114. [CrossRef]
18. Huang, Y.; Wang, L.; Guo, W.; Kang, Q.; Wu, Q. Chance Constrained Optimization in a Home Energy Management System. *IEEE Trans. Smart Grid* **2018**, *9*, 675–687. [CrossRef]
19. Mohandes, B.; El Moursi, M.S.; Wahbah, M.; El-Fouly, T.H.M. Renewable Energy Management System: Optimum Design and Hourly Dispatch. *IEEE Trans. Sustain. Energy* **2021**, *12*, 1615–1626. [CrossRef]
20. Elhadidy, M.A.; Shaahid, S.M. Optimal sizing of battery storage for hybrid wind diesel power systems. *Renew. Energy* **1998**, *7*, 6675. [CrossRef]
21. Chen, J.; Xiao, G.; Ferrari, M.L.; Yang, T.; Ni, M.; Cen, K. Dynamic simulation of a solar-hybrid microturbine system with experimental validation of main parts. *Renew. Energy* **2020**, *154*, 187–200. [CrossRef]
22. Wang, C.; Li, J. Power Management of a Stand-Alone Hybrid Wind-Microturbine Distributed Generation System. In Proceedings of the 2009 IEEE Power Electronics and Machines in Wind Applications, Lincoln, NE, USA, 24–26 June 2009. [CrossRef]
23. Hosseini, E.; Ghafoor, K.Z.; Emrouznejad, A.; Sadiq, A.S.; Rawat, D.B. Novel metaheuristic based on multiverse theory for optimization problems in emerging systems. *Appl. Intell.* **2021**, *51*, 3275–3292. [CrossRef]
24. Rosales Muñoz, A.A.; Grisales-Noreña, L.F.; Montano, J.; Montoya, O.D.; Perea-Moreno, A.-J. Application of the Multiverse Optimization Method to Solve the Optimal Power Flow Problem in Alternating Current Networks. *Electronics* **2022**, *11*, 1287. [CrossRef]
25. Symons, B.C.B.; Galvin, D.; Sahin, E.; Alexandrov, V.; Mensa, S. A practitioner's guide to quantum algorithms for optimisation problems. *J. Phys. A Math. Theor.* **2023**, *56*, 453001. [CrossRef]
26. Ajagekar, A.; You, F. Quantum computing for energy systems optimization: Challenges and opportunities. *Energy* **2019**, *179*, 76–89. [CrossRef]
27. Leonori, S.; Paschero, M.; Frattale Mascioli, F.M.; Rizzi, A. Optimization strategies for Microgrid energy management systems by Genetic Algorithms. *Appl. Soft Comput. J.* **2020**, *86*, 105903. [CrossRef]
28. Grisales-Noreña, L.F.; Montoya, O.D.; Ramos-Paja, C.A. An energy management system for optimal operation of BSS in DC distributed generation environments based on a parallel PSO algorithm. *J. Energy Storage* **2020**, *29*, 101488. [CrossRef]
29. Li, Y.; Wu, J. Optimum Integration of Solar Energy With Battery Energy Storage Systems. *IEEE Trans. Eng. Manag.* **2022**, *69*, 697. [CrossRef]

30. Sahu, P.; Verma, D.; Nema, S. Physical Design and Modelling of Boost Converter for Maximum Power Point Tracking in Solar PV systems. In Proceedings of the 2016 International Conference on Electrical Power and Energy Systems (ICEPES), Bhopal, India, 14–16 December 2016; IEEE: New York, NY, USA, 2016. [CrossRef]
31. Tiwari, R.; Krishnamurthy, K.; Neelakandan, R.B.; Padmanaban, S.; Wheeler, P.W. Neural Network Based Maximum Power Point Tracking Control with Quadratic Boost Converter for PMSG—Wind Energy Conversion System. *Electronics* **2018**, *7*, 20. [CrossRef]
32. Yang, L.-S.; Liang, T.-J. Analysis and Implementation of a Novel Bidirectional DC–DC Converter. *IEEE Trans. Ind. Electron.* **2012**, *59*, 10. [CrossRef]
33. Tan, N.M.L.; Abe, T.; Akagi, H. Design and Performance of a Bidirectional Isolated DC–DC Converter for a Battery Energy Storage System. *IEEE Trans. Power Electron.* **2012**, *27*, 1237. [CrossRef]
34. Gugulothu, R.; Nagu, B.; Pullaguram, D.; Chitti Babu, B. Optimal coordinated energy management strategy for standalone solar photovoltaic system with hybrid energy storage. *J. Energy Storage* **2023**, *67*, 107628. [CrossRef]
35. Manna, S.; Singh, D.K.; Akella, A.K.; Kotb, H.; Abo Ras, K.M.; Zawbaa, H.M.; Kamel, S. Design and implementation of a new adaptive MPPT controller for solar PV systems. *Energy Rep.* **2023**, *9*, 1818–1829. [CrossRef]
36. Guenounou, O.; Dahhou, B.; Chabour, F. Adaptive fuzzy controller based MPPT for photovoltaic systems. *Energy Convers. Manag.* **2014**, *78*, 843–850. [CrossRef]
37. Salem, A.A.; Nour Aldin, N.A.; Azmy, A.M.; Abdellatif, W.S.E. Implementation and validation of an adaptive fuzzy logic controller for MPPT of PMSG-based wind turbines. *IEEE Access* **2021**, *9*, 165690–165707. [CrossRef]
38. Chen, J.; Yao, W.; Zhang, C.; Ren, Y.; Jiang, L. Design of robust MPPT controller for grid-connected PMSG-Based wind turbine via perturbation observation based nonlinear adaptive control. *Renew. Energy* **2019**, *134*, 478–495. [CrossRef]
39. Sawle, Y.; Gupta, S.C.; Bohre, A.K. Optimal sizing of standalone PV/Wind/Biomass hybrid energy system using GA and PSO optimization technique. *Energy Procedia* **2017**, *117*, 690–698. [CrossRef]
40. Hosseinzadeh, M.; Salmasi, F.R. Robust Optimal Power Management System for a Hybrid AC/DC Micro-Grid. *IEEE Trans. Sustain. Energy* **2015**, *6*, 675. [CrossRef]
41. Eghtedarpour, N.; Farjah, E. Power Control and Management in a Hybrid AC/DC Microgrid. *IEEE Trans. Smart Grid* **2014**, *5*, 1234. [CrossRef]

sustainability

MDPI

Review

Sustainable Approaches to Microalgal Pre-Treatment Techniques for Biodiesel Production: A Review

Amarnath Krishnamoorthy, Cristina Rodriguez * and Andy Durrant

School of Engineering and Computing, University of the West of Scotland, Paisley PA1 2BE, UK
* Correspondence: cristina.rodriguez@uws.ac.uk

Abstract: Microalgae are a potential source of numerous nutritional products and biofuels. Their applications range from the food industry to the medical and fuel sectors and beyond. Recently, the conversion of biomass into biodiesel and other biofuels has received a lot of positive attention within the fossil fuel arena. The objective of biorefineries is to focus on utilising biomass efficiently to produce quality biofuel products by minimising the input as well as to reduce the use of chemical or thermal pre-treatments. Pre-treatment processes in biorefineries involve cell disruption to obtain lipids. Cell disruption is a crucial part of bioconversion, as the structure and nature of microalgae cell walls are complex. In recent years, many research papers have shown various pre-treatment methods and their advantages. The objective of this paper was to provide a comprehensive in-depth review of various recent pre-treatment techniques that have been used for microalgal biodiesel production and to discuss their advantages, disadvantages, and how they are applied in algal biorefineries.

Keywords: biofuel; microalgae; pre-treatment; biomass; cell disruption; lipids; biodiesel; renewable energy

1. Introduction

Natural resources are a significant part of the economic structures that meet the requirements of humanity. With the increasing human population, economic production is also constantly growing, which paves the way for research into the creation of new products and the innovation of current materials in an attempt to overcome the energy crisis. The energy crisis is one of the greatest current concerns for the world's stability and peace. Countries with developing economies that have limited natural resources need to secure fuel supplies. Fossil fuels, such as coal, petrol, natural gas, etc., have been viewed as fundamental energy sources [1], and they are used in very large amounts around the world. However, our long-term dependence on fossil fuels has challenged the lowering of greenhouse gases and has paved the way for global warming. The increase in the earth's overall temperature due to various human activities and natural causes has also contributed to the phenomenon of global warming. Some data have shown that the increase in the global temperature may result in increased health risks in future generations [2]. In order to retain clean ecosystems and maintain stability, renewable and eco-friendly biofuels are needed to replace fossil fuels [3]. These replacements are derived from natural resources, such as microalgae [4]. Algae comprise macroscopic and microscopic organisms, with some macroscopic organisms growing to a length of 10 m, and some microscopic organisms growing to a few micrometres in size.

Microalgae are considered to be a fascinating resource for industries, as they are helpful for producing multitudinous products because of their high growth rate, photosynthesis efficiency, and process optimisation. They have already been used in commercial industries, such as in animal feed, food, therapeutics, cosmetics, and biofuel [5–9]. The main advantages of culturing microalgae are that they can be cultured with minimal space, fewer nutrients, and minimal water (saline or brackish water) [10,11]. Microalgal cells are relatively small and are protected inside the cell wall. The Golgi apparatus within the cells contain certain products, and in some species, these products are to bound to the cell membranes. Due to these complex cell wall structures, cell disruption can be challenging [12,13]. Some microalgae species are easy to break down using a mild and effective cell disruption techniques, but this may not support large-scale production. Therefore, it is important to compare and choose feasible and energy-efficient cell disruption methods to obtain the highest standard of the extracted products, the most economical operating costs, and the highest lipid recovery rates.

In recent years, microalgae have gained more attention than most other sources of extractable biofuel. Biofuels are defined as fuels that are produced from agricultural or forestry materials or from the biodegradable parts of industrial waste [14]. Biofuel extraction has been calculated as producing 35 billion litres of fuel [15], with the USA, Brazil, and the European Union being the top producers [16]. Biodiesel is produced from vegetable oils [17], jatropha curcas [18], biobutanol [19], and algae [20]. Biodiesel can be generated from microalgal cell disruption using pre-treatment techniques that help to extract lipids. It is also clear that microalgae can produce large amounts of lipids. Table 1 shows some examples of algae species and the lipids extracted from them following pre-treatment.

The biological value of microalgal oil and biomass is due to their ability to synthesise a variety of elements, their growth capacity, and their capability to increase the efficiency of targeted bio-compounds using cultivation parameters and extraction methodologies [21,22]. Today, there is an increasing demand for algal biomass due to the increase in both traditional industries and microalgal applications. Estimates have shown that *Chlorella* and *Arthrospira* production has reached 2000–5000 tons and 6700–12,000 tons, respectively [23,24]. The higher initial costs of mass microalgal cultivation and the associated raw materials pose a problem for large-scale biomass production; however, reducing these costs and enhancing the economic productivity of microalgal lipids can be achieved using different techniques. The use of eco-friendly pre-treatment techniques such as mechanical [25], microwave [26], chemical [27,28], ultrasonic [29], high-pressure homogenisation [30] has been extensively studied. During cultivation, the use of genetic engineering processes to increase the recovery of lipids, proteins, and other valuable bioproducts is considered to be quite challenging [31]. This review paper focused on an analysis of research studies on the standard pre-treatment methods that are already in use as well as emerging techniques. All of the existing lipid extraction methods were analysed and compared using different species of microalgae.

2. Pre-Treatments

In various research studies from around the globe, techniques for pre-treating microalgae are still in development while researchers try to acquire more efficient lipid products (Table 1). The cell disruption technique involves breaking down the cells within the cell membranes to remove the intercellular products. It has also been shown that pre-treatment processes consume a lot of energy during cell disruption [32]. Some microalgal species are naturally good for efficient lipid extraction, but that does not apply to all types of algae. In order to produce significant cell disruption, the right pre-treatment method must be chosen to enhance the disruption efficiency [33]. Recent studies have shown that many new cell disruption techniques and methods are involved in the development of bioethanol and biofuels. The major obstacle to the use of biofuels has been the operational costs of large-scale production because the production of biofuels requires more input products (Figure 1) [34]. However, biofuels are emerging as non-toxic alternative fuel resources that are less harmful to the environment [20]. The use of biodiesel is increasing gradually, and as the demand rises, more pre-treatment methods and techniques are required [35].

Table 1. Lipids extracted after pre-treatment in microalgae.

Microalgae	Lipid Productivity (mg/L/day)	References
Phaeodactylum tricomutum	44.8	[36]
Chaetoceros muelleri	21.8	[37]
Skeletonema costatum	17.4	[37]
Botryococcus braunii	5.5	[38]
Dunaliella tertiolecta	60.6–69.8	[37]
Dunaliella sp.	33.5	[36]
Dunaliella salina	116	[36]
Nannochloris sp.	60.9–76.5	[37]
Nannochloropsis sp.	54.8	[38]
Nannochloropsis oculata	84.0–142.0	[36]
Scenedesmus sp.	40.8–53.9	[36]
Chlorella sp.	42.1	[38]
Chlorella vulgaris	11.2–40.0	[36]
Chlorella protothecoides	1214	[36]
Chlorella emersonii	10.3–50.0	[36]
Pavlova salina	49.4	[36]

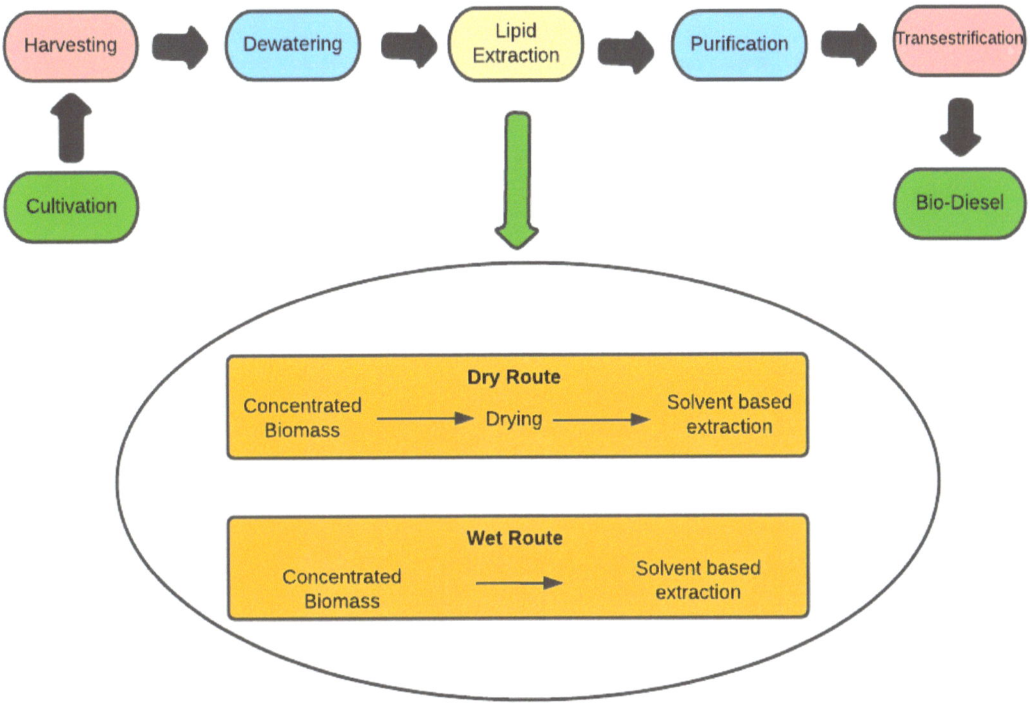

Figure 1. Approaches for converting microalgae to biodiesel.

A variety of pre-treatment techniques are currently used for cell disruption. These methods are mainly classified into mechanical, physical, thermal, chemical, biological, pulsed electric field, and combined techniques. Mechanical processes are the most widely used methods to reduce the rate of shear force required for cell wall rupture. In recent years, microwave, catalytic, bead beating, autoclaving, enzymatic, ultrasonic, autoclave, steam explosion, high-pressure homogenisation, high-speed homogenisation, and sonication methods have been studied for use in biodiesel applications and have shown good economic efficiency outcomes in large-scale production. It should be noted that the same microalgae can produce divergent lipid productivity results when using different pre-treatment techniques. As such, it is better to conduct a systematic evaluation on how the pre-treatment method influences the cell wall and cell size of a specific microalgae before choosing a certain technique [39]. When comparing the production processes of pre-treated biomass with non-pre-treated biomass, the energy balance favours the former [40].

2.1. Mechanical Pre-Treatment

Mechanical pre-treatment techniques involve the destruction of the cell wall using shear forces. Mechanical pre-treatment techniques can be classified into the categories of high-pressure homogenisation, high-speed homogenisation, and bead milling [41–43]. These methods are proven to extract lipids in a way that enhances large-scale biofuel production [44,45]. The main objective of these methods is to reduce cell wall particle size and crystallinity at the time of cell disintegration [46]. Mechanical pre-treatment methods have the advantage of preventing the cells from being contaminated and protect cell function from being damaged during cell rupture [47]. These methods become more effective when used in a combination, and they increase the cell surface area and produce more disruption efficiency [48].

2.1.1. High-Pressure Homogenisers

High-pressure homogenisers (HPHs) are specially made for emulsification techniques. This method is broadly used for the microalgae cell disruption process because of its continuous operation and scalability to generate wet biomass [49]. The HPH method has been shown to recover more microalgae lipids during cell rupture [50]. Different types of valve seat formats are available to optimise cell disruption efficiency [51]. The cells flow through the valve and strike the impact ring, exit the valve, and are discharged. Here, cell disruption is achieved through shear forces due to the impact caused, and hydrodynamic cavitation shows a pressure drop (Figure 2) [52]. HPHs show a higher possibility of obtaining wet microalgae concentrates with 25 W/W% solids for lipid recovery efficiently without consuming more energy [53]. In industries, pre-treatment methods using high-pressure homogenisers are used for cell disruption in seaweeds and yeast cells to improve lipid production [54]. An overview of research studies using this method is shown and discussed in Table 2.

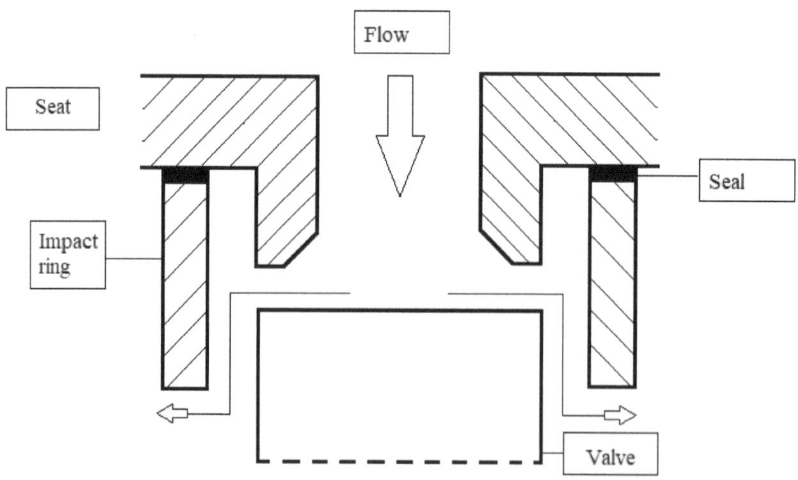

Figure 2. Schematic diagram of HPH value seat.

Table 2. Overview of previous research studies using high-pressure homogenisers.

Microalgae	Operating Conditions	Output Studied	Review	Reference
Nannochloris oculata	68.9 MPa and 310 MPa using nozzle diameters of 130 mm and 185 mm, respectively, 6 passes–100 mL	Efficiency increased.	Biodiesel production, total lipids.	[53]
	1%DCW, 125 MPa, 5 passes.	Efficiency of 200 (mg/g cell)	Biodiesel production, total lipids.	[55]
Chlorococcum sp.	0.85% DCW, 8 Mpa, 4 passes–200 mL	90% cell disruption achieved	Perfect cell count, total lipids.	[52]
Tetraselmissuecica	86 MPa. 0.85% DCW, 5 passes–200 mL	34.157 cell/mm^3 cell concentration	Perfect cell count, total lipids.	[52]
Auxenochlorella protothecoides	150 Mpa, 5 passes. Energy input of 1.5 MJ/kg dry weight–40 mL	Yields up to 35% (dry weight)	Perfect cell count, protein analysis	[56]
Chlorella vulgaris	150 Mpa, 5 passes. Energy input of 1.5 MJ/kg dry weight–40 mL	~25% (dry weight) protein release	Perfect cell count, protein analysis	[56]
Nannochloropsis sp.	150 Mpa, 1% DCW, nitrogen added, 6 passes–250 mL	90% protein achieved	Protein analysis	[57]
Chlorella saccharophila	200 to 1000 bar, t-butanol, ammonium sulphate.	Efficiency of 400 (mg/g cell)	Perfect cell count, total lipids.	[58]

The list of studies in the table shows that increasing the pressure and number of cycles will have a good impact on the lipid efficiency. Some studies suggest that lowering the dry cell weight and culture stress levels seems important but that modifying the nozzle diameter does not seem very effective [52]. Even though the use of HPHs is the most preferred method, they do have some disadvantages as well. When using a low DCW (0.01–0.85% *w*/*w*), the energy demand increases, and the hard cells become challenging to break. This indicates that HPH methods are not mild methods and are not acceptable for breaking fragile elements [59].

2.1.2. Bead Milling

Bead mills are widely used for lipid extraction during microalgae cell disruption. They provide good disruption efficiency in a single pass, and their industrial implementation values include temperature maintenance, easy operating procedures, large biomass set up, and easily available equipment [60]. Figure 3 shows the basic components of bead mills. They are classified into two types: agitated vessels and shaking vessels. Shaking vessels can be used to disrupt cells by vibrating the whole vessel. Agitated vessels use a spinning agitator that is filled with cell culture and beads. The cell disruption rate depends on size of the beads, the rigidity of the cell, and the biomass material of the microalgal cell [25]. It is theorised that after shear force is applied, the cells will be disrupted in the bead collision zones, and the energy will be transferred from the beads to the cells [61,62]. An overview of the literature is shown in Table 3 and discussed below.

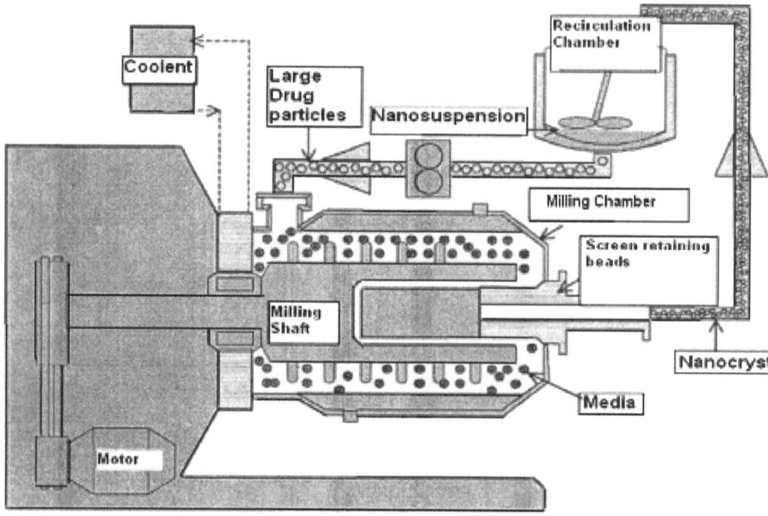

Figure 3. Schematic view of a wet bead milling process [63].

Table 3. Overview of the previous research studies using bead milling.

Microalgae	Operating Conditions	Output Studied	Review	Reference
Chlorella vulgaris	25 gDW L^{-1} biomass concentration, 2039 rpm, protease and cellulase (2% *v*/*w*, 1:1), 45 °C, 24 h–75 mL	75% lipid recovery (solid phase)	Perfect cell count, total lipids.	[64]
Nannochloropsis sp.	3 kw 0.5 beads, 4500 rpm/10 min–25 mL	Highest biomass concentration and COD reduction of 1.268 g/L and 71%, respectively	Perfect cell count, total lipids.	[65]
Chlorella vulgaris	Speed of the agitator set at 10 m^{-1} and a power of 24.5 kW for 90 min	95% increase in cell disruption	Perfect cell count, total lipids.	[66]
	3.3 kW, 0.40–0.50 mm beads, 10.7% dry cell weight–1.4 L	99% cell disintegration	Perfect cell count, total lipids	[67]
	(25–145 gDW kg^{-1}) and agitator speeds (6–12 m s^{-1})	97% cell disintegration	Perfect cell count, protein analysis	[68]
Nannochloropsis oculata	175 MPa, chloroform, methanol	Efficiency of 2.8 (mg/g cell)	Perfect cell count, total lipids.	[69]

Based on the case studies from the literature, various factors such as feed rate suspension, continuous operation, bead diameter, bead density, milling chamber design, biomass concentration, agitator design, agitator speed, bead filling, and the processing time of each batch affects the cell disruption rate during bead mill pre-treatment processes [70]. It can be also said that increasing the size of the beads will show more effective results than using small (0.5 mm) beads. The selected case studies also show that it is best to use low-density beads for low-viscosity media and high-density beads for high-viscosity media [42,67]. In spite of their advantages, their cons include their high energy demands and high operational costs, which makes this method less preferred for industries.

2.1.3. High-Speed Homogenisers

High-speed homogenisers (HSH) are devices that use a stirring mechanism that rotates at very high rpm and that consist of rotors and stators that are preferably assembled out of stainless steel. Cell disruption occurs when the cutting spindle rotates at a high speed, causing hydrodynamic cavitation and high shear forces inside microalgal cells by breaking their cell walls and extracting the intercellular elements from them. According to cell wall characteristics, operating conditions such as the homogenising speed, number of passes, and running period can be optimised to increase efficiency [71]. Additionally, other factors such as the microalgal species, dry cell concentration, and growth parameters influence the energy consumption and the efficiency of the pre-treatment process. A reduction in the biomass size due to the high pressure in the homogeniser causing a thermal effect on the sample results in the aggregation of the biocompounds and their release into to the aqueous media used as references [72–74]. The HSH technique is a very simple but very aggressive cell disruption technique that achieves effective results. The main advantage of this process is its short operating time and its potential to generate lipids and other compounds. Some of the research was conducted to increase the extraction yield using different species and biochemicals [75]. The main disadvantages of this technique are the high operational costs, the protein denaturation caused due to the shear force, and the increase thermal effect, and these make this technique less favourable for biorefinery industries [41]. An overview of studies from the literature using this method is shown in Table 4.

Table 4. Overview of previous research studies using HSH.

Microalgae	Operating Conditions	Output Studied	Review	Reference
Porphyridium cruentum	5500 rpm for 10 min	ω3-PUFA food products	Perfect cell count.	[72]
Synechococcus sp.	10,000 rpm for 1 min (5 cycles)	8.82% lipid recovery	Total lipids.	[74]
Laminaria digitata	150–500 bar for 15 min	20% lipid content	Perfect cell count, total lipids.	[76]
Chlorella sp.	12,000 rpm for 15 min	Lipid efficiency of 13.05%	Perfect cell count, total lipids.	[71]
Nannochloropsis sp.	Speed of 10,000 rpm–1 min–15 mL	Dry extraction yield of 75%	Total lipids.	[77]

2.2. *Physical Pre-Treatment*

Physical pre-treatments involve the application of mechanical forces such as shear force, microwaves, and ultrasound. Physical pre-treatment methods have advantages such as cost effectiveness, ease of commercialisation, and time saving. They consist of two major classifications: microwave and ultrasound methods.

2.2.1. Ultrasound Pre-Treatment

During ultrasonic pre-treatment, acoustic or sound energies of high-frequency waves are generated. By transmitting these shock waves into the cell wall, they cause cell disruption because of their high shear force. The pressure variation in these waves can produce cavitation within the cell [78,79]. The impact of ultrasound waves is mostly influenced by the cell wall structure and composition of the microalgae (Figure 4). Because of high temperature and pressure levels, the cavitation generates chemical reactions that are able to destroy organic matter and produce shear force, leading to the creation of H^+ and OH^- reactive radicals [80].

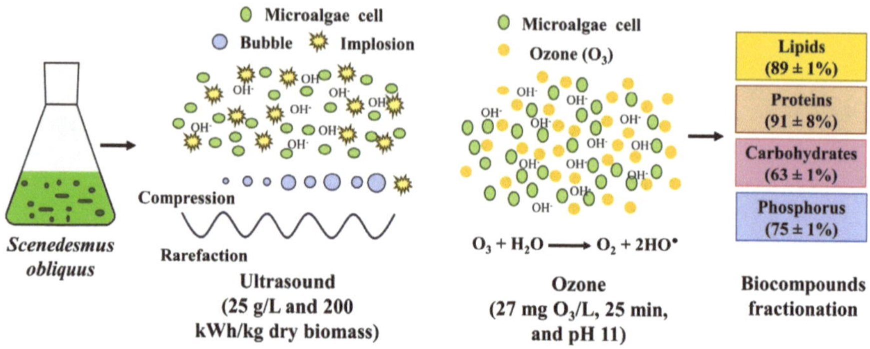

Figure 4. Graphic showing *Scenedesmus obliquus* during ultrasonic pre-treatment [81].

The ultrasound method is used in various applications, such as in olive mill wastewater, chicken and cattle manure, and sludge [82,83]. Microalgae biomass efficiency has been successfully increased to between 16% and 100% with high acoustic energy input. One study observed that there was no improvement in *spirulina maxima* when a semicontinuous reactor was used. This was mainly because of the characteristics of microalgae, which have a soft cell wall [84]. High temperatures should be avoided when using the ultrasound technique, as they result in the loss of volatile organics and reduce biomass production. This was suggested during a study with *Nannochloropsis salina*, which showed lower biomass yield when compared to raw biomass [85]. The schematic representation of the ultrasonic machine is shown in Figure 5. A review of case studies using this method is shown in Table 5.

Table 5. Overview of the previous research studies using ultrasonic methods.

Microalgae	Operating Conditions	Output Studied	Review	Reference
Botryococcus sp.	0.5%DCW, 5 min, 10 kHz–100 mL	8.8% lipid recovery	Total lipids.	[54]
Salvinia molesta.	5 min and frequency of 2 kHz–100 mL	19.7% increased lipid content	Perfect cell count, total lipids.	[86]
Chlorella vulgaris	20 kHz using 750 W for different times: 0, 5, 10, and 20 min, at 25 °C	23% lipid content	Perfect cell count, total lipids.	[63]
Chlorella sp.	20 kHz, 0.8 kW-h/L cells	Efficiency of 12.6 (mg/g cell)	Perfect cell count, total lipids.	[87]
Schizochytrium sp.	150 W, time for 30 min, with temperature 50 °C	Oil yields up to 93.76 (dry weight)	Total lipids.	[88]
Scenedesmus sp.	20 W and frequency 18 Hz for 5 s.	21.3% to 28.3% lipid yield	Perfect cell count, protein analysis	[64]

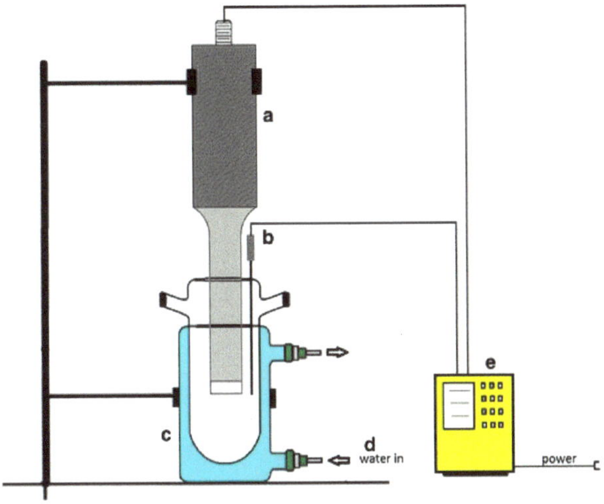

Figure 5. Schematic representation of ultrasonic system: (a) transducer, (b) thermometer, (c) ultrasonic reactor, (d) cryostat, (e) ultrasonic generator [89].

When pre-treating with ultrasonication, the working temperature significantly increases from around 50 to 90 °C, which also kills proteins and other intercellular elements [90]. The disadvantage of this method is its low disruption efficiency. Lipid quality can be increased with temperature control, but this decreases the efficiency slightly [66].

2.2.2. Microwave Techniques

Microwave techniques use optimal electromagnetic waves with frequencies ranging from 0.3 to 300 GHz that are used to heat localised areas. Here, the microwaves increase the kinetic energy of water molecules until they reach their boiling state [91]. Microwave treatments produce thermal radiation, and this effect is said to increase the temperature due to polarised macro molecules (Figure 6). These modules are aligned around the pole of the electromagnetic field, which is where hydrogen ions break down [26]. The pressure and the heat energy produced by microwaves cause damage to the cell wall and cell membranes [92]. A review of case studies using microwave techniques is shown in Table 6.

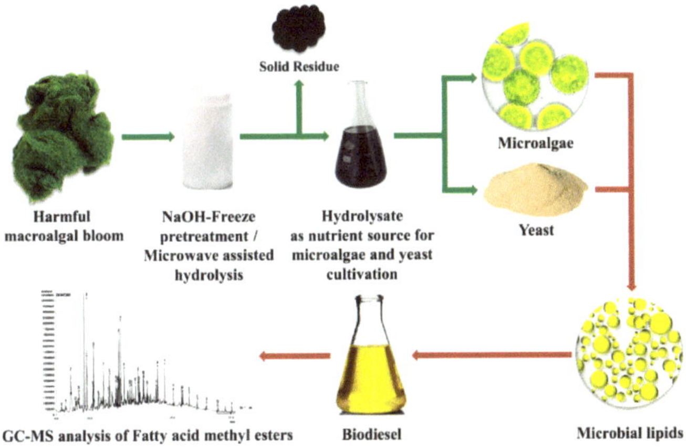

Figure 6. Picture of microwave-assisted treatment for microalgae [93].

Table 6. Overview of the previous research studies using microwave techniques.

Microalgae	Operating Conditions	Output Studied	Review	Reference
Nannochloropsis oceanica	Power of 1025 W and frequency of 245 MHz for 15 min	38.46% lipid production	Total lipids.	[94]
Yarrowia lipolytica	900 W power and a frequency of 245 MHz for 15 min	Lipid production of 8.18%	Total lipids.	[72]
Chlorella sp.	For 15 min, 450 W power. Biomass and methanol ratio of 1:12 (w/v), catalyst: KOH	32.18% lipid content	Biodiesel, total lipids.	[72]
	Power of 450 W, time of 60 min. Catalyst: 0.2 M H_2SO_4, 5 min	75.68% (FAME for biodiesel production)	Biodiesel, total lipids.	[95]
	2450 MHz and temperature of 100 °C, 5 min	Increased lipid efficiency.	Total lipids.	[96]
Botryococcus braunii	Power 1250 W and frequency 2450 MHz at 150 °C for 20 min	Enchanted lipid efficiency.	Biodiesel production, total lipids.	[73]
Nannochloropsis sp.	65 °C–25 min	42.22% dry biomass yield for biodiesel	Biodiesel production, total lipids.	[97]
	1.2 kW power and frequency of 2.45 GHz. 5–15 min.	Increased cell disruption efficiency.	Total lipids.	[98]
Nannochloropsis oculata	140 °C, 15 min	Lipid content increase: 6.25-fold	Perfect cell count, total lipids.	[99]

From the list of studies, the potential of using more specific energy can be more effective when using microwave techniques. Regardless, cell disruption occurs when water-based substances are observed by microwaves and are formed as head radicals [100]. Compared to normal heating, the microwaves will be uniform during temperature transmission. This method is shown to be more superior compared to the bead milling and ultrasound methods [101]. Even though these techniques have more advantages, they lack in terms of extraction yield the time required to use solvents. As such, the microwave method is well-suited for use as a mild microalgae cell disruption method [96].

2.3. Thermal Pre-Treatment

Thermal pre-treatment methods are techniques in which heat is added to the surface of algal biomass. This makes the microalgae disrupt the chemical bonds inside their cells, improving the solubilisation [102]. They provide high biomass yields and have low energy requirements when compared to other physical pre-treatment methods. Thermal pre-treatments are basically carried out by adding alkali or acidic chemicals to improve the cell disruption efficiency. Despite these high thermal properties, they may produce recalcitrant components that result in low biomass production and cannot be degraded anaerobically [103]. These methods are categorised into two types: steam explosion and autoclaving.

2.3.1. Steam Explosion

Steam explosion is an economical and effective method that is used in the processing of lignocellulosic components to improve the biomass efficiency. This method uses high temperatures ranging from about 160 °C to 260 °C (1.03–3.40 MPa) [104]. By using a catalyst such as NaOH or H_2SO_4, it is possible to obtain enhanced lipid efficiency [105]. Particle size, chemical composition, and shapes can be modified via explosive depressurisation and autohydrolysis [106,107]. In the search for the best method for cell disruption to enhance efficiency and to extract sugars and carbohydrates, the steam explosion technique is said to be perfect. The application of steam explosion with the acid catalyst method can efficiently extract more lipids [108]. Schematic diagram of steam explosion and a fractionation reactor is shown in Figure 7. A review of the case studies using steam explosion is shown in Table 7.

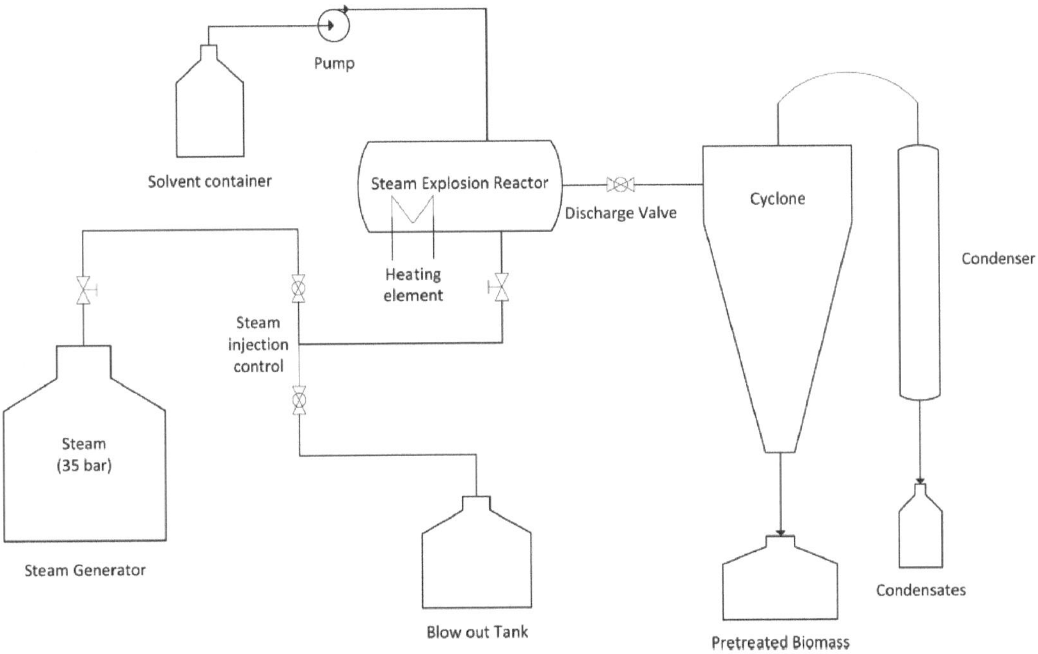

Figure 7. Schematic diagram of steam explosion and a fractionation reactor [109].

2.3.2. Autoclaving

This method is a heat transfer process that uses an absolute pressure of 0.3 MPa and a temperature of 121 °C for lipid extraction [110]. There are some studies that were conducted using a continuous reactor under a hydraulic retention time (HRT) of 15 to 20 days and at a temperature of 95 °C that show a positive energy balance. As such, it is clear there will be good biomass yield achieved by thermal pre-treatment methodologies when they are used in large-scale applications. A review of previous research studies is shown in Table 7 below.

Table 7. Overview of the previous research studies using steam explosion and autoclaving.

Microalgae	Operating Conditions	Output Studied	Review	Reference
Steam explosion				
Nannochloropsis gaditana sp.	150 °C for 5 min	Lipid recovery 0.3–3.6%	Total lipids.	[111]
Chlorella sorokiniana	120 °C for 5 min	17.9% and 18.2% lipid extraction	Total lipids.	[108]
Scenedesmus dimorphus	100 °C–130 °C	Enchanted solubilisation	Perfect cell count, total lipids.	[112]
Botryococcus braunii	90 °C, 10 min	Hydrocarbon (0.4% at 75 °C). 97.8 wt% recovery	Biodiesel production perfect cell count, total lipids.	[113]
Scenedesmus sp.	90 °C	Efficient cell disruption	Total lipids.	[114]
Nannochloropsis	30 °C and 60 °C	Biomass yield of 41%	Total lipids.	[114]
Chlorella pyrenoidosa	130 °C for 60 min–10 mL	2.1-fold increased lipid recovery	Total lipids.	[115]
Autoclaving				
Chlorella vulgaris	100 °C for 10 min–200 mL	15.4% lipid yield	Total lipids.	[116]
	100 °C, 1.5 MPa–5 min	Lipid content of 29.34%	Total lipids.	[54]
	121 °C with 0.1 MPa for 5 min	Lipid content of 24%	Total lipids.	[117]
Botryococcus sp.	125 °C with 1.5 MPa for 5 min	5.4–11.9% lipid recovery	Total lipids.	[54]

Balancing the energy of biomass production between energy consumed during production and energy harvested in the form of fuel is essential for biomass production to be cost effective. However, some reports suggest that using thermal pre-treatments with microalgae results in a negative balance [103]. When thermal methods are compared to other pre-treatment techniques such as physical and ultrasound pre-treatment methods, the energy that is consumed is comparatively low [114].

2.4. Chemical Treatments

Chemical treatments are ways of introducing chemical substances such as alkaline or organic solvents, detergents, chaotropes, antibiotics, hypochlorites, and chelating agents to enhance cell disruption. Usually, alkaline pre-treatment methods using alkaline compounds such as potassium, sodium hydroxide, and calcium at pH levels varying from 9 to 12 are applied for algal biomass. Acid pre-treatments are carried out by exposing H_2SO_4 and HCl at lower pH levels. Antibodies have the ability to extract lipids from cell membrane components by inhibiting them from the inside, whereas chelating agents cross-couple the cell membrane molecules to cause disruption. Detergents mix with membrane molecules, with the solvents in them dissolving and piercing the cell membrane and cell wall [52,118]. Oxidising agents such as hydrogen peroxide and ozone are used to disrupt cell walls. The energy required to enhance biomass is also too low compared to other methods such as physical or thermal ones [119]. Even though chemical pre-treatment methods are generally used for pre-treating cells, studies using them for microalgae biomass production are not as common as those using physical or thermal methods. The major problem with using chemical pre-treatments methods is that they are corrosive and toxic and may also produce inhibitory components. They may also lead to contamination [120].

Scenedesmus and *Chlorella* biomass showed pH improvements from 9 to 11 when chemical pre-treatment methods were used. By increasing the pH to 13, the microalgal cells were damaged because of the high alkalinity. It is also stated that low positive total energy values were attained in all cases [121]. Other studies concluded that treating *Chlorella* sp.

and Nannochloropsis sp. with different alkaline solutions has a negative effect, as these microalgae have robust cell wall conditions [122]. Using oxidising agents on microalgae looks challenging when compared to acid or alkali pre-treatment methods to generate biomass. Microalgae biomass is limited when it is pre-treated by ozonation. Applying ozone pre-treatments to the biomass improved energy efficiency from 6% to 66% at different stages [123]. Apart from these chemical methods, TiO_2 and hydroxyl radicals should also be studied to see if they increase biomass energy. These methods comprise different categories: solvent pre-treatments, catalytic pre-treatments, alkali and acid treatments, and enzymatic treatment.

2.4.1. Solvent Pre-Treatment

Biochemicals such as c-phycocyanin, astaxanthin, etc., are used as solvents to extract lipids. Studies report that some amine solvents can be used for cell disruption by modifying their polarity by adding them carbon dioxide. It can be said that algal biomass is highly influenced by amine solvents [124]. Additionally, not all amine solvents react with CO_2, and interaction depends on the polar compounds in the form of algal carbamates. To study this, more studies are needed to determine the properties of the reaction. In order to control the high energy consumption needed at the time of cell disruption and drying, switchable hydrophilicity solvents (SHSs) can be used. SHSs have the capacity to extract lipids by making contact with the lipids and organic solvents, thus increasing the extraction efficiency [125]. A review of the relevant research literature is shown in Table 8.

2.4.2. Catalytic Pre-Treatment

These methods comprise heterogeneous and homogeneous catalytic pre-treatment methods for biodiesel production [126]. Studies have noted that the use of a homogeneous catalyst in the process of biodiesel production results in advantages such as product purification, the reusability and recovery of the catalyst, lower water consumption, and less energy [127]. Research using 1 wt% of homogenous catalysts such as CH_3ONa, CH_3OK, NaOH, and KOH for biodiesel production with the addition of sunflower oil at a temperature of 60 °C for 3 h resulted in a biodiesel yield of 91.22% [128]. A review of previous research studies is shown in Table 8.

2.4.3. Enzymatic Treatment

This is a biochemical method that requires a mechanical technique, has a lower energy requirement, and can cause cells to rupture to achieve effective lipid production for biodiesel [40]. Cell efficiency can be improved by this technique, as the extraction process works with cellulose, alkaline protease, sanilase, and papain [129]. The process is non-flammable, inexpensive, and inert in nature, so it is very suitable for biodiesel production [130,131]. Additionally, enzymatic pre-treatment demonstrates an advantage when using rapeseed oil during the pre-treatment process, as it can separate microalgal strains and glycerol [132]. A review of previous research studies is shown in Table 8.

Table 8. Overview of the previous research studies using solvent, catalytic, and enzymatic treatment methods.

Microalgae	Operating Conditions	Output Studied	Review	Reference
Solvent treatment				
Chlorella vulgaris	Amine solvents (dimethylbutylamine, dipropylamine, ethylbutylamine, phenethylamine, and dimethylcyclohexylamine) + culture mixed in ratio of 1:1 with CO_2 treatment–50 mL	Lipid extraction yield of 9.16%	Total lipids.	[133]

Table 8. *Cont.*

Microalgae	Operating Conditions	Output Studied	Review	Reference
Chlorella sp.	Dimethylbenzylamine solvent, culture (1:1 ratio for 1 h extraction time)	Lipid extraction of 25.97,32 and 40.8%.	Total lipids.	[134]
Scenedesmus	Hexane: isopropanol (3:2) and solvent; culture (75:1) for 2 h extraction time.	FAMEs of 13% and total lipids, with polar FAME about 1.5% of total lipids	Biodiesel production, total lipids.	[135]
Nannochloropsis oceanica	TEPDA solvent: culture (1:4 ratio with 2 h extraction time)	98.2% lipid extraction efficiency.	Biodiesel production, total lipids.	[125]
Catalytic treatment				
Nannochloropsis sp.	Mixing 10% of Mg and Zr for 4 h at a temperature of 65 °C	Biodiesel potassium hydroxide yield of 28.0%.	Biodiesel production, total lipids.	[136]
Monoraphidium sp.	2 mL hexane and 5 mL of 20% saturated NaCl solution.	82.86% saponifiable components and 17.14% unsaponifiable components.	Total lipids.	[137]
Chlorella vulgaris	Dried in oven at 48 h at 100 °C	Lipid yield of 53.25%	Total lipids.	[138]
	220 °C, 2 h methanol per gram of biomass–8 mL	Biodiesel yield of 74.6%	Biodiesel production, total lipids.	[139]
Scenedesmus acutus	Dried in vacuum at 60 °C for 20 h.	~99 wt% hydrocarbons for biodiesel. 12.6% of extracted lipids.	Biodiesel production, total lipids.	[140]
Enzymatic treatment				
Rhodotorula glutinis	Adding glycerol, AA, and ChCl with 60 °C for 120 min, solid–liquid ratio is 1:20	Lipid yield improved by 32.1% and 54%	Total lipids.	[107]
Chlorella vulgaris	12 h hydrolysis by protease 2% (*v*/*w*) enzymes at 45 °C for 45 min and 12 h hydrolysis by cellulase (2% *v*/*w*) enzymes at 45 °C for 45 min	44% lipid yield.	Total lipids.	[64]
	Enzymatic hydrolysis was performed at pH 4.8 and 50 °C for 72 h.	1.10–1.69-fold and 85.3% hydrolysis yield.	Total lipids.	[141]
	Mixing sanilase and trypsin enzymes for hydrolysis.	30% lipid yield	Total lipids.	[142]
Nannochloropsis sp.	Enzymatic treatment at 50 °C for 30 min and a pH of 4.	90.0% lipid yield.	Total lipids.	[143]

2.5. Biological Pre-Treatment

Biological pre-treatments involve three factors: fungi, bacterial, and enzyme activity. These methods are considered to have low investment requirements, mild operating conditions, and less energy consumption and represent the best alternative to the aggressive mechanical techniques [144]. Lipases, glucanases, peptidases and glycosidases are also the most used enzyme classification methods to disrupt the cell wall. During the process, the enzymes mix inside a molecule inside the cell wall/membrane and break the bonds, resulting in cell disruption [145]. To enhance the algal biomass, an enzyme or mixture can be increased. The mixture mostly contains cellulose, starch-degrading enzymes, and hemicellulose [146]. Biological methods may be the best alternative to chemical and physical methods, as they avoid causing inhibitory problems; they are effective low-temperature alternative techniques to thermal pre-treatment methods [96]. The major disadvantage of biological methods is that they need 10–14 days, much longer than all of the other pre-treatment methods; they also need a big space to be carried out on an industrial scale. These methods can be used by themselves or can be combined with other pre-treatment techniques if the concentration of the recalcitrant compound is very high [147]. Some studies have shown that biological pre-treatment methods are mainly used for commercial enzymes and result in high methane output [148]. Biological pre-treatment methods also include algicidal pre-treatments, which consist of viruses, cyanobacteria, bacteria, and microalgae themselves. They have the capability to attack the extracellular compounds on the microalgae to extract the lipids [149]. The use of *Chlorella vulgaris* ESP-31 with bacterium *Flammeovirga yaeyamensis* for oil extraction over a 3-day-long pre-treatment process showed enzymes breakage in xylanase, amylase, and cellulase with a high lipid content of 21.5% [150]. Furthermore, a similar investigation of *Nannochloropsis* sp. biomass with different combinations of lysozyme, protease, cellulose, and pectinase enzymes showed a higher lipid content than when a single enzyme was used [143].

2.6. Supercritical Fluid Extraction

This method is said to be on the best and most effective techniques for lipid extraction processes and is eco-friendly [27,151]. This method requires the pressure and temperature to be increased more than the critical point to induce cell breakage. Substances such as CO_2, ammonia, methanol, and others can be used as supercritical extractants, and SC-CO_2 is the most commonly substance used due to its low cost, low temperature, and low pressure [152]. Lipids can be also directly transestrified into biodiesel using this method [153]. A study with the microalgae *Chlorella vulgaris* and *Nannochloropsis oculata* in combination with ethanol as a co-solvent resulted in cell disruption and extracted lipid percentages of 97% and 83% [154]. A similar study conducted by Viguera et al. [155] using *Chlorella protothecoides* microalgal species at 70 °C and 300 bars resulted in a higher lipid yield rate. SC-CO_2 along with *n*-hexane was used for lipid extraction in the microalgae *Schizochytrium* sp., and the results suggested that the lipid efficiency extracted from SC-CO_2 was more than the efficiency obtained from *n*-hexane [156]. Another study compared Bligh–Dyer with SC-CO_2 using *Scenedesmus obtusiusculus* and *Scenedesmus obliquus* at 12 Mpa and 20 °C and found that the lipid extraction rate was higher than 90% in the SC-CO_2 process, and the author suggested that this method is good for industrial purposes [157]. Using ethanol as a co-solvent in *Pavlova lutheri.* with various operating and extraction conditions, De Melo et al. achieved 3.5-fold and 7.9-fold higher extraction than fish oil [158]. The main problem with this method is the high price of the equipment and its limitations in large-scale applications.

2.7. Pulsed Electric Treatment

Pulsed electric and high-intensity field pulsed methods are techniques that use electric fields to disrupt the cell and produce lipid extraction. This produces electro-mechanical vibrations and an electric field that creates tension in the cell wall/membrane [159]. The high-strength power of ∼30 kV passes through the cell wall/membrane, and by increasing the power of the electric field by ∼2000 Hz, a large number of dissimilar electric charges pass over the dipolar molecules and break large elements, decreasing the complex molecular forms and piercing into the cell [160]. When the electric field exceeds a specific voltage, the inner pressure generated within the membrane creates an unequal amount of energy in an attempt to form unrepairable pores in the cell [161]. This pulse electric not only kills all of the cells in the membranes, but also attacks the molecular components within the cell. This technique also affects the nutritional products and the proteins due to the very high temperature [130]. This method has the advantage of being combined with other methods to achieve efficient cell disruption, but the solution that is added should be ion free. When treating marine algae with this pulse effect, the microalgae need to be prewashed and deionised to improve the ability of the pulse field to pass into them. This pre-treatment method produces mixed outputs. A study showed that less than 5% biogas was produced from the biomass [160]. Due to these problems, this method is not favoured by biorefineries.

2.8. Combined Pre-Treatment Methods

Combining various pre-treatment techniques can be used to reduce costs and enhance efficiency. Thermochemical pre-treatment methods are a blend of thermal and chemical methods, and when applied to *spirulina* biomass, they showed low biogas production, as these methods use toxic and chloride-based chemicals with a low pH [84]. Research using combined processes used the microwave and bead mills pre-treatment processes along with cell shattering via high-frequency shock waves [162,163]. The sonication method breaks the cell wall and reduces the size because of the cavitation effect [164]. Cell disruption occurred during bead beating in thee microalgal cells due to the high-speed spinning beads [80,82]. The amount of energy can be reduced for lipid extraction by combining ultrasonic and chemical cell disruption methods [117]. There are fewer studies on the use of combined methods compared to those highlighting the use of single pre-treatment methods, so future studies can focus on using combining methods for cell disruption. A list of previous studies is shown in Table 9.

Table 9. Overview of the previous research studies combining pre-treatment methods.

Microalgae	Treatment Type	Operating Conditions	Production Yield	References
Chlorella sp.	Homogenisation + thermal	84 MPa (123 °C and pH of 1.5, chloroform, methanol	Efficiency of 4.5 (mg/g cell)	[165]
Chlorella vulgaris	Microwave + solvent	700 W, 50 s–chloroform:methanol:water (2:2:1.9)	Lipid recovery of 31.70	[141]
Nannochloropsis oceanica	Microwave + diluted acid	140 °C, 25 min–H_2SO_4 (1% v/v)	Hydrogen yield of 183.9 mL/g TVS	[166]
Scenedesmus sp.	Thermal + alkaline	100 °C, 8 h–NaOH	Lipid extraction- 45.54 mL H_2/g (VS) of hydrogen	[167]
	Ultrasonication + solvent	30 kHz, 1 kW for 5 to 60 min–hexane, chloroform/methanol (1:1 v/v)	Efficiency of 0.144 to 0.72 (mg/g cell)	[168]

2.9. Other Latest Pre-Treatment Techniques

There is a need for novel techniques to decrease the recalcitrant properties of microalgae and influence biomass to disrupt high-value lipids, and new techniques are being implemented. Most researchers consider their work to be eco-friendly and appropriate for large-scale production. However, while their findings may be of good quality, they focus on industrial applications.

Updates to biological techniques with modifications were implemented in a recent study on lignocellulosic biomass, which showed that changing cellulose elements leads to increased enzyme hydrolysis, lower energy use, lower operational costs, and less hemicellulose loss [169,170].

The photocatalysis method uses light absorption to increase the temperature, creating a chemical reaction. *Chlorella vulgaris* treatment saw a mineralisation efficiency of 57% efficiency when using a light intensity of 4000 lux and a temperature of 25 ± 1 °C [171]. *Chlorella pyrenoidosa* and *Scenedesmus obliquus* treatment were investigated and resulted in decreased cell toxicity [171].

A study on cell disruption techniques resulted in advances in explosive decompression. This method uses propane, butane, or carbon dioxide for lipid extraction. *Haematococcus pluvialis* was suspended at a dry cell weight of 18.11%, and using explosive decompression, extraction increased from 72.3% to 92.6%. Because of the higher dry cell weight and the lower specific energy consumption, a high extraction yield was obtained [172].

Another study used pulsed arch technology in grape seeds to disrupt cells. This technology used high-electric energy discharge during a time phase and produced cavities within the cells due to the high temperature and pressure [173]. This is one among a number of highly aggressive methods, but it has still not been studied in microalgae. Perhaps this method could be modified in the future so that it could be used to disrupt cells with low electric energy, shear forces, and temperature to obtain good lipid productivity.

The autolysis extraction technique, a less explored method, represents a good approach for disrupting lipids. Cell disruption can be triggered by different atmosphere cues such as anoxia when there is an increase in temperature. *Chlamydomonas reinhardtii* [174] *Nannochloropsis gaditana* [175] were treated at temperatures of 50 °C and 38 °C, respectively, during a 24 h incubation period and showed cell breakage. This technique is considered for application because of its mild treatment conditions and low processing costs, even though this method seems to be slow.

Ionic liquid methods have been indicated to facilitate lipid extraction from wet microalgae. The influence of the ionic liquid 1-ethyl-3-methylimmidazolium methyl sulphate [EMIM][$MeSO_4$] was tested on *Nannochloropsis* sp. and resulted in a biodiesel yield of 40.9% [97]. Another study was conducted with 1 g of dry *Chlorella vulgaris* that had 4 mL of ionic liquid (1,3-dimethylimidazolium methyl phosphate) + 4 mL of methanol added to it, and it was treated for about 18 h at 65 °C. The results indicated the occurrence of cell disruption [176]. This method is also considered to be a mild process with slow results.

Research was conducted with *Nannochloropsis oculate* and *Scenedesmus dimorphus* using convectional ultrasonication at 100 W and a 20 kHz frequency and focused ultrasonication at 40 W and 3.2 MHz. The results indicated that higher efficiency and a better cell disruption efficiency can be obtained using focused ultrasonication than with convectional ultrasonication [177].

Pressurised liquid extraction is a technique that is also known as an accelerated solvent extraction technique. This method extracts the intercellular compounds in a shorter extraction time by utilizing a combination of pressure and temperature [178]. In addition, there have been several studies that have used supercritical water, propane, dimethyl ether, and n- butane as solvents [178,179].

3. Comparison and Discussion on Different Pre-Treatment Techniques

Microalgae are single-cell organisms. Their photosynthesis produces around 70% of the O_2 in the atmosphere. Not all microalgae are single-cell organisms, and some species of microalgae can grow as single cells or in colonies (according to colour) and take the shape of spheres or filaments [180]. During cell disruption, the inner forces cause temperature changes as well as cavitation, pressure, and molecular energy variations. The quality of the final product may be impacted by these events separately or collectively, depending on whether the contaminants are produced or the algal elements are degraded [181]. From the above methods, it is clear that mechanical and physical methods are considered to be more effective in enhancing lipid efficiency in large-scale applications. The cell disruption caused during high-speed homogenisation and bead milling works according to the principle of shear force, which is similar to the energy transfer caused by the current and waves effects observed in the microwave, ultrasound, and pulsed electric field techniques. Enhancing the efficiency of microalgae mainly relies on cell wall characteristics as well as on strain and operational parameters, including temperature, enzyme doses, and power input. Often, pre-treatment is the best way to improve biomass production with various ranges of efficiencies. Studies suggest that biomass yield can improve from 20% to 60% after the application of pre-treatment methods. Figure 8 shows the various microalgae that have been studied using different pre-treatment methods. Using a thermal pre-treatment method, *Botryococcus* sp. showed an improved lipid yield and increased biomass production [54].

The homogeniser and microwave methods were found to be more efficient compared to other methods. *Botryococcus* sp. showed lipid production of 28.6% and 28.1% when the microwave and homogeniser methods were used, respectively, and the bead beating method showed a higher lipid percentage when used with *Botryococcus braunii* compared to other pre-treatment techniques such as the French press, sonication, and homogeniser methods [164]. It should also be noted that the efficiency of the bead beating technique is not easily measured. *Chlorella vulgaris* showed high efficiency when the microwave method was applied and 7.9% lower efficiency compared to other mechanical methods. The microwave pre-treatment of *Scenedesmus* sp. resulted in high lipid efficiency when compared to other methods. However, the osmotic pre-treatment method is quite simple and produced similar output to the mechanical methods for *Scenedesmus* sp. and *Chlorella vulgaris*. A small problem with this method is that it requires a long pre-treatment time of 48 h [54]. A pre-treatment method similar to the microwave method that uses animal fats and vegetable oils has been studied and suggests that the microwave method is a simple, efficient, and easy pre-treatment technique. Furthermore, this research indicated the lipid extraction can be also easily measured and concluded that the microwave technique is the most applicable method for the large-scale production of microalgal biomass [163,182].

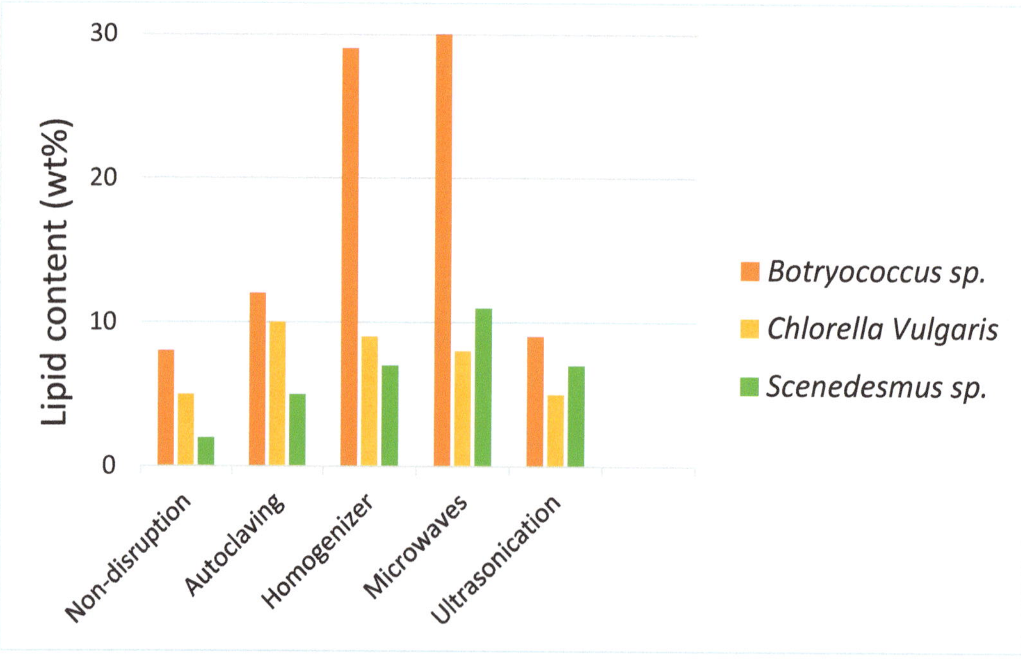

Figure 8. Schematic representation of lipid productivity of microalgae [54].

In various studies, modifying the pre-treatment methods and conditions has been seen as a strategy to enhance lipid production. The current studies in microalgal biofuel are mentioned because they highlight the ability of different methods to recover lipids [183]. The biomass production and the lipid recovery rate change from species to species and also vary across treatment methodologies. *Dunaliella salina* and *chaeoceros muelleri* were evaluated using osmotic shock by Lina et al. [184], where the effect of biomass and the water ratio was analysed with different levels of fluorescence ranges and timings. Many results were generated with different iterations, and it was concluded that the various results differ according to economic and efficiency factors. In a work by Gruber et al. [185], studies were conducted with methods such as microwave, ultrasonication, enzymatic, and wet milling using the microalgal biomass of *Chlorella vulgaris, Acutodesmus obliquus,* and *Chlorella emersonii* to study single and combined pre-treatment technique effects, and varying lipid-recovery rates and cost analysis were achieved. Another study by Francesso et al. [186] looked over the impacts of cell rupture using thermal, thermal hydrolysis, enzymatic, and ultrasound techniques using *Scenedesmus obliquus* and *Chlorella vulgaris.* In the initial set of experiments, the yields were shown to increase with the ultrasonic and enzymatic techniques compared to the untreated biomass, and then, when combining the thermal and thermal hydrolysis strategies, the yield percentages were lower. Additionally, apart from modifying and combining pre-treatment techniques, most of the studies focussed on cultivation process such as the selection of species [187], growth media [188], CO_2 [189], light [190], temperature [191], and nutrients [192]. Hence, the first suggestion for the microalgal pre-treatment is to determine the effects of the processing conditions, modifications, and pre-treatment techniques in combination to obtain an increase in the yield percentage functions. In a previous report using microalgae *Botryococcus braunii* pre-treated with thermal method for 140 °C for 10 min, a recovered lipid yield of 97.8 wt% was obtained [113]. In research using chemical methods and *Nannochloropsis oceanica* with diluted H_2SO_4 (1% v/v), the researcher obtained a hydrogen yield of 183.9 mL/g TVS [167]. Thermal pre-treatments are recommended compared to other methods such as ultrasound and biological pre-treatment methods [80]. For *Chlorella vulgaris*, enzymatic pre-treatment methods were more effective in enhancing biomass production. It also represents an energy-balanced pre-treatment method other than hydrothermal, thermal, and ultra-sound treatments. When thermal, ultrasound, hydrothermal, and microwave pre-treatments were compared using microalgae obtained from an open pond, a high lipid yield, organic matter solubilisation, and biomass concentration were found. It was also noted that thermal pre-treatment methods showed a positive energy balance because of energy gain [193]. Physical pre-treatment methods using ultrasound or microwaves are used to increase the biomass. The final quality of the product is related to the biochemical composition and morphology of the microalgae during cell disruption. The cell disruption effect was determined by Komaki et al. [194] by studying three various strains of *chlorella vulgaris*. The results indicated that digestibility occurred in one among them. Additionally, the selection of the extraction process has an impact on the final product. The summary of advantages and disadvantages of various pre-treatment techniques are shown in Table 10.

Table 10. The summary of advantages and limitations of various pre-treatment techniques.

Cell Rupture Method	Parameters Affecting Lipid Production	Advantages	Disadvantages	Reference
Mechanical	Design of the blade, number of passes, pressure, and speed of rotation.	Surface area increases. No inhibitory or toxic compounds. Easy to operate and commercialisable. Biomass is easy to handle.	Requires high energy. High capital and maintenance costs. Influence inert materials.	[25,195]
Ultrasonic	Power, time, and cycle number.	Extraction time and solvent consumption are reduced. Bulk medium of cell contents is reduced. No inhibitory or toxic compounds.	Requires high energy consumption. Scaling up is difficult. High capital and maintenance costs.	[90,196–198]
Microwave	Temperature, stirring, power, and time.	Less energy demand and solvent usage. Fast and uniform heating. Eco-friendly. High extraction yield.	Efficiency differs when solvents are volatile or nonpolar. Scaling up is difficult.	[163,199]

Table 10. *Cont.*

Cell Rupture Method	Parameters Affecting Lipid Production	Advantages	Disadvantages	Reference
Autoclaving	Temperature, thermal stress, and pressure.	High lipid content. Life span of the product of maintained. Lower energy demands.	Requires high energy consumption for industrial processes. Time consuming process and scaling up is difficult.	[25,200,201]
Steam Explosion	Temperature, thermal stress, microalgae species, and pressure.	No inhibitory or toxic compounds. Hazardous wastes can be reduced during lipid recovery. Low cost and commercialisable.	Efficiency is dependent on microalgae species. Requires high energy consumption for industrial processes. Time consuming process and scaling up is difficult. Energy costs for high temperatures.	[111,202,203]
Catalytic	Stirring, chemical concentration of KOH and NaOH.	Lower energy consumption. Hemicellulose solubilisation.	Expensive chemical cost. Toxic and inhibitory. Contamination during extraction.	[8,118]
Enzymatic	Enzymatic type, stirring.	Lower energy consumption. Higher lipid yield and speed process.	Expensive enzymatic cost. Agitation conditions.	[204,205]
Pulsed electric field treatment	Oscillation, time, microalgae type, growth phase conditions, and conductivity.	High lipid content. Non-inhibitory compounds. Speed and uniform cell disruption.	Requires high energy consumption for industrial process. High capital and maintenance costs.	[206]
Biological	Enzymes and combination of enzymes.	Energy demand is low. Non-inhibitory compounds.	Cross-contamination. High enzyme cost. Requires large space. Slow pre-treatment process.	[207]

For the process of biodiesel production, cell disruption and lipid extraction from microalgae have been studied and researched for years. The cell disruption pre-treatment techniques used for microalgae have both pros and cons depending on the energy consumption, type of microalgae, cost effectiveness, applicability, and efficiency. As discussed above, further advanced studies may be required to face the challenges of these methods and to bring them to reality.

4. Selection and Processing of Pre-Treatment Technique

The selection of a pre-treatment method and other processing steps is very crucial. These steps must be followed throughout the research process, from microalgae culturing to final processing. There are numerous choices for processing, and some steps may favour the discovery of new techniques.

1. Many iterations must be carried out for process development using published literature. New methods should be evaluated and analysed for lipid yield by considering the environmental impacts and cost factors of the method.
2. Improvement must be carried out according to successful studies using the available modern techniques, and the processing should be continued.
3. Energy and cost are very significant. Along with this, product evaluation is also important and can show an increase in process profitability.

4. Following the safety and legislation protocols is also essential. When using chemical methods, it is very important to consider safety and legislation factors.

5. Energy Consumption of Pre-Treatment Techniques

Even if pre-treatment techniques increase recovery, they demand a certain amount of energy for processing. The evaluation of cost effectiveness and energy consumption of different pre-treatment techniques should be economically relevant. The energy requirements should be viewed according to several factors, such as type of species, growth conditions, pre-treatment method, concentration, etc. For the mechanical and physical methods, the energy demands are ultimately high compared to other methods. Generally, for these methods, the energy consumption and cost effectiveness are the most influential conditions. However, for non-mechanical methods, the energy consumption depends on stirring, time, and temperature. Compared to biological enzymatic hydrolysis treatments, this technique requires very less energy but is only dependent on stirring [208]. By working on the industrial conditions, this method can be optimised in terms of its operating parameters to have a shorter working duration, which might increase its cost as well [209]. Compared the various mechanical methods such as the high-pressure homogenisation (HPH), high-speed homogenisation, microwave, ultrasonic, and bead milling techniques, studies conclude that HPH seems to be a more energy consuming technique [52–54] followed by the ultrasonication and microwave techniques [210]. Therefore, further studies have to be carried out to predict an actual energy demand that will help the specific pre-treatment technique chosen for industrial applications.

6. Key Challenges and Future Perspectives

Current applications in biofuel industries demand new economically and environmentally sustainable processes to overcome the demerits that they face. As such, large-scale and effective techniques for microalgal lipid extraction are required, and a considerable number of studies are needed. Future research should be focused on decreasing the energy consumption, overall cost effectiveness, adaptability, mildness, and recoverability of bio-products. As such, it is important to intensify approaches to minimise the cost and to utilise microalgal extracts to their full extent. In some species, genetic modifications that change their microbiological characteristics may pave a way to introduce new strains with enhanced outputs. Another challenge lies in contamination: it is better to carefully control the development of microorganisms such as bacteria, virus, and other predators. These can become a danger in growing cultures (affecting their growth) and might also reduce the efficiency [149,211]. As such, it is better to look for a potential new route to sort out this complex issue. Radio frequencies using certain magnetrons might cause rapid thermal effects and agitation, but this process has not been addressed in depth. However, a study on the non-thermal effects obtained using microwave radiation gives us a small idea about this process [212]. Pre-treatment techniques such as ultrasonic, microwave, and pulsed effect methods are currently being researched in combination with new approaches on a small scale, but if it possible to control their energy consumption, these methods could be used in large-scale industrial applications. A recent popular technique is the pulsed electric field technique, but this technique still being researched. The creation of new models and improving flow techniques are challenging to investigate. The HPH and bead mill methods are the most effective cell disruption techniques and are able to handle robust cell walls. However, cell walls are diverse, nano-biotechnology research may be promising for processing the weak cells. Alternatively, combining mechanical and non-mechanical methods, e.g., ultrasonic methods with enzymatic or chemical methods, could reduce energy demands. Studies are required for up-and-coming methods such as explosive decompression and cationic polymer-coated membrane treatment to obtain their cell disruption ability for lipid extraction. A method for collecting generic data about different lipid efficiencies should be produced, this could help us to achieve higher level of understanding and would encourage researchers to discover new mild techniques.

7. Conclusions

Considering the pre-treatment techniques reviewed in this paper, the studies shown here attempt to provide cost-effective solutions for the increasing universal energy demands. Complex cell wall/membrane structures may remain as an obstacle to biodiesel production; however, the application of pre-treatment methods will improve the quality of feedstock yields by disrupting the microalgal cells and extracting lipids. Depending on the type of microalgal species, the cell wall/membrane structure acts as a main factor that influences microalgae solubilisation and has various efficiency outcomes. Although most of the pre-treatment techniques were found to have positive attributes in biodiesel production, after checking numerous ongoing research studies, it is well-understood that there is no best methodology for the application of pre-techniques for microalgal lipid extraction because each and every method has both pros and cons for different microalgal species, and the lipid productivity percentage relies on the microalgal species being considered and their characteristics during cell rupture. In large-scale applications, energy and cost requirements are denoted as main indicators, and the discussed techniques are not always feasible for biorefineries due to the high energy consumption and operational costs. The major aspects of industrial microalgae lipid extraction techniques are their universality, energy efficiency, selectivity, mildness, and controllability. Recent research shows that mechanical techniques are optimal for industrial lipid extraction, but they consume a large amount of energy. Non-mechanical techniques may have a low energy demand, but the final quality may be low, and they have a long pre-treatment time. Studies related to biological pre-treatment during on-site enzyme production and enzyme immobilisation indicate that the cost of pure enzymes is high and non-recyclable. Therefore, further studies must focus on reducing the cost of biological pre-treatment methods. Thermal, biological, and chemical pre-treatment methods have been found to produce a higher energy balance than microwave and ultrasonic methods. Microwave and ultrasonic methods lack in biomass yield enhancement and are easily commercialisable. Capital investment in ultrasound and microwave pre-treatment methods is considerably higher than investment in biological and chemical pre-treatment methods. Energy-demanding techniques such as the bead mill or high-pressure homogeniser techniques are mostly preferred for large-scale applications. Future studies can place importance on oxidative pre-treatment techniques as well pulsed and laser electric arc techniques for cell disruption technologies for biodiesel production. These can be very useful for increasing the potential capabilities of microalgal biomass. Hence, it can be concluded that pre-treatment techniques must aim to provide an enhanced lipid efficiency with a reduced energy demand to produce a sustainable energy source.

Author Contributions: A.K., conceptualisation, formal analysis, writing—original draft, writing—review and editing, visualisation; C.R., conceptualisation, writing—review and editing, supervision; A.D., conceptualisation, writing—review and editing, supervision. All authors have read and agreed to the published version of the manuscript.

Funding: This research received no external funding.

Institutional Review Board Statement: Not applicable.

Informed Consent Statement: Not applicable.

Data Availability Statement: Not applicable.

Conflicts of Interest: The authors declare no conflict of interest.

References

1. Chandrasekhar, K.; Lee, Y.-J.; Lee, D.-W.; Chandrasekhar, K.; Lee, Y.-J.; Lee, D.-W. Biohydrogen Production: Strategies to Improve Process Efficiency through Microbial Routes. *Int. J. Mol. Sci.* **2015**, *16*, 8266–8293. [CrossRef] [PubMed]
2. Watts, N.; Amann, M.; Arnell, N.; Ayeb-Karlsson, S.; Belesova, K.; Berry, H.; Bouley, T.; Boykoff, M.; Byass, P.; Cai, W.; et al. The 2018 report of the Lancet Countdown on health and climate change: Shaping the health of nations for centuries to come. *Lancet* **2018**, *392*, 2479–2514. [CrossRef]

3. Schenk, P.M.; Thomas-Hall, S.R.; Stephens, E.; Marx, U.C.; Mussgnug, j.; Posten, C.; Kruse, O.; Hankamer, B. Second Generation Biofuels: High-Efficiency Microalgae for Biodiesel Production. *BioEnergy Res.* **2008**, *1*, 20–43. [CrossRef]
4. Kadier, A.; Simayi, Y.; Abdeshahian, P.; Azman, N.F.; Chandrasekhar, K.; Kalil, M.S. A comprehensive review of microbial electrolysis cells (MEC) reactor designs and configurations for sustainable hydrogen gas production. *Alex. Eng. J.* **2016**, *55*, 427–443. [CrossRef]
5. Brennan, L.; Owende, P. Biofuels from microalgae-A review of technologies for production, processing, and extractions of biofuels and co-products. *Renew. Sustain. Energy Rev.* **2010**, *14*, 557–577. [CrossRef]
6. Chisti, Y. Biodiesel from microalgae. *Biotechnol. Adv.* **2007**, *25*, 294–306. [CrossRef]
7. de Stefano, L.; de Stefano, M.; de Tommasi, E.; Rea, I.; Rendina, I. A natural source of porous biosilica for nanotech applications: The diatoms microalgae. *Phys. Status Solidi (C) Curr. Top. Solid State Phys.* **2011**, *8*, 1820–1825. [CrossRef]
8. Harun, R.; Danquah, M.K. Enzymatic hydrolysis of microalgal biomass for bioethanol production. *Chem. Eng. J.* **2011**, *168*, 1079–1084. [CrossRef]
9. Hazar, H.; Aydin, H. Performance and emission evaluation of a CI engine fueled with preheated raw rapeseed oil (RRO)-diesel blends. *Appl. Energy* **2010**, *87*, 786–790. [CrossRef]
10. Tsukahara, K.; Sawayama, S. Liquid fuel production using microalgae. *J. Japan Pet. Inst.* **2005**, *48*, 251–259. [CrossRef]
11. Roberts, G.W.; Fortier, M.O.P.; Sturm, B.S.M.; Stagg-Williams, S.M. Promising pathway for algal biofuels through wastewater cultivation and hydrothermal conversion. *Energy Fuels* **2013**, *27*, 857–867. [CrossRef]
12. Gerken, H.G.; Donohoe, B.; Knoshaug, E.P. Enzymatic cell wall degradation of Chlorella vulgaris and other microalgae for biofuels production. *Planta* **2013**, *237*, 239–253. [CrossRef]
13. Scholz, M.J.; Weiss, T.L.; Jinkerson, R.E.; Jing, J.; Roth, R.; Goodenough, U.; Posewitz, M.C.; Gerken, H. Ultrastructure and composition of the Nannochloropsis gaditana cell wall. *Eukaryot. Cell* **2014**, *13*, 1450–1464. [CrossRef] [PubMed]
14. Dufey, A.; International Institute for Environment and Development. *Biofuels Production, Trade and Sustainable Development: Emerging Issues*; IIED: London, UK, 2006.
15. Olson, D.L.; Rosacker, K. Crowdsourcing and open source software participation. *Serv. Bus.* **2013**, *7*, 499–511. [CrossRef]
16. Balat, M. An Overview of Biofuels and Policies in the European Union. *Energy Sources Part B Econ. Plan. Policy* **2007**, *2*, 167–181. [CrossRef]
17. Shay, E.G. Diesel fuel from vegetable oils: Status and opportunities. *Biomass Bioenergy* **1993**, *4*, 227–242. [CrossRef]
18. Becker, K.; Makkar, H.P.S. Jatropha curcas: A potential source for tomorrow' s oil and biodiesel. *Lipid Technol.* **2008**, *20*, 104–107. [CrossRef]
19. Dürre, P. Biobutanol: An attractive biofuel. *Biotechnol. J.* **2007**, *2*, 1525–1534. [CrossRef]
20. Roessler, P.G.; Brown, L.M.; Dunahay, T.G.; Heacox, D.A.; Jarvis, E.E.; Schneider, J.C.; Talbot, S.G.; Zeiler, K.G. Genetic Engineering Approaches for Enhanced Production of Biodiesel Fuel from Microalgae. *ACS Symp. Ser.* **1994**, *566*, 255–270. [CrossRef]
21. Li-Beisson, Y.; Thelen, J.J.; Fedosejevs, E.; Harwood, J.L. The lipid biochemistry of eukaryotic algae. *Prog. Lipid Res.* **2019**, *74*, 31–68. [CrossRef] [PubMed]
22. Sun, X.M.; Ren, L.J.; Zhao, Q.Y.; Ji, X.J.; Huang, H. Microalgae for the production of lipid and carotenoids: A review with focus on stress regulation and adaptation. *Biotechnol. Biofuels* **2018**, *11*, 1–16. [CrossRef]
23. Barkia, I.; Saari, N.; Manning, S.R. Microalgae for High-Value Products Towards Human Health and Nutrition. *Mar. Drugs* **2019**, *17*, 304. [CrossRef]
24. Lever, J.; Brkljaca, R.; Kraft, G.; Urban, S. Natural Products of Marine Macroalgae from South Eastern Australia, with Emphasis on the Port Phillip Bay and Heads Regions of Victoria. *Mar. Drugs* **2020**, *18*, 142. [CrossRef]
25. Halim, R.; Rupasinghe, T.W.T.; Tull, D.L.; Webley, P.A. Mechanical cell disruption for lipid extraction from microalgal biomass. *Bioresour. Technol.* **2013**, *140*, 53–63. [CrossRef] [PubMed]
26. Patil, P.; Reddy, H.; Muppaneni, T.; Ponnusamy, S.; Sun, Y.; Dailey, P.; Cooke, P.; Patil, U.; Deng, S. Optimization of microwave-enhanced methanolysis of algal biomass to biodiesel under temperature controlled conditions. *Bioresour. Technol.* **2013**, *137*, 278–285. [CrossRef]
27. Patil, P.D.; Dandamudi, K.P.R.; Wang, J.; Deng, Q.; Deng, S. Extraction of bio-oils from algae with supercritical carbon dioxide and co-solvents. *J. Supercrit. Fluids* **2018**, *135*, 60–68. [CrossRef]
28. Meullemiestre, A.; Breil, C.; Abert-Vian, M.; Chemat, F. Microwave, ultrasound, thermal treatments, and bead milling as intensification techniques for extraction of lipids from oleaginous Yarrowia lipolytica yeast for a biojetfuel application. *Bioresour. Technol.* **2016**, *211*, 190–199. [CrossRef]
29. Passos, F.; Astals, S.; Ferrer, I. Anaerobic digestion of microalgal biomass after ultrasound pretreatment. *Waste Manag.* **2014**, *34*, 2098–2103. [CrossRef]
30. Clarke, A.; Prescott, T.; Khan, A.; Olabi, A.G. Causes of breakage and disruption in a homogenizer. *Appl. Energy* **2010**, *87*, 3680–3690. [CrossRef]
31. Maltseva, Y.M.K.; Maltsev, Y.; Maltseva, K.; Khmelnytskyi, B. Fatty acids of microalgae: Diversity and applications. *Rev. Environ. Sci. Bio/Technol.* **2021**, *20*, 515–547. [CrossRef]
32. Chen, Z.; Wang, L.; Qiu, S.; Ge, S. Determination of Microalgal Lipid Content and Fatty Acid for Biofuel Production. *Bio. Med Res. Int.* **2018**, *2018*, 1052036. [CrossRef]
33. Dixon, C.; Wilken, L.R. Green microalgae biomolecule separations and recovery. *Bioresour. Bioprocess.* **2018**, *5*, 14. [CrossRef]

34. Yoo, G.; Park, W.K.; Kim, C.W.; Choi, Y.E.; Yang, J.W. Direct lipid extraction from wet Chlamydomonas reinhardtii biomass using osmotic shock. *Bioresour. Technol.* **2012**, *123*, 717–722. [CrossRef]
35. Engler, C.R. *Disruption of Microbial Cells*; Pergamon Press: Oxford, UK, 1985; Volume 2, pp. 305–324.
36. Deshmukh, S.; Kumar, R.; Bala, K. Microalgae biodiesel: A review on oil extraction, fatty acid composition, properties and effect on engine performance and emissions. *Fuel Processing Technol.* **2019**, *191*, 232–247. [CrossRef]
37. Sajjadi, B.; Chen, W.Y.; Raman, A.A.A.; Ibrahim, S. Microalgae lipid and biomass for biofuel production: A comprehensive review on lipid enhancement strategies and their effects on fatty acid composition. *Renew. Sustain. Energy Rev.* **2018**, *97*, 200–232. [CrossRef]
38. Ferreira, G.F.; Pinto, L.F.R.; Filho, R.M.; Fregolente, L.v. A review on lipid production from microalgae: Association between cultivation using waste streams and fatty acid profiles. *Renew. Sustain. Energy Rev.* **2019**, *109*, 448–466. [CrossRef]
39. Medeiros, H.; Kirleis, A.W.; Vetter, R.J. Microwave Inactivation of Thioglucosidase in Crambe. *J. Am. Oil Chem. Soc.* **1978**, *55*, 679–682. [CrossRef]
40. Mercer, P.; Armenta, R.E. Developments in oil extraction from microalgae. *Eur. J. Lipid Sci. Technol.* **2011**, *113*, 539–547. [CrossRef]
41. Engler, C.R. Disruption of Microbial Cells. *Princ Biotechnol Eng. Consid* **1985**, *2*, 305–324.
42. Schütte, H.; Kroner, K.H.; Hustedt, H.; Kula, M.R. Experiences with a 20 litre industrial bead mill for the disruption of microorganisms. *Enzym. Microb. Technol.* **1983**, *5*, 143–148. [CrossRef]
43. Shirgaonkar, I.Z.; Lothe, R.R.; Pandit, A.B. Comments on the mechanism of microbial cell disruption in high-pressure and high-speed devices. *Biotechnol. Prog.* **1998**, *14*, 657–660. [CrossRef]
44. Kargi, F.; Moo-Young, M. *Transport Phenomena in Bioprocesses*; Pergamon Press: Oxford, UK, 1985.
45. Tedesco, S.; Lochlainn, D.M.; Olabi, A.G. Particle size reduction optimization of Laminaria sbiomass for enhanced methane production. *Energy* **2014**, *76*, 857–862. [CrossRef]
46. Greenwell, H.C.; Laurens, L.M.L.; Shields, R.J.; Lovitt, R.W.; Flynn, K.J. Placing microalgae on the biofuels priority list: A review of the technological challenges. *J. R. Interface R. Soc.* **2010**, *06*, 703–726. [CrossRef] [PubMed]
47. Montgomery, D. 'Helping the Guards': Illegal displays and blueshirt criminality, 1932–1936. *Eire-Ireland* **2014**, *49*, 22–43. [CrossRef]
48. Spiden, E.M.; Scales, P.J.; Kentish, S.E.; Martin, G.J.O. Critical analysis of quantitative indicators of cell disruption applied to Saccharomyces cerevisiae processed with an industrial high pressure homogenizer. *Biochem. Eng. J.* **2013**, *70*, 120–126. [CrossRef]
49. Ekpeni, L.E.N.; Benyounis, K.Y.; Nkem-Ekpeni, F.F.; Stokes, J.; Olabi, A.G. Underlying factors to consider in improving energy yield from biomass source through yeast use on high-pressure homogenizer (hph). *Energy* **2015**, *81*, 74–83. [CrossRef]
50. Hoare, M.; Dunnill, P. Biochemical engineering challenges of purifying useful proteins. *Philos. Trans. R. Soc. Lond. Ser. B Biol. Sci.* **1989**, *324*, 497–507. [CrossRef]
51. Moore, E.K.; Hoare, M.; Dunnill, P. Disruption of baker's yeast in a high-pressure homogenizer: New evidence on mechanism. *Enzym. Microb. Technol.* **1990**, *12*, 764–770. [CrossRef]
52. Halim, R.; Harun, R.; Danquah, M.K.; Webley, P.A. Microalgal cell disruption for biofuel development. *Appl. Energy* **2012**, *91*, 116–121. [CrossRef]
53. Lee, A.K.; Lewis, D.M.; Ashman, P.J. Disruption of microalgal cells for the extraction of lipids for biofuels: Processes and specific energy requirements. *Biomass Bioenergy* **2012**, *46*, 89–101. [CrossRef]
54. Lee, J.-Y.; Yoo, C.; Jun, S.-Y.; Ahn, C.-Y.; Oh, H.-M. Comparison of several methods for effective lipid extraction from microalgae. *Bioresour. Technol.* **2010**, *101*, S75–S77. [CrossRef]
55. Shene, C.; Monsalve, M.T.; Vergara, D.; Lienqueo, M.E.; Rubilar, M. High pressure homogenization of Nannochloropsis oculata for the extraction of intracellular components: Effect of process conditions and culture age. *Eur. J. Lipid Sci. Technol.* **2016**, *118*, 631–639. [CrossRef]
56. Papachristou, I.; Silve, A.; Jianu, A.; Wüstner, R.; Nazarova, N.; Müller, G.; Frey, W. Evaluation of pulsed electric fields effect on the microalgae cell mechanical stability through high pressure homogenization. *Algal Res.* **2020**, *47*, 101847. [CrossRef]
57. Grimi, N.; Dubois, A.; Marchal, L.; Jubeau, S.; Lebovka, N.I.; Vorobiev, E. Selective extraction from microalgae Nannochloropsis susing different methods of cell disruption. *Bioresour. Technol.* **2014**, *153*, 254–259. [CrossRef] [PubMed]
58. Mulchandani, K.; Kar, J.R.; Singhal, R.S. Extraction of Lipids from Chlorella saccharophila Using High-Pressure Homogenization Followed by Three Phase Partitioning. *Appl. Biochem. Biotechnol.* **2015**, *1613*, 176–1626. [CrossRef] [PubMed]
59. Scopus—Document Details—Evaluation of Nutritional Value and Safety of the Green Microalgae Chlorella Vulgaris Treated with Novel Processing Methods | Signed in. Available online: https://www.scopus.com/record/display.uri?eid=2-s2.0-79961163938&origin=inward&txGid=50f30500ddba8978d392e1c26c85bc87&featureToggles=FEATURE_NEW_DOC_DETAILS_EXPORT:1 (accessed on 4 March 2022).
60. Bierau, H.; Zhang, Z.; Lyddiatt, A. Direct process integration of cell disruption and fluidised bed adsorption for the recovery of intracellular proteins. *J. Chem. Technol. Biotechnol.* **1999**, *74*, 208–212. [CrossRef]
61. Melendres, A.v.; Unno, H.; Shiragami, N.; Honda, H. A concept of critical velocity for cell disruption by bead mill. *J. Chem. Eng. Jpn.* **1992**, *25*, 354–356. [CrossRef]
62. Bunge, F.; Pietzsch, M.; Müller, R.; Syldatk, C. Mechanical disruption of Arthrobacter sDSM 3747 in stirred ball mills for the release of hydantoin-cleaving enzymes. *Chem. Eng. Sci.* **1992**, *47*, 225–232. [CrossRef]
63. Nekkanti, V.; Vabalaboina, V.; Pillai, R. Drug Nanoparticles—An Overview. In *The Delivery of Nanoparticles*; InTech: London, UK, 2012. [CrossRef]

64. Alavijeh, R.S.; Karimi, K.; Wijffels, R.H.; van den Berg, C.; Eppink, M. Combined bead milling and enzymatic hydrolysis for efficient fractionation of lipids, proteins, and carbohydrates of Chlorella vulgaris microalgae. *Bioresour. Technol.* **2020**, *309*, 123321. [CrossRef] [PubMed]
65. Emparan, Q.; Jye, Y.S.; Danquah, M.K.; Harun, R. Cultivation of Nannochloropsis smicroalgae in palm oil mill effluent (POME) media for phycoremediation and biomass production: Effect of microalgae cells with and without beads. *J. Water Process. Eng.* **2020**, *33*, 101043. [CrossRef]
66. Schwenzfeier, A.; Wierenga, P.A.; Gruppen, H. Isolation and characterization of soluble protein from the green microalgae Tetraselmis sp. *Bioresour. Technol.* **2011**, *9121*, 102–9127. [CrossRef] [PubMed]
67. Doucha, J.; Lívanský, K. Influence of processing parameters on disintegration of Chlorella cells in various types of homogenizers. *Appl. Microbiol. Biotechnol.* **2008**, *81*, 431–440. [CrossRef]
68. Postma, P.R.; Miron, T.L.; Olivieri, G.; Barbosa, M.J.; Wijffels, R.H.; Eppink, M.H.M. Mild disintegration of the green microalgae Chlorella vulgaris using bead milling. *Bioresour. Technol.* **2015**, *184*, 297–304. [CrossRef]
69. Taleb, A.; Kandiliana, R.; Toucharda, R.; Montalescota, V.; Rinaldia, T.; Tahab, S.; Takachec, H.; Marchala, L.; Legranda, J.; Pruvosta, J. Screening of freshwater and seawater microalgae strains in fully controlled photobioreactors for biodiesel production. *Bioresour. Technol.* **2016**, *218*, 480–490. [CrossRef]
70. Huang, O.; Bandyopadhyay, A.; Bose, S. Influence of processing parameters on PZT thick films. *Mater. Sci. Eng. B Solid-State Mater. Adv. Technol.* **2005**, *116*, 19–24. [CrossRef]
71. Singh, S.; Meena, P.; Saharan, V.K.; Bhoi, R.; George, S. Enhanced lipid recovery from chlorella sBiomass by green approach: A combination of ultrasonication and homogenization pre-treatment techniques (hybrid method) using aqueous deep eutectic solvents. *Mater. Today Proc.* **2022**, *57*, 179–186. [CrossRef]
72. Bernaerts, T.M.M.; Panozzo, A.; Doumen, V.; Foubert, I.; Gheysen, L.; Goiris, K.; Moldenaers, P.; Hendrickx, M.E.; Van Loey, A.M. Microalgal biomass as a (multi)functional ingredient in food products: Rheological properties of microalgal suspensions as affected by mechanical and thermal processing. *Algal Res.* **2017**, *25*, 452–463. [CrossRef]
73. Zhou, L.; Guan, Y.; Bi, J.; Liu, X.; Yi, J.; Chen, Q.; Wu, X.; Zhouet, M. Change of the rheological properties of mango juice by high pressure homogenization. *LWT—Food Sci. Technol.* **2017**, *82*, 121–130. [CrossRef]
74. Wang, M.; Yin, Z.; Zeng, M. Microalgae as a promising structure ingredient in food: Obtained by simple thermal and high-speed shearing homogenization. *Food Hydrocoll.* **2022**, *131*, 107743. [CrossRef]
75. Balasubramanian, R.K.; Doan, T.T.Y.; Obbard, J.P. Factors affecting cellular lipid extraction from marine microalgae. *Chem. Eng. J.* **2013**, *215–216*, 929–936. [CrossRef]
76. Malafronte, L.; Yilmaz-Turan, S.; Krona, A.; Martinez-Sanz, M.; Vilaplana, F.; Lopez-Sanchez, P. Macroalgae suspensions prepared by physical treatments: Effect of polysaccharide composition and microstructure on the rheological properties. *Food Hydrocoll.* **2021**, *120*, 106989. [CrossRef]
77. Wang, G.; Wang, T. Characterization of lipid components in two microalgae for biofuel application. *JAOCS J. Am. Oil Chem. Soc.* **2012**, *89*, 135–143. [CrossRef]
78. Chisti, Y.; Moo-Young, M. Disruption of microbial cells for intracellular products. *Enzym. Microb. Technol.* **1986**, *8*, 194–204. [CrossRef]
79. Shi, M.; Wang, F.; Lan, P.; Zhang, Y.; Zhang, M.; Yan, Y.; Liu, Y. Effect of ultrasonic intensity on structure and properties of wheat starch-monoglyceride complex and its influence on quality of norther-style Chinese steamed bread. *LWT* **2021**, *138*, 110677. [CrossRef]
80. Alzate, M.E.; Muñoz, R.; Rogalla, F.; Fdz-Polanco, F.; Pérez-Elvira, S.I. Biochemical methane potential of microalgae: Influence of substrate to inoculum ratio, biomass concentration and pretreatment. *Bioresour. Technol.* **2012**, *123*, 488–494. [CrossRef] [PubMed]
81. Scenedesmus Dimorphus and Scenedesmus Quadricauda: Two Potent Indigenous Microalgae Strains for Biomass Production and CO2 Mitigation—A Study on Their Growth Behavior and Lipid Productivity under Different Concentration of Urea as Nitrogen Source. | Semantic Scholar. Available online: https://www.semanticscholar.org/paper/Scenedesmus-dimorphus-and-Scenedesmus-quadricauda-%3A-Goswami-Kalita/98746c87fce372e744d9ba4071fb23867f6abd7d (accessed on 16 May 2022).
82. González-Fernández, C.; Sialve, B.; Bernet, N.; Steyer, J.P. Comparison of ultrasound and thermal pretreatment of Scenedesmus biomass on methane production. *Bioresour. Technol.* **2012**, *110*, 610–616. [CrossRef]
83. Zhang, X.; Mi, T.; Gao, W.; Wu, Z.; Yuan, C.; Cui, B.; Dai, Y.; Liu, P. Ultrasonication effects on physicochemical properties of starch–lipid complex. *Food Chem.* **2022**, *388*, 133054. [CrossRef] [PubMed]
84. Samson, R.; Leduy, A. Influence of mechanical and thermochemical pretreatments on anaerobic digestion ofSpirulina maxima algal biomass. *Biotechnol. Lett.* **1983**, *5*, 671–676. [CrossRef]
85. Schwede, S.; Kowalczyk, A.; Gerber, M.; Span, R. Influence of Different Cell Disruption Techniques on Mono Digestion of Algal Biomass. Available online: https://www.semanticscholar.org/paper/Influence-of-Different-Cell-Disruption-Techniques-Schwede-Kowalczyk/ad21f21a945afe30cb56c09e8866b467db460611 (accessed on 10 July 2022).
86. Mubarak, M.; Shaija, A.; Suchithra, T.V. Ultrasonication: An effective pre-treatment method for extracting lipid from Salvinia molesta for biodiesel production. *Resour. -Effic. Technol.* **2016**, *2*, 126–132. [CrossRef]
87. Natarajan, R.; Ang, W.M.R.; Chen, X.; Voigtmann, M.; Lau, R. Lipid releasing characteristics of microalgae species through continuous ultrasonication. *Bioresour. Technol.* **2014**, *158*, 7–11. [CrossRef]

88. Hao, X.; Suo, H.; Peng, H.; Xu, P.; Gao, X.; Du, S. Simulation and exploration of cavitation process during microalgae oil extracting with ultrasonic-assisted for hydrogen production. *Int. J. Hydrog. Energy* **2021**, *46*, 2890–2898. [CrossRef]
89. Ji, J.; Wang, J.; Li, Y.; Yu, Y.; Xu, Z. Preparation of biodiesel with the help of ultrasonic and hydrodynamic cavitation. *Ultrasonics* **2006**, *44*, e411–e414. [CrossRef] [PubMed]
90. Borthwick, K.A.J.; Coakley, W.T.; McDonnell, M.B.; Nowotny, H.; Benes, E.; Gröschl, M. Development of a novel compact sonicator for cell disruption. *J. Microbiol. Methods* **2005**, *60*, 207–216. [CrossRef] [PubMed]
91. Rodriguez, C.; Alaswad, A.; Mooney, J.; Prescott, T.; Olabi, A.G. Pre-treatment techniques used for anaerobic digestion of algae. *Fuel Processing Technol.* **2015**, *138*, 765–779. [CrossRef]
92. Choi, I.; Cho, S.J.; Chun, J.K.; Moon, T.W. Extraction yield of soluble protein and microstructure of soybean affected by microwave heating. *J. Food Processing Preserv.* **2006**, *30*, 407–419. [CrossRef]
93. Kumar, V.; Arora, N.; Pandey, S.; Jaiswal, K.K.; Nanda, M.; Vlaskin, M.S.; Chauhan, P.K. Microwave-assisted pretreatment of harmful algal blooms for microbial oil-centered biorefinery approach. *Biomass Convers. Biorefinery* **2020**, *12*, 3097–3105. [CrossRef]
94. Cheng, J.; Huang, R.; Yu, T.; Li, T.; Zhou, J.; Cen, K. Biodiesel production from lipids in wet microalgae with microwave irradiation and bio-crude production from algal residue through hydrothermal liquefaction. *Bioresour. Technol.* **2014**, *151*, 415–418. [CrossRef]
95. Biodiesel Production from Dry Microalga Biomass by Microwave-Assisted In-Situ Transesterification | MATEC Web of Conferences. Available online: https://www.matec-conferences.org/articles/matecconf/abs/2018/15/matecconf_rsce2018_06005/matecconf_rsce2018_06005.html (accessed on 6 January 2022).
96. Zheng, H.; Yin, J.; Gao, Z.; Huang, H.; Ji, X.; Dou, C. Disruption of chlorella vulgaris cells for the release of biodiesel-producing lipids: A comparison of grinding, ultrasonication, bead milling, enzymatic lysis, and microwaves. *Appl. Biochem. Biotechnol.* **2011**, *164*, 1215–1224. [CrossRef]
97. Wahidin, S.; Idris, A.; Yusof, N.M.; Kamis, N.H.H.; Shaleh, S.R.M. Optimization of the ionic liquid-microwave assisted one-step biodiesel production process from wet microalgal biomass. *Energy Convers. Manag.* **2018**, *171*, 1397–1404. [CrossRef]
98. Iqbal, J.; Theegala, C. Microwave assisted lipid extraction from microalgae using biodiesel as co-solvent. *Algal Res.* **2013**, *2*, 34–42. [CrossRef]
99. Biller, P.; Friedman, C.; Ross, A.B. Hydrothermal microwave processing of microalgae as a pre-treatment and extraction technique for bio-fuels and bio-products. *Bioresour. Technol.* **2013**, *136*, 188–195. [CrossRef] [PubMed]
100. Amarni, F.; Kadi, H. Kinetics study of microwave-assisted solvent extraction of oil from olive cake using hexane. Comparison with the conventional extraction. *Innov. Food Sci. Emerg. Technol.* **2010**, *11*, 322–327. [CrossRef]
101. Pasquet, V.; Chérouvrier, J.; Farhat, F.; Thiéry, V.; Piot, J.; Bérard, J.; Kaas, R.; Serive, B.; Patrice, T.; Cadoret, J.; et al. Study on the microalgal pigments extraction process: Performance of microwave assisted extraction. *Process. Biochem.* **2011**, *46*, 59–67. [CrossRef]
102. Aboulfoth, A.; El Gohary, E.H.; El Monayeri, O.D. Effect of thermal pretreatment on the solubilization of organic matters in a mixture of primary and waste activated sludge. *J. Urban Environ. Eng.* **2015**, *9*, 82–88. [CrossRef]
103. Mahdy, A.; Mendez, L.; Ballesteros, M.; González-Fernández, C. Enhanced methane production of Chlorella vulgaris and Chlamydomonas reinhardtii by hydrolytic enzymes addition. *Energy Convers. Manag.* **2014**, *85*, 551–557. [CrossRef]
104. Velásquez, J.A.; Ferrando, F.; Salvadó, J. Effects of kraft lignin addition in the production of binderless fiberboard from steam exploded Miscanthus sinensis. *Ind. Crops Prod.* **2003**, *18*, 17–23. [CrossRef]
105. Schell, D.; Nguyen, Q.; Tucker, M.; Boynton, B. Pretreatment of softwood by acid-catalyzed steam explosion followed by alkali extraction. *Appl. Biochem. Biotechnol. Part A Enzym. Eng. Biotechnol.* **1998**, *17*, 70. [CrossRef]
106. Romaní, A.; Garrote, G.; Ballesteros, I.; Ballesteros, M. Second generation bioethanol from steam exploded Eucalyptus globulus wood. *Fuel* **2013**, *111*, 66–74. [CrossRef]
107. Chen, W.H.; Pen, B.L.; Yu, C.T.; Hwang, W.S. Pretreatment efficiency and structural characterization of rice straw by an integrated process of dilute-acid and steam explosion for bioethanol production. *Bioresour. Technol.* **2011**, *102*, 2916–2924. [CrossRef]
108. Lorente, E.; Farriol, X.; Salvadó, J. Steam explosion as a fractionation step in biofuel production from microalgae. *Fuel Process. Technol.* **2015**, *131*, 93–98. [CrossRef]
109. Matsakas, L.; Nitsos, C.; Raghavendran, V.; Yakimenko, O.; Persson, G.; Olsson, E.; Rova, U.; Olsson, L.; Christakopoulos, P. A novel hybrid organosolv: Steam explosion method for the efficient fractionation and pretreatment of birch biomass. *Biotechnol. Biofuels* **2018**, *11*, 160. [CrossRef]
110. Olofsson, M.; Lamela, T.; Nilsson, E.; Bergé, J.-P.; Del Pino, V.; Uronen, P.; Legrand, C. Combined effects of nitrogen concentration and seasonal changes on the production of lipids in Nannochloropsis oculate. *Mar. Drugs* **2014**, *12*, 1891–1910. [CrossRef]
111. Nurra, C.; Torras, C.; Clavero, E.; Ríos, S.; Rey, M.; Lorente, E.; Farriol, X.; Salvadó, J. Biorefinery concept in a microalgae pilot plant. Culturing, dynamic filtration and steam explosion fractionation. *Bioresour. Technol.* **2014**, *163*, 136–142. [CrossRef]
112. Chandra, T.S.; Suvidha, G.; Mukherji, S.; Chauhan, V.S.; Vidyashankar, S.; Krishnamurthi, K.; Sarada, R.; Mudliar, S.N. Statistical optimization of thermal pretreatment conditions for enhanced biomethane production from defatted algal biomass. *Bioresour. Technol.* **2014**, *162*, 157–165. [CrossRef] [PubMed]
113. Kita, K.; Okada, S.; Sekino, H.; Imou, K.; Yokoyama, S.; Amano, T. Thermal pre-treatment of wet microalgae harvest for efficient hydrocarbon recovery. *Appl. Energy* **2010**, *87*, 2420–2423. [CrossRef]
114. Marsolek, M.D.; Kendall, E.; Thompson, P.L.; Shuman, T.R. Thermal pretreatment of algae for anaerobic digestion. *Bioresour. Technol.* **2014**, *151*, 373–377. [CrossRef]

115. Song, W.; Huang, R.; Guo, H.; Yin, C.; Wang, C.; Cheng, J.; Yang, W. Characterization of wet microalgal cells pretreated with steam for lipid extraction. *Chin. J. Chem. Eng.* **2021**, *37*, 114–120. [CrossRef]
116. Silva, A.P.F.d.; Costa, M.C.; Lopes, A.C.; Neto, E.F.A.; Leitão, R.C.; Mota, C.R.; Santos, A.B.d. Comparison of pretreatment methods for total lipids extraction from mixed microalgae. *Renew. Energy* **2014**, *63*, 762–766. [CrossRef]
117. Prabakaran, P.; Ravindran, A.D. A comparative study on effective cell disruption methods for lipid extraction from microalgae. *Lett. Appl. Microbiol.* **2011**, *53*, 150–154. [CrossRef] [PubMed]
118. Miranda, J.R.; Passarinho, P.C.; Gouveia, L. Pre-treatment optimization of Scenedesmus obliquus microalga for bioethanol production. *Bioresour. Technol.* **2012**, *104*, 342–348. [CrossRef] [PubMed]
119. Kendir, E.; Ugurlu, A. A comprehensive review on pretreatment of microalgae for biogas production. *Int. J. Energy Res.* **2018**, *42*, 3711–3731. [CrossRef]
120. Gonzalez-Fernandez, C.; Barreiro-Vescovo, S.; de Godos, I.; Fernandez, M.; Zouhayr, A.; Ballesteros, M. Biochemical methane potential of microalgae biomass using different microbial inocula. *Biotechnol. Biofuels* **2018**, *11*, 184. [CrossRef]
121. Cho, S.; Park, S.; Seon, J.; Yu, J.; Lee, T. Evaluation of thermal, ultrasonic and alkali pretreatments on mixed-microalgal biomass to enhance anaerobic methane production. *Bioresour. Technol.* **2013**, *143*, 330–336. [CrossRef] [PubMed]
122. Bohutskyi, P.; Betenbaugh, M.J.; Bouwer, E.J. The effects of alternative pretreatment strategies on anaerobic digestion and methane production from different algal strains. *Bioresour. Technol.* **2014**, *155*, 366–372. [CrossRef] [PubMed]
123. Cardeña, R.; Moreno, G.; Bakonyi, P.; Buitrón, G. Enhancement of methane production from various microalgae cultures via novel ozonation pretreatment. *Chem. Eng. J.* **2017**, *307*, 948–954. [CrossRef]
124. Jessop, P.G.; Subramaniam, B. Gas-expanded liquids. *Chem. Rev.* **2007**, *2666*, 107–2694. [CrossRef]
125. Guo, H.; Cheng, J.; Mao, Y.; Qian, L.; Yang, W.; Park, J.Y. Synergistic effect of ultrasound and switchable hydrophilicity solvent promotes microalgal cell disruption and lipid extraction for biodiesel production. *Bioresour. Technol.* **2022**, *343*, 126087. [CrossRef]
126. Atadashi, I.M.; Aroua, M.K.; Aziz, A.R.A.; Sulaiman, N.M.N. The effects of catalysts in biodiesel production: A review. *J. Ind. Eng. Chem.* **2013**, *26*, 25. [CrossRef]
127. Vicente, G.; Martínez, M.; Aracil, J. Integrated biodiesel production: A comparison of different homogeneous catalysts systems. *Bioresour. Technol.* **2004**, *92*, 297–305. [CrossRef]
128. Li, Y. One-step production of biodiesel from Nannochloropsis son solid base Mg-Zr catalyst. *Appl. Energy* **2011**, *3313*, 88–3317. [CrossRef]
129. Taher, H.; Al-Zuhair, S.; Al-Marzouqi, A.H.; Haik, Y.; Farid, M. Effective extraction of microalgae lipids from wet biomass for biodiesel production. *Biomass Bioenergy* **2014**, *66*, 159–167. [CrossRef]
130. Kondo, M.; Rezaei, K.; Temelli, F.; Goto, M. On-line extraction—Reaction of canola oil with ethanol by immobilized lipase in SC-CO 2. *Ind. Eng. Chem. Res.* **2002**, *41*, 5770–5774. [CrossRef]
131. Tewari, Y.B.; Ihara, T.; Phinney, K.W.; Mayhew, M.P. A thermodynamic study of the lipase-catalyzed transesterification of benzyl alcohol and butyl acetate in supercritical carbon dioxide media. *J. Mol. Catal. B Enzym.* **2004**, *304*, 131–136. [CrossRef]
132. Besa, E.C. New opportunities in a personalized approach to the preleukemic phase of myelodysplastic syndrome and acute myelogenous leukemia. *Trends Green Chem.* **2018**, *4*, 1–45. [CrossRef]
133. Choi, O.K.; Lee, J.W. CO2-triggered switchable solvent for lipid extraction from microalgal biomass. *Sci. Total Environ.* **2022**, *819*, 153084. [CrossRef] [PubMed]
134. Anto, S.; Premalatha, M.; Mathimani, T. Tertiary amine as an efficient CO2 switchable solvent for extracting lipids from hypersaline microalgae. *Chemosphere* **2022**, *288*, 132442. [CrossRef]
135. Lage, S.; Willfors, A.; Hörnberg, A.; Gentili, F.G. Impact of organic solvents on lipid-extracted microalgae residues and wastewater sludge co-digestion. *Bioresour. Technol. Rep.* **2021**, *16*, 100850. [CrossRef]
136. Wimmer, Z.; Zarevúcka, M. A Review on the Effects of Supercritical Carbon Dioxide on Enzyme Activity. *Int. J. Mol. Sci.* **2010**, *11*, 233–253. [CrossRef] [PubMed]
137. Delmiro, T.M.; Calixto, G.Q.; Viegas, C.V.; Melo, D.M.A.; Viana, G.A.C.M.; Mendes, L.B.B.; Braga, R.M. Renewable aromatic hydrocarbons from flash catalytic pyrolysis of Monoraphidium slipid extract. *Bioresour. Technol. Rep.* **2021**, *15*, 100799. [CrossRef]
138. Raheem, A.; Cui, X.; Mangi, F.H.; Memon, A.A.; Ji, G.; Cheng, B.; Dong, W.; Zhao, M. Hydrogen-rich energy recovery from microalgae (lipid-extracted) via steam catalytic gasification. *Algal Res.* **2020**, *52*, 102102. [CrossRef]
139. Felix, C.; Ubando, A.; Madrazo, C.; Gue, I.H.; Sutanto, S.; Tran-Nguyen, P.L.; Go, A.W.; Ju, Y.; Culaba, A.; Chang, J.; et al. Non-catalytic in-situ (trans) esterification of lipids in wet microalgae Chlorella vulgaris under subcritical conditions for the synthesis of fatty acid methyl esters. *Appl. Energy* **2019**, *248*, 526–537. [CrossRef]
140. Santillan-Jimenez, E.; Pace, R.; Marques, S.; Morgan, T.; McKelphin, C.; Mobley, J.; Crocker, M. Extraction, characterization, purification and catalytic upgrading of algae lipids to fuel-like hydrocarbons. *Fuel* **2016**, *180*, 668–678. [CrossRef]
141. Lee, S.Y.; Cho, J.M.; Chang, Y.K.; Oh, Y.K. Cell disruption and lipid extraction for microalgal biorefineries: A review. *Bioresour. Technol.* **2017**, *244*, 1317–1328. [CrossRef]
142. Liang, K.; Zhang, Q.; Cong, W. Enzyme-Assisted Aqueous Extraction of Lipid from Microalgae. *J. Agric. Food Chem.* **2012**, *601*, 1177–11776. [CrossRef]
143. Wu, C.; Xiao, Y.; Lin, W.; Li, J.; Zhang, S.; Zhu, J.; Rong, J. Aqueous enzymatic process for cell wall degradation and lipid extraction from Nannochloropsis sp. *Bioresour. Technol.* **2017**, *223*, 312–316. [CrossRef]

144. Harrison, S.T.L. Bacterial cell disruption: A key unit operation in the recovery of intracellular products. *Biotechnol. Adv.* **1991**, *9*, 217–240. [CrossRef]
145. McKenzie, H.A.; White, F.H. Lysozyme and α-lactalbumin: Structure, function, and interrelationships. *Adv. Protein Chem.* **1991**, *41*, 173–315. [CrossRef] [PubMed]
146. Lucy; Montgomery, F.R.; Bochmann, G. Pretreatment of Feedstock for Enhanced Biogas Production. 2014. Available online: https://www.iea-biogas.net/files/daten-redaktion/download/Technical%20Brochures/pretreatment_web.pdf (accessed on 10 July 2022).
147. Kumar, P.; Barrett, D.M.; Delwiche, M.J.; Stroeve, P. Methods for Pretreatment of Lignocellulosic Biomass for Efficient Hydrolysis and Biofuel Production. *Ind. Eng. Chem. Res.* **2009**, *48*, 3713–3729. [CrossRef]
148. Fu, C.C.; Hung, T.C.; Chen, J.Y.; Su, C.H.; Wu, W.T. Hydrolysis of microalgae cell walls for production of reducing sugar and lipid extraction. *Bioresour. Technol.* **2010**, *101*, 8750–8754. [CrossRef] [PubMed]
149. Demuez, M.; González-Fernández, C.; Ballesteros, M. Algicidal microorganisms and secreted algicides: New tools to induce microalgal cell disruption. *Biotechnol. Adv.* **2015**, *33*, 1615–1625. [CrossRef] [PubMed]
150. Chen, C.Y.; der Bai, M.; Chang, J.S. Improving microalgal oil collecting efficiency by pretreating the microalgal cell wall with destructive bacteria. *Biochem. Eng. J.* **2013**, *81*, 170–176. [CrossRef]
151. Abrahamsson, V.; Cunico, L.P.; Andersson, N.; Nilsson, B.; Turner, C. Multicomponent inverse modeling of supercritical fluid extraction of carotenoids, chlorophyll A, ergosterol and lipids from microalgae. *J. Supercrit. Fluids* **2018**, *139*, 53–61. [CrossRef]
152. Nagappan, S.; Devendran, S.; Tsai, P.C.; Dinakaran, S.; Dahms, H.U.; Ponnusamy, V.K. Passive cell disruption lipid extraction methods of microalgae for biofuel production—A review. *Fuel* **2019**, *252*, 699–709. [CrossRef]
153. Lee, S.; Posarac, D.; Ellis, N. An experimental investigation of biodiesel synthesis from waste canola oil using supercritical methanol. *Fuel* **2012**, *91*, 229–237. [CrossRef]
154. Obeid, S.; Beaufils, N.; Camy, S.; Takache, H.; Ismail, A.; Pontalier, P.Y. Supercritical carbon dioxide extraction and fractionation of lipids from freeze-dried microalgae Nannochloropsis oculata and Chlorella vulgaris. *Algal Res.* **2018**, *34*, 49–56. [CrossRef]
155. Viguera, M.; Marti, A.; Masca, F.; Prieto, C.; Calvo, L. The process parameters and solid conditions that affect the supercritical CO2 extraction of the lipids produced by microalgae. *J. Supercrit. Fluids* **2016**, *113*, 16–22. [CrossRef]
156. Zinnai, A.; Sanmartin, C.; Taglieri, I.; Andrich, G.; Venturi, F. Supercritical fluid extraction from microalgae with high content of LC-PUFAs. A case of study: Sc-CO2 oil extraction from Schizochytrium sp. *J. Supercrit. Fluids* **2016**, *116*, 126–131. [CrossRef]
157. Lorenzen, J.; Gl, N.; Tippelt, M.; Stege, A.; Qoura, F.; Sohling, U.; Brück, T. Extraction of microalgae derived lipids with supercritical carbon dioxide in an industrial relevant pilot plant. *Bioprocess Biosyst. Eng.* **2017**, *40*, 911–918. [CrossRef] [PubMed]
158. De Melo, M.M.R.; Sapatinha, M.; Pinheiro, J.; Lemos, M.F.L.; Bandarra, N.M.; Batista, I.; Paulo, M.C.; Coutinho, J.; Saraiva, J.A.; Portugal, I.; et al. Supercritical CO_2 extraction of Aurantiochytrium sbiomass for the enhanced recovery of omega-3 fatty acids and phenolic compounds. *J. CO2 Util.* **2020**, *38*, 24–31. [CrossRef]
159. Sheng, J.; Vannela, R.; Rittmann, B.E. Evaluation of cell-disruption effects of pulsed-electric-field treatment of Synechocystis PCC 6803. *Environ. Sci. Technol.* **2011**, *45*, 3795–3802. [CrossRef]
160. Salerno, M.B.; Lee, H.-S.; Parameswaran, P.; Rittmann, B.E. Using a pulsed electric field as a pretreatment for improved biosolids digestion and methanogenesis. *Water Environ. Res.* **2009**, *81*, 831–839. [CrossRef]
161. Schoenbach, K.H.; Joshi, R.P.; Stark, R.H.; Dobbs, F.C.; Beebe, S.J. Bacterial decontamination of liquids with pulsed electric fields. *IEEE Trans. Dielectr. Electr. Insul.* **2000**, *7*, 637–645. [CrossRef]
162. Cravotto, G.; Boffa, L.; Mantegna, S.; Perego, P.; Avogadro, M.; Cintas, P. Improved extraction of vegetable oils under high-intensity ultrasound and/or microwaves. *Ultrason. Sonochem.* **2008**, *15*, 898–902. [CrossRef] [PubMed]
163. Virot, M.; Tomao, V.; Ginies, C.; Visinoni, F.; Chemat, F. Microwave-integrated extraction of total fats and oils. *J. Chromatogr. A* **2008**, *1196–1197*, 57–64. [CrossRef] [PubMed]
164. Lee, S.J.; Yoon, B.D.; Oh, H.M. Rapid method for the determination of lipid from the green alga Botryococcus braunii. *Biotechnol. Tech.* **1998**, *12*, 553–556. [CrossRef]
165. Spiden, E.M.; Scales, P.J.; Yap, B.H.J.; Kentish, S.E.; Hill, D.R.A.; Martin, G.J.O. The effects of acidic and thermal pretreatment on the mechanical rupture of two industrially relevant microalgae: Chlorella sand Navicula sp. *Algal Res.* **2015**, *7*, 5–10. [CrossRef]
166. Xia, A.; Cheng, J.; Lin, R.; Lu, H.; Zhou, J.; Cen, K. Comparison in dark hydrogen fermentation followed by photo hydrogen fermentation and methanogenesis between protein and carbohydrate compositions in Nannochloropsis oceanica biomass. *Bioresour. Technol.* **2013**, *138*, 204–213. [CrossRef] [PubMed]
167. Yang, Z.; Guo, R.; Xu, X.; Fan, X.; Li, X. Enhanced hydrogen production from lipid-extracted microalgal biomass residues through pretreatment. *Int. J. Hydrog. Energy* **2010**, *35*, 9618–9623. [CrossRef]
168. Adam, F.; Abert-Vian, M.; Peltier, G.; Chemat, F. 'Solvent-free' ultrasound-assisted extraction of lipids from fresh microalgae cells: A green, clean and scalable process. *Bioresour. Technol.* **2012**, *114*, 457–465. [CrossRef]
169. Zabed, H.M.; Akter, S.; Yun, J.; Zhang, G.; Awad, F.N.; Qi, X.; Sahu, J.N. Recent advances in biological pretreatment of microalgae and lignocellulosic biomass for biofuel production. *Renew. Sustain. Energy Rev.* **2019**, *105*, 105–128. [CrossRef]
170. Gaurav, N.; Sivasankari, S.; Kiran, G.S.; Ninawe, A.; Selvin, J. Utilization of bioresources for sustainable biofuels: A Review. *Renew. Sustain. Energy Rev.* **2017**, *73*, 205–214. [CrossRef]
171. Lu, Z.; Xu, Y.; Peng, L.; Liang, C.; Liu, Y.; Ni, B.J. A two-stage degradation coupling photocatalysis to microalgae enhances the mineralization of enrofloxacin. *Chemosphere* **2022**, *293*, 133523. [CrossRef]

172. Eppink, M.H.M.; Olivieri, G.; Reith, H.; van den Berg, C.; Barbosa, M.J.; Wijffels, R.H. From current algae products to future biorefinery practices: A review. *Adv. Biochem. Eng. Biotechnol.* **2019**, *166*, 99–123. [CrossRef] [PubMed]
173. Boussetta, N.; Lesaint, O.; Vorobiev, E. A study of mechanisms involved during the extraction of polyphenols from grape seeds by pulsed electrical discharges. *Innov. Food Sci. Emerg. Technol.* **2013**, *19*, 124–132. [CrossRef]
174. Production and Characterization of Algae Extract from Chlamydomonas Reinhardtii. Available online: https://www.scielo.cl/scielo.php?script=sci_arttext&pid=S0717-34582014000100003 (accessed on 12 January 2014).
175. Halim, R.; Hill, D.R.; Hanssen, E.; Webley, P.A.; Blackburn, S.; Grossman, A.R.; Posten, C.; Martin, G.J. Towards sustainable microalgal biomass processing: Anaerobic induction of autolytic cell-wall self-ingestion in lipid-rich Nannochloropsis slurries. *Green Chem.* **2019**, *21*, 2967–2982. [CrossRef]
176. Orr, V.C.A.; Plechkova, N.v.; Seddon, K.R.; Rehmann, L. Disruption and Wet Extraction of the Microalgae Chlorella vulgaris Using Room-Temperature Ionic Liquids. *ACS Sustain. Chem. Eng.* **2016**, *4*, 591–600. [CrossRef]
177. Wang, M.; Yuan, W.; Jiang, X.; Jing, Y.; Wang, Z. Disruption of microalgal cells using high-frequency focused ultrasound. *Bioresour. Technol.* **2014**, *153*, 315–321. [CrossRef]
178. Ruiz-Domínguez, M.C.; Fuentes, J.L.; Mendiola, J.A.; Cerezal-Mezquita, P.; Morales, J.; Vílchez, C.; Ibáñez, E. Bioprospecting of cyanobacterium in Chilean coastal desert, Geitlerinema smolecular identification and pressurized liquid extraction of bioactive compounds. *Food Bioprod. Process.* **2021**, *128*, 227–239. [CrossRef]
179. Wang, Q.; Oshita, K.; Takaoka, M.; Shiota, K. Influence of water content and cell disruption on lipid extraction using subcritical dimethyl ether in wet microalgae. *Bioresour. Technol.* **2021**, *329*, 124892. [CrossRef]
180. Microalgae: How Tiny, Single-Celled Organisms Can Revolutionise the World—ADVISIAN. Available online: https://www.advisian.com/en/global-perspectives/microalgae-revolutionise-the-world (accessed on 1 January 2020).
181. Khan, M.I.; Shin, J.H.; Kim, J.D. The promising future of microalgae: Current status, challenges, and optimization of a sustainable and renewable industry for biofuels, feed, and other products. *Microb. Cell Factories* **2018**, *17*, 36. [CrossRef]
182. Mahesar, S.A.; Sherazi, S.T.H.; Abro, K.; Kandhro, A.; Bhanger, M.I.; van de Voort, F.R.; Sedman, J. Application of microwave heating for the fast extraction of fat content from the poultry feeds. *Talanta* **2018**, *75*, 1240–1244. [CrossRef]
183. Han, S.F.; Jin, W.B.; Tu, R.J.; Wu, W.M. Biofuel production from microalgae as feedstock: Current status and potential. *Crit. Rev. Biotechnol.* **2015**, *35*, 255–268. [CrossRef] [PubMed]
184. González-González, L.M.; Astals, S.; Pratt, S.; Jensen, P.D.; Schenk, P.M. Impact of osmotic shock pre-treatment on microalgae lipid extraction and subsequent methane production. *Bioresour. Technol. Rep.* **2019**, *7*, 100214. [CrossRef]
185. Gruber-Brunhumer, M.R.; Jerney, J.; Zohar, E.; Nussbaumer, M.; Hieger, C.; Bromberger, P.; Bochmann, G.; Jirsa, F.; Schagerl, M.; Obbard, J.P.; et al. Associated effects of storage and mechanical pre-treatments of microalgae biomass on biomethane yields in anaerobic digestion. *Biomass Bioenergy* **2016**, *93*, 259–268. [CrossRef]
186. Ometto, F.; Quiroga, G.; Pšenička, P.; Whitton, R.; Jefferson, B.; Villa, R. Impacts of microalgae pre-treatments for improved anaerobic digestion: Thermal treatment, thermal hydrolysis, ultrasound and enzymatic hydrolysis. *Water Res.* **2014**, *65*, 350–361. [CrossRef] [PubMed]
187. Okoro, V.; Azimov, U.; Munoz, J.; Hernandez, H.H.; Phan, A.N. Microalgae cultivation and harvesting: Growth performance and use of flocculants—A review. *Renew. Sustain. Energy Rev.* **2019**, *1093*, 11564. [CrossRef]
188. Mehrabadi, A.; Craggs, R.; Farid, M.M. Biodiesel production potential of wastewater treatment high rate algal pond biomass. *Bioresour. Technol.* **2016**, *221*, 222–233. [CrossRef]
189. Singh, S.P.; Singh, P. Effect of CO2 concentration on algal growth: A review. *Renew. Sustain. Energy Rev.* **2014**, *38*, 172–179. [CrossRef]
190. Zhu, L.D.; Li, Z.H.; Hiltunen, E. Strategies for Lipid Production Improvement in Microalgae as a Biodiesel Feedstock. *BioMed Res. Int.* **2016**, *2016*, 8792548. [CrossRef]
191. Dickinson, S.; Mientus, M.; Frey, D.M.; Amini-Hajibashi, A.; Ozturk, S.; Shaikh, F.; Sengupta, D.; El-Halwagi, M.M. A review of biodiesel production from microalgae. *Clean Technol. Environ. Policy* **2017**, *19*, 637–668. [CrossRef]
192. Cordeiro, R.S.; Vaz, I.C.D.; Magalhães, S.M.S.; Barbosa, F.A.R. Effects of nutritional conditions on lipid production by cyanobacteria. *An Da Acad. Bras. De Ciências* **2021**, *89*, 2017–2031. [CrossRef] [PubMed]
193. Passos, I.F.F.; Carretero, J. Comparing pretreatment methods for improving microalgae anaerobic digestion: Thermal, hydrothermal, microwave and ultrasound. *Chem. Eng. J.* **2015**, *279*, 667–672. [CrossRef]
194. Komaki, H.; Yamashita, M.; Niwa, Y.; Tanaka, Y.; Kamiya, N.; Ando, Y.; Furuse, M. The effect of processing of Chlorella vulgaris: K-5 on in vitro and in vivo digestibility in rats. *Anim. Feed. Sci. Technol.* **1998**, *70*, 363–366. [CrossRef]
195. Cho, S.C.; Choi, W.; Oh, S.; Lee, C.; Seo, Y.; Kim, J.; Song, C.; Kim, G.; Lee, S.; Kang, D.; et al. Enhancement of lipid extraction from marine microalga, Scenedesmus associated with high-pressure homogenization process. *J. Biomed. Biotechnol.* **2012**, *2012*, 359432. [CrossRef] [PubMed]
196. Chandler, D.P.; Brown, J.; Bruckner-Lea, C.J.; Olson, L.; Posakony, G.J.; Stults, J.R.; Valentine, N.B.; Bond, L.J. Continuous spore disruption using radially focused, high-frequency ultrasound. *Anal. Chem.* **2001**, *84*, 3773–3789. [CrossRef]
197. Taylor, M.T.; Belgrader, P.; Furman, B.J.; Pourahmadi, F.; Kovacs, G.T.A.; Northrup, M.A. Lysing bacterial spores by sonication through a flexible interface in a microfluidic system. *Anal. Chem.* **2001**, *73*, 492–496. [CrossRef]
198. Fykse, E.M.; Olsen, J.S.; Skogan, G. Application of sonication to release DNA from Bacillus cereus for quantitative detection by real-time PCR. *J. Microbiol. Methods* **2003**, *55*, 1–10. [CrossRef]

199. Pan, X.; Niu, G.; Liu, H. Comparison of microwave-assisted extraction and conventional extraction techniques for the extraction of tanshinones from Salvia miltiorrhiza bunge. *Biochem. Eng. J.* **2002**, *12*, 71–77. [CrossRef]
200. Ryckebosch, E.; Muylaert, K.; Foubert, I. Optimization of an Analytical Procedure for Extraction of Lipids from Microalgae. *J. Am. Oil. Chem. Soc.* **2012**, *89*, 189–198. [CrossRef]
201. Luo, J.M.; Xiao, X.; Luo, S.L. Biosorption of cadmium(II) from aqueous solutions by industrial fungus Rhizopus cohnii. *Trans. Nonferrous Met. Soc. China* **2010**, *20*, 1104–1111. [CrossRef]
202. Ko, C.H.; Wang, Y.N.; Chang, F.C.; Lee, C.Y.; Chen, W.H.; Hwang, W.S. Energy analysis of bioethanols produced from dendrocalamus latiflorus. *Chem. Eng. Trans.* **2013**, *34*, 109–114. [CrossRef]
203. Cheng, J.J.; Timilsina, G.R. Status and barriers of advanced biofuel technologies: A review. *Renew. Energy* **2011**, *36*, 3541–3549. [CrossRef]
204. González-Fernández, C.; Ballesteros, M. Linking microalgae and cyanobacteria culture conditions and key-enzymes for carbohydrate accumulation. *Biotechnol. Adv.* **2012**, *30*, 1655–1661. [CrossRef]
205. Ghosh, A.; Khanra, S.; Mondal, M.; Halder, G.; Tiwari, O.N.; Saini, S.; Bhowmick, T.K.; Gayen, K. Progress toward isolation of strains and genetically engineered strains of microalgae for production of biofuel and other value added chemicals: A review. *Energy Convers. Manag.* **2016**, *113*, 104–118. [CrossRef]
206. Brown, R.E.; Bartoletti, D.C.; Harrison, G.I.; Gamble, T.R.; Bliss, J.G.; Powell, K.T.; Weaver, J.C. Multiple-pulse electroporation: Uptake of a macromolecule by individual cells of Saccharomyces cerevisiae. *J. Electroanal. Chem.* **1992**, *2343*, 235–245. [CrossRef]
207. Ehimen, E.A.; Holm-Nielsen, J.-B.; Poulsen, M.; Boelsmand, J.E. Influence of different pre-treatment routes on the anaerobic digestion of a filamentous algae. *Renew. Energy* **2013**, *50*, 476–480. [CrossRef]
208. Córdova, O.; Passos, F.; Chamy, R. Enzymatic Pretreatment of Microalgae: Cell Wall Disruption, Biomass Solubilisation and Methane Yield Increase. *Appl. Biochem. Biotechnol.* **2019**, *189*, 787–797. [CrossRef] [PubMed]
209. Scopus—Document Details—Null Signed in. Available online: https://www.scopus.com/record/display.uri?eid=2-s2.0-85098242260&origin=inward&txGid=01fba7859a81fc45a17d6ee7a4eea580&featureToggles=FEATURE_NEW_DOC_DETAILS_EXPORT:1 (accessed on 15 June 2022).
210. McMillan, J.R.; Watson, I.A.; Ali, M.; Jaafar, W. Evaluation and comparison of algal cell disruption methods: Microwave, waterbath, blender, ultrasonic and laser treatment. *Appl. Energy* **2013**, *103*, 128–134. [CrossRef]
211. He, S.; Fan, X.; Luo, S.; Katukuri, N.R.; Guo, R. Enhanced the energy outcomes from microalgal biomass by the novel biopretreatment. *Energy Convers. Manag.* **2017**, *135*, 291–296. [CrossRef]
212. Chen, Z.; Li, Y.; Wang, L.; Liu, S.; Wang, K.; Sun, J.; Xu, B. Evaluation of the possible non-thermal effect of microwave radiation on the inactivation of wheat germ lipase. *J. Food Process. Eng.* **2017**, *40*, e12506. [CrossRef]

sustainability

MDPI

Review

Solar Energy: Applications, Trends Analysis, Bibliometric Analysis and Research Contribution to Sustainable Development Goals (SDGs)

Khaled Obaideen [1], Abdul Ghani Olabi [1,2,*], Yaser Al Swailmeen [1], Nabila Shehata [3], Mohammad Ali Abdelkareem [1,4,*], Abdul Hai Alami [1], Cristina Rodriguez [5] and Enas Taha Sayed [4]

1 Sustainable Energy & Power Systems Research Centre, RISE, University of Sharjah, Sharjah P.O. Box 27272, United Arab Emirates
2 Mechanical Engineering and Design, School of Engineering and Applied Science, Aston University, Aston Triangle, Birmingham B4 7ET, UK
3 Environmental Science and Industrial Development Department, Faculty of Postgraduate Studies for Advanced Sciences, Beni-Suef University, Beni-Suef 65211, Egypt
4 Chemical Engineering Department, Faculty of Engineering, Minia University, Minya 61517, Egypt
5 School of Computing, Engineering, and Physical Sciences, University of the West of Scotland, Paisley PA1 2BE, UK
* Correspondence: aolabi@sharjah.ac.ae (A.G.O.); mabdulkareem@sharjah.ac.ae (M.A.A.)

Abstract: Over the past decade, energy demand has witnessed a drastic increase, mainly due to huge development in the industry sector and growing populations. This has led to the global utilization of renewable energy resources and technologies to meet this high demand, as fossil fuels are bound to end and are causing harm to the environment. Solar PV (photovoltaic) systems are a renewable energy technology that allows the utilization of solar energy directly from the sun to meet electricity demands. Solar PV has the potential to create a reliable, clean and stable energy systems for the future. This paper discusses the different types and generations of solar PV technologies available, as well as several important applications of solar PV systems, which are "Large-Scale Solar PV", "Residential Solar PV", "Green Hydrogen", "Water Desalination" and "Transportation". This paper also provides research on the number of solar papers and their applications that relate to the Sustainable Development Goals (SDGs) in the years between 2011 and 2021. A total of 126,513 papers were analyzed. The results show that 72% of these papers are within SDG 7: Affordable and Clean Energy. This shows that there is a lack of research in solar energy regarding the SDGs, especially SDG 1: No Poverty, SDG 4: Quality Education, SDG 5: Gender Equality, SDG 9: Industry, Innovation and Infrastructure, SDG 10: Reduced Inequality and SDG 16: Peace, Justice and Strong Institutions. More research is needed in these fields to create a sustainable world with solar PV technologies.

Keywords: Sustainable Development Goals (SDGs); bibliometric analysis; solar energy; large-scale solar PV power plants; residential applications of solar PV; green hydrogen; water desalination

1. Introduction

The sun is a huge sphere of gases emitting light and energy towards our planet, and this energy received from the sun is a form of radiation best known as electromagnetic radiation [1]. It can be measured in the wavelength range of a measuring device, and it is called solar irradiance. The units used to measure this solar irradiance are W/m^2 in the SI system of measuring units [2] and are the radiant energy emitted into the surrounding environment integrated over a specified time. Figure 1 shows the average of this irradiance around the planet. Sun irradiation, solar exposure, solar insolation or insolation and several other terms are all used for this combined solar irradiance.

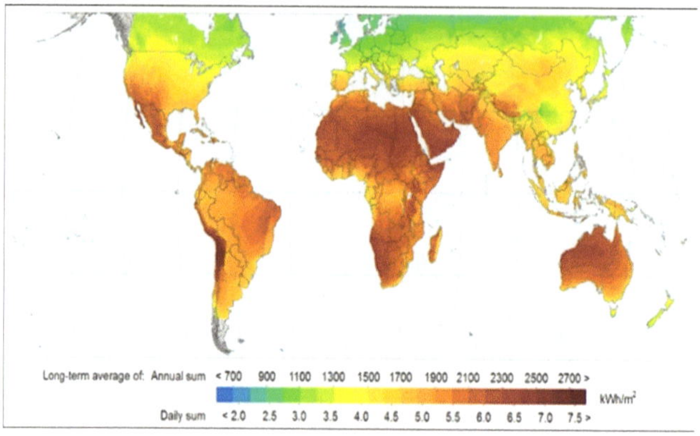

Figure 1. Long-term average irradiance around planet earth (from www.smartenergyconsulting.com, accessed on 13 July 2022).

Solar energy is the term for the energy collected from solar irradiance, and this energy can be in the form of heat (thermal energy), a chemical change or process or even pure electrical energy (electricity) [3,4]. The overall amount of solar energy that strikes the earth is much greater than all its current and future energy demands, so this highly distributed source could meet all of our energy demands if it is properly utilized. Solar energy, in contrast to common sources of energy such as coal, petroleum and natural gas, has become one of the most prominent and environmentally safe energy sources recently, which indicates that it will last millions if not billions of years. The sun is not just a tremendously strong energy source; its light is by far the most abundant source of energy received by the earth. However, its intensity at the surface is quite modest, which is mostly due to the distance between the earth and the sun, causing a vast radial dispersion of radiation along the way. The earth's atmosphere and clouds absorb or scatter more than half of the incoming sunlight [5], resulting in a small extra loss. More than half of the sunlight received from the sun is called visible light, with the remaining consisting of infrared, ultraviolet and other kinds of electromagnetic radiation. The amount of raw energy received from the sun is more than enough to cover all the planet's energy demands thousands of times over [6], so this is why solar energy possesses an incurable potential that needs to be explored thoroughly. Unfortunately, however, despite the fact that solar energy is a free and available almost everywhere on the planet, the high cost of collecting this energy and then converting it and storing it restricts its use in many regions all around the globe. Solar radiation can be turned into thermal energy (heat) or electrical energy, with the former being more straightforward [7], as the heat emitted from the sun is used directly for heating purposes for a long time. However, solar heat (thermal energy) can be used in so many different applications and ways in our current time, as the technology for solar collectors and dryers has improved drastically, as can be seen from the literature [4,8–11].

Renewable energy has proven its ability to supply considerable amounts of energy for all sorts of different locations. Solar, wind and hydroelectric energy supply and consumption should be favored, especially given greater knowledge of the detrimental environmental implications of oil and gas production. The need for renewable energy is expanding all around the world [12], as renewable energy adoption is crucial for the power industry all around the globe, enabling us to finally achieve sustainable energy development in our regions. Solar energy plays a major role in this development, as it can even challenge other renewable energy sources. Table 1 below shows a comparison between solar energy and other renewables [13].

Table 1. Advantages and disadvantages of some renewable energy resources.

Energy	Advantages	Disadvantages
Biomass	• Very versatile in terms of fuel used • Reduces waste • Carbon-neutral	• Presence of impurities • Requires a lot of space • Can cause deforestation
Geothermal energy	• Has huge potential • Effective for heating/cooling purposes • Eco-friendly	• Very restricted in terms of location • High cost
Hydro-energy	• Low emissions • Can generate power during day or night • Safe	• Can have potential consequences on some sea habitats • Expensive
Solar energy	• The most sustainable source, as the sun will last for a long time • Completely clean • No noise generated • Simple • Low-maintenance	• Not available during the night • Highly dependent on the weather • Storage issues • High initial cost
Wind energy	• Cost-effective • Clean energy • Provides jobs	• Can pose a threat to some kinds of wildlife • Very noisy • Wind is not available all the time • High initial cost

This paper aims to provide knowledge about the different applications of solar PV systems and about how they assist with the accomplishment of the Sustainable Development Goals (SDGs). A bibliometric study is also conducted to understand the publication trends of "solar" and its discussed applications, namely "Large Scale Solar PV", "Residential Solar PV", "Green Hydrogen", "Water Desalination" and "Transportation" within the SDGs, while shedding light on what the SDGs lack/require in research in the fields of solar energy and determining the most frequent keywords in each of these fields.

2. Solar PV: The State of the Art

2.1. Fundamentals of Solar PV

Solar photovoltaic (PV) cells directly take sunlight and convert it into electrical energy [14,15], and this phenomenon is known as the PV effect. The science behind these solar PV cells is quite simple; a semiconductor such as silicon is used alongside metal or another semiconductor [16]. As sunlight strikes the junction between them, electrons obtain enough energy to be activated and freed from the semiconductor, thus moving to the other side in one direction, resulting in a small electric voltage in the cell. A single PV cell produces a small amount of power of usually around 2 W [17]. Small solar PV cells are commonly used in small/low-power devices, such as calculators, watches and flashlights. However, these cells can be combined and connected to create a single module commonly known as a solar PV panel. Solar PV panels can be used individually or connected together (in series or parallel) to form an array of panels that produces much larger amounts of electricity [18].

A solar PV system comprises several components, and the PV panel/array is only one of them. Figure 2 shows some other various components that are used, such as a battery in the case of off-grid PV systems used to store the produced energy, alongside a charge controller to regulate and control the charge and to protect the battery. An inverter is also used convert the DC (direct current) produced by the solar panels into the AC (alternating current), which is used in all houses.

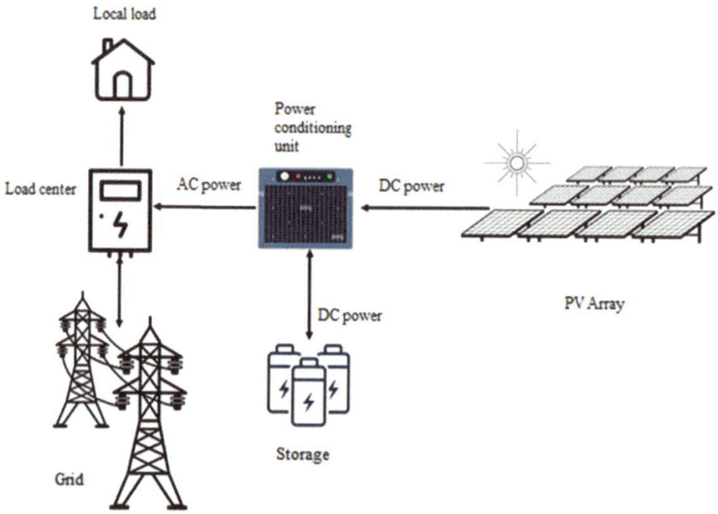

Figure 2. Schematic of an on-grid solar system.

Commonly used solar PV panels have an energy efficiency percentage ranging between 16% and 22%, which is quite low, forcing the use of a higher number of panels/cells to produce enough power output to support the grid [19]. Much research is being conducted to increase the output efficiency of solar PV cells, and their technology has advanced in the past few years, which, in return, is very promising.

2.2. Different Generations of Solar PV

Solar PV technology can be categorized into three main sets [20], which are commonly known as the first, second and third generations of solar PV, determined by their technologies and market entry time. It is very important to understand and distinguish between different types of solar PV cells when choosing one over the other, as this can be influenced by numerous factors such as the cost of the cells, efficiency, materials and even geographical location. Table 2 presents a summary of the different types and generations of solar PV technology [21–32].

2.2.1. The First Generation of Solar PV Technology

The vast majority of today's solar PV cells use silicon as a semiconducting material [33], mainly monocrystalline silicon or polycrystalline silicon. These cells are known for their high efficiency rates and long lifespan, making them a very valuable option for residential and commercial use.

Monocrystalline Silicon Solar Cells

The most common types of solar cells and the oldest of them are monocrystalline silicon solar cells. These cells comprise a single pure silicon crystal. Because of their purity, these cells are characterized by a high efficiency rate (usually above 20%), but their purity is also the reason why they are one of the most expensive cells [34]. However, huge improvements in solar PV technologies over the past years, as well as the drop in raw silicon prices and the rise of newer PV cells, have caused their prices to drop significantly. It is also worth mentioning that silicon-based PV cells are more sensitive to higher temperatures, which can lead to a drop in their overall efficiency.

Polycrystalline Silicon Solar Cells

These cells are produced with a faster method compared with the previous one. By melting raw silicon together [35], PV cells that are made using this way are cheaper in comparison to mono-silicon cells, but they are also a little less efficient and have a shorter lifespan, which are important factors to be considered.

2.2.2. The Second Generation of Solar PV Technology

The second generation of solar PV cells refers to what is known as thin film solar cells [36]. They are typically made from very thin layers (only a few micrometers thick) of polycrystalline semiconducting material. These solar cells are cheap and very easy to construct because less of the material is used for their production. They are also characterized by their outstanding flexibility and light weight, which make them optimal for portable needs. Thin-film solar cells are also very common in buildings and small PV systems.

Amorphous Silicon Solar Cells

Amorphous silicon solar cells are one of the main thin film cells that are commonly used today; these cells are found in very small devices, such as calculators. Usually made with a one-micrometer thickness [37], these very thin cells are very flexible and great for curved surface usage, and they also function very well in low light. However, their disadvantage is their low efficiency rate of around only 7%.

Cadmium Telluride Solar Cells

Cadmium telluride solar cells have gained huge popularity in PV technology, as they can be produced expeditiously with very low prices. In addition to that, these cells have good light absorption and minimal energy losses, making them a great option overall, but the shortcoming of this technology is that cadmium is a known toxic material, which has made many people concerned regarding the use of this type of solar PV panel [16].

Copper Indium Gallium Diselenide Solar Cells

Copper indium gallium diselenide solar cells are another type of thin film solar cells [38] in use. The technology behind such cells is quite complex, and their materials are hard to work with, making it very difficult to withstand the competition against silicon-based PV cells. Copper indium gallium diselenide solar PV cells have higher costs compared with amorphous silicon and cadmium telluride cells.

2.2.3. The Third Generation of Solar PV Technology

Solar PV technology has advanced tremendously since its first generation appeared, and it is still advancing every day as new research and experiments are conducted in this field all around the globe, leading to a new generation of solar PV cells that is still mostly under research and development. This makes them unavailable for commercial use [33]. There are many different types and technologies of what can be classified as a third-generation solar PV cell. One example is an organic PV (OPV) cell, in which an organic semiconductor is used as the light absorbent layer, and the organic semiconductor is usually either a very small molecular or a polymer. Another type of third-generation PV cell is called a concentrated PV (CPV) cell. They work like any conventional PV cell, but they consist of multiple junctions and can reach a very high efficiency level of up to 40% [39]. In addition to that, CPV cells can withstand high temperatures and can still be three times more efficient than traditional solar PV systems. To reach such high efficacy rates, CPV systems are integrated with lenses, tracking devices and cooling systems, which lead to large capital costs in most cases. CPV cells are already available on a large scale, with a capacity of 27 kW in Malaysia [40], 480 kW in USA [41] and 75 kW in Japan [42], as well as in several sites around the world.

Table 2. Comparison of different generations of solar PV cells.

Generation	Type	Efficiency	Advantages	Disadvantages
First Generation	Monocrystalline silicon	Up to 24% [43]	• High efficiency • Long lifetime	• High cost
	Polycrystalline silicon	13–20% [43]	• Lower cost	• Lower efficiency
Second Generation	Amorphous silicon	5–10% [14]	• Lower cost • Flexible • Ease of production	• Shorter lifetime • Lower efficiency
	Cadmium Telluride	18–22% [44]	• Lower cost • High absorption	• Toxic
	Copper Indium Gallium Diselenide	15–22% [45]	• Higher heat resistance	• Higher cost
Third Generation	Organic PV	Up to 17% [46]	• Lightweight • Eco-friendly	• Lower efficiency • Shorter lifetime
	Concentrated PV	40% [47]	• Very high efficiency • Can withstand high temperatures	• Very high cost • Must be integrated with solar tracking systems and cooling devices to reach high efficiency

3. Applications of Solar PV Technology

The pursuit of a green and sustainable energy source has quickly become one of the world's greatest challenges, especially in this modern era. This has mainly occurred as a result of the abrupt debilitation of conventional energy resources, such as oil and fossil fuels, as well as global environmental climate changes, such as global warming and the ever-increasing demand for power and energy. Among various sustainable and renewable energy technologies, solar PV cells are the most common and mature green energy systems [48] used to counter the world's growing need for energy. Like any other energy technology, solar PV cells are not perfect and face many challenges when integrated. However, with their fast advancement in technology, solar PV cells have managed to be used in various applications in our modern time, with some systems producing as little as a few watts (W) and others producing megawatts (MW) of power every day.

3.1. *Large-Scale Solar PV Power Plants*

Solar PV technology has gained wide popularity in the last few years in large-scale power generation, making it a key factor in the power sector around the globe. Solar PV power plants (also known as solar farms) function very similarly to how a small domestic PV system does; however, solar PV power plants are usually connected to the power grid and are known as grid-connected centralized PV plants [49,50]. These power plants feed a large bulk of energy to the associated grid through the centralized generation of energy. A PV energy generation system is usually considered a power plant when its total power output is no less than 1 MW. However, some power plants are rated with a power generation output of dozens of MW to several hundred MW [51]. Such systems are designed and regulated with varying complexity regarding their environmental surroundings and geographical location in order to obtain the highest possible performance, all while maintaining a desirable balance between the performance and the cost of the system. For instance, in 2022, the United Arab Emirates had solar PV power plants of a total capacity of 950 MW [52].

3.1.1. Components of Solar PV Power Plants

PV power plants consist of several components, as shown in Figure 3, such as the cells, mounting, connections (both mechanical and electrical) and many others [53]. The appropriate selection of these components plays a major role in the design of the system, and the most important components are shown below.

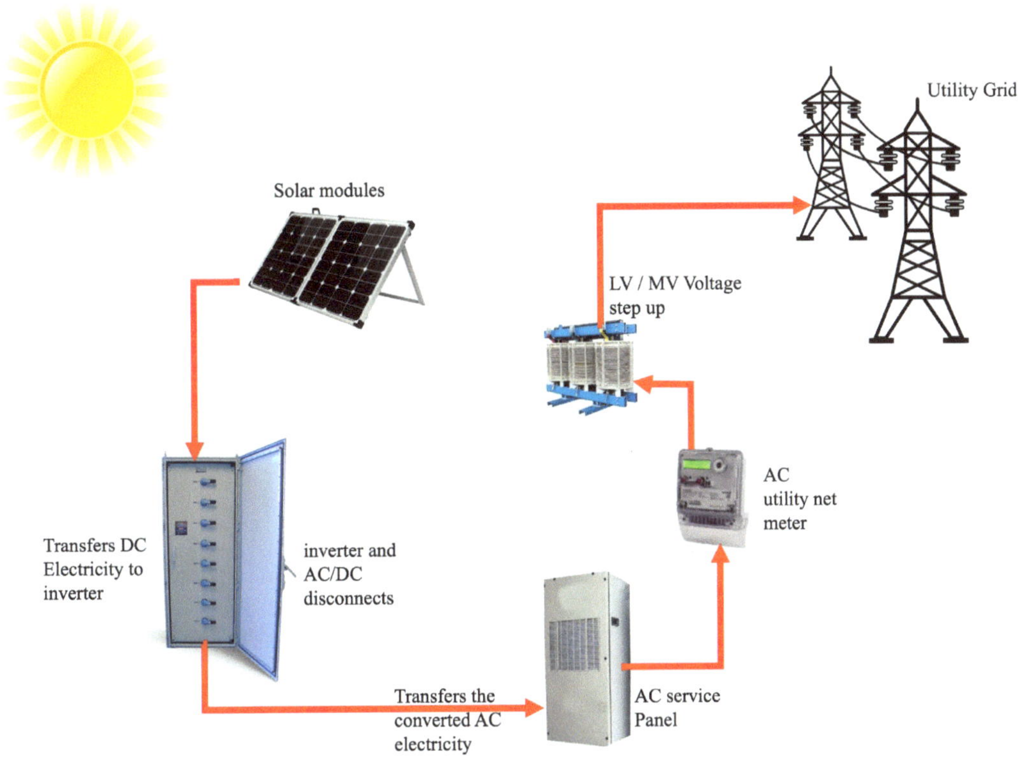

Figure 3. Main components of a solar power plant.

PV Modules

Solar PV power plants comprise a large number of PV cells that are connected together, thus making PV modules which, in return, are also connected together, making PV strings that are able to generate the desired power output. Solar PV modules are the core of a PV plant, and their sole function is to take the solar irradiance coming directly from the sun and convert it into electrical energy through this process, known as the PV effect. The generated electrical output is in the form of DC current. Solar PV modules have the highest cost of components in a PV power plant [54], which is usually due to the large number of modules that are used.

The site of the PV power plant plays a major role in determining what kind of solar PV module to use, and there are many important factors [55] to consider here regarding the surrounding environment. The three most important environmental factors are: (1) the temperature, (2) distance to shore and (3) dust and snow.

Mounting and Tracking

In a PV power plant, a mounting structure is required for the panels to keep them attached and supported in a secure manner while keeping them facing the correct direction for maximum power generation. These structures can be either a fixed system or a tracking system. In a fixed mounting system, the solar PV panels are oriented in a fixed direction, with a fixed tilt angle all the time. It is crucial to choose the correct angle and direction in order to obtain the maximum power output possible from the total solar irradiance on the solar panels. Unlike fixed mounting systems, tracking systems allow the panels to follow the sun as it moves during the day, and solar tracking systems are essentially the only parts that move in a solar PV power plant. Tracking systems help solar PV panels generate more energy than fixed systems due to them following the direction of direct sunlight, and trackers can help PV power plants generate up to 37% more energy output than systems without trackers [56].

Inverters

An inverter is used to convert DC electricity coming from the solar panels into AC electricity to feed into the grid. Inverters are classified into two main types: central inverters and string inverters [57]. For central inverters, a whole block (multiple strings) of solar panels is connected to a single central inverter, thus converting a huge amount of power, whereas string inverters use a separate inverter for each string of panels.

Transformers

These devices are critical, and they are required in PV power plants. As the output voltage coming from the inverters is usually lower than that of the AC grid, this is where transformers are used to increase the voltage to match the required voltage of the grid so that, when the plant is connected to the distribution grid, electricity can be directly pumped into it. When the plant is connected to a transmission network, the grid transformer is necessary to increase the voltage even more.

Grid Interface

The power is transferred into the grid network from this point. The power plant's substation also includes necessary grid interface switch-gear, such as circuit breakers and disconnects that are used to shut down the system in cases of faults, as well as generation and supply metering equipment for the PV power plant's safety and isolation.

3.2. *Residential Applications for Solar PV Systems*

Solar PV systems are not limited to utility-scale power generation; solar PV systems have gained usage in many households all across the world, proving that solar power is an effective source of energy. Solar PV modules have grown in popularity in recent years due to the numerous advantages and opportunities they provide [19]. A rooftop solar PV system, as shown in Figure 4, refers to a household or building that has solar PV panels installed and mounted on its roof for electrical energy generation purposes. Rooftop-mounted solar PV systems are very modest in terms of size and output compared to utility-scale solar PV power plants with capacities in the MW range. Residential rooftop solar PV systems normally have a capacity of 5–20 kW [58], whereas commercial rooftop PV systems can easily reach 100 KW to 1 MW depending on the size of the building [59], the available space and the number of installed solar modules. Larger industrial-scale PV systems ranging from 1 to 10 MW [60] can be installed on big rooftops, such as large factory roofs, etc. Although rooftop solar PV systems can vary in terms of design and used components, all of these systems require a base set of components, which are PV solar panels used to convert sunlight into electricity, mounts, clamps used to fix and install the modules on a set of racks, inverters for DC/AC conversion and batteries for storage (if required), cables and other possible accessories.

There are two main ways for households to adopt solar PV into them, either by using a grid-connected PV system or a stand-alone PV system [61]. Grid-connected systems heavily depend on a nearby power grid to remain connected to a power source at all times. This means that, if the building requires more power than the solar panels have produced, if a problem occurs in the PV system that prevents it from generating electricity, or even if power is used during the night when there is no sunlight, the grid can provide the necessary power. Grid-connected PV systems are also great for areas that usually have cloudy and dusty weather situations. However, if a residential solar panel system manages to produce more electricity in a day than that which the household uses, the excess energy can then be transferred to the nearby power grid, which provides support to it and relieves the load of the grid during peak hours, creating a solid network of energy between residential areas and their associated grid. This is why grid-connected solar power systems account for the majority of rooftop PV installations, with so many governments and utility companies adopting this method. Additionally, they provide attractive Net Metering and Feed-in Tariff programs that can help consumers [62] offset their power expenses. Grid-connected solar PV systems have many advantages, but what makes them unique is the fact that they are less complicated to install. They do not require a battery for storage, as the extra energy goes directly to the grid, making these systems provide the benefit of maximizing the usage of generated energy due to the absence of storage losses. However, stand-alone PV systems have no direct connection to the power grid. As shown in Figure 5, these systems are built to take the electricity generated by the solar panels and store it inside batteries, and the stored electricity from these batteries is used to power all the gadgets connected to it. Stand-alone systems are usually utilized in places where there is no access to the grid [63], but that is not the only case, as many people today use these systems in conjunction with the grid. The advantage of these systems lies in their ability to use them in the nighttime when solar panels do not generate any power. However, stand-alone PV systems are often more expensive than grid-connected PV systems due to the high cost of batteries [64,65]. Until now, the most popular kind of batteries for residential PV has been lead–acid batteries, which are a form of electro-chemical storage [66]. They has solely been employed in off-grid solar applications because of their exorbitant pricing, but with power rates rising in many areas around the globe and battery prices falling, viable battery storage solutions for grid-connected homes are developing.

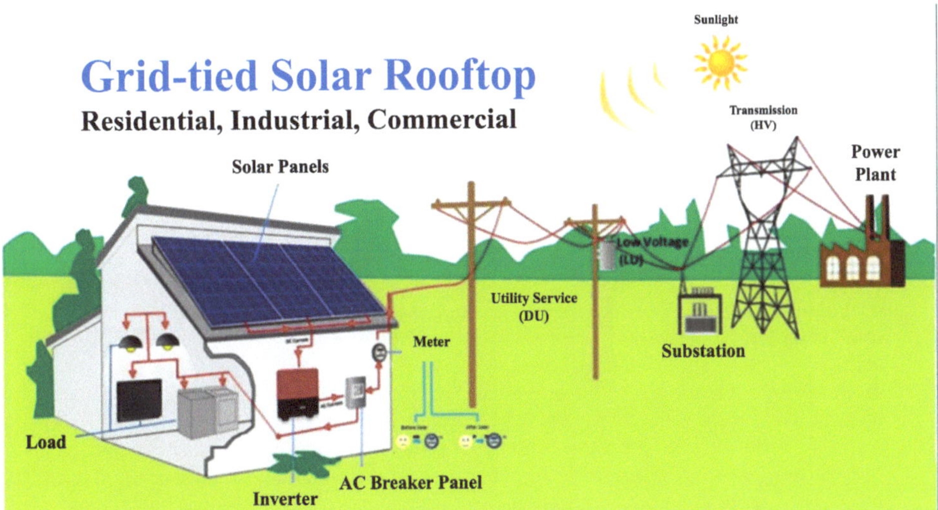

Figure 4. Grid-connected solar rooftop system (from www.solarideatspl.com, accessed on 13 July 2022).

From the standpoint of global climate change mitigation and sustainable energy supply, this breakthrough is critical. Residential battery storage systems with solar PV and smart inverter technologies can modify this paradigm, allowing users to not only adjust when they use power but also to lower the amount of electricity they need from the utility grid or to even completely unplug from it.

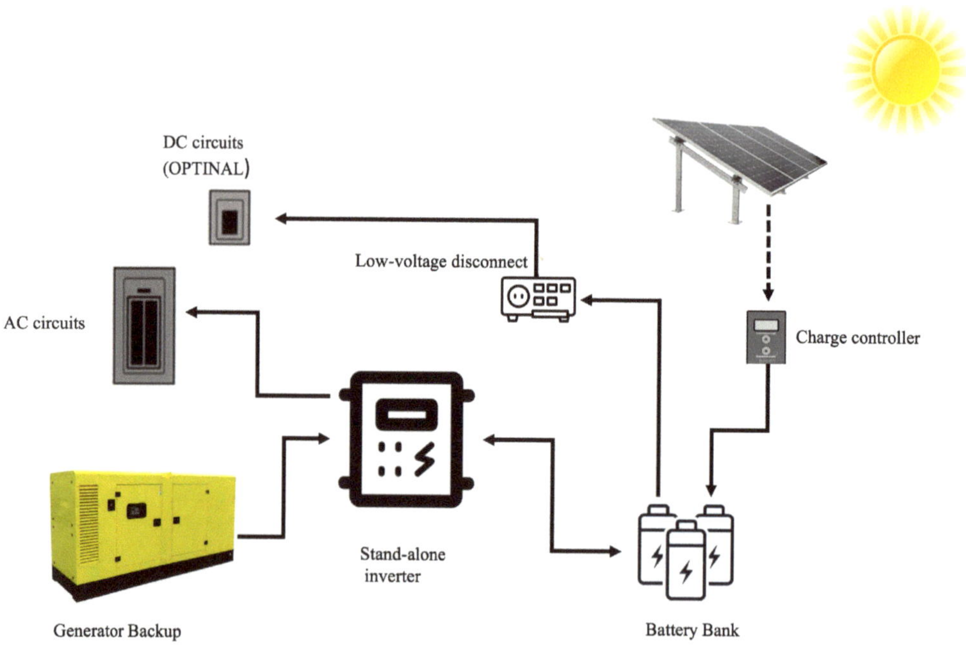

Figure 5. Rooftop stand-alone PV system with batteries.

3.2.1. PV-Powered "Heating, Ventilation and Air Conditioning" (HVAC) Units

In hot and humid areas, such as the Middle East, air conditioning systems are critical for ensuring thermal comfort in all interior spaces. Air conditioning, which includes cooling and de-humidification, especially during summertime [67], has experienced a significant rise in power demand due to the widespread usage of air conditioning equipment, which has eventually led to high stress levels on the grid, especially during peak hours. As a result of the high match between the diurnal cooling load and the availability of solar irradiance, using solar PV systems for domestic air conditioning is a very appealing application, as it has the power to supply the chosen indoor area with the needed cooling capacity. Solar PV cooling systems can employ electricity generated by PV cells from residential PV systems to power a traditional vapor compression air conditioning system. Despite PV systems being simple, their applications have been hampered for years by the high cost of solar PV modules alongside the rest of the accompanying air conditioning equipment. However, with the rapid decrease in the cost of solar PV modules and the rapid increase in energy usage for cooling/heating purposes in recent years, more attention has been drawn to these technologies. There are many advantages of a PV-powered air conditioner, and the most important of them can be seen in its very simple structure, operational stability and rapid reaction to load variations during the day. PV-powered air conditioners also have the ability to offer warmth in the winter under certain climate conditions [68,69], because, in many cases, traditional residential air conditioners can act as heat pumps. Because both the heating/cooling demand and the electricity generated from a PV system vary during the day and are not always the same, whenever the PV system generates extra power, it can be stored in the battery of a stand-alone residential PV system or can even be transferred to the grid with a grid-connected residential PV system [68,70]. Air conditioning energy usage is now at a larger proportion of the total building energy consumption in hot climate countries. In the UAE, it is estimated that the cooling energy used in buildings and homes exceeds 50%, as shown in Figure 6. This is why PV-powered air conditioning systems installed on a large scale might be a useful way to relieve the load on the electric grid during peak hours, saving not just energy but also a huge sum of money over time.

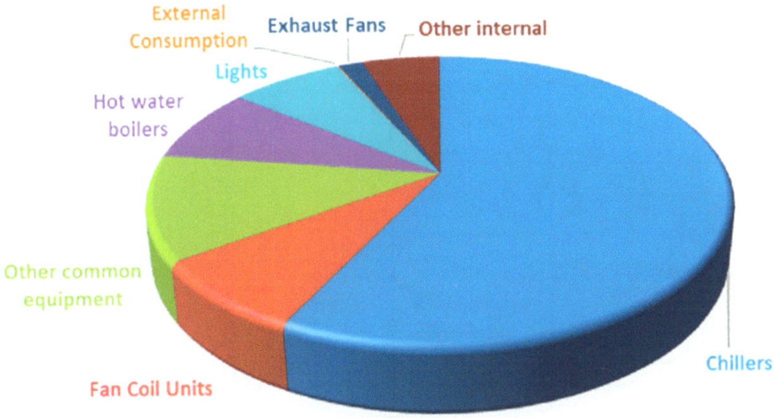

Figure 6. Energy consumption of residential equipment in the UAE [71], open access.

PV-powered air conditioners have three main types based on what kind of electricity they use to function [72,73], which are DC solar air conditioners, AC solar air conditioners and hybrid solar air conditioners. DC solar air conditioners work using direct current (DC), and these air conditioners can be either connected directly to solar PV panels or connected to a battery system charged by PV panels. These air conditioners run completely off the grid, as they cannot function with an alternating current (AC) supply. DC solar air conditioners are quite simple and easy to use; however, if they are used during the nighttime, it should be ensured that there is enough battery storage capacity. In contrast, AC solar air conditioners (shown in Figure 7) use alternating current (AC) as the driving power and require the presence of an inverter. These air conditioners are the most common of the three. AC solar air conditioners may or may not use a battery, but their main advantage is that they can be connected to the grid and run on it either partially or fully. As the name implies, hybrid solar air conditioners function using DC and AC supply, as they are connected to both the solar system and the grid unit at the same time, making them an overall solution that works with any system available, switching from DC to AC or the opposite with total ease. Although these air conditioners work with inverters, they are not necessary components.

3.2.2. PV-Powered Water Pumps

Water is essential for human survival because it is required not only for drinking purposes but also for everyday domestic use and even large-scale applications such as irrigation and hydro-power generation. This is why water plays an important part in any country's growth, and even though fresh water is plentiful, it is not always readily available in certain places. Research indicates that, by 2025, around three billion people will not have access to potable water [74] around the globe. This will raise the need for a pump to take water from deep below earth's surface, where it is needed. Water pumps have been used for decades. However, they are commonly diesel-powered pumps, but due to the unfortunate increase in the price of fuel in the global marketplace, toxic pollutants from its combustion, the cost of maintenance and its limited lifespan, an alternative is highly needed. Different renewable energy sources can be utilized for water pumping, but solar is gaining a lot of popularity because it is accessible almost everywhere on the planet, even in distant locations, reducing reliance on the grid and diesel for pump drives [75]. The concept is very simple. Solar energy is clean and abundant, and solar PV panels turn sunlight directly into usable electrical energy, which is, in return, used to power the water pump directly or via an inverter. Solar-powered water pumps have various benefits over traditional pumping systems [76], such as the fact that gasoline and diesel engines require expensive fuel for long periods of time and pollute the air and noise in clean residential areas, and solar water pumps are ecologically benign and do not require extra fuel to operate, making them cost-effective, environmentally friendly and dependable, with less maintenance and a longer life span.

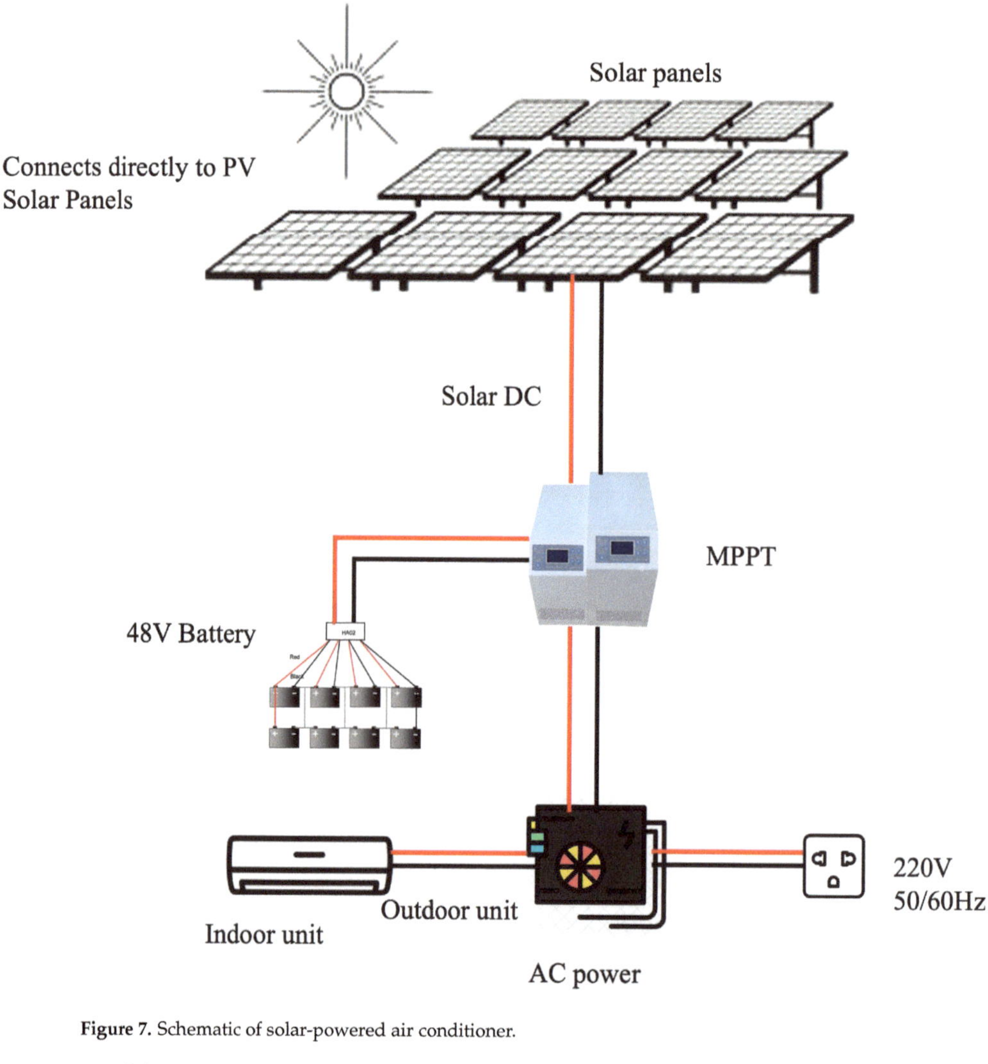

Figure 7. Schematic of solar-powered air conditioner.

Solar water pumps can be developed or designed in a variety of ways, but the majority of these systems function with the same common method, which is to employ PV panels to power water pumps. The system consists of a solar PV system integrated with a power conditioning unit, a hydraulic water pump and a storage tank, as shown in Figure 8. PV panels are sometimes installed with a tracking system for increased efficiency, but it is not a necessity. In addition, because the produced energy is predominantly DC and the pump is mostly AC, an inverter must be utilized in order to convert the output energy to AC before it can power the water pump. The water pump can be submerged in the water or positioned at ground level, depending on what type of solar water pump is used [77]. The water pump delivers the water upward from a water reservoir to a storage tank high above, the reason for which is to allow the storage tank to use gravity to naturally pump water back down whenever needed at no extra cost. The head of the pump is used for the height difference between the tank and the water reservoir level, which is a crucial consideration when choosing a pump to deliver water from the reservoir to the tank.

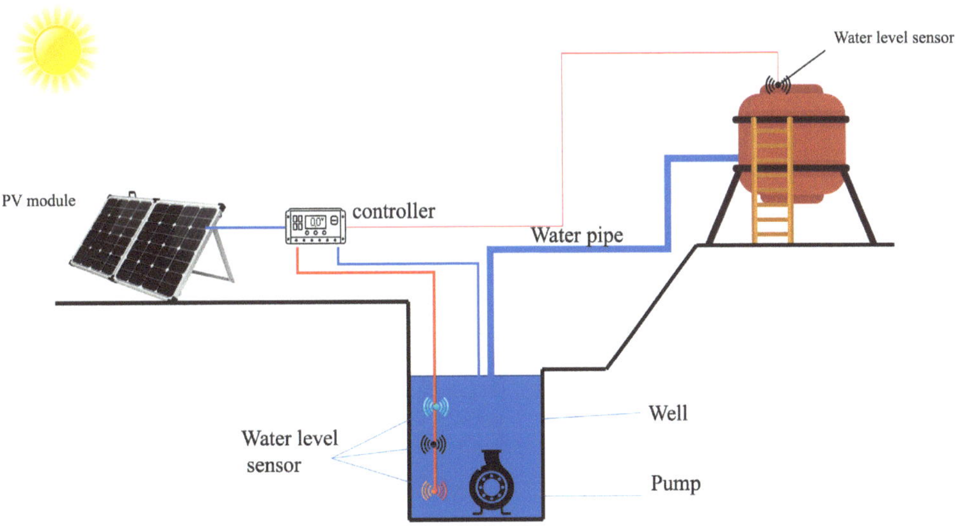

Figure 8. Schematic of a solar-powered water pump.

3.3. Green Hydrogen

Energy storage technologies are undoubtedly a critical and important component of modern-day energy sources, particularly those of which are considered renewable, such as wind and solar. The reason is that renewable energy sources do, in fact, require a storage system to ensure a reliable power supply due to their intermittent nature [78]. Storage systems enable a solution for the maximum demand problem, and from an energy standpoint, energy storage technologies can take numerous shapes and forms. Figure 9 shows the different ways that energy can be stored. Anything that allows for the storage of energy, such as batteries, can be employed as an energy storage system in practice. However, not all storage systems are devices, as green hydrogen is considered a great way to gather and store energy in a safe and clean manner [79].

Hydrogen fuel is a long-term clean energy carrier that can play an important part in today's global energy situation, as it is highly capable of reducing the CO_2 footprints [80] of the transportation and manufacturing industries. Hydrogen fuel is rarely found in its natural condition (H_2) on our planet, and traces of it can be found in our atmosphere, the majority of which is combined with oxygen to form water. Thus, H_2 must be manufactured artificially by humans. Today, hydrogen is primarily derived from hydrocarbons, which, in return, emit a significant quantity of CO_2 into the environment [78]. However, other methods are available, such as water electrolysis. Water electrolysis is a typical method for producing pure hydrogen, in which electricity splits water into its two counterparts, hydrogen and oxygen, in a safe controlled process. Figure 10 shows a simple illustration of water electrolysis. Hydrogen, like gasoline, is an extremely combustible type of fuel; however, unlike gasoline, which spreads horizontally, hydrogen flames soar upward due to its buoyancy, as it is far lighter than the air around it.

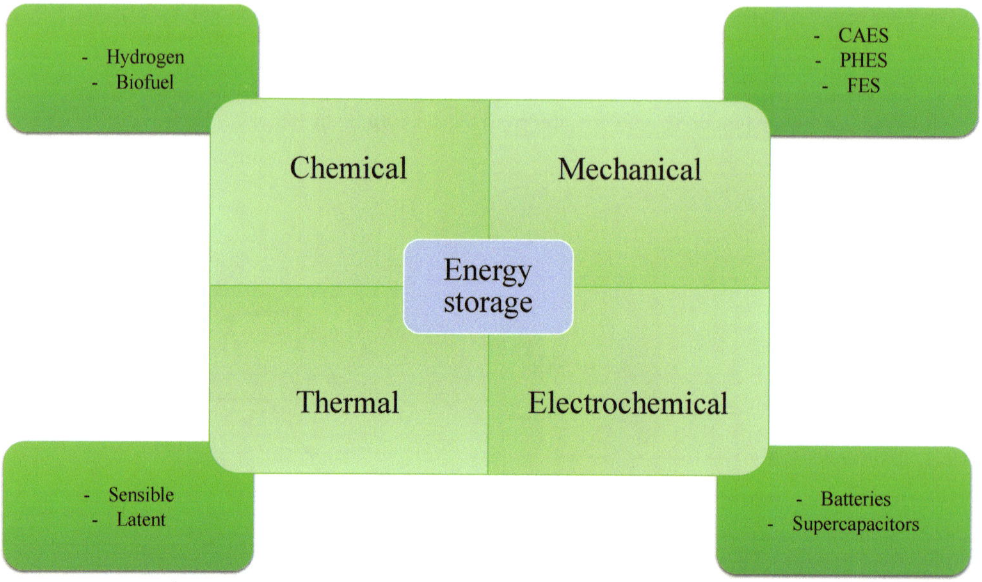

Figure 9. Types of energy storage routes.

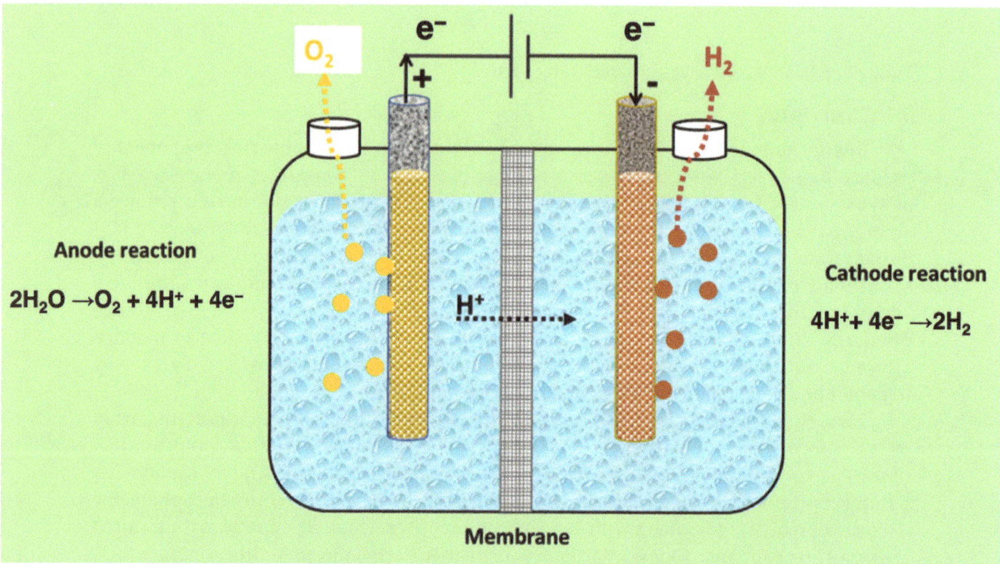

Figure 10. Simple water electrolysis.

Hydrogen can be classified into three main types depending on how it is extracted. The first and most common type of hydrogen is known as gray hydrogen [81], which is considered the least renewable kind of hydrogen because it is obtained from natural gas and is manufactured using fossil fuels. The process is conducted via reforming natural gas, which is a method of altering the molecular structure of hydrocarbons through a catalytic chemical reaction. Methane, which is a major constituent in natural gas, is then mixed with steam at a very high temperature to create hydrogen and carbon dioxide. Gray hydrogen is quite cheap, and it is widely utilized in the chemical sector to create fertilizers and refine oils. However, its shortcoming is the huge amount of CO_2 produced from this process, which is ten times more than the amount of H_2 produced, thus giving it the name gray hydrogen. The second type of hydrogen available is known as blue hydrogen [82], which is produced using the exact method used for producing gray hydrogen, with the difference being how the carbon that is generated as a result is handled. CO_2 produced here is not released to the atmosphere, but it is captured instead and stored in a process called "Carbon Capture and Storage", or CCS for short [83]. Blue hydrogen is now gaining more traction as a viable alternative because it emits less CO_2 than gray hydrogen, making it more environmentally friendly. However, it does not completely reduce carbon emissions into the environment, as it just stores it, making it not free of cost and causing several challenges for blue hydrogen. The third and final type of hydrogen is called green hydrogen [84]. This type of hydrogen is truly a step closer to achieving true sustainability, as it is produced in a completely different manner compared to the previous types of hydrogen. Green hydrogen is produced using water electrolysis, which is a process we discussed earlier, where electrical energy is applied to water in order to separate hydrogen and oxygen molecules from each other. The electricity for this procedure comes from renewable sources [85,86] rather than fossil fuels; thus, the name "green hydrogen" was chosen for it. This method is unique, as it produces a complete loop of sustainable energy with no hazardous gases produced at any step in the manufacturing process, making it the ultimate aim in the hydrogen fuel sector. Several variables are responsible in determining the wholesale cost of green hydrogen, such as, but not limited to, the total expense of the electrical power that is produced from renewable and sustainable energy sources to initiate the electrolysis process. Additionally, the cost of the electrolysis itself, in which hydrogen is produced from water, should be addressed, as it is a critical factor as well.

In 2021, water electrolysis was found to provide only about 4% of the much-needed hydrogen, with the remaining 96% coming primarily from non-renewable sources [87], such as fossil fuels, leaving behind the enormous potential to replace fossil fuels if technological limitations can be overcome. By building a strong connection between energy consumption and production capacity in both decentralized and centralized systems, hydrogen can improve the overall low-carbon energy system reliability. However, even though all the environmental and economic securities of hydrogen fuel in end-use applications seem very promising, the implementation of hydrogen manufacturing techniques, generation and distribution networks, and sales services is difficult, especially with the high costs of electricity. This is why investigations into using solar energy as a prime source for generating required energy are critical to address these challenges for an industrialized system in manufacturing solar green hydrogen.

Figure 11 shows how a solar–electrolyzer system is connected. PV cells and an electrolyzer are linked together in a series of connections, thus resulting in the same operational current and voltage for both of these devices [88]. As result, the current and voltage during this process's operation are determined by the crossing point between them. In PV–electrolysis wired systems, the PV junction is commonly electrically connected to the catalysts of the system. The strength of these systems is not only how they run as a standalone system but also how they can be integrated into residential solar PV systems with total ease, as whenever a residential solar PV rooftop system produces more power than its required residential demand, the excess can then be fed into a water electrolysis cell to produce green hydrogen. Usually, batteries are used in off-grid solar systems to store excess energy for later use, but, as is known, the limited lifespan and poor storage capacity of traditional batteries render them unsuitable in long-term sustainable energy storage, which is why renewable energy storage in the form of hydrogen is much more dependable and efficient than standard battery storage.

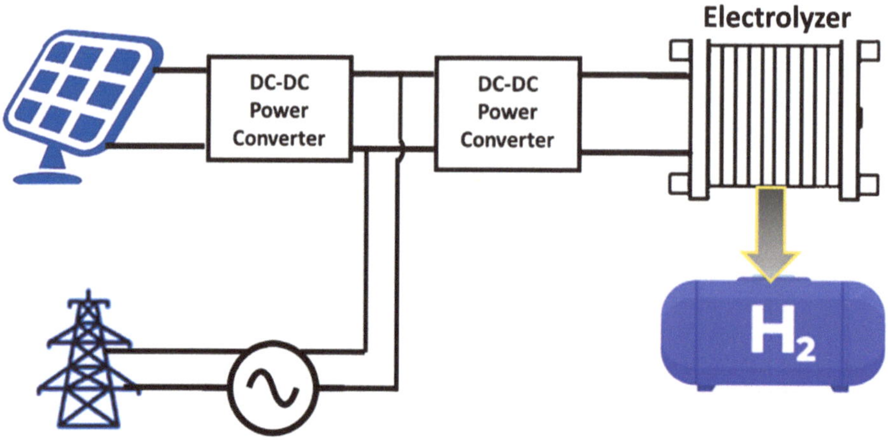

Figure 11. Schematic of solar–electrolyzer system with a storage tank [89], open access.

Producing hydrogen from the excess electricity of residential buildings and homes and utilizing it as an energy storage bank to create electrical energy when the sun is not present is very cost-effective due to lower fuel cell pricing. Solar-produced green hydrogen may be used in a variety of applications on both industrial and residential levels, as well as transportation, making PV green hydrogen an extraordinary idea. Subsequently, solar PV systems can play a critical role in the globe's forthcoming energy sector. This is attributed to solar-to-hydrogen being a way of creating green energy growth, and hydrogen is a long-term energy vessel that can replace non-renewable energy sources while lowering CO_2 emissions, hence mitigating global climate change. Hydrogen generation using ubiquitous renewable solar energy and plentiful water is an eco-conscious answer to the world's growing energy needs and requirements that also provides needed long-term energy stability.

3.4. Water Desalination

Water desalination is a process in which salt is eliminated from salty water to produce potable water [90]. Water desalination provides a great number of advantages, including the capacity to easily utilize any available water resource, such as seawater. It also provides the ability to secure the collected water in dry and coastal locations using techniques that can handle a variety of operating circumstances. However, water desalination comes with several drawbacks as well, such as the required energy for the process, as water desalination requires a lot of energy. There is also the threat of released salts and fuels from the process, as they can impact the associated environment and the living creatures within it. In addition, as is obvious, the system requires a high cost.

One category of water desalination technologies is called membrane separation. This technology was once restricted to municipal water treatment [91] in the past, but as the technology developed over time and new membrane types were created, its applications further expanded to include not only the water industry but also other high-return industries, such as beverage purification processes and chemical processes regarding enzymes and separation. In order to create potable water, this method employs a reasonably permeable membrane to transport either water or salt between two separated zones with various concentrations. The idea of combining solar PV with water desalination systems is a very luxurious one, especially in locations where there is an abundance of both solar irradiance and salt water. There are three main processes that allow the integration of solar PV as a source of power in water desalination systems; these processes are known as reverse osmosis (RO), electrodialysis reversal and membrane distillation [92].

RO is a water filtration method that focuses on separating all the ions, salts and other undesirable particles from saline water in order to produce drinking water [93]. This method is known for using a semi-permeable membrane for the separation. RO is similar to normal osmosis, in which there are two liquids separated by a membrane and the less concentrated liquid tends to go through the membrane to the liquid with the highest concentration in order to eventually become even. However, in RO, pumps are used to apply pressure to counteract osmotic pressure, which, in return, results in forcing all the water to cross the membrane, leaving behind all the salts and solid particles on the other side [94–96]. RO has proven that it can remove not only salts but also a wide range of dissolved materials as well as biological species such as bacteria from water, making RO a method utilized in both global industrial operations as well as potable water production. RO is quite different from filtration, as in filtration, the membrane in use has pores in with a size of 0.01 μm and is dependent on the size of the particles in general. RO uses a membrane of 0.001 μm and depends largely on the pressure applied to the side of the membrane [97,98].

Electrodialysis reversal desalination (EDR) is also a membrane desalination process like RO [99]. However, it differs quite a lot regarding how the salts are separated from the water, as in this process, an electrical current is used to separate all kinds of salt ions from the water through the membrane. This system's main components consist of a desalination unit, a selective membrane, a pump and a power source unit. The electrodes in the desalination unit are submerged in the saltwater solution, which then produces a current that moves the ions through the membrane separating them from the water. The process is then reversed in a periodic manner every set amount of time by altering the polarity of the electrodes, allowing higher amounts of water to be recovered. The electrodes can be made from different kinds of metals, such as niobium and titanium, and are usually coated with platinum coatings, and the membranes of such systems usually last for 5–7 years [100]. Electrodialysis reversal desalination has some advantages over RO, such as a high water recovery rate, which is mainly due to the reverse in polarity as well as the fact that the water can be treated without any chemicals. However, these systems have the disadvantage of the desalination process always being kept under a certain current density limit in order to keep the total efficiency of the system high and prevent the ionization of water.

Finally, there is membrane distillation (MD), which is a hydrophobic membrane-based thermally driven water desalination technique [101]. In this process, a vapor pressure difference is created between the two faces of the membrane pores, which, in turn, drives this desalination process and allows the volatile solution components to transfer their mass and heat. Membrane desalination's simplicity, along with the fact that it can utilize waste heat or even other renewable energy sources, such as solar and geothermal energy, makes it a promising water desalination technique that can be used in conjunction with other processes in hybrid or integrated systems. The vapor pressure differential created by a temperature difference across the membrane provides the driving power for the membrane desalination process. Membrane desalination can be kept at a significantly lower temperature of around 30–60 °C compared to conventional thermal distillation methods because the driving force is not purely thermal [102]. In addition, due to surface tension in the process, the hydrophobic property of the membrane prevents water molecules from entering. Water vapor, on the other hand, produces a pressure differential and passes through the membrane pore system before condensing on the opposite cold side of the membrane. Although membrane desalination systems use heat, solar PV modules can be used as a power source for these systems, as shown in Figure 12, by providing electricity to power up the heaters in the MD system.

3.5. Transportation

The transportation sector presently accounts for roughly 20% of worldwide CO_2 emissions, which come from fuel combustion and internal combustion engines in most of today's road vehicles [103]. Unfortunately, road transportation emissions are continuing to rise as time passes, which is why electric vehicles (EVs) are becoming a more appealing option for reducing CO_2 and other hazardous air pollutants in this industry [104]. An electric vehicle is one that is propelled by one or more electric motors, meaning that EVs can run directly on electricity. Electric vehicles can be powered by a collector system that uses electricity from outside the vehicle, or they can be self-powered with a battery. Figure 13 shows that electric vehicles are not only pure electricity machines, as hybrid EVs exist [105], which can run on both electricity and conventional fossil fuels exactly like internal combustion engine vehicles. EVs are not limited to small cars only, as road and rail vehicles, underwater watercrafts, electric airplanes and spacecrafts are all examples of electric vehicles. There are already more than five million EVs on the road, nearly 67% of which are pure battery EVs. EVs also have better efficiency than conventional cars and can be converted to using low-carbon electricity for fuel. In 2018, EVs accounted for approximately 2.5% of the new vehicle market and are expected to reach 15% by 2030 [106].

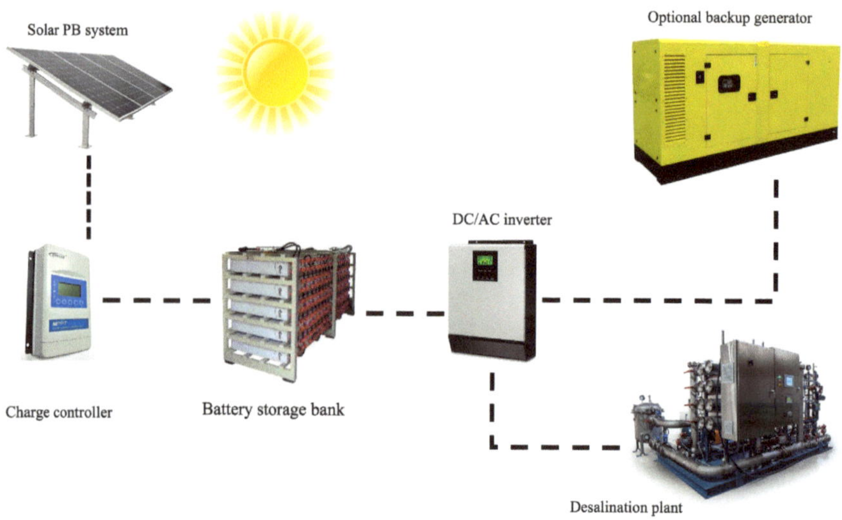

Figure 12. Schematic of solar water desalination system.

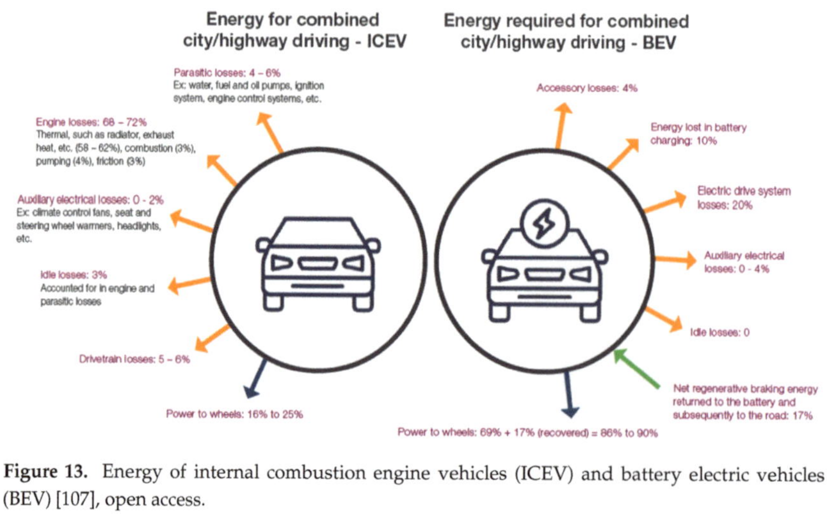

Figure 13. Energy of internal combustion engine vehicles (ICEV) and battery electric vehicles (BEV) [107], open access.

In most cases, the local grid provides the power needed to charge an electric vehicle, but the reason why EVs are becoming very popular and attractive is because they can run on renewable energy sources, especially solar energy, as it is free fuel for the car to use. The adoption of specialized solar PV charging stations, as shown in Figure 14, is highly capable of reducing CO_2 emissions from road vehicles greatly, as they limit the use of fossil fuels at the bare minimum. Solar-powered EVs are also more than capable of cutting emissions, as they enable a decrease in local grid overloading [108], especially in peak hours, and enhanced grid flexibility, which are further advantages of the synergy between solar PV and EV technologies. The constantly lowering battery prices that are driving EV adoption also allow for the addition of batteries to store collected solar energy during the day for later charging in PV-powered charging stations. This is a great feature, as it also helps in boosting the grid's resilience to the inherent intermittent nature of solar energy in general. PV cells can also be directly incorporated into the electric car's body so that they can also be used to charge EVs, which are known as PV-integrated EVs [109]. However, given the diverse orientations of main vehicle surfaces and the vehicle's ever-changing exposure to the sun during the day, owing to factors such as location at different times of the day and frequent shade, estimating the potential usefulness of PV systems in boosting EV range is extremely challenging, which is why they are not implemented as much. Solar PV EVs can be helpful in powering up secondary systems and extra functions in vehicles, such as lighting systems or even possibly passenger cooling systems within the car.

Figure 14. Schematic of a solar charger for EVs.

From the aforementioned applications, it is clear that the applications of solar energy play a significant role in deciding proper solar technology. However, it is also important to mention that the that the selection of sustainable solar energy technology varies from one place to another according to meteorological parameters; therefore, they must be considered before selecting the proper solar technology for a specific application.

4. Methodology

The first step for conducting this analysis was to consolidate our search to a database. The SCOPUS database was thus chosen. For the search parameters, we searched the required topic, such as "solar", which was our first part of this analysis. Then, some restrictions were added to the search, such as the date range, the language and the type of paper. After obtaining the results, this search was combined with each of the United

Nations (UNs) Sustainable Development Goals (SDGs) one at a time. The process was then repeated for all five applications discussed in this study. All the data were transferred to Microsoft Excel, where they were analyzed and cleaned, removing things such as repeated words, etc. These data were then finally ready for the bibliometric analysis and data presentation. Figure 15a,b show a simplified version of the process.

(**a**)

(**b**)

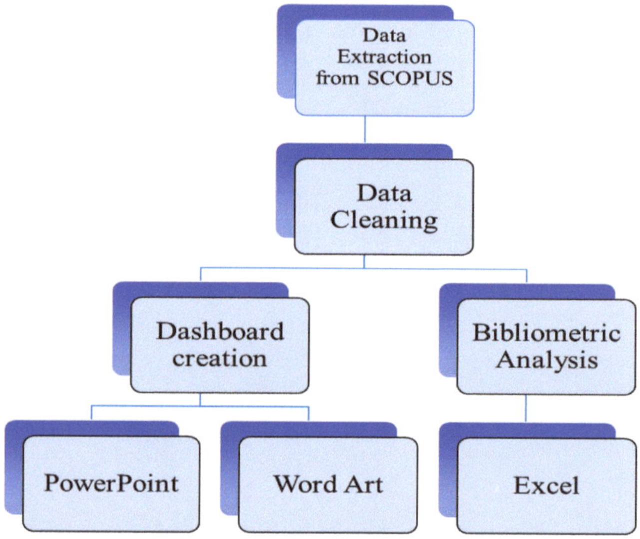

Figure 15. Data extraction process of "solar" from SCOPUS (**a**) and analysis of data process (**b**).

5. Results and Discussions

This section describes the bibliometric analysis of the solar photovoltaic systems in relation to the SDGs. For investigating and evaluating vast amounts of scientific data, a bibliometric analysis is a well-liked and great technique, as it allows exploring the subtleties and points of interest of a particular field's developed history in depth over a period of time while highlighting different important aspects. The aim was to utilize bibliometric techniques regarding solar energy and its applications discussed in this study in combination with the SDGs [52], as shown in Figure 16, in order to shed light on the research gap related to the contribution of solar energy to the SDGs.

The SCOPUS online database was used to collect all the scientific papers published regarding "solar" from 2011 to 2021, and the total number of documents extracted from the database was 276,989. These data were then imported to Microsoft Excel for statistical analysis purposes and data cleaning. Afterwards, the SDGs were implemented one by one alongside "solar" in the SCOPUS database search, eventually resulting in a total of 126,513 documents, which was less than half of the papers discussing solar in the selected years. The same procedure was also conducted for the five applications of solar discussed in this paper, which are as follows: large-scale solar PV power plants, residential applications of solar PV, green hydrogen, water desalination and transportation. Figure 17 shows an overview of the distribution of solar papers by the type of SDG, and from this chart, it can be clearly seen that SDG 7 (affordable and clean energy) was, by far, the most dominant in solar research, with 91,708 documents in total, rounding up to over 72% of the total documents, followed by SDG 13 (Climate action) with 11.46%. The contribution of solar research in the other SDGs was weak. This was also true for the other five applications, and SDG 7-related papers had the highest percentage in each of them. It can also be noticed that SDG 4 (quality education) was the lowest in solar research with merely 32 documents, which was only 0.03% of the total documents. The most interesting finding was that, in general, most published papers focused on technical perspectives of solar and on the environmental pillar of sustainable development. The other pillars of sustainable development (social and economic) were less analyzed in the literature.

Figure 16. Sustainable Development Goals (SDGs) (from www.sdgs.un.org, accessed on 13 July 2022).

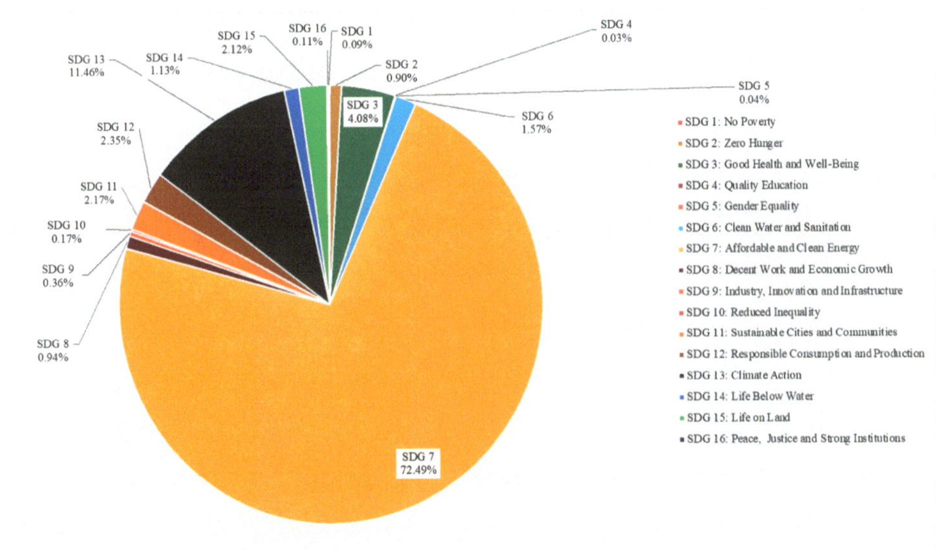

Figure 17. Solar papers regarding SDGs between 2011 and 2021.

Figure 18 compares the research papers based on the related SDGs for each year. It can be noticed that the number of the research papers increased from year to year, indicating that the SDGs received more attention regarding solar applications from year to year. In 2011, only 5179 documents were released, whereas in 2021, that number increased by almost four times to reach 20,392 documents. However, it is clear that some

SDGs lacked enough research in solar energy; for example, SDG 4 and SDG 5 had only two documents released each in the year 2011, whereas in 2021, they had four and five documents released, respectively.

Application 3, green hydrogen, had the lowest contribution among the other applications, with only 201 documents published. Furthermore, there were no papers published regarding SDGs 1, 3, 4, 5, 8, 10 and 16. This could be related to the fact that the study of green hydrogen is quite new. In all applications, SDG 7 (Affordable and Clean Energy) was clearly dominant in the number of documents published.

In application 4 (water desalination), SDG 6 (Clean Water and Sanitation) was the second highest, with 89 documents published out of 723. For application 5 (transportation), SDG 11 (Sustainable Cities and Communities) was the second highest, with 73 documents out of 1898. From these obtained data, it can be easily seen that, in the field of solar energy, the majority of SDG research was concentrated heavily within SDG 7 (Affordable and Clean Energy), as it is directly related to the energy sector. The data lacked enough research in the vast majority of the remaining SDGs, especially SDGs 1, 4, 5, 9, 10 and 16.

These results indicate that solar energy, in general, needs more research in terms of SDGs. In reality, solar energy does not only contribute to the SDGs by providing clean and cheap energy, but it also affects all the other SDGs. Climate change has been a global issue for a long time, causing extreme damage to countless environments on the planet mainly due to the high usage of fossil fuels. By utilizing and investing in solar energy, it is possible to achieve low emission societies with healthy land, water and air quality, thus accomplishing SDGs 13, 14 and 15. Through solar energy adoption, not only can it reduce emissions and carbon footprints, but it can also lead to significant economic development. One way of achieving this economic development is through the creation of new employment. Solar energy also offers potential for additional economic activity, which is another benefit. These actions are crucial for promoting growth from the ground up. As new business models appear, entrepreneurs can invest in them, laying the groundwork for the generation of power. Supply problems can also call for a much more centralized structure.

During the COVID-19 pandemic [110,111], the importance of the internet in gaining access to education was demonstrated. During that year, SDG 4—quality education—suffered, mostly in areas without access to power and the internet. Schools may also have not had access to electricity. Without it, learning materials and learning time are reduced. However, the provision of the required electricity can keep students at school for a longer period of time as a result of solar panels in schools. These locations can now have computers and internet connections installed so that kids can study more. Additionally, access to basic necessities such as food, water and health care is now made feasible as a result of solar energy. Farmers rely on solar-powered irrigation systems in the contemporary world to help them produce more food. When there is a surplus, food security improves, hunger declines and thus SDG 2 is addressed. Solar energy also helps to accomplish the SDGs in the area of health care. The sun is a clean energy source that does not contaminate the atmosphere. Because of this, most nations that use this technology have clean air and inhabitants who are not prone to illness or bad health because of pollution. Modernizing and enhancing the efficiency of healthcare facilities is another application for solar energy. In addition to keeping the lights on and life-saving equipment running, solar-powered health care can also chill vaccines and perform many other life-saving tasks. More and more establishments must install solar panels in order to fulfill SDG 3. They offer affordable and dependable electricity. Health institutions can reduce their power costs and save money that can be put back into the building. Additionally, SDG 5 strives to empower all women and girls in order to achieve gender equality. When there is access to power, children in developing countries, especially girls, can pursue an education. Energy also encourages women to pursue their own businesses. The presence of solar energy in the neighborhood helps to level the playing field for women and girls. This is how solar energy can lay the basic groundwork for achieving all SDGs, proving how important it is for our future.

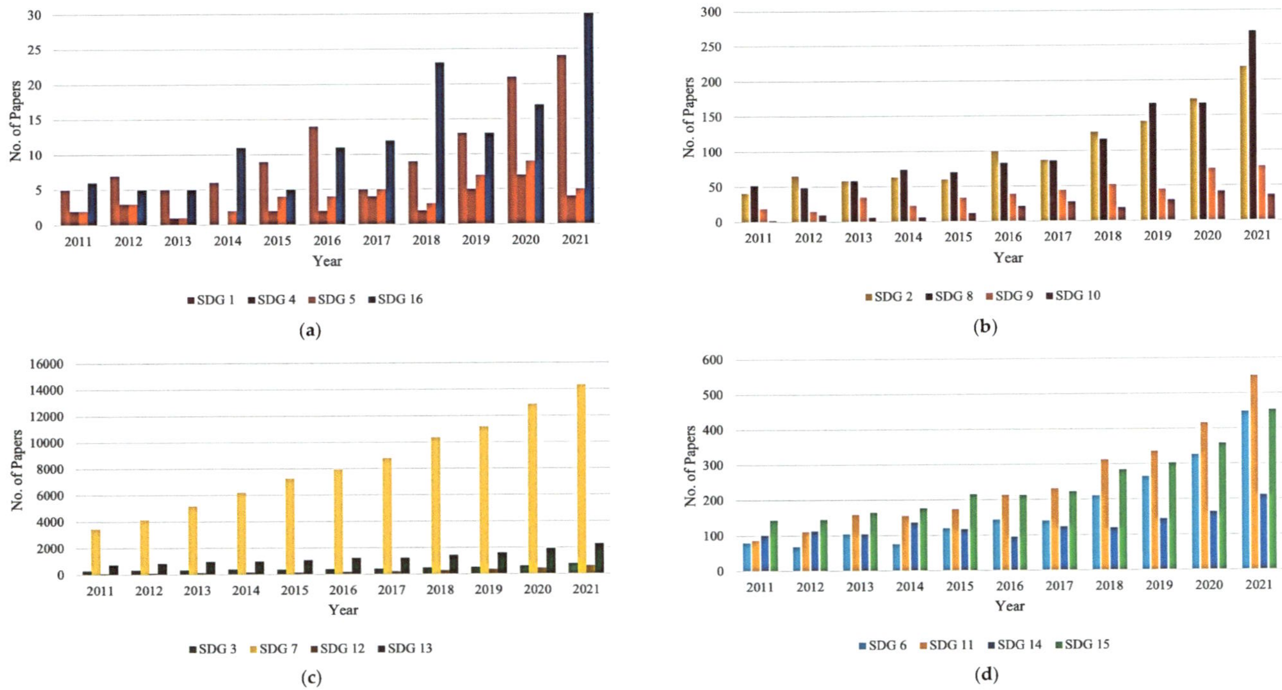

Figure 18. Solar SDG-related research papers by year: (**a**) SDG1, SDG 4, SDG 5 and SDG 16; (**b**) SDG2, SDG 8, SDG9 and SDG 10; (c) SDG3, SDG 7, SDG 12 and SDG 13; (**d**) SDG6, SDG 11, SDG 14 and SDG 15. A similar analysis, i.e., investigating the number of research papers related to SDGs, was conducted for the main five applications of solar (i.e., "Large-Scale Solar PV", "Residential Solar PV", "Green Hydrogen", "Water De-salination" and "Transportation"). For application 1, which is large-scale solar PV power plants with a capacity of more than 1 MW, the total number of documents including SDGs was 4451 and was the highest among the five applications, and like the rest, SDG 7 had the highest percentage of documents with 84%. The second was SDG 13 with 6% of the total documents, and SDG 4 came last with zero documents.

Figure 19 shows how solar energy relates to the three pillars of sustainable development, indicating why more research and the implementation of solar energy plans in all sectors are required for a secure, sustainable and reliable future.

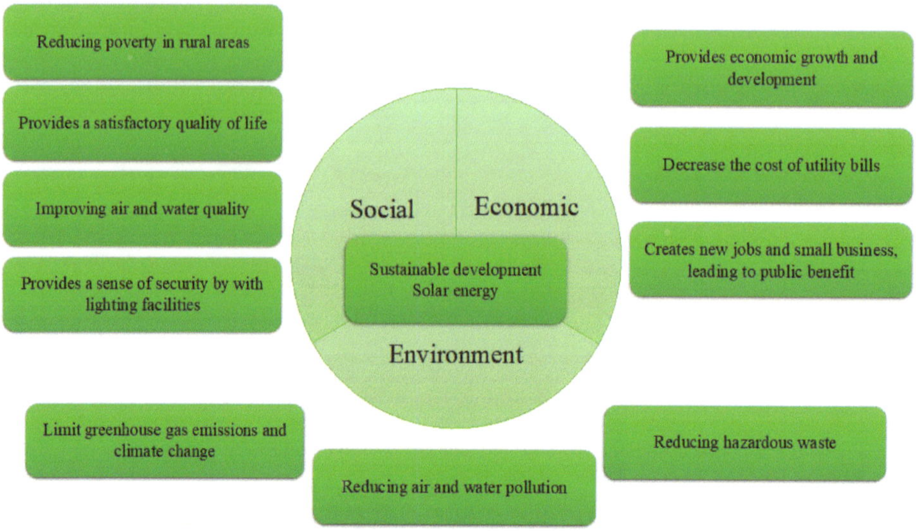

Figure 19. Solar energy and the three pillars of sustainable development.

Moreover, from the data collected and the bibliometric analysis, the keywords from all papers and documents were also extracted. This allowed us to find and explore the points of interest in each of the various analysis fields. This analysis focused on the results of SDG 7 only in each application, because it was the SDG with the greatest number of documents. In the solar analysis, the highest occurring keyword was "Solar Power Generation" with 23,169 documents, and the lowest was "Photocatalytic Activity" with 1166 documents. For applications 1 (Power Plants) and 2 (Residential), the highest occurring keyword was "Photovoltaic Cells" with 1633 and 563 documents, respectively, and the lowest was "Solar Heating" with 59 documents for application 1 and "Hydrogen" with 23 documents for application 2. For application 3, (Green Hydrogen) the highest was "Hydrogen Production" with 141 documents, and the lowest was "current" with only 3 documents. For applications 4 and 5 (Water Desalination and Transportation) the highest occurring keywords were "Desalination" and "Electric Vehicles" with 475 and 625 documents, respectively, and the lowest were "Water Conservation" and "Electric Load Management" with 11 and 27 documents, respectively. The obtained results for the keywords are logical and make sense, especially the highest occurring keywords in each analysis, as they are directly connected to the topic itself. Figure 20a–f shows the dashboards made for solar and the five applications that are discussed in this paper using the data collected above. These dashboards offer a synopsis of the most significant findings and results obtained throughout the analysis that was carried out. The results that are presented in these dashboards include the total number of publications, the number of patents that were discovered, the percentage of open-access articles, the number of publications that were discovered each year, the top three countries in terms of the number of publications that were discovered and, finally, a word cloud that was generated from all of the keywords that were extracted from the data. In Figure 20a, it can be seen from the data presented that there was a huge number of patents registered (310,009) in the field of solar energy between 2011 and 2021. According to what is shown in the graph, in the next few years, there will likely be an increase in the number of publications that are related to solar energy. China is the country that is responsible for the publication of the most papers, with a total of 66,823. The USA came in second place with 57,266 published papers, and India came in third place with 24,632 published papers. The vast disparity in the number of publications between the leading two countries (China and USA) and India suggests that China and the USA will most likely continue to be the leading two countries in terms of research in solar energy.

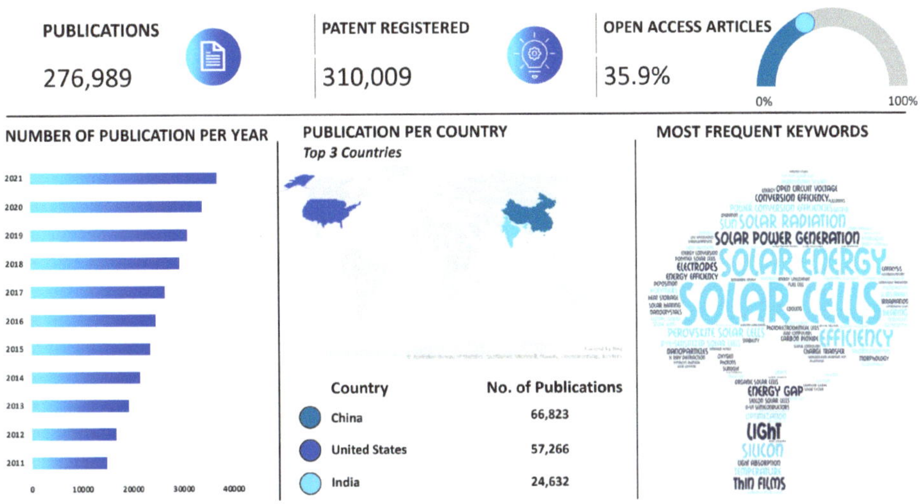

(a)

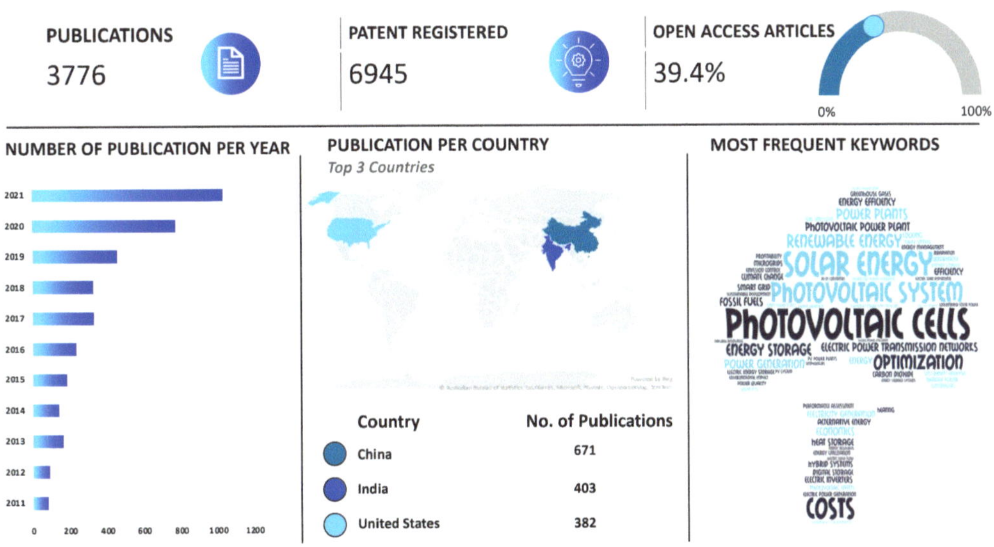

(b)

Figure 20. *Cont.*

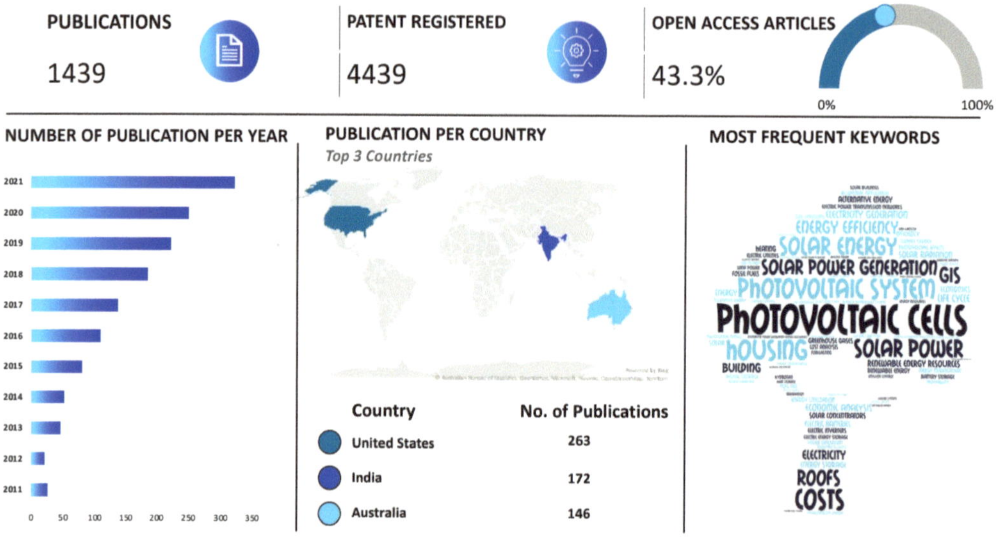

(c)

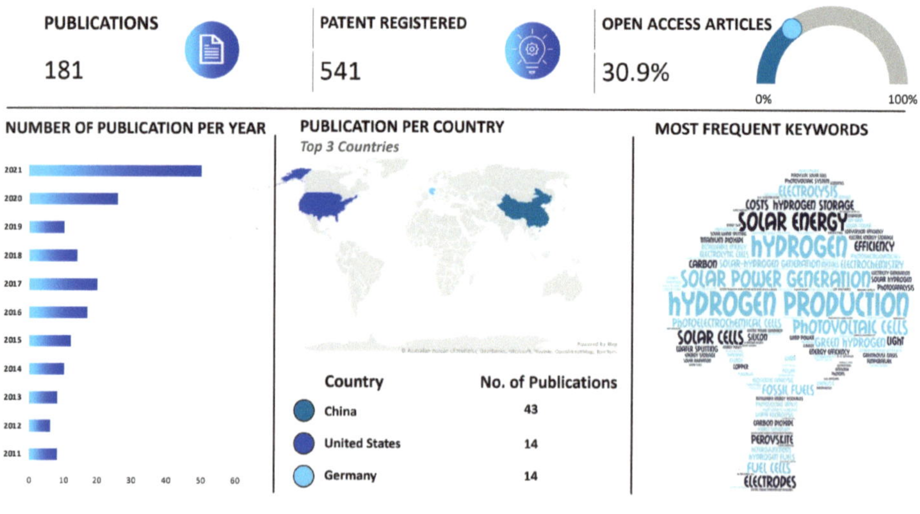

(d)

Figure 20. *Cont.*

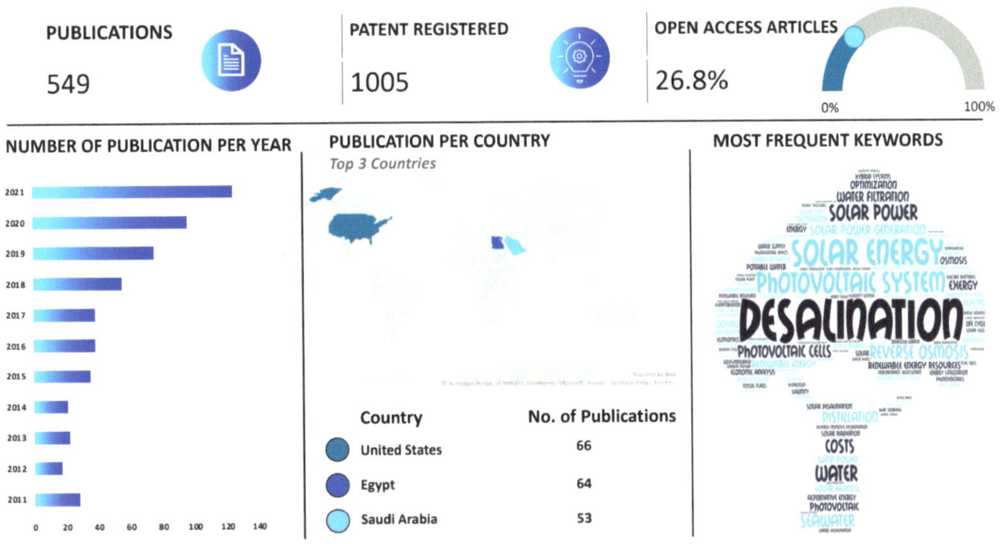

(e)

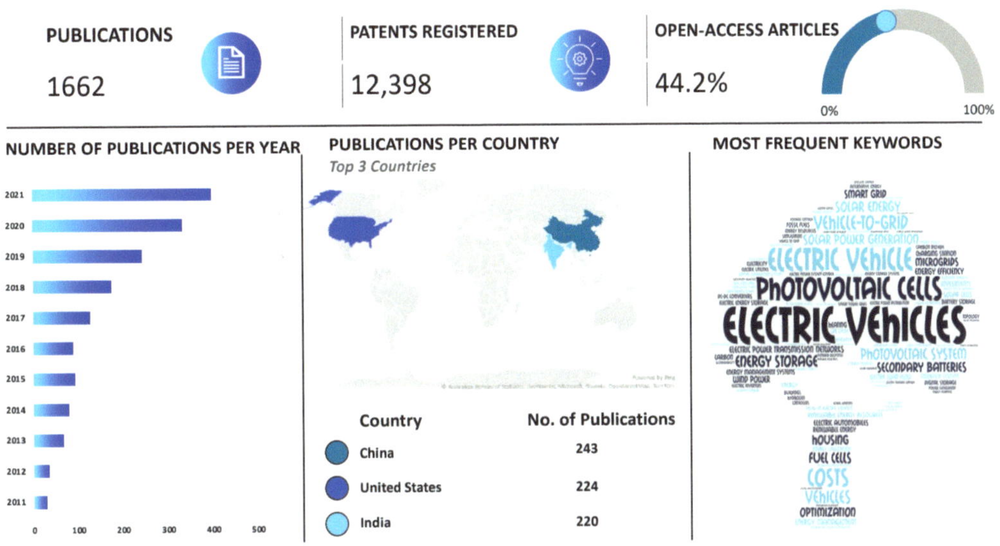

(f)

Figure 20. (**a**) Dashboard for solar; (**b**) Dashboard for large-scale solar power generation; (**c**) Dashboard for residential solar; (**d**) Dashboard for green hydrogen; (**e**) Dashboard for water desalination; (**f**) Dashboard for transportation.

As can be seen in the word cloud that was generated, the term "solar cells" appears most frequently in these results. The findings that are presented in Figure 20b are associated with the first application discussed in this paper, which is solar power generation on a large scale. It was observed that, in the years 2020 and 2021, there was a noticeable increase in the number of publications. This was quite a reasonable result given that the world witnessed an increase in the number of solar power plants in the most recent years. With a total of 671, 403, and 382 published papers, China, India and the USA, respectively, were the top three countries in terms of the number of publications. In Figure 20c, it can be seen that the third dashboard focuses on residential PV, and it is important to note that the number of studies conducted on this subject is also growing. The USA, India and Australia are found to be the top three countries in this category, with a total of 263, 172, and 146, respectively. According to these findings, a greater number of people in the USA are considering the possibility of installing PV solar panels on the roofs of their homes and other buildings. This work includes a dashboard for green hydrogen, which can be found in Figure 20d. The data make it very clear that China was the leading country regarding research on the production of green hydrogen from solar PV, with 43 published papers, and the USA and Germany came in second place with only 14 publications each. This is a clear indication of China's dominance in this area. Although 43 publications may appear to be a very low number when compared with other results when viewed in isolation, this may be due to the fact that green hydrogen is still a relatively new topic that has only recently gained popularity. Figure 20e shows the dashboard for the solar water desalination system. As depicted in the figure, the total number of papers published each year has been steadily growing, particularly in the recent years. The USA, Egypt and Saudi Arabia were the top three countries represented here, each with a total of 53 publications, 66 publications and 64 publications, respectively. The fact that two of the top three countries were located in the MENA region (i.e., the Middle East and North Africa) demonstrates that the concept of desalinating water by utilizing solar energy has the potential to gain an enormous amount of popularity in the MENA region in the following years. The last dashboard in this paper, as shown in Figure 20f, is related to transportation (which concerns solar PV and electric vehicles (EVs)). The dashboard analysis shows that the total number of publications was found to be 1662 published papers, with 12,398 patents registered. Out of the 1662 publications, 44.2% were open-access articles, which accounted for 734 papers, allowing people from all around the world to reach these papers free of charge. It can be seen from the chart that, from 2011 to 2015, the number of published papers slightly increased with each passing year; however, in 2016, the number of publications slightly dropped for an unknown reason. Then, from 2017 and onwards, the number of publications went back to increasing, especially in the last few years. This result is in accordance with the strategic plan for switching from fossil-fuel-based transportation to electric transportation. People are switching to electric vehicles from normal combustion engine cars, and several countries are promoting the use of electric vehicles and hybrid vehicles that run on both electricity and combustion in order to minimize the environmental impacts of combustion engine vehicles. The top three countries in the number of published papers for this topic were found to be China in first place with 243 published papers, then the USA with 224 published papers and finally India in third place with 220 published papers. As can be seen, the numbers are quite close to each other, and this can indicate that these three countries could be, in the future, the leading countries in terms of research regarding solar PV and electric vehicles. Some of the most recurring keywords were found to be "electric vehicles", "photovoltaic cells", "vehicle to grid" and "secondary batteries". Even though the six dashboards below do not cover all the results and findings in this paper, they still present a summary for the readers of some of the most important results that were found during this bibliometric analysis for solar and the five chosen applications.

6. Conclusions and Recommendations

Solar energy is one of the most prominent renewable energy sources. Solar energy holds great potential for solving many global issues, such as climate change, securing fresh water and securing a green energy source. This work discusses the state of the art of solar photovoltaics systems, including their fundamentals, different generations of the solar PV and large-scale applications of solar PV systems, such as residential applications, green hydrogen production, water desalination and transportation. Then, a bibliometric analysis is described regarding solar PV systems in relation to the United Nations (UNs) Sustainable Development Goals (SDGs). A total of 276,989 documents available on the SCOPUS database were analyzed, and the main conclusions of this study can be summarized as follows:

- Solar PV systems applied on a commercial scale in power plants with a power capacity of more than 1 MW and can reach the Giga scale.
- Solar PV power desalination systems are available worldwide on a commercials scale, and their potential is increasing over time. Significant progress has been realized in the application of solar PV in the production of green hydrogen and green buildings as well as electric vehicles in the transportation sector.
- The bibliometric study demonstrates that there has been a notable increase in the amount of research conducted on solar energy over the past decade.
- The majority of solar research is concentrated within SDG 7, which is directly related to clean energy, while lacking in many other SDGs, such as SDG 6: Clean Water and Sanitation, SDG 1: No Poverty, SDG 4: Quality Education, SDG 5: Gender Equality, SDG 9: Industry, Innovation and Infrastructure, SDG 10: Reduced Inequality and SDG 16: Peace, Justice and Strong Institutions.
- China, India and the USA were the primary contributors of published research, and the rest of the world contributed a significantly lower number of publications.
- This study proves that more research is needed in the solar energy sector regarding the SDGs, as solar energy impacts not only SDG 7 but all other SDGs, whether directly or indirectly. Such studies are crucial for securing a sustainable future for current and future generations.
- Within the five chooses application of solar energy, a ranking of publications was as follows: (1) large-scale solar power generation, (2) transportation, (3) residential applications, (4) water desalination and (5) green hydrogen production.
- The vast majority of the research was focused mainly on the technical aspects of solar energy, whereas other important aspects received significantly less attention.
- There is a significant lack of research in relation to SDG 1: No Poverty, SDG 4: Quality Education, SDG 5: Gender Equality and SDG 16: Peace, Justice and Strong Institutions.
- The results of the bibliometric indicate that there was a total of 276,989 published studies, in addition to 310,009 patents that were registered.
- Overall, there was a good percentage of research published in open-access journals, which shows good financial support from different stakeholders.
- The keywords "electric vehicles", "photovoltaic cells", "vehicle to grid" and "secondary batteries" were found to be among the most frequently used keywords.

Based on the funding of this research, a key policy priority should therefore be to explore the impact of the solar energy on all the SDGs, especially SDG 1: No Poverty, SDG 4: Quality Education, SDG 5: Gender Equality and SDG 16: Peace, Justice and Strong Institutions. Moreover, there is a need for more research in the rest of the world, other than China, India and the USA, to better understand the solar impact on SDG achievements. Furthermore, policymakers should focus on understating the social and economic aspect of solar energy, such as the total number of jobs created by solar projects, the impact on gross domestic product (GDP) and gender inequality in solar projects. In addition, policymakers should allocate a larger amount of their funds to new areas of solar energy applications, such as green hydrogen and electric vehicles. Another implication of our study is that policymakers should ensure that there is alignment between funded research and the priority SDGs and SDG targets within the country. Finally, policymakers are recommended to push toward increasing the percentage of open-access research.

Author Contributions: Conceptualization, K.O., Y.A.S., N.S. and E.T.S.; methodology, K.O., A.G.O. and Y.A.S.; formal analysis, N.S., A.H.A. and C.R.; investigation, A.G.O., M.A.A., A.H.A. and E.T.S.; resources, A.G.O., M.A.A. and C.R.; data curation, K.O. and Y.A.S.; writing—original draft preparation, K.O., A.G.O., Y.A.S., N.S., M.A.A., A.H.A., C.R. and E.T.S.; draft preparation, K.O., A.G.O., Y.A.S., N.S., M.A.A., A.H.A., C.R. and E.T.S. All authors have read and agreed to the published version of the manuscript.

Funding: This work was supported by the University of Sharjah, Project No. 19020406129.

Institutional Review Board Statement: Not applicable.

Informed Consent Statement: Not applicable.

Data Availability Statement: Not applicable.

Conflicts of Interest: The authors declare no conflict of interest.

References

1. Bhuvaneswari, C.; Rajeswari, R.; Kalaiarasan, C. Analysis of solar energy based street light with auto tracking system. *Int. J. Adv. Res. Electr. Electron. Instrum. Eng.* **2013**, *2*, 3422–3428.
2. Michael, P.; Johnston, D.; Moreno, W. A conversion guide: Solar irradiance and lux illuminance. *J. Meas. Eng.* **2020**, *8*, 153–166. [CrossRef]
3. Mekhilef, S.; Saidur, R.; Safari, A. A review on solar energy use in industries. *Renew. Sustain. Energy Rev.* **2011**, *15*, 1777–1790. [CrossRef]
4. Gupta, A.; Das, B.; Biswas, A.; Mondol, J. Sustainability and 4E analysis of novel solar photovoltaic-thermal solar dryer under forced and natural convection drying. *Renew. Energy* **2022**, *188*, 1008–1021. [CrossRef]
5. Al Dashti, H. *The Effect of the Chemical Composition of the Atmosphere on the Climate of State of Kuwait*; Cairo University: Giza, Egypt, 2012.
6. Barsanti, L.; Gualtieri, P. Is exploitation of microalgae economically and energetically sustainable? *Algal Res.* **2018**, *31*, 107–115. [CrossRef]
7. Ashok, S. *Solar Energy*; Encyclopedia Britannica: Chicago, IL, USA, 2020.
8. Gupta, A.; Das, B.; Biswas, A.; Mondol, J. An environmental and economic evaluation of solar photovoltaic thermal dryer. *Int. J. Environ. Sci. Technol.* **2022**, *19*, 10773–10792. [CrossRef]
9. Jha, P.; Das, B.; Gupta, R. Energy matrices evaluation of a conventional and modified partially covered photovoltaic thermal collector. *Sustain. Energy Technol. Assess.* **2022**, *54*, 102610. [CrossRef]
10. PraveenKumar, S.; Agyekum, E.; Qasim, M.; Alwan, N.; Velkin, V.; Shcheklein, S. Experimental assessment of thermoelectric cooling on the efficiency of PV module. *Int. J. Renew. Energy Res. (IJRER)* **2022**, *12*, 1670–1681.
11. Agyekum, E.; PraveenKumar, S.; Alwan, N.; Velkin, V.; Shcheklein, S.; Yaqoob, S. Experimental investigation of the effect of a combination of active and passive cooling mechanism on the thermal characteristics and efficiency of solar PV module. *Inventions* **2021**, *6*, 63. [CrossRef]
12. Kochtcheeva, L. Renewable energy: Global challenges. *E-International Relations*, 27 May 2016.
13. Sipahutar, R.; Bernas, S.; Imanuddin, M. Renewable energy and hydropower utilization tendency worldwide. *Renew. Sustain. Energy Rev.* **2013**, *17*, 213–215.
14. Parida, B.; Iniyan, S.; Goic, R. A review of solar photovoltaic technologies. *Renew. Sustain. Energy Rev.* **2011**, *15*, 1625–1636. [CrossRef]
15. Ahmadi, A.; Das, B.; Ehyaei, M.; Esmaeilion, F.; El Haj Assad, M.; Jamali, D.; Koohshekan, O.; Kumar, R.; Rosen, M.; Negi, S.; et al. Energy, exergy, and techno-economic performance analyses of solar dryers for agro products: A comprehensive review. *Sol. Energy* **2021**, *228*, 349–373. [CrossRef]
16. Bagher, A.; Vahid, M.; Mohsen, M. Types of solar cells and application. *Am. J. Opt. Photonics* **2015**, *3*, 94–113. [CrossRef]
17. Anku, N.; Adu-Gyamfi, D.; Kankam, A.; Takyi, A.; Amponsah, R. A model for photovoltaic module optimization. *J. Mech. Eng. Autom* **2015**, *5*, 72–79.
18. Koussa, M.; Cheknane, A.; Hadji, S.; Haddadi, M.; Noureddine, S. Measured and modelled improvement in solar energy yield from flat plate photovoltaic systems utilizing different tracking systems and under a range of environmental conditions. *Appl. Energy* **2011**, *88*, 1756–1771. [CrossRef]
19. Hosenuzzaman, M.; Rahim, N.; Selvaraj, J.; Hasanuzzaman, M.; Malek, A.; Nahar, A. Global prospects, progress, policies, and environmental impact of solar photovoltaic power generation. *Renew. Sustain. Energy Rev.* **2015**, *41*, 284–297. [CrossRef]
20. Allouhi, A.; Rehman, S.; Buker, M.; Said, Z. Up-to-date literature review on Solar PV systems: Technology progress, market status and R&D. *J. Clean. Prod.* **2022**, *362*, 132339.
21. Obaideen, K.; Abdelkareem, M.; Wilberforce, T.; Elsaid, K.; Sayed, E.; Maghrabie, H.; Olabi, A. Biogas role in achievement of the sustainable development goals: Evaluation, Challenges, and Guidelines. *J. Taiwan Inst. Chem. Eng.* **2022**, *131*, 104207. [CrossRef]

22. Chiu, J.-S.; Zhao, Y.-M.; Zhang, S.; Wuu, D.-S. The role of laser ablated backside contact pattern in efficiency improvement of mono crystalline silicon PERC solar cells. *Sol. Energy* **2020**, *196*, 462–467. [CrossRef]
23. Jiang, L.; Cui, S.; Sun, P.; Wang, Y.; Yang, C. Comparison of monocrystalline and polycrystalline solar modules. In Proceedings of the 2020 IEEE 5th Information Technology and Mechatronics Engineering Conference (ITOEC), Chongqing, China, 12–14 June 2020; pp. 341–344.
24. Becker, C.; Amkreutz, D.; Sontheimer, T.; Preidel, V.; Lockau, D.; Haschke, J.; Jogschies, L.; Klimm, C.; Merkel, J.; Plocica, P. Polycrystalline silicon thin-film solar cells: Status and perspectives. *Sol. Energy Mater. Sol. Cells* **2013**, *119*, 112–123. [CrossRef]
25. Corcelli, F.; Ripa, M.; Ulgiati, S. End-of-life treatment of crystalline silicon photovoltaic panels. An emergy-based case study. *J. Clean. Prod.* **2017**, *161*, 1129–1142. [CrossRef]
26. Lin, L.; Xu, X.; Chu, C.; Majeed, M.; Yang, J. Mesoporous Amorphous Silicon: A Simple Synthesis of a High-Rate and Long-Life Anode Material for Lithium-Ion Batteries. *Angew. Chem.* **2016**, *55*, 14063–14066. [CrossRef] [PubMed]
27. Kang, H. Crystalline silicon vs. amorphous silicon: The significance of structural differences in photovoltaic applications. *IOP Conf. Ser. Earth Environ. Sci.* **2021**, *726*, 012001. [CrossRef]
28. Gessert, T.; Ali, S. 1.19-cadmium telluride photovoltaic thin film: CdTe. In *Comprehensive Renewable Energy*; Elsevier: Oxford, UK, 2012; pp. 423–438.
29. Nikolić, D.; Bojić, M.; Skerlić, J.; Radulović, J.; Taranović, D. A review of non-silicon and new photovoltaics technology for electricity generation. *Zb. Međunarodne Konf. O Obnov. Izvorima Električne Energ.* **2018**, *2*, 1–7.
30. Yao, H.; Wang, J.; Xu, Y.; Zhang, S.; Hou, J. Recent Progress in Chlorinated Organic Photovoltaic Materials. *Acc. Chem. Res.* **2020**, *53*, 822–832. [CrossRef]
31. Chemisana, D. Building integrated concentrating photovoltaics: A review. *Renew. Sustain. Energy Rev.* **2011**, *15*, 603–611. [CrossRef]
32. Shanks, K.; Senthilarasu, S.; Mallick, T. Optics for concentrating photovoltaics: Trends, limits and opportunities for materials and design. *Renew. Sustain. Energy Rev.* **2016**, *60*, 394–407. [CrossRef]
33. Sharma, S.; Jain, K.; Sharma, A. Solar cells: In research and applications—A review. *Mater. Sci. Appl.* **2015**, *6*, 1145. [CrossRef]
34. Goetzberger, A.; Hebling, C.; Schock, H.-W. Photovoltaic materials, history, status and outlook. *Mater. Sci. Eng. R Rep.* **2003**, *40*, 1–46. [CrossRef]
35. El Chaar, L.; El Zein, N. Review of photovoltaic technologies. *Renew. Sustain. Energy Rev.* **2011**, *15*, 2165–2175. [CrossRef]
36. Badawy, W. A review on solar cells from Si-single crystals to porous materials and quantum dots. *J. Adv. Res.* **2015**, *6*, 123–132. [CrossRef] [PubMed]
37. Beaucarne, G. Silicon thin-film solar cells. *Adv. OptoElectronics* **2007**, *2007*, 36970. [CrossRef]
38. Mesquita, D.d.B.; Silva, J.L.d.S.; Moreira, H.S.; Kitayama, M.; Villalva, M.G. A review and analysis of technologies applied in PV modules. In Proceedings of the 2019 IEEE PES Innovative Smart Grid Technologies Conference-Latin America (ISGT Latin America), Gramado, Brazil, 15–18 September 2019; pp. 1–6.
39. Pérez-Higueras, P.; Muñoz, E.; Almonacid, G.; Vidal, P. High Concentrator PhotoVoltaics efficiencies: Present status and forecast. *Renew. Sustain. Energy Rev.* **2011**, *15*, 1810–1815. [CrossRef]
40. Tan, M.-H.; Chong, K.-K.; Wong, C.-W. Optical characterization of nonimaging dish concentrator for the application of dense-array concentrator photovoltaic system. *Appl. Opt.* **2014**, *53*, 475–486. [CrossRef]
41. Sala, G.; Anton, I.; Arboiro, J.; Luque, A.; Camblor, E.; Mera, E.; Gasson, M.; Cendagorta, M.; Valera, P.; Friend, M. The 480 kWp EUCLIDESTM-THERMIE Power plant: Installation, Set-up and First Results. In Proceedings of the Sixteenth European Photovoltaic Solar Energy Conference, Glasgow, UK, 1–5 May 2020; pp. 2072–2077.
42. Matsushima, T.; Setaka, T.; Muroyama, S. Concentrating solar module with horizontal reflectors. *Sol. Energy Mater. Sol. Cells* **2003**, *75*, 603–612. [CrossRef]
43. Ameur, A.; Berrada, A.; Loudiyi, K.; Adomatis, R. Performance and energetic modeling of hybrid PV systems coupled with battery energy storage. In *Hybrid Energy System Models*; Elsevier: Amsterdam, The Netherlands, 2021; pp. 195–238.
44. Todorov, T.; Bishop, D.; Lee, Y. Materials perspectives for next-generation low-cost tandem solar cells. *Sol. Energy Mater. Sol. Cells* **2018**, *180*, 350–357. [CrossRef]
45. Ramanujam, J.; Singh, U. Copper indium gallium selenide based solar cells–a review. *Energy Environ. Sci.* **2017**, *10*, 1306–1319. [CrossRef]
46. Ma, Q.; Jia, Z.; Meng, L.; Zhang, J.; Zhang, H.; Huang, W.; Yuan, J.; Gao, F.; Wan, Y.; Zhang, Z. Promoting charge separation resulting in ternary organic solar cells efficiency over 17.5%. *Nano Energy* **2020**, *78*, 105272. [CrossRef]
47. Jakhar, S.; Soni, M.; Gakkhar, N. Historical and recent development of concentrating photovoltaic cooling technologies. *Renew. Sustain. Energy Rev.* **2016**, *60*, 41–59. [CrossRef]
48. Hyder, F.; Sudhakar, K.; Mamat, R. Solar PV tree design: A review. *Renewable Sustain. Energy Rev.* **2018**, *82*, 1079–1096. [CrossRef]
49. Verma, A.; Singhal, S. Solar PV performance parameter and recommendation for optimization of performance in large scale grid connected solar PV plant—Case study. *J. Energy Power Sources* **2015**, *2*, 40–53.
50. Praveenkumar, S.; Agyekum, E.; Kumar, A.; Ampah, J.; Afrane, S.; Amjad, F.; Velkin, V. Techno-economics and the identification of environmental barriers to the development of concentrated solar thermal power plants in India. *Appl. Sci.* **2022**, *12*, 10400. [CrossRef]

51. Mansouri, N.; Lashab, A.; Sera, D.; Guerrero, J.; Cherif, A. Large photovoltaic power plants integration: A review of challenges and solutions. *Energies* **2019**, *12*, 3798. [CrossRef]
52. Obaideen, K.; AlMallahi, M.N.; Alami, A.; Ramadan, M.; Abdelkareem, M.; Shehata, N.; Olabi, A. On the contribution of solar energy to sustainable developments goals: Case study on Mohammed bin Rashid Al Maktoum Solar Park. *Int. J.* **2021**, *12*, 100123. [CrossRef]
53. Lumby, B. *Utility-Scale Solar Photovoltaic Power Plants: A Project Developer's Guide*; The World Bank: Bretton Woods, NH, USA, 2015.
54. Benda, V.; Černá, L. PV cells and modules–State of the art, limits and trends. *Heliyon* **2020**, *6*, e05666. [CrossRef]
55. Aghaei, M.; Eskandari, A.; Vaezi, S.; Chopra, S. Solar PV power plants. In *Photovoltaic Solar Energy Conversion*; Elsevier: Amsterdam, The Netherlands, 2020; pp. 313–348.
56. Seme, S.; Štumberger, B.; Hadžiselimović, M.; Sredenšek, K. Solar photovoltaic tracking systems for electricity generation: A review. *Energies* **2020**, *13*, 4224. [CrossRef]
57. Díez-Mediavilla, M.; Dieste-Velasco, M.; Rodríguez-Amigo, M.; García-Calderón, T.; Alonso-Tristán, C. Performance of grid-tied PV facilities based on real data in Spain: Central inverter versus string system. *Energy Convers. Manag.* **2014**, *86*, 1128–1133. [CrossRef]
58. Sahu, A.; Yadav, N.; Sudhakar, K. Floating photovoltaic power plant: A review. *Renew. Sustain. Energy Rev.* **2016**, *66*, 815–824. [CrossRef]
59. Ghosh, S.; Nair, A.; Krishnan, S. Techno-economic review of rooftop photovoltaic systems: Case studies of industrial, residential and off-grid rooftops in Bangalore, Karnataka. *Renew. Sustain. Energy Rev.* **2015**, *42*, 1132–1142. [CrossRef]
60. Trattner, A.; Höglinger, M.; Macherhammer, M.; Sartory, M. Renewable hydrogen: Modular concepts from production over storage to the consumer. *Chem. Ing. Tech.* **2021**, *93*, 706–716. [CrossRef]
61. Shankarappa, N.; Ahmed, M.; Shashikiran, N.; Naganagouda, D. Solar photovoltaic systems–applications & configurations. *Int. Res. J. Eng. Technol.* **2017**, *4*, 1851–1855.
62. Tomar, V.; Tiwari, G. Techno-economic evaluation of grid connected PV system for households with feed in tariff and time of day tariff regulation in New Delhi–A sustainable approach. *Renew. Sustain. Energy Rev.* **2017**, *70*, 822–835. [CrossRef]
63. Mamaghani, A.; Escandon, S.; Najafi, B.; Shirazi, A.; Rinaldi, F. Techno-economic feasibility of photovoltaic, wind, diesel and hybrid electrification systems for off-grid rural electrification in Colombia. *Renew. Energy* **2016**, *97*, 293–305. [CrossRef]
64. Marino, C.; Nucara, A.; Panzera, M.; Pietrafesa, M.; Pudano, A. Economic comparison between a stand-alone and a grid connected PV system vs. grid distance. *Energies* **2020**, *13*, 3846. [CrossRef]
65. Praveenkumar, S.; Gulakhmadov, A.; Kumar, A.; Safaraliev, M.; Chen, X. Comparative Analysis for a Solar Tracking Mechanism of Solar PV in Five Different Climatic Locations in South Indian States: A Techno-Economic Feasibility. *Sustainability* **2022**, *14*, 11880. [CrossRef]
66. Agnew, S.; Dargusch, P. Effect of residential solar and storage on centralized electricity supply systems. *Nat. Clim. Change* **2015**, *5*, 315–318. [CrossRef]
67. Gugulothu, R.; Somanchi, N.; Banoth, H.; Banothu, K. A review on solar powered air conditioning system. *Procedia Earth Planet. Sci.* **2015**, *11*, 361–367. [CrossRef]
68. Li, Y.; Zhang, G.; Lv, G.; Zhang, A.; Wang, R. Performance study of a solar photovoltaic air conditioner in the hot summer and cold winter zone. *Sol. Energy* **2015**, *117*, 167–179. [CrossRef]
69. Liu, Z.; Li, A.; Wang, Q.; Chi, Y.; Zhang, L. Performance study of a quasi grid-connected photovoltaic powered DC air conditioner in a hot summer zone. *Appl. Therm. Eng.* **2017**, *121*, 1102–1110. [CrossRef]
70. Praveenkumar, S.; Gulakhmadov, A.; Agyekum, E.; Alwan, N.; Velkin, V.; Sharipov, P.; Safaraliev, M.; Chen, X. Experimental study on performance enhancement of a photovoltaic module incorporated with CPU heat pipe—A 5E analysis. *Sensors* **2022**, *22*, 6367. [CrossRef]
71. Afshari, A.; Nikolopoulou, C.; Martin, M. Life-Cycle Analysis of Building Retrofits at the Urban Scale—A Case Study in United Arab Emirates. *Sustainability* **2014**, *6*, 453–473. [CrossRef]
72. Charadi, S.; Salbi, A.; Redouane, A.; El Hasnaoui, A. Smart hybrid AC-DC distribution system for solar electric house: Case of an air conditioner system. In Proceedings of the International Conference on Sustainable Renewable Energy Systems and Applications, Tebessa, Algeria, 4–5 December 2019; pp. 1–4.
73. Tanaka, K.; Makino, Y.; Sakoguchi, E.; Takeoka, A.; Kuwano, Y. Residential solar-powered air conditioners. *Electr. Eng. Jpn.* **1994**, *114*, 123–133. [CrossRef]
74. Postel, S. Water and world population growth. *Am. Water Work. Assoc.* **2000**, *92*, 131–138. [CrossRef]
75. Verma, S.; Mishra, S.; Chowdhury, S.; Gaur, A.; Mohapatra, S.; Soni, A.; Verma, P. Solar PV powered water pumping system–A review. *Mater. Today Proc.* **2021**, *46*, 5601–5606. [CrossRef]
76. Aliyu, M.; Hassan, G.; Said, S.; Siddiqui, M.; Alawami, A.; Elamin, I. A review of solar-powered water pumping systems. *Renew. Sustain. Energy Rev.* **2018**, *87*, 61–76. [CrossRef]
77. Sashidhar, S.; Reddy, V.; Fernandes, B. A single-stage sensorless control of a PV-based bore-well submersible BLDC motor. *IEEE J. Emerg. Sel. Top. Power Electron.* **2018**, *7*, 1173–1180. [CrossRef]
78. Raza, S.; Janajreh, I.; Ghenai, C. Sustainability index approach as a selection criteria for energy storage system of an intermittent renewable energy source. *Appl. Energy* **2014**, *136*, 909–920. [CrossRef]

79. Ramadan, M. A review on coupling Green sources to Green storage (G2G): Case study on solar-hydrogen coupling. *Int. J. Hydrog. Energy* **2021**, *46*, 30547–30558. [CrossRef]
80. Hosseini, S.; Wahid, M. Hydrogen from solar energy, a clean energy carrier from a sustainable source of energy. *Int. J. Energy Res.* **2020**, *44*, 4110–4131. [CrossRef]
81. Park, C.; Koo, M.; Woo, J.; Hong, B.; Shin, J. Economic valuation of green hydrogen charging compared to gray hydrogen charging: The case of South Korea. *Int. J. Hydrog. Energy* **2022**, *47*, 14393–14403. [CrossRef]
82. Howarth, R.; Jacobson, M. How green is blue hydrogen? *Energy Sci. Eng.* **2021**, *9*, 1676–1687. [CrossRef]
83. Olabi, A.; Obaideen, K.; Elsaid, K.; Wilberforce, T.; Sayed, E.; Maghrabie, H.; Abdelkareem, M. Assessment of the pre-combustion carbon capture contribution into sustainable development goals SDGs using novel indicators. *Renew. Sustain. Energy Rev.* **2022**, *153*, 111710. [CrossRef]
84. Atilhan, S.; Park, S.; El-Halwagi, M.; Atilhan, M.; Moore, M.; Nielsen, R. Green hydrogen as an alternative fuel for the shipping industry. *Curr. Opin. Chem. Eng.* **2021**, *31*, 100668. [CrossRef]
85. Khan, M.; Al-Shankiti, I.; Ziani, A.; Idriss, H. Demonstration of green hydrogen production using solar energy at 28% efficiency and evaluation of its economic viability. *Sustain. Energy Fuels* **2021**, *5*, 1085–1094. [CrossRef]
86. Praveenkumar, S.; Agyekum, E.; Ampah, J.; Afrane, S.; Velkin, V.; Mehmood, U.; Awosusi, A. Techno-economic optimization of PV system for hydrogen production and electric vehicle charging stations under five different climatic conditions in India. *Int. J. Hydrog. Energy* **2022**, *47*, 38087–38105. [CrossRef]
87. Vidas, L.; Castro, R. Recent developments on hydrogen production technologies: State-of-the-art review with a focus on green-electrolysis. *Appl. Sci.* **2021**, *11*, 11363. [CrossRef]
88. Kim, J.; Hansora, D.; Sharma, P.; Jang, J.; Lee, J. Toward practical solar hydrogen production–an artificial photosynthetic leaf-to-farm challenge. *Chem. Soc. Rev.* **2019**, *48*, 1908–1971. [CrossRef]
89. Tebibel, H.; Labed, S. System Design and Optimal Management Strategy for Photovoltaic Energy and Hydrogen Production. In *Progress in Clean Energy*; Springer: Berlin/Heidelberg, Germany, 2015; pp. 513–523.
90. Compain, P. Solar energy for water desalination. *Procedia Eng.* **2012**, *46*, 220–227. [CrossRef]
91. Shivajirao, P.A. Treatment of distillery wastewater using membrane technologies. *Int. J. Adv. Eng. Res. Stud.* **2012**, *1*, 275–283.
92. Gorjian, S.; Ghobadian, B.; Ebadi, H.; Ketabchi, F.; Khanmohammadi, S. *Applications of Solar PV Systems in Desalination Technologies, in: Photovoltaic Solar Energy Conversion*; Academic Press Elsevier: Cambridge, MA, USA, 2020; pp. 237–274.
93. Qasim, M.; Badrelzaman, M.; Darwish, N.; Darwish, N.; Hilal, N. Reverse osmosis desalination: A state-of-the-art review. *Desalination* **2019**, *459*, 59–104. [CrossRef]
94. Curto, D.; Franzitta, V.; Guercio, A. A review of the water desalination technologies. *Appl. Sci.* **2021**, *11*, 670. [CrossRef]
95. Idrees, M. Performance analysis and treatment technologies of reverse osmosis plant–a case study. *Case Stud. Chem. Environ. Eng.* **2020**, *2*, 100007. [CrossRef]
96. Lutchmiah, K.; Verliefde, A.; Roest, K.; Rietveld, L.; Cornelissen, E. Forward osmosis for application in wastewater treatment: A review. *Water Res.* **2014**, *58*, 179–197. [CrossRef]
97. Wlodarczyk, R.; Kwarciak-Kozlowska, A. Treatment of waterborne pathogens by reverse osmosis. In *Wlodarczyk Renata*; Kwarciak-Kozlowska, A., Ed.; Elsevier: Amsterdam, The Netherlands, 2020; pp. 57–80.
98. Obaideen, K.; Shehata, N.; Sayed, E.; Abdelkareem, M.; Mahmoud, M.; Olabi, A. The Role of Wastewater Treatment in Achieving Sustainable Development Goals (SDGs) and sustainability guideline. *Energy Nexus* **2022**, *7*, 100112. [CrossRef]
99. Chao, Y.-M.; Liang, T. A feasibility study of industrial wastewater recovery using electrodialysis reversal. *Desalination* **2008**, *221*, 433–439. [CrossRef]
100. Kucera, J. *Desalination: Water from Water*; John Wiley Sons: Hoboken, NJ, USA, 2019.
101. Drioli, E.; Ali, A.; Macedonio, F. Membrane distillation: Recent developments and perspectives. *Desalination* **2015**, *356*, 56–84. [CrossRef]
102. Peng, P.; Fane, A.; Li, X. Desalination by membrane distillation adopting a hydrophilic membrane. *Desalination* **2005**, *173*, 45–54. [CrossRef]
103. Albuquerque, F.; Maraqa, M.; Chowdhury, R.; Mauga, T.; Alzard, M. Greenhouse gas emissions associated with road transport projects: Current status, benchmarking, and assessment tools. *Transp. Res. Procedia* **2020**, *48*, 2018–2030. [CrossRef]
104. Liu, L.; Kong, F.; Liu, X.; Peng, Y.; Wang, Q. A review on electric vehicles interacting with renewable energy in smart grid. *Renew. Sustain. Energy Rev.* **2015**, *51*, 648–661. [CrossRef]
105. Arif, S.; Lie, T.; Seet, B.; Ayyadi, S.; Jensen, K. Review of electric vehicle technologies, charging methods, standards and optimization techniques. *Electronics* **2021**, *10*, 1910. [CrossRef]
106. Jungmeier, G.; Canella, L.; Elgowainy, A.; Ehrenberger, S.; Pérez, G.B.; Roye, P.-O.; Lim, O. Evaluation of the Environmental Benefits of the Global EV-Fleet in 40 Countries–A LCA Based Estimation in IEA HEV. In Proceedings of the EVS32 Symposium, Lyon, France, 19–22 May 2019.
107. Market Snapshot: Battery Electric Vehicles Are Far More Fuel Efficient than Vehicles with Internal Combustion Engines. 2021. Available online: https://www.cer-rec.gc.ca/en/data-analysis/energy-markets/market-snapshots/2021/market-snapshot-battery-electric-vehicles-are-far-more-fuel-efficient-than-vehicles-with-internal-combustion-engines.html (accessed on 22 October 2022).

108. Rodriguez, A.S.; de Santana, T.; MacGill, I.; Ekins–Daukes, N.; Reinders, A. A feasibility study of solar PV–powered electric cars using an interdisciplinary modeling approach for the electricity balance, CO2 emissions, and economic aspects: The cases of The Netherlands, Norway, Brazil, and Australia. *Prog. Photovolt. Res. Appl.* **2020**, *28*, 517–532. [CrossRef]
109. Khan, S.; Ahmad, A.; Ahmad, F.; Shemami, M.S.; Alam, M.S.; Khateeb, S. A comprehensive review on solar powered electric vehicle charging system. *Smart Sci.* **2018**, *6*, 54–79. [CrossRef]
110. Olabi, V.; Wilberforce, T.; Elsaid, K.; Sayed, E.; Abdelkareem, M. Impact of COVID-19 on the Renewable Energy Sector and Mitigation Strategies. *Chem. Eng. Technol.* **2022**, *45*, 558–571. [CrossRef]
111. Elsaid, K.; Olabi, V.; Sayed, E.; Wilberforce, T.; Abdelkareem, M. Effects of COVID-19 on the environment: An overview on air, water, wastewater, and solid waste. *J. Environ. Manag.* **2021**, *292*, 112694. [CrossRef]

sustainability

MDPI

Article

Optimisation of Process Parameters to Maximise the Oil Yield from Pyrolysis of Mixed Waste Plastics

Farjana Faisal [1], Mohammad Golam Rasul [1], Ashfaque Ahmed Chowdhury [2,*] and Md Islam Jahirul [1]

[1] Fuel and Energy Research Group, School of Engineering and Technology, Central Queensland University, Rockhampton, QLD 4702, Australia; f.faisal@cqu.edu.au (F.F.); m.rasul@cqu.edu.au (M.G.R.); m.j.islam@cqu.edu.au (M.I.J.)

[2] Fuel and Energy Research Group, School of Engineering and Technology, Central Queensland University, Gladstone, QLD 4680, Australia

* Correspondence: a.chowdhury@cqu.edu.au

Abstract: The study sought to optimise process parameters of thermal pyrolysis of mixed waste plastic (MWP) to maximise pyrolytic oil yield. High-density polyethylene (HDPE), polypropylene (PP), and polystyrene (PS) were used as feedstocks for pyrolysis. Response surface methodology (RSM) and Box–Behnken design (BBD) were used to optimise the pyrolysis process. The optimisation was carried out by varying three independent variables, namely, reaction temperature (460–540 °C), residence time (30–150 min), and size of MWP feedstock (5–45 mm), to increase the liquid oil yield. A BBD matrix was used to generate the design of the experiments, and 15 experiments were conducted. The highest liquid oil yield of 75.14 wt% was obtained by optimising the operating parameters, which were a reaction temperature of 535.96 °C, a reaction time of 150 min, and a feedstock particle size of 23.99 mm. A model was developed to determine the relationships among the independent variables, and analysis of variance (ANOVA) was used to investigate their impact on maximising oil yield. ANOVA results showed that the temperature and residence time had the maximum impact on oil yield, followed by feedstock size. Physicochemical analysis of the properties of the plastic pyrolytic oil (PPO) revealed that the crude PPO obtained from the MWP had higher water (0.125 wt%) and sulfur content (5.12 mg/kg) and lower flash point (<20 °C) and cetane index (32), which makes it unsuitable for use as an automobile fuel. However, these issues can be resolved by upgrading the PPO using different posttreatment techniques, such as distillation and hydrotreatment.

Keywords: waste to energy; response surface methodology; Box-Behnken design; plastic pyrolytic oil optimisation; surface and contour plots; physicochemical properties

1. Introduction

Technological advancements and improved living standards are driving higher demand for petroleum-based fuels. Power generation, automobiles, agriculture, industries, and domestic useable machinery rely on petroleum fuels. Alternative energy sources are receiving increased attention to satisfy growing energy demands. Plastics have become an integral part of our daily life due to their durability, affordability, versatility, and lightweight nature [1]. These materials find extensive application across diverse sectors including packaging, aerospace, consumer goods, electronics, construction, transportation, automotive, biomedical, textiles, leisure, and more [1]. As the amount of plastic production and use in everyday life is increasing, their amount in residues and subsequently in municipal solid waste (MSW) is also rising, which is a matter of growing concern because of the very slow degradability of plastics.

Many researchers have turned to using plastic waste to produce plastic diesel as a means of reducing dependency on depleting fossil fuels and managing plastic waste problems more efficiently. At present, 70% of municipal solid waste (MSW) is disposed of in landfills, 19% is recycled, and a mere 11% is converted into energy [1]. From 1950 to 2016, global plastic production increased from 1.5 million tons to 335 million tons [2]. Plastic waste can have a long-lasting impact on the environment, taking hundreds of years to decompose and often ending up in waterways, landfills, and hazardous waste stockpiles. As plastic degrades through various environmental conditions like wear and tear, exfoliation, fragmentation, abrasion, gradual degradation, etc., it can cause a range of environmental problems, including release of microplastics [3,4]. Microplastics have a small particle size (less than 5 mm) and their degradation rate is slow. They possess high stability and can endure in the atmosphere for extended durations [4]. Rough estimates suggest that approximately 10–20 million tons of plastic waste enter the oceans annually [5]. The influence of plastic waste on marine life is significant and far-reaching, given its persistence in the environment for decades to even centuries.

The world's oceans currently host over 5 trillion plastic fragments floating on their surfaces. These fragments come in different sizes and shapes and weigh more than 250,000 tons [6]. According to a report from 2019, China was responsible for producing the highest amount of plastic material, accounting for approximately 31% of global production. The second highest plastic producers were the NAFTA countries (USA, Canada, and Mexico), contributing to 19% of global plastic production [7]. Given that crude oil serves as a basis for plastic production, there is potential to convert waste plastic into energy [8]. The production of plastics consumes 4% of the fossil fuels produced worldwide [8]. To fuel the operations of plastic manufacturing industries, an extra 4% is necessary for power generation [5]. Landfilling and incineration are commonly used methods for eliminating waste plastics. However, landfilling requires a massive amount of land outside of human habitats, which makes obtaining new sites for landfills increasingly challenging. These landfill sites cause serious health and environmental concerns, including groundwater contamination, heightened greenhouse gas (GHG) emissions, potential fire and explosion risks, threats to human health, and sanitation issues [9].

Incineration has the potential to recover some energy from plastic wastes. Nevertheless, it generates pollutants like nitrous and sulfur oxides, light hydrocarbons, dust, and dioxins, which can damage human health and degrade air quality during the process [10]. Pyrolysis is an advanced thermal treatment (ATT) method that can effectively manage waste plastic in a sustainable way. Pyrolysis is a thermochemical recycling technique that transforms waste plastics into liquid oil, along with valuable byproducts like flammable gases and char. It is a thermal degradation process that involves moderate heat and pressure in an inert environment, breaking down long-chain hydrocarbons into shorter ones. Compared to incineration, pyrolysis is preferred as it yields lower levels of toxic and greenhouse gases, reducing emissions like nitrogen oxides, sulfur oxides, and other harmful gases. The liquid product generated through pyrolysis boasts a higher heating value. However, the output yield of the pyrolysis process hinges on various factors: temperature, heating rate, reactor design, residence time, feedstock volume, moisture content, particle size, and elemental composition. This paper concentrates on optimising three key parameters: temperature, residence time, and feedstock particle size. There has been limited exploration using response surface methodology (RSM) to fine-tune the waste plastic pyrolysis process. In a study by Selvaganapathy et al. [11], the process parameters were optimised to increase the yield of liquid fuel in the pyrolysis of waste polystyrene. A pyrolytic reaction with a 1 kg capacity was used, and the experiments were designed using a central composite design (CCD) matrix in RSM with Design Expert 12 statistical programming software. For optimisation, four parameters were chosen: reaction temperature (ranging from 350 to 500 °C), polystyrene size (2 to 6 mm), weight of polystyrene feedstock (ranging between 250 to 750 gm), and retention time (10 to 90 min), with the aim to enhance liquid oil yield. They executed 30 experiments to ascertain the liquid oil yield, matching these results with values obtained via numerical simulations using a quadratic model. ANOVA was employed to discern the connection between independent variables and their influence on liquid oil yield. Their conclusion highlighted temperature and reaction time as the most critical factors for maximising liquid oil production.

In another study, RSM was utilised to optimise the process parameters in catalytic pyrolysis of waste HDPE to maximise liquid oil yield in a 300 mL capacity reactor [12]. Three parameters were chosen for optimisation—reaction temperature, mass ratio between HDPE and modified catalyst, and acidity of the modified catalyst—using face central composite design (FCCD). Under optimum conditions (reaction temperature of 450 °C, catalyst-to-waste HDPE ratio at 1:4, and a catalyst acidity of 0.341), they recorded a substantial 78.7 wt% yield of liquid product. Furthermore, they attained a notably high determination coefficient of R^2, affirming the model's reliability.

Ayodele et al. [13] also used RSM to study the influence of operating parameters on the oil yield from waste plastic pyrolysis. They considered four operating factors, plastic-type (PS, LDPE and PP, HDPE, PET, and PVC), temperature (300–740 °C), heating rate (3–25 °C/min), and reaction time (20–150 min). They documented that the temperature and type of plastic used had greater influence on the liquid oil yield from pyrolysis compared with the other two parameters studied. They observed that the liquid product obtained under optimised conditions had a higher heating value and was in the range of hydrocarbons (C10–C25). Pinto et al. [14] also investigated the effect of experimental operating conditions on liquid oil yield from pyrolysis of a mixture of recycled plastic (80 wt%), scrap tire (10 wt%), and pine (10 wt%) in an autoclave with a capacity of 1 L, by using RSM. The recycled plastic they used comprised PE, PS, and PP, and they considered three factors for optimising liquid oil yield—temperature, reaction time, and initial pressure—using an experimental factorial design. The maximum liquid yield obtained was 91.3 wt% for optimum conditions of 350 °C, 30-min reaction time, and 0.2 MPa initial pressure. Miranda et al. [15] also used RSM for the thermal pyrolysis of mixed waste plastic (PE, PP and PS) and rubber tires. They optimised the process parameters to maximise the liquid oil yield and used 70 wt% mixed waste plastic and 30 wt% rubber tires. Three important process parameters were chosen for optimisation: temperature, reaction time and initial pressure. The optimum values were 370 °C, 15 min, and 0.48 MPa, respectively. The experimental deviation of 0.95% resulted in a highest liquid oil yield of 81.3 wt%. In a recently published study, Prabha et al. [16] conducted optimisation of the pyrolysis process using single-use waste PP grocery bags. They used a laboratory scale semi-batch type pyrolyzer with only 1 kg capacity. The process parameters considered were reaction temperature (400–550 °C), nitrogen flow rate (5–20 mL min^{-1}), and substrate feed rate (0.25–1.5 kg h^{-1}). However, the process parameters considered in this present study are vital and have more influence on the pyrolysis product yield. That research has several differences with this present study, as here a pilot scale batch type pyrolytic reactor that provides more reliable data was used. Moreover, a mixture of HDPE, PP, and PS was used in this study and more characterisation parameters have been considered. Dutta et al. [17] also conducted optimisation of the thermal pyrolysis of virgin and waste HDPE in a small lab-scale batch pyrolytic reactor. They documented maximum liquid yield as 69.33% for optimum operating conditions, which is lower than was achieved in this present study (75.14%). Moreover, they considered only two independent variables, temperature and virgin to waste plastic ratio, whereas in this study three vital pyrolysis process parameters, which are temperature, reaction time, and size of the feedstock have been considered.

Some researchers have used response surface methodology to optimise the process parameters of the pyrolysis process for biofuel production. For example, Jahirul et al. [18] studied process optimisation by response surface methodology (RSM) for producing biodiesel from nonedible beauty leaves. Their optimisation approach was based on the Box-Behnken design in RSM and considered three main factors (reaction temperature, catalyst concentration, and methanol to oil molar ratio). Both linear and full quadratic regression models were developed to analyse the results. ANOVA was conducted to determine the significance of factors and their interaction, and optimum values were obtained. Li et al. [19] also used RSM to optimise process parameters to maximise bio-oil yield using a microwave-assisted pyrolysis process from cotton stalks. In that study, they chose the Box-Behnken design approach. For the optimisation study, three parameters were chosen: reaction temperature, reaction time, and microwave power. They conducted 17 experiments following the software-generated experimental design. It was concluded that a 32.47 wt% bio-oil yield can be achieved by using the optimum values of the process parameters, which are 1800 W as the microwave power, a 24 min reaction time, and a 550 °C reaction temperature. Gupta et al. [20] used a combination of RSM and an artificial neural network (ANN) to optimise biofuel production from the pyrolysis of pine needles. They chose three factors, temperature (400–700 °C), inert gas flow rate (100–250 mL/min), and heating rate (10–50 °C/min), at three different levels of −1, 0, and 1 for optimisation and conducted a total of 20 experimental runs planned by the software. They used RSM in Design Expert 12 statistical software to determine the influence and significance of process parameters on three responses, increasing bio-oil yield, and decreasing biochar and gas yield, using a second-order polynomial equation. For improved prediction of the pyrolysis process output, ANN was used. They documented that the most influential variable is temperature, which maximises pyrolytic product yield. They concluded that the predicted optimum bio-oil yield is 51.11 wt% and 51.70 wt %, by using RSM and ANN modelling for values of parameters such as temperature 552.06 °C, inert gas flow rate 164.40 mL/min, and heating rate 50 °C/min [20].

This research article has several novelties compared with the existing research articles and knowledge. A literature review revealed that insufficient numbers of researchers have studied the optimisation of process parameters from mixed waste plastic pyrolysis. Among the very few studies where the optimisation of mixed waste plastic pyrolysis process parameters was discussed, only a very small size pyrolyzer was used. Experimental results obtained using small pyrolyzers (<1 L) might not be as accurate as the present study, where a 20-L capacity pyrolyzer was used. The pyrolyzer used in this study also has an automatic proportional–integral–derivative (PID) control unit that provides proper and accurate control over the operating parameters.

Furthermore, research papers where complete characterisation of the fuel properties of crude oil obtained under optimum operating conditions has been conducted and compared with standard values for automobile standard diesel fuels are rare. In this paper, optimisation of the pyrolysis process parameters from mixed waste plastics was conducted by using RSM to obtain the maximum liquid oil yield. After conducting some prior experiments, three waste plastics were selected, HDPE, PP, and PS, with a 1:1:1 ratio to favour liquid fuel production. The main objective of the present study was to maximise the output of liquid oil yield by optimising three important parameters: reaction temperature (460–540 °C), retention time (30–150 min), and MWP feedstock particle size (5–45 mm), by using RSM. The use of RSM can reduce the number of experiments to a great extent and help predict pyrolysis product yields. The model developed in this work is applicable for optimising the mixed waste plastic pyrolysis process and can be used by other researchers, with small adjustments based on their operating conditions.

2. Materials and Methodology

2.1. Preparing the Feedstock for Pyrolysis

MWP samples were sourced from the local municipal solid waste (MSW) site in Rockhampton, Australia. Among the MWP mix, HDPE, PS, and PP samples were segregated for this specific study based on their resin identification codes. Typically, high- and low-density polyethylene (HDPE and LDPE), polypropylene (PP), and polystyrene (PS) collectively constitute a substantial fraction of overall plastic waste, accounting for 46%, 16%, and 16%, respectively [21]. Among the other common waste plastics, PVC and PET are worth mentioning. However, neither of these are particularly suitable for pyrolysis experiments and have several disadvantages. For example, PET increases char and gas production but reduces the production of liquid pyrolytic oil if used in smaller amounts in catalytic pyrolysis and contains heteroatoms [22,23]. PVC generates hazardous phosgene and HCl gases during pyrolysis, affects catalytic activity in catalytic pyrolysis because of the presence of chlorine or coke deposition, and at low temperatures (250–320 °C), dichlorination occurs [24,25]. Waste plastic water tanks were used as the source of HDPE feedstock. They were shredded in an industrial-scale shredder. Waste PP was sourced from used and broken plastic chairs, food containers and packaging, bottles, jars, hot beverage cups, etc., and shredded in the same lab facility. The shredded HDPE and PP feedstocks were further shredded (second shredder) into smaller sizes (5 mm and 25 mm) in the CQU North Rockhampton campus laboratory. Waste polystyrene foam boxes were used as the source of polystyrene (PS). PS foams underwent cutting into smaller pieces in the CQU North Rockhampton campus laboratory. Figure 1, below, reveals the three types of feedstocks used for the pyrolysis experiments and the second shredder used for shredding them in the present study.

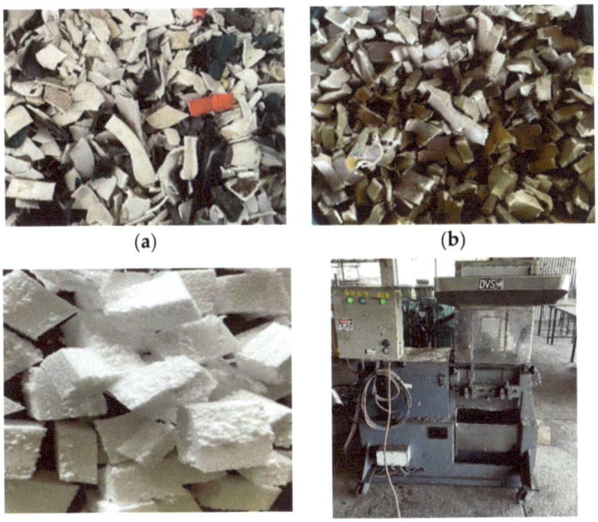

(**a**) (**b**) (**c**) (**d**)

Figure 1. Shredded waste plastic feedstocks: (**a**) PP, (**b**) HDPE, (**c**) PS, and (**d**) shredder.

The HDPE, PP, and PS feedstocks were further shredded into two other sizes of approximately 5 mm and 25 mm in the shredder to satisfy the experimental design plan obtained using Minitab software 21.1.0. No pretreatment of the feedstocks was conducted other than manually removing some dirt and dust before shredding the feed materials. For preparing the feedstock, HDPE-, PP-, and PS-type waste plastics were used in equal proportions, i.e., 1:1:1. Figure 2 shows the three different sizes of the mixed feedstock used for the pyrolysis optimisation experiments.

Figure 2. MWP feedstocks of three different sizes: 5 mm, 25 mm, and 45 mm.

The literature review also mentioned shredding of waste plastics before using them for pyrolysis [23,26,27].

2.2. Experimental Set up and Pyrolysis of MWP

The mixed waste plastics (MWP) were thermally pyrolyzed in a 20 L pilot-scale batch reactor situated at the Waste-to-Energy Laboratory on the Central Queensland University campus in Rockhampton, Australia [28]. Figure 3 displays both the physical representation of the reactor and a schematic diagram illustrating the pyrolysis reactor setup. The pyrolytic reactor system comprises a reactor chamber, a condenser, an oil collection unit, a char collector, a mechanical stirrer, a stirrer motor, a nitrogen gas cylinder, and a PID controller. The reactor walls are heated by electric heating coils that surround the vessel and are insulated. This pyrolytic plant can operate both as a batch reactor and an Auger reactor. However, the pyrolysis of MWP was conducted in a batch process by feeding shredded mixed plastic chips (HDPE, PP, and PS) of different sizes directly into the reactor through the feeding hopper. K-type thermocouples were installed at different points inside the reactor chamber to measure the temperature. The temperature and heating rate were managed by an automatic PID controller unit. Nitrogen gas was employed to purge the system, eliminating residual air and oxygen, and it had its own pressure regulator valve. The reactor system's pressure was kept at 30 KPa, to ensure that the system operated under atmospheric or nearly atmospheric conditions, and a pressure gauge was used to monitor it. Additionally, a pressure relief valve was included in the system to eliminate excess pressure in case of excessive pressure build-up. The reactor system also had a condenser to condense the vapour after it left the reactor. An external chiller unit was utilised to control the condenser's temperature, and the chiller unit's temperature range could be adjusted between −5 °C and 20 °C. However, for the present study, −5 °C was selected. Ethylene glycol was utilised to cool the chiller unit, ensuring the water temperature remained below 0 °C while circulating through the condenser.

The pyrolytic reactor chamber received 1.5 kg of processed mixed waste plastic feedstocks—comprising HDPE, PS, and PP at a ratio of 1:1:1—through a feeding hopper situated atop the primary reactor chamber. The pyrolysis experiments employed a fixed-bed batch pyrolysis reactor (Figure 3a), as depicted. The system, illustrated in Figure 3b, was loaded with shredded plastic samples, sealed, and checked for air leaks using high-pressure nitrogen. To ensure airtightness, the system was purged with nitrogen gas from a nitrogen bottle for 15 min, eliminating any residual oxygen. The chiller unit was activated, followed by the heater/electric power, set to a heating rate of 6 °C/min, while the pyrolysis temperature was regulated using the PID controller. In this study, the pyrolysis temperature varied between 460, 500, and 540 °C, depending on the design obtained from the Minitab statistical design software for the optimisation studies. The mechanical stirrer was turned on to agitate the feedstock continuously. The feedstock was thermally decomposed at the provided temperature, producing vapour/gas and char. Vapours generated within the heated reactor chamber were channeled through a shell-and-tube condenser unit, where the condensable vapours formed liquid oil. The non-condensable vapours, referred to as syngas, were released through the exit valve. After concluding the experiment, the electric heater was deactivated, enabling the gradual cooling of the reactor. The plastic crude oil was then collected from the bottom of the oil tank, while the solid residue (char) was directly gathered from the bottom outlet of the pyrolytic reactor vessel. In this study, 1.5 kg of mixed waste plastic feedstock with a 1:1:1 ratio of HDPE, PP, and PS was processed. The feedstock was loaded into the pyrolytic reactor chamber using a feeding hopper situated at the top of the main chamber. In conducting the experiments, a fixed-bed batch pyrolysis reactor, depicted in Figure 3a, was utilised. Meanwhile, Figure 3b illustrates the schematic diagram of the pyrolysis reactor system employed in this study. After filling the reactor with processed (shredded) waste plastic samples and sealing it, the system underwent high-pressure nitrogen testing to check for leaks. Upon confirming the airtightness of the system, a 15 min purge with nitrogen gas was conducted to eliminate any remaining oxygen. Following the nitrogen purging, the sequence involved activating the chiller unit and subsequently turning on the electric heater.

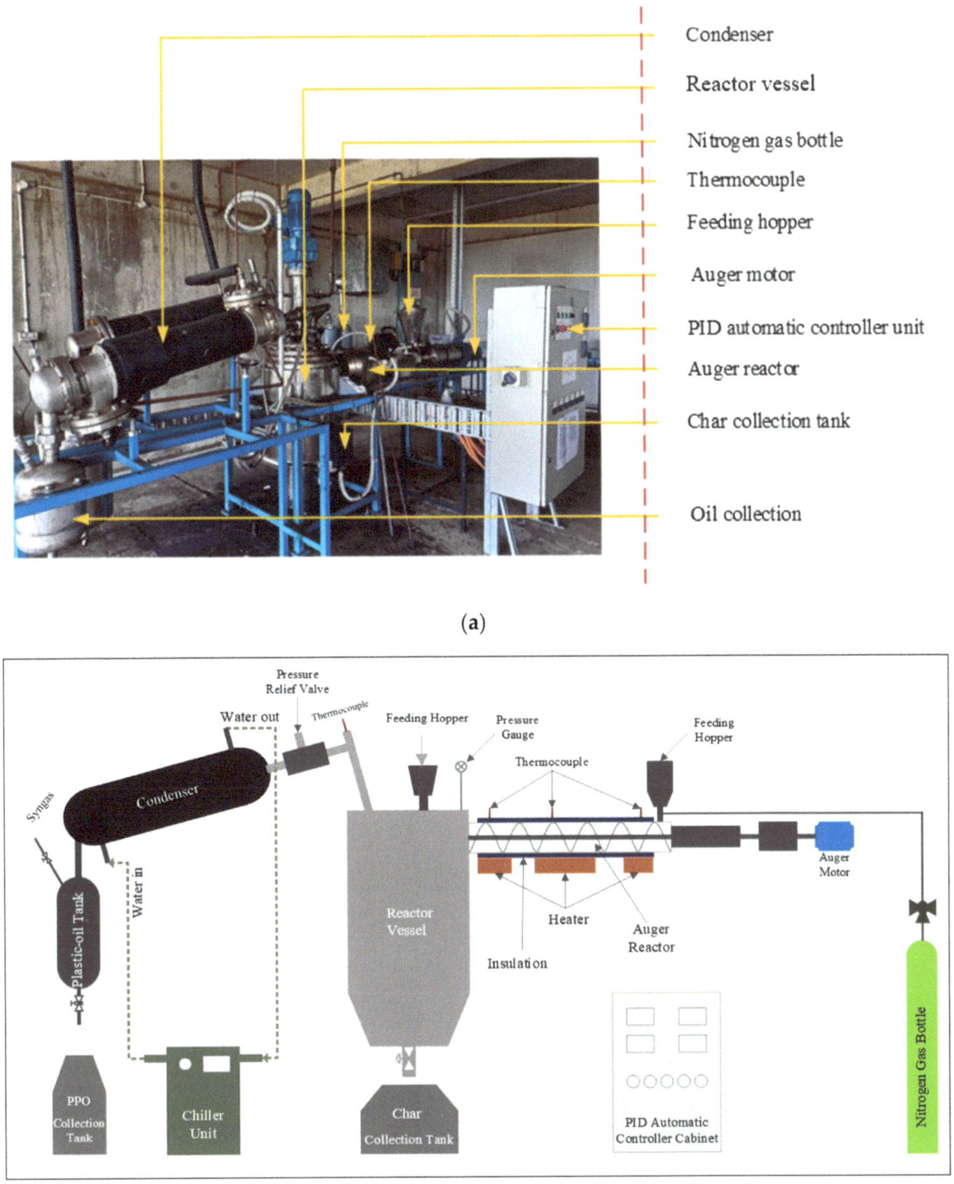

(**a**)

(**b**)

Figure 3. (**a**) Plastic pyrolytic plant, (**b**) process schematic diagram.

The heating rate was set to 6 °C/min, and the pyrolysis temperature was varied (460, 500, and 540 °C) based on the experimental design obtained from the Minitab statistical design software. The mechanical stirrer was turned on to agitate the feedstock continuously. The feedstock underwent thermal decomposition at the provided temperature, producing vapour/gas and char. The vapours produced in the heated reactor chamber were directed through a shell-and-tube condenser unit, where the condensable vapours were gathered as liquid oil. Non-condensable vapours or gases, known as syngas, were released through the exit valve. When the experiment was finished, the electric heater was turned off to facilitate the gradual cooling of the reactor. Following that, the plastic crude oil was obtained from the bottom of the oil tank, while the solid residue (char) was collected directly from the lower outlet of the pyrolytic reactor vessel.

The yield of crude PPO (CPPO), char and gas can be calculated from the formula below [29,30]:

$$Y_0 = \frac{m_0 \times 100\%}{m_f}$$

$$Y_c = \frac{m_c \times 100\%}{m_f}$$

$$Y_g = 100 - (Y_0 + Y_c)$$

where Y_0, Y_c and Y_g represent the yields of CPPO, char, and syngas derived from pyrolysis of MWP, respectively, m_0 and m_c are the masses of CPPO and char, respectively (in kg), and m_f the total mass of feedstock used for every experiment is 1.5 kg. The production of syngas was calculated by adding the production of CPPO and char and subtracting them from 100, as the reactor had no mass flowmeter to measure the mass of syngas. In this study, priority was given to the yield of CPPO from the pyrolysis process, and the produced char and syngas were not analysed. However, char and syngas are two valuable pyrolysis products and can be used in many sectors, for example, to produce power and heat and upgrade soil quality.

By-Products of MWP Pyrolysis Process

(1) Gas

Composition of gas produced from pyrolysis of waste plastic depends on the composition of the feedstock material. The main components of gases obtained from pyrolysis of different types of plastic feedstocks are methane, hydrogen, propane, propene, ethane, ethene, and butane [31], but PET produces carbon dioxide and carbon monoxide along with those and PVC produces hydrogen chloride [32]. Gases produced from pyrolysis of waste plastic include methane (34%), hydrogen (20%), nitrogen (17.5%), CO (13%), CO_2 (11%), ethylene (3%), and oxygen (1.5%) [33]. As the temperature of the pyrolysis process increases, the amount of gas production increases [34]. It was documented that pyrolysis of individual plastics produced more gas than in mixture when pyrolysis was carried out for LDPE, PS, and their mixtures at 350 °C [35]. Long residence time and high temperature maximise gas production in the pyrolysis process [36].

The gas produced from the pyrolysis process has higher heating value and can be used in boilers for heating or in gas turbines for the generation of electricity without any flue gas treatment [37]. However, Joo and Guin. [38] mentioned that produced gas from thermal pyrolysis is not suitable as a fuel source and needs refining before use. The gases formed in the pyrolysis process consist of methane (6.6%), ethane ethylene (10.6%), propane (7.4%), propylene (29.1%), isobutane (1.9%), n-butane (0.9%), C4 (unsaturated) (25.6%), Iso C_5–n-C_5 (0.1%), C_5+higher (15.3%), hydrogen (2.5%), and CO/CO_2 (<400 ppm) [39,40]. Kaimal & Vijayabalan [41] also mentioned that gas produced from pyrolysis of waste plastic consists of propylene, isobutane, ethane, and small amounts of methane. After separation from other gas components, ethene and propene can be used as chemical feedstock for the production of polyolefins [42].

(2) Char

Char is the unburnt plastic left in the reactor after the pyrolysis process and is considered as a by-product [43]. Char formation maximises with a slow heating rate at very low temperatures, decreased feed rate, and long residence time, whereas char formation is low in fast pyrolysis [42,44]. Char production can also be reduced by using a microporous catalyst [45]. However, catalytic activity reduces when char is produced due to decreasing active sites [46]. Char is produced in small amounts (1–1.3 g in 1 kg plastic) during the pyrolysis process [47,48]. Char mainly consists of volatile matter and fixed carbon (97%) with small amount of moisture and ash [49]. Char has low sulfur content, making it convenient to use as a fuel [42]. Char can be used in different environmental applications like heavy metal adsorption from municipal and industrial wastewater and toxic gases [50]. A proximate analysis of the solid residue produced in waste plastic pyrolysis was conducted [51] and it was found that higher heating value can be obtained. Char has the potential to be used for an energy source in boilers or as a feedstock for activated carbon [37], in road surfacing, as a building material [44], and as an absorbent in water treatment for removing heavy metals through upgrading treatment [42].

(3) Wax Analysis

Wax can sometimes be produced from pyrolysis of waste plastic. Polyolefin polymers are considered as valuable products and a potential fuel source because of having high elemental carbon. Wax is a residue of the pyrolysis process caused by devolatilisation and has carbon range between C20–C50 [52]. The wax produced has elemental carbon ranging from 81.04 to 87.67% and can be used as high-carbon-based material.

Environmental impact assessment of the pyrolysis process: Recycling rates of waste plastics are very low and most of them find their way into landfills and pollute the environment. Pyrolysis of waste and contaminated plastics can reduce the harmful impact on the environment to a great extent by reducing their global warming impact. Garcia et al. [53] conducted an environmental impact assessment of contaminated waste plastics in three different scenarios, one where pyrolysis was conducted with combustion of the char and the other two where pyrolysis of the char with carbon dioxide and with potassium hydroxide was considered. In all three scenarios, the environmental impact was found to be less than landfilling. Hermanns et al. [54] found that global warming impacts can be reduced by about 60% to 94% compared with incineration through pyrolysis when considering a life-cycle assessment of mixed waste plastics. Moreover, Jeswani et al. [55] also conducted life-cycle assessment of mixed plastic wastes and found that pyrolysis has about 50% lower climate change impact and life-cycle energy use than energy recovery from MPW.

3. Design of Experiments (DoE) Using RSM

Selecting an appropriate experimental design is a critical step in DoE methodology. To optimise the liquid oil yield of the thermal pyrolysis process of MWP, this study employed response surface methodology (RSM) to identify the most influential operating parameters. RSM is a statistical analysis technique capable of optimising or minimising a dependent variable, which is determined by various independent variables. The RSM design procedure encompasses three facets: design, modeling, and optimisation. It is employed to enhance the efficiency of manufacturing processes and formulations [14,15,56]. A literature review observed that many researchers have utilised RSM to optimise process parameters [15,30,57]. RSM can predict a particular response with the help of creating a conceptual model. In RSM, useful equations and effective visualisation are established with the help of the developed model for optimising the response of the dependent variable. The flow diagram of the processes of RSM used in this study for model development and validation is shown in Figure 4.

The first step in RSM is to select an appropriate design of experiment (DoE). In the single-factor optimisation method, one parameter is varied at a time, keeping the other factors constant, unlike in RSM, where interactions among variable process parameters can also be considered [58]. The two main design techniques in RSM, central composite design and Box-Behnken design, were used for this study. RSM has the advantage of efficiently reducing the number of experiments and making it possible to change all the variables simultaneously and not one at a time [15]. Response surface methodology (RSM) is an amalgamation of mathematical and statistical techniques. It is used to optimise and analyse responses obtained during experiments with multiple variables. It is very suitable in cases of modelling, analysing, and optimising studies where multiple factors impact the response of interest. The reason for choosing Box–Behnken design over central composite design is the simplicity of the design and its effectiveness. One of the research articles concludes that the BBD method is better than other RSM designs because of its efficacy, after comparing BBD with several other RSM designs [59]. Furthermore, BBD offers the advantage of excluding combinations where all components are simultaneously at their maximum or minimum levels, thus preventing unacceptable results by averting tests conducted under extreme conditions. Box-Behnken design can consider three levels for each factor considered. First- and second-order coefficients can be estimated efficiently using BBD, but they cannot include runs from a factorial experiment.

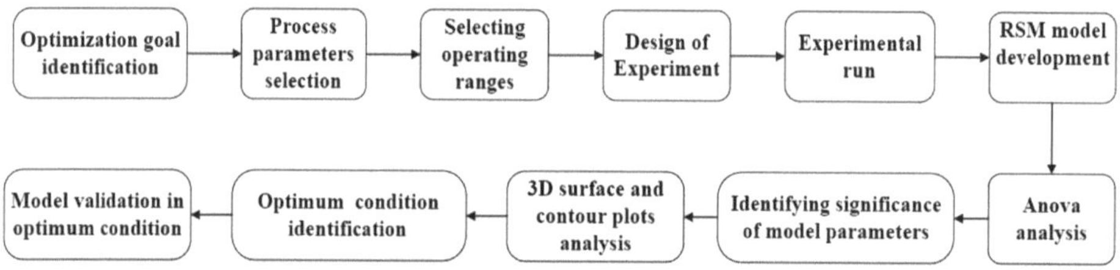

Figure 4. Flow diagram of the processes of RSM for model development and validation.

For conducting DoE, selecting the most important factors affecting the response of interest and the ranges along which they can vary was needed. According to previous experiments conducted by the authors of this present study and the literature review, three vital factors were chosen for optimising MWP pyrolysis studies: temperature, residence time, and feedstock particle size. The literature suggests that these three factors significantly impact the product yield (predominantly liquid oil) from the thermal pyrolysis process. Other factors significantly affect the pyrolytic oil yield, such as varying the amount of catalyst used, stirrer speed, and nitrogen flow rate. However, in this study, only thermal pyrolysis was considered and not catalytic pyrolysis, so no catalyst was used. The stirrer speed was not considered, as from some preliminary tests conducted by the researchers, it was found that varying stirrer speed and feedstock volume do not significantly impact pyrolytic oil yield in terms of pilot-scale thermal pyrolysis. The nitrogen flow rate was also not considered, as nitrogen was used as an inert gas in the experiment and only to purge the system to ensure that experiments could run without oxygen. After purging the system with nitrogen for 15 min, the nitrogen supply was stopped and was not used during the experiment. These are the reasons for not exploring those factors (type and amount of catalyst used, stirrer speed, nitrogen flow rate) within the optimisation study in the present work.

Selecting an appropriate design of experiment (DoE) is a critical step that significantly influences the reliability and meaningfulness of the study's results. It determines the probable relationship among various factors (which is three for this study, n = 3) and builds a response surface for all of them. Therefore, a 2^n factorial design was not considered. The reason for not choosing a 3^n factorial design was to reduce the cost, time, and complexity of conducting 27 experimental runs. The Box–Behnken design was selected as the DoE for the RSM other than the 3^n factorial design, to reduce the time and cost associated with extra experiments [30,60].

In the application of RSM, the correlation between the response variable and the independent variables can be represented by the equation below:

$$Y = f(A,\ B,\ C)$$

where Y = liquid oil yield from MWP pyrolysis (response variable), while A, B, and C are independent variables representing temperature, residence time, and feed amount, respectively. The factors were investigated at three different levels: central (0), low (−1), and high (+1).

The formula [30] provided below was used to establish the total number of experiments:

$$T = p2 + p + q$$

where T = total number of experimental runs, p = number of factors, and q = number of the centre point repeated.

There were 15 experiments with three centre-point trials (q = 3); three factors and three levels were chosen for the present optimisation study. The Minitab 16 statistical software package (Version 20.2, USA) was used to randomly generate the run orders of the experiments, shown in design Table 1, and a total of 15 runs were generated. The main aim of the experimental design was to optimise the process parameters to maximise the liquid oil yield (Y) from the pyrolysis of MWP. For statistical analysis, a second-order polynomial model was developed using the Box-Behnken design of the RSM to determine the effect of the three chosen process parameters on the liquid oil yield. Fitting the results to a quadratic polynomial model enables effective prediction of the system response, in this case, the liquid oil yield. It is a way to capture the non-linear relationships and optimise the experimental conditions for the desired outcome.

Table 1. Range and level of independent factors chosen for the optimisation study.

Variables	Unit	Symbol Coded	Range and Levels		
			−1	0	1
Temperature	°C	A	460	500	540
Residence time	min	B	30	90	150
Feedstock size	mm	D	5	25	45

The form of the second-order polynomial model for predicting liquid oil yield is as follows [30]:

$$Y = \beta_0 + \beta_1 X_1 + \beta_2 X_2 + \beta_3 X_3 + \beta_{1,2} X_1 X_2 + \beta_{1,3} X_1 X_3 + \beta_{2,3} X_2 X_3 + \beta_{1,1} {X_1}^2 + \beta_{2,2} {X_2}^2 + \beta_{3,3} {X_3}^2$$

Were Y = predicted liquid oil yield, β_0 = constant coefficient, β_1 = linear coefficient, β_2 = quadratic coefficient, β_3 = interaction coefficient, and X_1, X_2, and X_3 are coded values of the independent variables.

Regression coefficients provide valuable insights into the extent of change in the response variable for each unit change in the independent variable, while holding all other independent variables constant. They are key indicators of the relationships within the model. This statistical control is essential in distinguishing the impact of one variable from all the others present in the model. Table 1 clearly describes the parameters and their range for the experimental study. Table 2 illustrates the actual and predicted yield of plastic pyrolytic oil from MWP pyrolysis and the experimental design matrix.

Table 2. Experimental design matrix along with actual and predicted product yields of PPO.

Run	Temp. (°C)	Residence Time (min)	Feedstock Size (mm)	Liquid Oil Yield (%)	
				Actual	Predicted
1	500 (0)	150 (1)	5 (−1)	61.09	60.4563
2	500 (0)	90 (0)	25 (0)	55.19	54.0533
3	460 (−1)	150 (1)	25 (0)	31.65	34.0075
4	460 (−1)	90 (0)	5 (−1)	25.67	23.9462
5	500 (0)	30 (−1)	45 (1)	19.91	20.5438
6	540 (1)	90 (0)	5 (−1)	50.53	50.6387
7	500 (0)	90 (0)	25 (0)	53.90	54.0533
8	540 (1)	90 (0)	45 (1)	47.30	49.0238
9	540 (1)	150 (1)	25 (0)	74.47	74.9950
10	500 (0)	30 (−1)	5 (−1)	21.94	24.1888
11	500 (0)	150 (1)	45 (1)	59.57	57.3212
12	540 (1)	30 (−1)	25 (0)	28.31	25.9525
13	460 (−1)	30 (−1)	25 (0)	10.53	10.0050
14	460 (−1)	90 (0)	45 (1)	18.89	18.7812
15	500 (0)	90 (0)	25 (0)	53.07	54.0533

4. Statistical Analysis of the Developed Model

The optimisation process of MWP pyrolysis was statistically analysed using RSM, with the assistance of ANOVA. The aim of the analysis was to identify the process parameters that yielded better results. The actual yield of PPO was calculated through practical experiments, while the predicted PPO yield values were obtained via Minitab software. The fitness of the developed model was assessed using ANOVA. The impact of various factors was determined by the F value and the ratio of the mean square of treatments to the mean square of the error, in conjunction with the coefficient values. A factor is deemed statistically significant when the F value is sufficiently large, and the p value is less than 0.05. In addition to p and F values, various correlation coefficients, such as R^2, predicted R^2 (R^2 pred), and adjusted R^2 (R^2 adj.), along with degrees of freedom, were considered to further validate the accuracy of the model. Moreover, 3D surface plots and contour plots were drawn to determine and analyse the effects of independent factors on the yield of PPO.

5. Results and Discussion

5.1. Optimisation of Liquid Oil Yield

As mentioned earlier, in the present study, three parameters were considered for optimisation: temperature, residence time, and feedstock particle size, and experimental design was conducted using BBD in response surface methodology. Each of the parameters was assessed at three different points. All the points were analysed to identify the topics producing higher amounts of pyrolytic liquid oil. The temperature was varied from 460 °C to 540 °C, the reaction time was changed from 30 min to 150 min, and the feedstock particle size was varied from 5 mm to 45 mm for the optimisation study. According to Table 2, Run 9 yielded the highest liquid oil, while Run 13 had the lowest liquid oil yield among the experimental design matrices. The table includes the actual yield of plastic oil along with the predicted yield. The difference between the actual and predicted values was minimal and negligible.

Figure 5 shows the graphical representation of the experimental vs. predicted liquid oil yield from the pyrolysis of MWP. The experimental yield of liquid oil varied in the range of 10.53 wt% to 74.47 wt%, and the predicted yield also matches closely with the experimental data, which shows the consistency of the developed model. R^2 was obtained as 0.995, representing a reliable relationship between the independent and dependent variables. The response variable means at different factor levels were compared, to verify factor importance via ANOVA assessments.

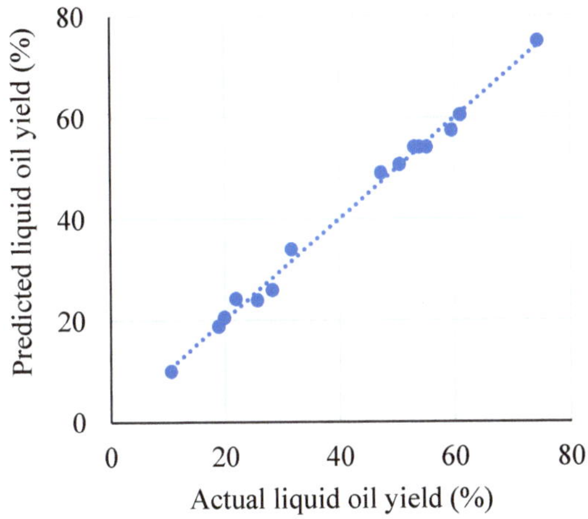

Figure 5. Experimental vs. predicted yield of PPO.

Table 3 illustrates the elemental composition analysis of MWP. The data reveal that MWP has higher carbon and hydrogen contents, which are 79.9% and 12.6%, respectively. The proximate analysis also shows higher volatile matter content in waste plastics, 92.90%, and a small amount of moisture and fixed carbon content.

Table 3. Ultimate and proximate analysis of MWP [61].

Analysis	Property	Amount (wt%)
Ultimate	C	79.9
	H	12.6
	S	0.00
	N	0.00
	O	5.1
	Chlorine	1.13
	HHV (MJ/kg)	44.4
Proximate	Moisture content	1.37
	Volatile matter	92.90
	Fixed carbon	1.14

The model includes a polynomial equation representing the relationships between the dependent (response) and independent variables. Y is the pyrolytic oil yield, A is the temperature, B is the residence time, and D is the feedstock particle size.

$$\text{Liquid oil yield, Y (wt\%)} = -1828 + 7.232\ A - 0.683\ B + 0.230\ D$$
$$-0.007139\ A \times A - 0.001775\ B \times B - 0.01759\ D \times D + 0.002608\ A \times B + 0.00111\ A \times D + 0.00011\ B \times D$$

The ANOVA results are presented in Table 4. Terms with a p value below 0.05 are considered statistically significant. Therefore, it can be said that A, B, A^2, B^2, D^2, and A $\times$ B are considered as the significant terms of the developed model. In addition, it can also be concluded that among all the terms, A, B, and A^2 have the highest significance in relation to the amount of liquid oil yield.

5.2. Analysis of Variance (ANOVA)

Analysis of variance was conducted to determine the statistical fit of the developed model.

Table 4 shows that the p value of the model is less than 0.0001, and the F value is 93.52, which provides evidence supporting the developed model's statistical fitness. The quadratic model obtained fits effectively in predicting the response, evidenced by a high determination coefficient of R^2 (99.41%). When the difference between two correlations, R^2 pred and R^2 adj., is smaller than 0.2, then it is considered that they are in reasonable agreement. Thus, in this developed model, R^2(pred) and R^2(adj.) values can be regarded as in good agreement.

Table 4. Analyzing the quadratic model for PPO yield through ANOVA.

Source	DF	Adj SS	Adj MS	F Value	*p* Value	Remark
Model	9	5190.04	576.67	93.52	0.0000	Significant
A	1	1620.80	1620.80	262.84	0.0000	Significant
B	1	2667.79	2667.79	432.62	0.0000	Significant
D	1	22.98	22.98	3.73	0.1114	Not significant
A^2	1	481.68	481.68	78.11	0.0003	Significant
B^2	1	150.84	150.84	24.46	0.0043	Significant
D^2	1	182.69	182.69	29.63	0.0028	Significant
A × B	1	156.75	156.75	25.42	0.0040	Significant
A × D	1	3.15	3.15	0.51	0.5067	Not significant
B × D	1	0.07	0.07	0.01	0.9222	Not significant
Lack-of-Fit	3	28.55	9.52	8.34	0.1090	Not significant
Pure Error	2	2.28	1.14			
Total	14	5220.88				

R-sq = 99.41%; R-sq(adj) = 98.35%; R-sq(pred) = 91.15%.

5.3. Interactions of the Process Parameters on PPO Yield

This study examined the influence of three distinct process factors—temperature, residence time, and feedstock particle size—on the liquid oil yield from thermal pyrolysis or MWP. The interaction between these factors was also examined using 3D surface plots and 2D contour plots. To determine the optimal liquid yield, two parameters were varied at a time while the third was kept constant at its average value. The contour plots served as a visual representation to depict the relationship between each factor's response and the experimental levels. They also graphically displayed the regression equation [12]. Figure 6a,b illustrate the impact of temperature and residence time on the liquid oil yield from the thermal pyrolysis of MWP, with the feedstock size held constant at 25 mm.

Figure 6 demonstrates that the yield of plastic pyrolytic oil is influenced by the combined effects of temperature and residence time. The liquid oil yield was noted to rise with the elevation of both temperature and residence time. The increase in liquid oil yield with reaction temperature is predictable because the MWP molecules receive more energy at higher temperatures and form monomers by breaking more intermolecular bonds of the macromolecules [57]. Moreover, previous research also mentioned an increase in liquid oil yield at increased residence time due to increased heat energy being transported to the MWP molecules and helping them decompose more. Figure 6 shows that at temperatures higher than 500 °C, the liquid oil yield increased with increasing residence time. Moreover, when the reaction temperature was lower than approximately 500 °C, even at the highest residence time, the liquid oil yield was not in the higher range.

From the ANOVA results documented in Table 4, it can be noted that the F values for temperature (A) and residence time (B) were 262.84 and 432.62, respectively, for individual terms and 78.11 and 24.46 for their square terms. The *p* values for both the individual and squared terms are below 0.05, indicating that their influence on the liquid oil yield is statistically significant. The highest amount of liquid was 74.47% at a temperature of 540 °C and 150 min residence time. The liquid oil yield decreased after approximately 536 °C, while the residence times were longer. The reason could be the formation of more gases that are not condensable at increased temperatures by secondary cracking [12]. Thus, choosing the higher temperature range of 540 °C was appropriate. Nevertheless, there was no observed decrease in the production of liquid oil yield for the residence time of 150 min. This behaviour denotes that pyrolytic oil continued to be produced at a residence time of

150 min at temperatures above 500 °C. The MWP pyrolysis model is significantly affected by reaction time and temperature at the 95% confidence level.

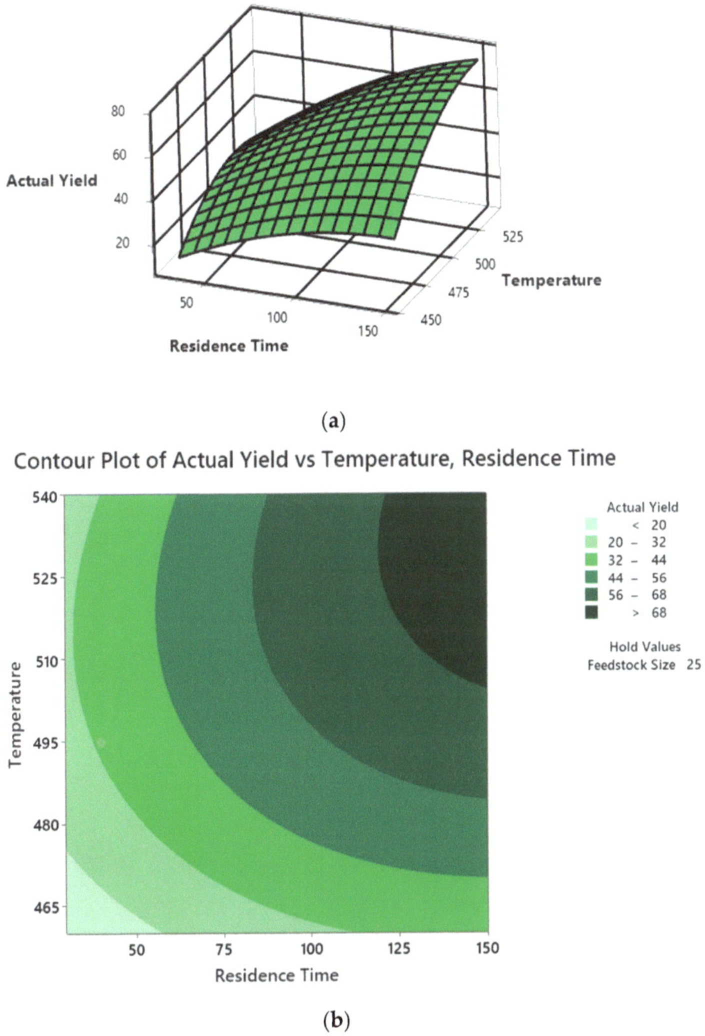

Figure 6. (**a**) The 3D surface and (**b**) contour plot for PPO yield illustrate the combined impact of temperature and residence time while maintaining a constant feedstock particle size of 25 mm.

In Figure 7a, the three-dimensional surface plots depict the empirical model's variation with respect to two factors, temperature and feedstock particle size, within the considered experimental range and at a fixed residence time of 90 min. The above figure shows that the liquid oil yield increases with increasing temperature. Figure 7 shows that at temperatures higher than 495 °C, the liquid oil yield increases sharply for both small and large particle sizes of the feedstock. At temperatures above 500 °C, a higher liquid oil yield was only achieved when the feedstock size was approximately 24 mm. From the nature of the graph, it can be said that after reaching a specific higher temperature, the liquid oil yield starts to decrease, because of the secondary cracking of volatiles. This phenomenon increases the yield of syngas [62]. The liquid oil yield was lower when the feedstock size was too low in the 5 to 20 mm range and for larger particle sizes greater than approximately 25 mm. Figure 7 also shows that particle size also has some impact on the pyrolytic oil yield. For larger size feedstock particles, heat transfer reduces as the rate of heat and mass transfer to the feedstock depend on its shape, size, and homogeneity. Therefore, the temperature in the feedstock reduces and thus the yield of PPO [30]. However, the p value for feedstock particle size (single term) is 0.111, more significant than 0.05, which means that its effect on liquid oil yield is insignificant. Selvaganapathy et al. [11] also observed a less significant effect of feedstock particle size on liquid oil yield from pyrolysis of waste PS.

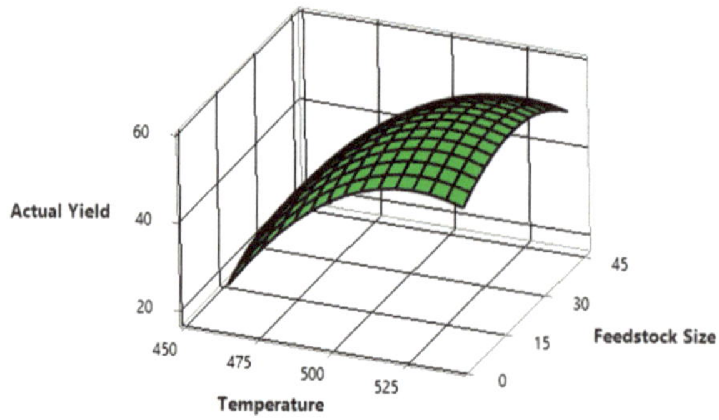

(**a**)

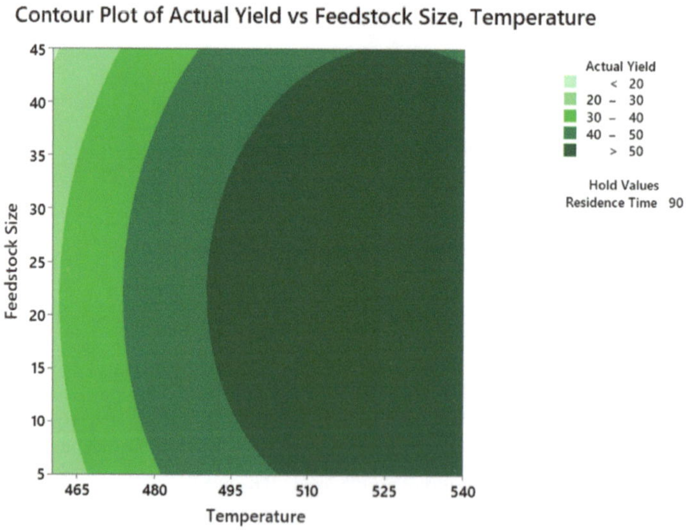

(**b**)

Figure 7. (**a**) The 3D surface and (**b**) contour plot for PPO yield showing the combined effect of temperature and feedstock particle size, keeping residence time fixed at mean value (90 min).

The combined effect of residence time and feedstock particle size can be demonstrated from the 3D surface and 2D contour plots in Figure 8a,b. It is evident from the contour plot that when the residence time increases, the liquid oil yield increases as well, especially after approximately 80 min. However, after reaching a certain highest yield with the increase of residence time, further liquid PPO yield starts to decrease as most of the feedstock is pyrolyzed by then. Particle size did not significantly affect liquid oil yield ($p < 0.05$) in either single or interacting terms. However, when the feedstock particle size was very small (less than 20 mm) and too large (greater than 30 mm), the PPO yield was relatively small. Thus, the optimum yield was found when the feedstock particle size was close to the mean value (25 mm).

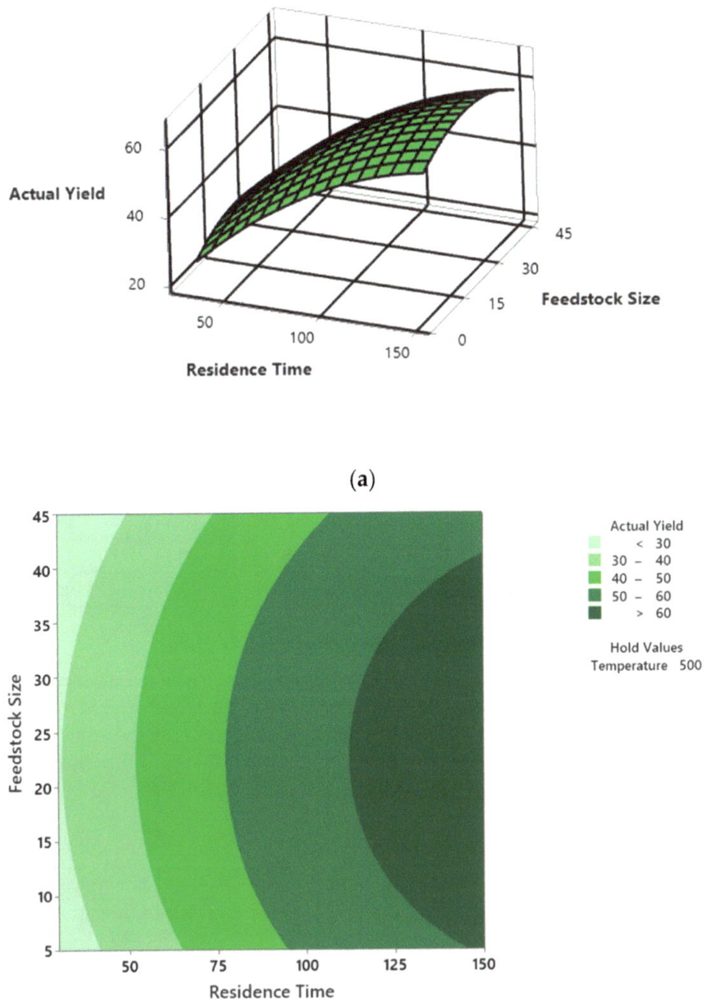

Figure 8. (**a**) The 3D surface and (**b**) contour plot for PPO yield, while maintaining the temperature constant at 500 °C, demonstrating the combined impact of residence time and feedstock particle size.

The process parameters were optimised by keeping the independent variables' values within the stated range and maximising the yield of PPO. Table 5 shows the ranges for the independent and dependent variables for the optimisation study and the maximum and minimum amounts of liquid oil obtained from the experiments.

Table 5. Independent and dependent variable ranges for the optimisation study.

Parameters	Objective	Minimum Threshold	Maximum Boundary
Temperature (°C)	Within range	460	550
Residence time (min)	Within range	30	150
Feedstock particle size (mm)	Within range	5	45
Yield of PPO (%)	Maximise	10.53	74.47

To attain a peak liquid plastic oil yield of 75.14%, the optimal operational conditions were a temperature of 535.96 °C, a residence time of 150 min, and a feedstock particle size of 23.99 mm. The actual values of the operating parameters chosen for the optimisation study varied slightly from the abovementioned values due to the limitation of the pyrolysis instrument. The parameters chosen for the optimisation studies were a temperature of 536 °C, a residence time of 150 min, and a feedstock particle size of 24 mm. Three experiments were conducted under the optimum operating conditions to validate the model. Table 6 displays both the experimental and predicted values of the PPO yield under optimal conditions.

Table 6. Experimental and predicted yield of PPO under optimum operating conditions.

Run	Temp. (°C)	Residence Time (min)	Particle Size (mm)	Exp. PPO Yield (%)	Predicted PPO Yield (%)	Error (%)
1	536	150	24	74.77	75.14	0.49
2	536	150	24	75.92	75.14	1.03
3	536	150	24	75.98	75.14	1.11
Avg.	536	150	24	75.55	75.14	0.88

5.4. *Characterisation of PPO under Optimum Conditions*

The PPO acquired under optimum operating conditions underwent characterisation. Characterisation involved conducting fuel property tests and Fourier transform infrared spectroscopy (FTIR). Standard test methods were employed to determine the liquid fuel's density, kinematic viscosity, flash point, calorific value, water content, lubricity, sulfur content, and cetane index, and are listed in Table 7.

Table 7. The standards for conducting property tests of the PPO.

Fuel Properties	Unit	Testing Methods	Equipment Used
Flash point	°C	D93B	Pensky Martins closed-cup apparatus (Koehler instrument company, INC., Bohemia, NY, USA)
Calorific value	kJ/kg	D4809	Parr 6200 calorimeter (Parr Instrument Company, Chicago, IL, USA)
Density	kg/L	D4052	DM40 Mettler Toledo density meter (METTLER TOLEDO, Columbus, OH, USA)
Cetane index (CI)	-	D4737 A	Ignition quality tester (CFR Engines, Ottawa, ON, Canada)
Lubricity	µm	D6079	Ducom lubricity meter (Ducom instruments, New York, NY, USA)
Kinematic viscosity (KV)	cSt	D7042	Stabinger viscometer SVM 3000 (Anton Paar, Gurugram, India)
Sulfur content	(mg/kg)	D5453	XPLORER sulfur analyser (TE Instrument, Delft, The Netherlands)
Water content	wt%	D2709	Centrifuge sigma (Thermo Fisher Scientific, Waltham, MA, USA)

The properties of the crude PPO were measured and compared with ASTM, EN, and Australian standard diesel and are described in Table 8.

Table 8. Comparison of the fuel properties of crude PPO with three different diesel fuel standards.

Properties	ASTM Std.	EN Std.	Aus. Std. Diesel	Crude PPO	CPPO [63]	CPPO [64]	CPPO [65]
HHV (MJ/kg)	42–46	-	44–46	44.15	43.58	37.72	45.24
KV (mm^2/s)	1.9–4.1	2–4.5	2–4.5	1.318	2.3	1.66	3.11
CI	40 min	46 min	46 min	32	66.7	-	46.7
Density@15 °C (g/cm^2)	-	0.82–0.84	0.82–0.85	0.832	0.8027	0.899	0.823
Flash point (°C)	52 min	55 min	61.5 min.	<20	48	35	54
Water content (wt. %)	0.05	0.02 max	0.02 max.	0.125	0.051	-	-
Sulfur content (mg/kg)	15 max	10 max	10 max.	5.12	42.9	-	-
Lubricity (WSD, µm)	520 max	460 max	460 max.	220	-	-	-

Table 8 shows that the HHV of the produced crude PPO under the optimum operating conditions satisfies two of the fuel standards. However, the HHV obtained by Khatha et al. [64] was 37.72 MJ/kg and lower than the described value for diesel fuel in all the standards. The kinematic viscosity measured for the produced crude PPO was lower than the value mentioned in all the diesel fuel standards. Khatha et al. [64] also found a low KV (1.66 mm^2/s) for the crude PPO produced in their study. The cetane index for the crude PPO in this study was found to be low 32 and failed to satisfy any of the diesel fuel test standards. Nevertheless, the water content, sulfur content, lubricity, and density values agree well with all the diesel testing standards. The literature review includes reports of higher sulfur content in crude PPO; for example, Bezergianni et al. [63] documented 42.9 mg/kg of sulfur from PPO obtained from waste plastics, which does not satisfy any of the diesel fuel property measurement standards. However, the flash point for the crude PPO produced under optimised pyrolytic conditions in this study was very low, at less than 20 °C, and using this in an engine would be a fire hazard. Bezergianni et al. [63] and Janyalertadun et al. [65] also documented lower flash points for crude PPO compared with the standards. Finally, it can be concluded that the crude PPO produced from the pyrolysis of MWP had low KV, CI, and flash points. At the moment, these factors hinder its suitability as a readily applicable fuel in automobile engines. However, posttreatment processes, such as distillation and hydrotreatment, can significantly improve these parameters. The researchers in this study conducted distillation and hydrotreatment of crude PPO obtained under optimum operating conditions and analysed the fuel properties of distilled and hydrotreated PPO. Distillation was conducted with a pilot-scale vacuum distillation unit (iFischer AutoDest 800/860 AC, i-Fischer Engineering, Germany). Gasoline distillation temperatures were set at <170 °C, diesel distillation temperatures ranged from 170 °C to 380 °C, with lube oil and residue distillation temperatures > 380 °C. After completing the distillation run, the diesel fraction was collected and analysed. The distilled diesel fraction was hydrotreated afterwards, which is a vital refinery process to improve the quality of the pyrolytic oil by removing heteroatoms and saturating alkenes [66,67]. Hydrotreatment was carried out in this study to catalyse hydrodesulfurization (HDS), nitrogen removal (HDN), and hydrodeoxygenation (HDO) from the distilled diesel fraction of the plastic pyrolytic oil.

Experiments conducted by the researchers in the present study have shown that properties such as water content, sulfur content, cetane index, lubricity, etc., improve significantly after post-treating the crude PPO. For example, the cetane index was found to be 32 for PPO from waste plastics, which was very low and does not satisfy the minimum requirements of the applied diesel fuel standards; however, CI increased to 48 in distilled plastic diesel and further to 59 in hydrotreated plastic diesel, which is well above the minimum requirement. It was also observed that DPD and HPD satisfied the quality specifications of Australian standard diesel standards. Kinematic viscosity of CPPO increased from 1.32 to 2.72 mm^2/s after distillation; after hydrotreatment, it reached 2.95 mm^2/s. Reduction in sulfur content from 5.12 mg/kg to 2.83 mg/kg was achieved when waste plastics PPO was hydrotreated, satisfying the diesel standards. The lubricity of DPD and HPD also satisfies the Australian diesel standards.

Detail analysis of the distillation and hydrotreatment processes and the physicochemical properties of the crude, distilled, and hydrotreated pyrolytic oil has been discussed in another research paper by the current authors [28].

5.5. FTIR Analysis of PPO

Fourier transform infrared spectroscopy (FTIR) plays a crucial role as an analytical technique for the identification of distinct functional groups within pyrolytic oil. Figure 9 illustrates the FTIR spectrum obtained via analysing the PPO produced by thermal pyrolysis of MWP under optimum operating conditions. The FTIR technique works by interacting pyrolytic oil with infrared light, which causes the chemical bonds to undergo stretching, absorption, and contracting of infrared radiation at a specific wavelength [12]. FTIR analysis can identify different functional groups by analysing the peak intensities of PPO.

In Figure 9, the peak at approximately wavenumber 2924.65 is probably due to the presence of alkanes with C–H stretching vibrations with a strong appearance, and the peak detected at approximately wavenumber 2856.35 cm^{-1} also represents alkanes due to C–H stretching. Selvaganapathy et al. [11] also documented the presence of alkanes at wavenumbers 2919.22 cm^{-1} and 2854.01 cm^{-1} with C–H stretching with a strong appearance, for PPO obtained from the pyrolysis of polystyrene. The relatively smaller peaks at 3082.93 cm^{-1} probably represent the presence of functional group alkenes because of =C–H stretching. The wavenumber of 1642.26 cm^{-1} is perhaps due to the presence of alkene functional groups due to C=C stretching [12]. The peak with a strong appearance at 1454.31 cm^{-1} could be due to the C–H bending and scissoring of alkane functional groups. Moreover, an alkane group was detected at 1376.86 cm^{-1}, probably because of CH_3C–H bending in the medium appearance. The peak found at 965.27 cm^{-1} is probably because of the vibration of C–H bending and represents the functional group alkene. The wavenumber 888.66 cm^{-1} exhibits the presence of an alkene functional group due to =C–H bending in a strong appearance. Moreover, an alkene functional group was detected at a frequency of approximately 775.23 cm^{-1}, which might result from C–H bending vibrations. The presence of alkanes, alkenes, aromatics, etc., suggests the potential of using PPO as a liquid hydrocarbon [30,68]. These results are comparable to the findings of research articles by Kumar and Singh [9] and Selvaganapathy et al. [11], where FTIR analysis was conducted for PPO obtained from the pyrolysis of HDPE and PS plastics, respectively, under optimum conditions [11,12]. In conclusion, it can be affirmed that the diverse chemical compounds found in pyrolytic oil derived from waste plastics can be readily categorised through the analysis of FTIR spectra. Categorisation means FTIR analysis can identify specific functional groups, molecular bonds, or characteristic absorption bands present in the spectra of the oil samples.

For example, the presence of various chemical groups like alkanes, aromatics, alkenes, etc., can be identified by FTIR spectra. Moreover, comparison among different oil samples is possible by FTIR analysis, based on the intensity and position of absorption peaks corresponding to different fu nctional groups. These characterisation results can be useful for identifying the influence of different pyrolysis process parameters on the chemical composition of the produced pyrolytic oils and thus can be used for optimisation studies as well.

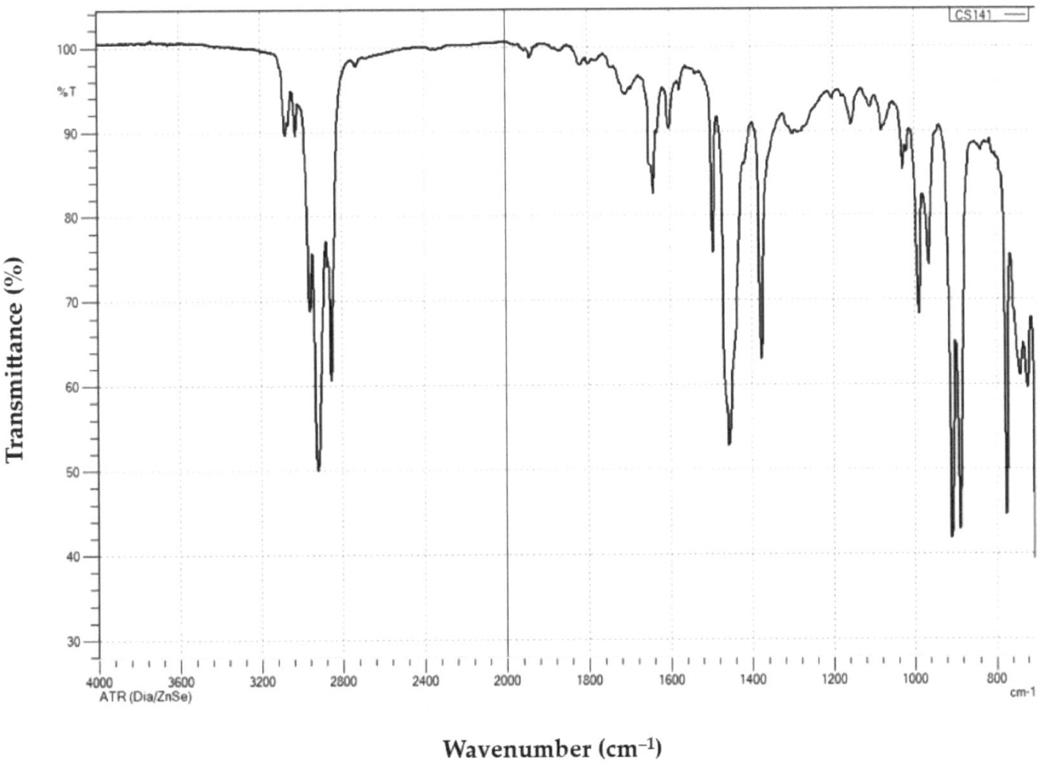

Figure 9. FTIR spectra of PPO under optimum conditions.

6. Conclusions

In this study, thermal pyrolysis of MWP to produce liquid pyrolytic oil was conducted. To maximise the yield of liquid oil, a combination of RSM and BBD was used to optimise three vital experimental operating parameters: reaction temperature, feedstock particle size, and reaction time. A quadratic model was developed and the validation of the model was conducted under the optimum operating conditions achieved from the model. The experimental average maximum yield of plastic pyrolytic oil (PPO) was found to be 75.55%, with an average error of 0.88%. The experimental and predicted liquid yields were close, indicating the accuracy of the model. The quadratic model is a strong fit for predicting the liquid oil yield response, as indicated by its high coefficient of determination R^2 (99.41%). The researchers found that temperature and reaction time had a greater impact on liquid oil yield compared with feedstock particle size, based on ANOVA analysis. The model can be used by other researchers, with a few modifications depending on the experimental conditions. The oil produced has properties comparable to petroleum fuels and thus has the potential to be used as automobile fuel. However, the crude PPO has a lower CI, KV, and flash point and higher sulfur and water content than the standard values for automobile diesel according to ASTM, EN, and Australian standards. Therefore, additional processing is required for crude PPO to make it suitable for use as a direct replacement fuel in automobile engines.

Author Contributions: Conceptualization, F.F. and A.A.C.; methodology, F.F.; software, F.F.; validation, F.F. and M.I.J.; formal analysis, F.F. and M.I.J.; investigation, F.F.; resources, A.A.C.; data curation, F.F. and M.I.J.; writing—original draft preparation, F.F.; writing—review and editing, M.G.R.; A.A.C. and M.I.J.; visualization, F.F.; supervision, M.G.R., A.A.C. and M.I.J.; project administration, M.G.R.

and M.I.J.; funding acquisition, M.G.R., and M.I.J. All authors have read and agreed to the published version of the manuscript.

Funding: This research was funded by CRC-P project with the project number CRCPEIGHT000194, which focuses on Australian Standard Diesel from Mixed Waste Plastics, and the Advanced Queensland Industry Research Fellowship project with the project number RSH/6017, which aims to optimise renewable fuel production from mixed waste for a more affordable and environmentally friendly Queensland.

Institutional Review Board Statement: Not applicable.

Informed Consent Statement: Not applicable.

Data Availability Statement: Data is contained within the article.

Conflicts of Interest: The authors declare no conflict of interest.

References

1. Nanda, S.; Berruti, F. Thermochemical conversion of plastic waste to fuels: A review. *Environ. Chem. Lett.* **2021**, *19*, 123–148. [CrossRef]
2. Statistica. Global Plastic Production 1950–2020. 2021. Available online: https://www.statista.com/statistics/282732/global-production-of-plastics-since-1950/ (accessed on 25 January 2024).
3. Zhang, K.; Shi, H.; Peng, J.; Wang, Y.; Xiong, X.; Wu, C.; Lam, P.K. Microplastic pollution in China's inland water systems: A review of findings, methods, characteristics, effects, and management. *Sci. Total Environ.* **2018**, *630*, 1641–1653. [CrossRef] [PubMed]
4. Padervand, M.; Lichtfouse, E.; Robert, D.; Wang, C. Removal of microplastics from the environment. A review. *Environ. Chem. Lett.* **2020**, *18*, 807–828. [CrossRef]
5. Gourmelon, G. Global plastic production rises, recycling lags. *Vital Signs* **2015**, *22*, 91–95.
6. Eriksen, M.; Lebreton, L.C.; Carson, H.S.; Thiel, M.; Moore, C.J.; Borerro, J.C.; Galgani, F.; Ryan, P.G.; Reisser, J. Plastic pollution in the world's oceans: More than 5 trillion plastic pieces weighing over 250,000 tons afloat at sea. *PLoS ONE* **2014**, *9*, e111913. [CrossRef]
7. Statistica. Plastic Material Production Worldwide by Region 2019. 2019. Available online: https://journals.plos.org/plosone/article?id=10.1371/journal.pone.0111913 (accessed on 25 January 2024).
8. Kumar, R.; Mishra, M.; Singh, S.; Kumar, A. Experimental evaluation of waste plastic oil and its blends on a single cylinder diesel engine. *J. Mech. Sci. Technol.* **2016**, *30*, 4781–4789. [CrossRef]
9. Kumar, A.; Sharma, M. GHG emission and carbon sequestration potential from MSW of Indian metro cities. *Urban Clim.* **2014**, *8*, 30–41. [CrossRef]
10. Lee, K.-H.; Shin, D.-H. Characteristics of liquid product from the pyrolysis of waste plastic mixture at low and high temperatures: Influence of lapse time of reaction. *Waste Manag.* **2007**, *27*, 168–176. [CrossRef] [PubMed]
11. Selvaganapathy, T.; Muthuvelayudham, R.; Jayakumar, M. Process parameter optimization study on thermolytic polystyrene liquid fuel using response surface methodology (RSM). *Mater. Today Proc.* **2020**, *26*, 2729–2739. [CrossRef]
12. Kumar, S.; Singh, R.K. Optimization of process parameters by response surface methodology (RSM) for catalytic pyrolysis of waste high-density polyethylene to liquid fuel. *J. Environ. Chem. Eng.* **2014**, *2*, 115–122. [CrossRef]
13. Ayodele, T.R.; Durodola, O.; Ogunjuyigbe, A.S.; Munda, J.L. Effects of operating factors on the bio-oil produced from pyrolysis of plastic wastes using response surface methodology. In Proceedings of the 2019 IEEE PES/IAS PowerAfrica, Abuja, Nigeria, 20–23 August 2019; pp. 527–532.
14. Pinto, F.; Paradela, F.; Gulyurtlu, I.; Ramos, A.M. Prediction of liquid yields from the pyrolysis of waste mixtures using response surface methodology. *Fuel Process. Technol.* **2013**, *116*, 271–283. [CrossRef]
15. Miranda, M.; Pinto, F.; Gulyurtlu, I.; Cabrita, I.; Nogueira, C.; Matos, A. Response surface methodology optimization applied to rubber tyre and plastic wastes thermal conversion. *Fuel* **2010**, *89*, 2217–2229. [CrossRef]
16. Prabha, B.; Ramesh, D.; Sriramajayam, S.; Uma, D. Optimization of Pyrolysis Process Parameters for Fuel Oil Production from the Thermal Recycling of Waste Polypropylene Grocery Bags Using the Box–Behnken Design. *Recycling* **2024**, *9*, 15. [CrossRef]
17. Dutta, N.; Mondal, P.; Gupta, A. Optimization of process parameters using response surface methodology for maximum liquid yield during thermal pyrolysis of blend of virgin and waste high-density polyethylene. *J. Mater. Cycles Waste Manag.* **2022**, *24*, 1182–1193. [CrossRef]
18. Jahirul, M.I.; Koh, W.; Brown, R.J.; Senadeera, W.; O'Hara, I.; Moghaddam, L. Biodiesel production from non-edible beauty leaf (*Calophyllum inophyllum*) oil: Process optimization using response surface methodology (RSM). *Energies* **2014**, *7*, 5317–5331. [CrossRef]
19. Li, X.; Wang, B.; Wu, S.; Kong, X.; Fang, Y.; Liu, J. Optimizing the conditions for the microwave-assisted pyrolysis of cotton stalk for bio-oil production using response surface methodology. *Waste Biomass Valorization* **2017**, *8*, 1361–1369. [CrossRef]

20. Gupta, S.; Patel, P.; Mondal, P. Biofuels production from pine needles via pyrolysis: Process parameters modeling and optimization through combined RSM and ANN based approach. *Fuel* **2022**, *310*, 122230. [CrossRef]
21. Vasile, C.; Brebu, M.; Sakata, Y.; Pakdel, H.; Roy, C.; Miranda, R. Solid waste treatment by pyrolysis methods. *J. Environ. Prot. Ecol.* **2002**, *3*, 230e235.
22. Thorat, P.; Warulkara, S.; Sathonea, H. Pyrolysis of waste plastic to produce Liquid Hydroocarbons. *Adv. Polym. Sci. Technol. Int. J.* **2013**, *3*, 14–18.
23. Sembiring, F.; Purnomo, C.W.; Purwono, S. Catalytic Pyrolysis of Waste Plastic Mixture. In *IOP Conference Series: Materials Science and Engineering*; IOP Publishing: Bristol, UK, 2018; p. 012020.
24. Lopez-Urionabarrenechea, A.; De Marco, I.; Caballero, B.; Laresgoiti, M.; Adrados, A. Catalytic stepwise pyrolysis of packaging plastic waste. *J. Anal. Appl. Pyrolysis* **2012**, *96*, 54–62. [CrossRef]
25. Singh, R.; Ruj, B.; Sadhukhan, A.; Gupta, P.; Tigga, V. Waste plastic to pyrolytic oil and its utilization in CI engine: Performance analysis and combustion characteristics. *Fuel* **2020**, *262*, 116539. [CrossRef]
26. Almohamadi, H.; Alamoudi, M.; Ahmed, U.; Shamsuddin, R.; Smith, K. Producing hydrocarbon fuel from the plastic waste: Techno-economic analysis. *Korean J. Chem. Eng.* **2021**, *38*, 2208–2216. [CrossRef]
27. Miandad, R.; Barakat, M.; Aburiazaiza, A.S.; Rehan, M.; Ismail, I.; Nizami, A. Effect of plastic waste types on pyrolysis liquid oil. *Int. Biodeterior. Biodegrad.* **2017**, *119*, 239–252. [CrossRef]
28. Faisal, F.; Rasul, M.; Chowdhury, A.A.; Schaller, D.; Jahirul, M. Uncovering the differences: A comparison of properties of crude plastic pyrolytic oil and distilled and hydrotreated plastic diesel produced from waste and virgin plastics as automobile fuels. *Fuel* **2023**, *350*, 128743. [CrossRef]
29. Song, Q.; Zhao, H.-y.; Xing, W.-l.; Song, L.-h.; Yang, L.; Yang, D.; Shu, X. Effects of various additives on the pyrolysis characteristics of municipal solid waste. *Waste Manag.* **2018**, *78*, 621–629. [CrossRef]
30. Hasan, M.; Rasul, M.; Jahirul, M.; Khan, M. Fast pyrolysis of macadamia nutshell in an auger reactor: Process optimization using response surface methodology (RSM) and oil characterization. *Fuel* **2023**, *333*, 126490. [CrossRef]
31. Williams, P.T.; Williams, E.A. Interaction of plastics in mixed-plastics pyrolysis. *Energy Fuels* **1999**, *13*, 188–196. [CrossRef]
32. López, A.; de Marco, I.; Caballero, B.M.; Adrados, A.; Laresgoiti, M.F. Deactivation and regeneration of ZSM-5 zeolite in catalytic pyrolysis of plastic wastes. *Waste Manag.* **2011**, *31*, 1852–1858. [CrossRef]
33. Kalargaris, I.; Tian, G.; Gu, S. Combustion, performance and emission analysis of a DI diesel engine using plastic pyrolysis oil. *Fuel Process. Technol.* **2017**, *157*, 108–115. [CrossRef]
34. López, A.; De Marco, I.; Caballero, B.; Laresgoiti, M.; Adrados, A.; Torres, A. Pyrolysis of municipal plastic wastes II: Influence of raw material composition under catalytic conditions. *Waste Manag.* **2011**, *31*, 1973–1983. [CrossRef]
35. Onwudili, J.A.; Insura, N.; Williams, P.T. Composition of products from the pyrolysis of polyethylene and polystyrene in a closed batch reactor: Effects of temperature and residence time. *J. Anal. Appl. Pyrolysis* **2009**, *86*, 293–303. [CrossRef]
36. Prabir, B. *Biomass Gasification and Pyrolysis: Practical Design and Theory*; Elsevier Inc.: Amsterdam, The Netherlands, 2010.
37. Fernández, Y.; Arenillas, A.; Menéndez, J.Á. *Microwave Heating Applied to Pyrolysis*; INTECH Open Access Publisher: London, UK, 2011.
38. Joo, H.; Guin, J.A. Continuous upgrading of a plastics pyrolysis liquid to an environmentally favorable gasoline range product. *Fuel Process. Technol.* **1998**, *57*, 25–40. [CrossRef]
39. Mani, M.; Subash, C.; Nagarajan, G. Performance, emission and combustion characteristics of a DI diesel engine using waste plastic oil. *Appl. Therm. Eng.* **2009**, *29*, 2738–2744. [CrossRef]
40. Mani, M.; Nagarajan, G.; Sampath, S. Characterisation and effect of using waste plastic oil and diesel fuel blends in compression ignition engine. *Energy* **2011**, *36*, 212–219. [CrossRef]
41. Kaimal, V.K.; Vijayabalan, P. A detailed study of combustion characteristics of a DI diesel engine using waste plastic oil and its blends. *Energy Convers. Manag.* **2015**, *105*, 951–956. [CrossRef]
42. Sharuddin, S.D.A.; Abnisa, F.; Daud, W.M.A.W.; Aroua, M.K. A review on pyrolysis of plastic wastes. *Energy Convers. Manag.* **2016**, *115*, 308–326. [CrossRef]
43. Miandad, R.; Barakat, M.; Aburiazaiza, A.S.; Rehan, M.; Nizami, A. Catalytic pyrolysis of plastic waste: A review. *Process Saf. Environ. Prot.* **2016**, *102*, 822–838. [CrossRef]
44. Jung, S.-H.; Cho, M.-H.; Kang, B.-S.; Kim, J.-S. Pyrolysis of a fraction of waste polypropylene and polyethylene for the recovery of BTX aromatics using a fluidized bed reactor. *Fuel Process. Technol.* **2010**, *91*, 277–284. [CrossRef]
45. Lee, K.-H. Effects of the types of zeolites on catalytic upgrading of pyrolysis wax oil. *J. Anal. Appl. Pyrolysis* **2012**, *94*, 209–214. [CrossRef]
46. Miskolczi, N.; Ateş, F.; Borsodi, N. Comparison of real waste (MSW and MPW) pyrolysis in batch reactor over different catalysts. Part II: Contaminants, char and pyrolysis oil properties. *Bioresour. Technol.* **2013**, *144*, 370–379. [CrossRef]
47. López, G.; Olazar, M.; Artetxe, M.; Amutio, M.; Elordi, G.; Bilbao, J. Steam activation of pyrolytic tyre char at different temperatures. *J. Anal. Appl. Pyrolysis* **2009**, *85*, 539–543. [CrossRef]
48. Williams, P.T. Yield and composition of gases and oils/waxes from the feedstock recycling of waste plastic. In *Feedstock Recycling and Pyrolysis of Waste Plastics: Converting Waste Plastics into Diesel and Other Fuels*; Wiley: Hoboken, NJ, USA, 2006; pp. 285–313.
49. Jamradloedluk, J.; Lertsatitthanakorn, C. Characterization and utilization of char derived from fast pyrolysis of plastic wastes. *Procedia Eng.* **2014**, *69*, 1437–1442. [CrossRef]

50. Heras, F.; Jimenez-Cordero, D.; Gilarranz, M.A.; Alonso-Morales, N.; Rodriguez, J. Activation of waste tire char by cyclic liquid-phase oxidation. *Fuel Process. Technol.* **2014**, *127*, 157–162. [CrossRef]
51. Syamsiro, M.; Saptoadi, H.; Norsujianto, T.; Noviasri, P.; Cheng, S.; Alimuddin, Z.; Yoshikawa, K. Fuel oil production from municipal plastic wastes in sequential pyrolysis and catalytic reforming reactors. *Energy Procedia* **2014**, *47*, 180–188. [CrossRef]
52. Arabiourrutia, M.; Elordi, G.; Lopez, G.; Borsella, E.; Bilbao, J.; Olazar, M. Characterization of the waxes obtained by the pyrolysis of polyolefin plastics in a conical spouted bed reactor. *J. Anal. Appl. Pyrolysis* **2012**, *94*, 230–237. [CrossRef]
53. Garcia-Garcia, G.; Martín-Lara, M.; Calero, M.; Blázquez, G. Environmental impact of different scenarios for the pyrolysis of contaminated mixed plastic waste. *Green Chem.* 2024. *advance article*. [CrossRef]
54. Hermanns, R.; Kraft, A.; Hartmann, P.; Meys, R. Comparative life cycle assessment of pyrolysis–recycling Germany's sorted mixed plas-tic waste. *Chem. Ing. Tech.* **2023**, *95*, 1259–1267. [CrossRef]
55. Jeswani, H.; Krüger, C.; Russ, M.; Horlacher, M.; Antony, F.; Hann, S.; Azapagic, A. Life cycle environmental impacts of chemical recycling via pyrolysis of mixed plastic waste in comparison with mechanical recycling and energy recovery. *Sci. Total Environ.* **2021**, *769*, 144483. [CrossRef] [PubMed]
56. Cornell, J. *How to Apply Response Surface Methodology*; ASQC: Milwaukee, WI, USA, 1990; Volume 8, p. 3.
57. Wong, S.L.; Ngadi, N.; Amin, N.A.S.; Abdullah, T.A.T.; Inuwa, I.M. Pyrolysis of low density polyethylene waste in subcritical water optimized by response surface methodology. *Environ. Technol.* **2016**, *37*, 245–254. [CrossRef] [PubMed]
58. Azargohar, R.; Dalai, A. Production of activated carbon from Luscar char: Experimental and modeling studies. *Microporous Mesoporous Mater.* **2005**, *85*, 219–225. [CrossRef]
59. Ferreira, S.C.; Bruns, R.; Ferreira, H.S.; Matos, G.D.; David, J.; Brandão, G.; da Silva, E.P.; Portugal, L.; Dos Reis, P.; Souza, A. Box-Behnken design: An alternative for the optimization of analytical methods. *Anal. Chim. Acta* **2007**, *597*, 179–186. [CrossRef] [PubMed]
60. Kuehl, R.O. *Designs of Experiments: Statistical Principles of Research Design and Analysis*; Duxbury Press: New York, NY, USA, 2000.
61. Cho, M.-H.; Jung, S.-H.; Kim, J.-S. Pyrolysis of mixed plastic wastes for the recovery of benzene, toluene, and xylene (BTX) aromatics in a fluidized bed and chlorine removal by applying various additives. *Energy Fuels* **2010**, *24*, 1389–1395. [CrossRef]
62. Moreira, R.; dos Reis Orsini, R.; Vaz, J.M.; Penteado, J.C.; Spinacé, E.V. Production of biochar, bio-oil and synthesis gas from cashew nut shell by slow pyrolysis. *Waste Biomass Valorization* **2017**, *8*, 217–224. [CrossRef]
63. Bezergianni, S.; Dimitriadis, A.; Faussone, G.-C.; Karonis, D. Alternative diesel from waste plastics. *Energies* **2017**, *10*, 1750. [CrossRef]
64. Khatha, W.; Ekarong, S.; Somkiat, M.; Jiraphon, S. Fuel Properties, Performance and Emission of Alternative Fuel from Pyrolysis of Waste Plastics. In *IOP Conference Series: Materials Science and Engineering*; IOP Publishing: Bristol, UK, 2020; p. 012001.
65. Janyalertadun, A.; Santaweesuk, C.; Sanongraj, S. Fuel production, performance, and emission of a CI engine using waste plastics oil. *World J. Eng.* **2017**, *14*, 114–120. [CrossRef]
66. Bezergianni, S.; Kalogianni, A.; Dimitriadis, A. Catalyst evaluation for waste cooking oil hydroprocessing. *Fuel* **2012**, *93*, 638–641. [CrossRef]
67. Belbessai, S.; Azara, A.; Abatzoglou, N. Recent Advances in the Decontamination and Upgrading of Waste Plastic Pyrolysis Products: An Overview. *Processes* **2022**, *10*, 733. [CrossRef]
68. Bakhtyari, A.; Makarem, M.; Rahimpour, M. Light olefins/bio-gasoline production from biomass. In *Bioenergy Systems for the Future*; Elsevier: Amsterdam, The Netherlands, 2017; pp. 87–148.

sustainability

MDPI

Article

Optimal Parameter Determination of Membrane Bioreactor to Boost Biohydrogen Production-Based Integration of ANFIS Modeling and Honey Badger Algorithm

Hegazy Rezk [1,*], A. G. Olabi [2], Mohammad Ali Abdelkareem [2,3], Abdul Hai Alami [2] and Enas Taha Sayed [3]

1 Department of Electrical Engineering, College of Engineering in Wadi Alddawasir, Prince Sattam bin Abdulaziz University, Wadi Alddawasir 11991, Saudi Arabia
2 Sustainable Energy & Power Systems Research Centre, RISE, University of Sharjah, Sharjah P.O. Box 27272, United Arab Emirates
3 Faculty of Engineering, Minia University, Minia 61111, Egypt
* Correspondence: hr.hussien@psau.edu.sa

Abstract: Hydrogen is a new promising energy source. Three operating parameters, including inlet gas flow rate, pH and impeller speed, mainly determine the biohydrogen production from membrane bioreactor. The work aims to boost biohydrogen production by determining the optimal values of the control parameters. The proposed methodology contains two parts: modeling and parameter estimation. A robust ANIFS model to simulate a membrane bioreactor has been constructed for the modeling stage. Compared with RMS, thanks to ANFIS, the RMSE decreased from 2.89 using ANOVA to 0.0183 using ANFIS. Capturing the proper correlation between the inputs and output of the membrane bioreactor process system encourages the constructed ANFIS model to predict the output performance exactly. Then, the optimal operating parameters were identified using the honey badger algorithm. During the optimization process, inlet gas flow rate, pH and impeller speed are used as decision variables, whereas the biohydrogen production is the objective function required to be maximum. The integration between ANFIS and HBA boosted the hydrogen production yield from 23.8 L to 25.52 L, increasing by 7.22%.

Keywords: membrane bioreactor; biohydrogen; optimization; modeling

1. Introduction

The energy demand is increasing globally due to the quick expansion of urbanization and the economy. Fossil fuel utilization is contributing to greenhouse gas emissions and global climate change [1]. Global warming, air pollutions and acid rains are examples of environmental challenges produced by a severe and long-term reliance on fossil fuels [2]. Renewable energy resources including hydro [3], wind [4,5], tidal [6], solar [7], biomass [8], etc. are sustainable and clean energy sources that can be used instead of fossil fuels. Biofuel is a liquid, solid or gaseous fuel produced from biomass resources, including organic wastes such as bioethanol, biohydrogen and biogas [9]. Thereby, biofuel is a promising solution for simultaneous waste management and energy generation.

Hydrogen has developed as a new energy source in addition to its traditional function as an industrial feedstock, primarily for producing methanol, ammonia and petroleum refining [10]. Hydrogen's incredible efficiency and low pollution make it attractive for use in various other contexts, including militarized equipment, power generation and transportation [11]. Biohydrogen has been mentioned as a future technology that could be used to switch to cleaner energy and make transportation systems that are more environmentally friendly [12]. Thermochemical conversion (such as gasification and combustion) [13], biochemical conversion (such as anaerobic digestion and dark fermentation) [14] or the use of MEC "microbial electrolysis cell" are the most common methods for converting biobased substances into biohydrogen [15].

Wastewater is water left over after freshwater, raw water, drinking water or salty water has been used in various ways. Wastewater is also water that has been used for farming, household, industrial or commercial activities, stormwater/surface runoff and any sewer infiltration or inflow. Adsorption, filtration, membrane processes, advanced oxidation and biological processes are some of the techniques used to treat wastewater [16,17]. Among these processes, biological processes are one of the best, most practical and most successful options for secondary treatment. The biological approach uses less energy than other methods of producing hydrogen since it may be carried out at mild operating conditions [18]. Therefore, fermentation-based biohydrogen production is more practical, affordable and straightforward and has drawn increased attention in recent years [19,20].

Biohydrogen is produced using various reactor technologies, including bubble columns [21], membrane reactors [22] and stirred tanks [23]. Membrane bioreactors (MBRs) of multiple designs, sizes and styles have lately come into prominence as efficient systems for producing hydrogen and treating sewage [24,25]. The Anaerobic MBRs "AnMBRs", a new biohydrogen production technique, are constructed by consolidating traditional anaerobic fermenters with membrane technology [26,27].

MGSBR "membrane gas separation-based bioreactor" is constructed by incorporating gas separation membranes into a hydrogen-producing reactor (i.e., stirred tank reactor) as an alternative of the purifier filter [28]. The admixture from MGSBR is split into two streams named retentate (which is made up of products with no energy value, such as N2 and CO2) and permeate (which is full of hydrogen and whose purity relies on the membrane selectivity) [28].

Hydrogen was produced from expired soft drinks using a packed-filter bioreactor [29]. Different parameters, including the packed filter position in the bioreactor, were investigated at dimensionless heights (h/H) of 1/4, 2/4, 3/4 and 4/4, and the concentrations of the feeding substrate of 2.51, 3.48, 5.34, 10.19 and 19.51 g total sugar/L, were also investigated. With a definite concentration "20 g total sugar/L" and hydraulic retention time (1 h), the results revealed that a packed filter located at 2/4 has the highest hydrogen generation; *Clostridium tyrobutyricum*, *Bifidobacterium crudilactis* and *Clostridium* sp. were the major bacteria strains. This bioreactor was able to produce steady hydrogen. The best possible biohydrogen yield was achieved by online monitoring of a dark fermenter using an experimentally validated heuristic optimization technique.

Due to the system's complexity, several factors, including the mass transfer coefficient, membrane properties and permeability and impeller rotational speed, impact how well the MGSBR performs. Furthermore, it is difficult to set up all the variables that affect the process and make it successful. Therefore, it appears essential to use computational models to predict and subsequently test many parameters that affect the system's operation [30–32]. The glucose concentration and consumption in the bioreactor were estimated using a Luenberger observer linked to a super-twisting observer [33]. The maximum output and best input flow rate of inlet glucose concentration might be calculated using a nonlinear programming (NLP) optimization problem. A super-twisting controller with the perfect feed flow rate set as its starting value monitored this maximum production. Three different types of non-toxic porous materials, including PN "porous nylon", CPE "chlorinated polyethene" and LS "loofah sponge" were used as fixed-bed materials in a batch anaerobic digestion bioreactor for hydrogen production [34]. The CPE-based bioreactor immobilized the richest density of microorganisms, keeping the pH level stable in the best range and assisting the hydrogen conversion via the butyrate pathway.

Rothan et al. [35] used RSM to increase hydrogen production by ammonia decomposition. Three parameters are considered: electrical voltage, ammonia flow rate and upstream pressure. The maximum hydrogen yield of 93.438% has been achieved with voltage of 110 V, NH_3 flow rate of 30 L/h and upstream pressure of 0 kPa [35]. In the same direction, using RSM, Liu et al. [36] simultaneously optimized the production of volatile fatty acids and biohydrogen from food waste. The total solid content, pH and reaction time are considered to increase biohydrogen [36]. The maximum hydrogen of 46.03 mL g^{-1} has been obtained at 7100 g L^{-1} and 3d, respectively, for total solid content, pH and reaction time.

Adaptive Network Fuzzy Inference System, or simply ANFIS, is one of the artificial intelligence tools that can grasp the trend of the model from its experimental input–output data. Unlike normal mathematical tools, ANFIS expresses the functionality between the output and the inputs in the form of some IF-THEN fuzzy rules. The proposed methodology for the case under investigation contains two phases. The first phase is building an accurate model using ANFIS to simulate the membrane bioreactor process in terms of three controlling input parameters. The second phase is optimal parameter identification process using honey badger algorithm. During the optimization process, there are three controlling parameters. The impeller speed, pH and gas flow rate are used as decision variables, whereas the objective function is boosting the biohydrogen productivity.

The contribution of the paper can be summarized by the following points:

- Constructing an accurate ANFIS model of membrane bioreactor.
- For the first time, the honey badger algorithm is applied to determine the impeller speed, pH and gas flow rate.
- Boosting the biohydrogen productivity from membrane bioreactor.

The rest of the paper is organized following this order: brief description of the dataset is presented in Section 2. In Section 3, the proposed methodology has been explained. The results are discussed in Section 4. Finally, the main findings are outlined in Section 5.

2. Dataset

The bioreactor design employed in the current investigation was modeled after the experimental configuration as in [37]. A 5 L stirred-tank reactor operated under batch and batch recycled flow modes was used for producing biohydrogen. Using anaerobic-activated sludge, sucrose of 35 g/L was employed as the sole substrate for biohydrogen generation. The bioreactor was embedded with turbine "Rushton" and a six-bladed impeller, and an exterior diameter of 8.7 cm was utilized to preserve uniformity. Table 1 summarizes the bioreactor setup's geometric features; more details are displayed in [38].

Table 1. Summary of the bioreactor setup's geometric features.

Variables		
Impeller Speed (rpm)	**Inlet Gas Flow Rate (m^3/s)**	**pH**
40	2×10^{-5}	5
120	3×10^{-5}	6
200	4×10^{-5}	7

According to the literature [37,38], the reactor was kept at 35.0 °C, and pH of 5.5 was maintained using NaOH. A sparger at the bottom of the reactor supplied the recycling flow that was used to aerate the bioreactor.

3. Suggested Method

The suggested method includes two phases: ANFIS modeling and parameter identification.

3.1. ANFIS Based Model

For the ANFIS model, the nonlinear mapping of the inputs is attained by membership functions (MFs) in the fuzzification layer. Generating the rules of ANFIS, assessing the rules' outputs and combining the fired rules to acquire the output happens in the inference engine phase [39]. Ultimately, the output is converted from fuzzy form to its crisp value in the defuzzification layer. In spite of there being several MF forms and defuzzification methods, the gaussian shape and weight average are the best nominated, respectively.

For the ANFIS model, the relations between the inputs and outputs are modeled using an IF-THEN rule. The following relation can be presented as an example of the ANFIS rule [40].

IF m is M, and n is N, THEN z = f(m, n).

Where m and b are the inputs, and z is the output, M and N are the MFs of m and N, respectively.

The fuzzy rule demonstrates that the output z is a function of the inputs m and n.

The output f can be estimated using the two rules' outputs, f_1 and f_2, as follows:

$$f = \widetilde{\omega}_1 f_1 + \widetilde{\omega}_2 f_2 \tag{1}$$

Evaluating $\widetilde{\omega}_1 g_1(x,y)$ and $\widetilde{\omega}_2 g_2(x,y)$

$$\widetilde{\omega}_1 = \frac{\omega_1}{\omega_1 + \omega_2} \tag{2}$$

$$\widetilde{\omega}_2 = \frac{\omega_2}{\omega_1 + \omega_2} \tag{3}$$

$$\omega_1 = \mu_{A_1} * \mu_{B_1} \tag{4}$$

$$\omega_2 = \mu_{A_2} * \mu_{B_2} \tag{5}$$

$\mu_{A_1}, \mu_{A_2}, \mu_{B_1} and\ \mu_{B_2}$ are the MF values of the two inputs.

3.2. Honey Badger Algorithm

The honey badger algorithm is one of the most recent approaches that has been presented by Hashim et al. [41]. The algorithm was inspired from the smart foraging behavior of honey badgers. The HBA is characterized by exploration/exploitation balance; this merit enhances the ability of the approach to solve complex optimization problems with multi-local solutions. The honey badger determines the location of its prey via walking slowly and continuously using mouse-sniffing skills. It begins with locating the approximate site of the prey by digging and eventually catching it. A Honeyguide bird can locate hives but cannot obtain the honey. This makes a connection between them, as the bird leads the badger to hives and helps it open the hives via its claws; finally, they enjoy the teamwork reward. Therefore, the food source can be located via either a digging phase or a honey phase. The approach begins by defining a population of probable solutions with size N; this is formulated as follows:

$$x_i = lb_i + r_1(ub_i - lb_i) \tag{6}$$

where x_i represents the position of ith honey badger; lb_i and ub_i denote the lower and upper limits of the search space, and r_1 is a random number in range [0, 1]. After that, the algorithm calculates the prey smell intensity, which refers to the prey concentration strength and its distance from the badger. It can be calculated by the inverse square low as follows:

$$I_i = r_2 \times \frac{S}{4\pi {d_i}^2} \tag{7}$$

where r_2 denotes random number in the range [0, 1]; S and d_i represent the concentration strength (prey location) and distance between the ith badger and the prey, respectively. They can be calculated as follows:

$$S = (x_i - x_{i+1})^2 \tag{8}$$

$$d_i = x_{prey} - x_i \tag{9}$$

where x_{prey} is the position of prey. The third step followed in HBA is updating the density factor (α); it controls the transition process from exploration to exploitation; it can be calculated as follows:

$$\alpha = C \times exp\left(\frac{-t}{t_{max}}\right) \tag{10}$$

where C is a constant greater than unity. The HBA employs a parameter named flag (F) to alert the direction of the algorithm search, which helps validate high chances for agents to scan the search space thoroughly. The positions of agents are updated via the digging phase and the honey phase; in the first one, the honey badger moves in cardioid shape via the following equation:

$$x_{new} = x_{prey} + F \times \beta \times I \times x_{prey} + F \times r_3 \times \alpha \times d_i \times |cos(2\pi r_4) \times (1 - cos(2\pi r_5))| \tag{11}$$

where β is the badger's ability to acquire food; $\beta \geq 1$, r_3, r_4 and r_5 are random numbers in range [0, 1]; the value of F is assigned according to the following formula:

$$F = \begin{cases} 1 & if\ r_6 \leq 0.5 \\ -1 & else \end{cases} \tag{12}$$

where r_6 represents a random number in the range [0, 1]. On the other hand, the honey is conducted as follows:

$$x_{new} = x_{prey} + F \times r_7 \times \alpha \times d_i \tag{13}$$

where r_7 represents the random number in the range [0, 1].

4. Results and Discussion

4.1. Modeling

The considered dataset employed to construct the ANFIS model contains 20 measured samples. Such data are divided into two parts. The first part used to train the model contains 14 points, whereas the remaining is used to test the model. To train the model, the LSE and backpropagation are used in the forward and backward paths, respectively. For generating the rules, the SC has been considered. Finally, the model was trained to achieve a minimum RMSE value. The numerical assessment of the constructed model is presented in Table 2.

Table 2. Statistical metrics of membrane bioreactor model.

RMSE			Squared-*R*		
Train	**Test**	**All**	**Train**	**Test**	**All**
0.0057	0.0322	0.0183	1.0	0.9999	0.9999

Referring to Table 2, using ANFIS, the RMSE values are 0.0057 and 0.0322 for training and testing data. The Squared-*R* values are 1.00 and 0.9999, respectively, for training and testing. Compared with RMS, thanks to ANFIS, the RMSE decreased from 2.89 using ANOVA to 0.0183 using ANFIS. The low RMSE and the high Squared-*R* using ANFIS demonstrate an effective modeling stage. Figure 1 presents the three inputs and single output structure of the ANFIS model. The outlines of the gaussian shape MFs are presented in Figure 2.

Figure 3 indicates the three-dimensional plots with contours of the inputs against the output. The peak value of the output goes to the dark red, but the smallest value goes to the dark blue.

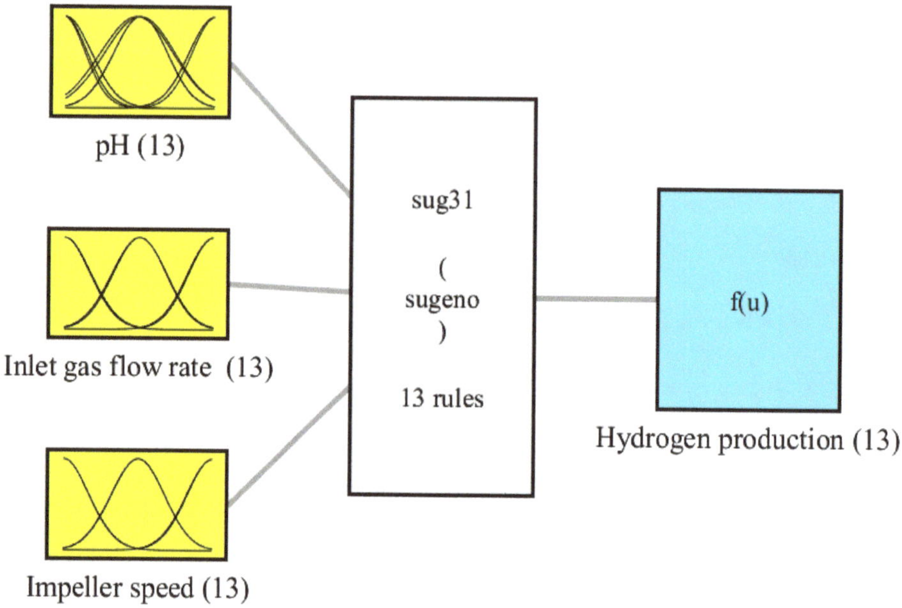

Figure 1. Arrangement of ANFIS model for membrane bioreactor.

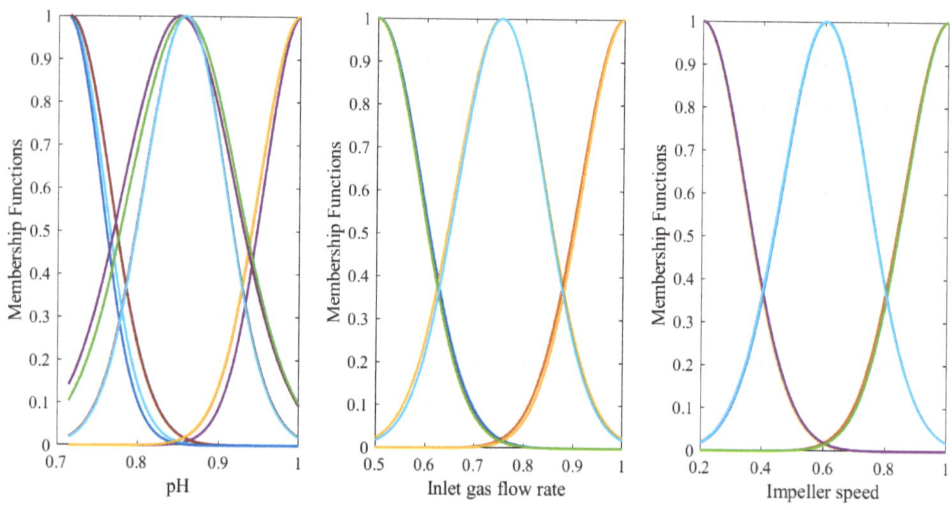

Figure 2. Membership functions of ANFIS model for membrane bioreactor.

Figure 3 illustrates the influence of the interaction of two parameters on hydrogen production, i.e., pH and flow rate (Figure 3a), pH and impeller speed (Figure 3b) and flow rate and impeller speed (Figure 3c). As depicted in Figure 3a, the increase in the pH value from 5 to 7 results in improving the hydrogen productivity, especially at a pH value of around 6.5 at the whole range of the flow rates, while the increase in the flow rate from 20 mL/s to 30 mL/s has a positive effect on the hydrogen production. In contrast, the further increase in the flow rate to 40 mL/s decreased hydrogen production. The positive effect of the pH on the biohydrogen from 5 to around 6.5 would be related to the positive effect of the pH on the growth rate of the microorganisms responsible for the hydrogen production; however, the little bit of decrease in the hydrogen production beyond pH more than 6.5 could be related to the directing the bio-reaction for production other products rather than the biohydrogen [42]. Furthermore, at low pH values, the activity of the (Fe–Fe) hydrogenase enzyme is inhibited, and thus biohydrogen decreases at low pH values, especially lower than 5 [43,44]. The effect of the increase in the impeller speed and the pH was seen in Figure 3b. As depicted from the figure, the increase in the impeller speed from 40 to 120 resulted in improving the biohydrogen yield, while the further increase in the impeller speed resulted in decreasing the biohydrogen production. The pH effect is similar to that discussed in Figure 3b, while the pH effect is clear at a higher impeller rate than that at a low impeller rate. The improvement in the biohydrogen production by increasing the impeller speed to 120 rpm would be related to improved mass transfer [45,46], while at a higher rotation speed of more than 12 rpm had a negative effect on the biohydrogen production that could be related to the negative effect of the rotation speed at the intense agitation and turbulent flow conditions (Re > 10^3) on the metabolic activity of the butyric acid pathway [46–48]. As depicted in Figure 3c, the increase in the gas flow rate from 20 to around 30 mL/min improved the biohydrogen production, especially at high impeller speed. It decreased with the further increase in the impeller's speed. The increase in the recycled gas flow rate from 20 to 30 mL/s would increase the formation of gas bubbles in the bioreactor media, thus increasing the interfacial area between the liquid and the gas. Such behavior at a higher speed of the impeller would significantly improve the mass transfer and thus the biohydrogen production [45]. However, a higher recycled gas flow rate beyond 30 mL/s would decrease the metabolic activity of the butyric acid pathway, thus decreasing the biohydrogen production, as discussed above [46–48].

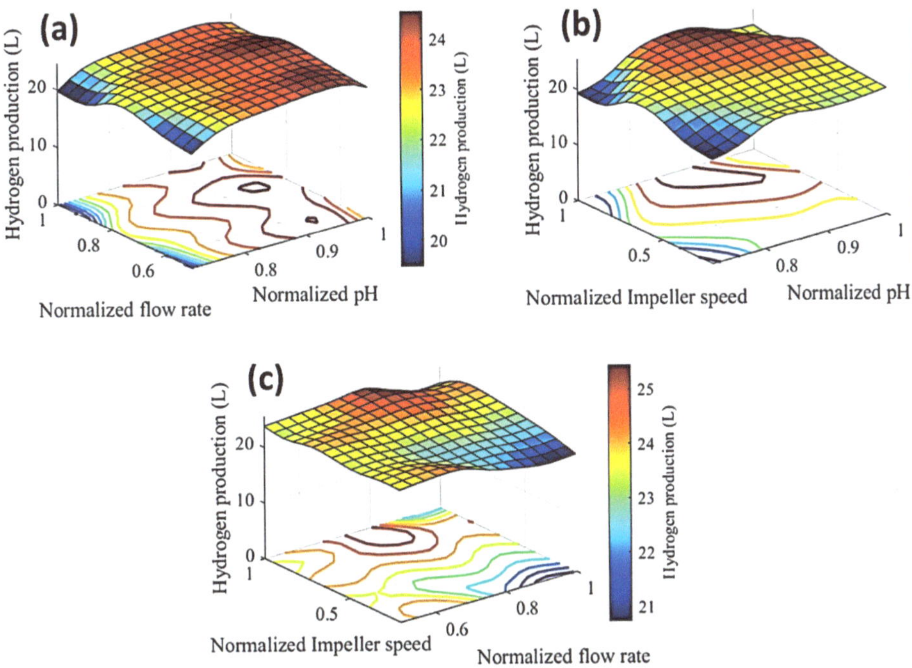

Figure 3. Three-dimensional plot of controlling parameters; (**a**) flow rat and pH, (**b**) pH and impeller speed and (**c**) impeller speed and the flow rate.

Catching the proper correlation between the inputs and output of the membrane bioreactor process system encourages the constructed ANFIS model to predict the output performance entirely. This is clear from the plotting of the ANFIS model's predicted outputs versus the CFD data as explained in Figure 4. The great fit between the calculated and CFD data can be examined. The predictions' plots across the 100% precision line are also displayed in Figure 5.

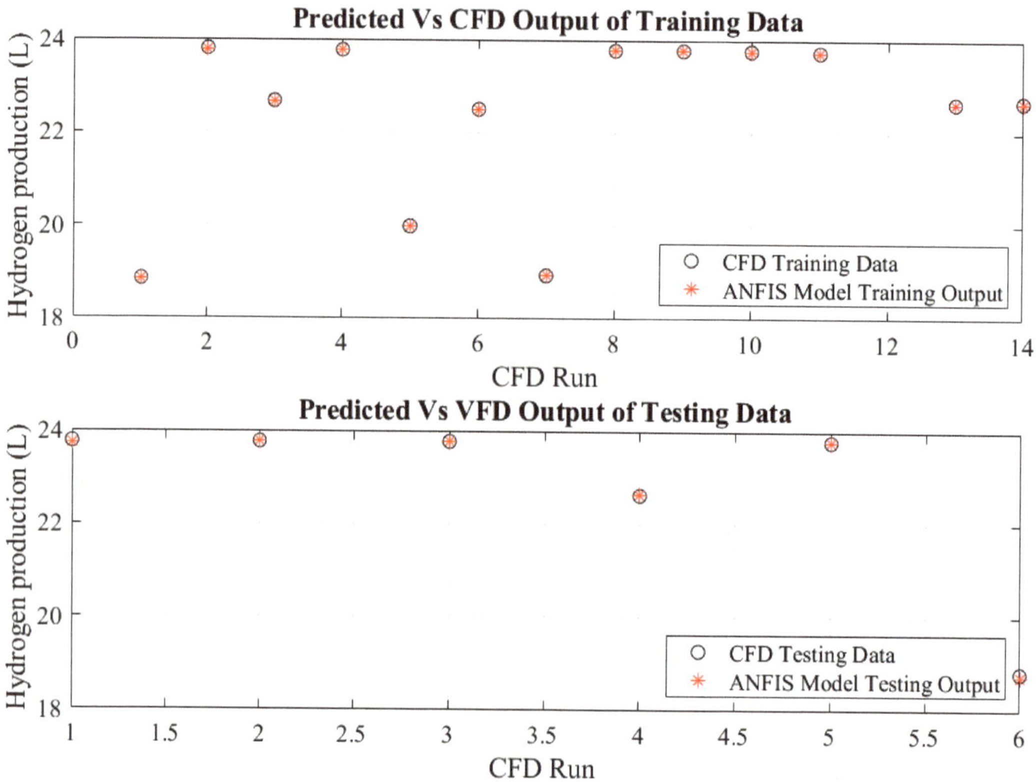

Figure 4. Predicted versus CFD data of ANFIS model.

4.2. Optimization Results

This section aims to define the best values of pH, flow rate and impeller speed to increase biohydrogen production. Consequently, after constructing a dependable ANFIS model, HBA has been used to estimate the best values for three controlling parameters. The problem statement of the objective function can express as follows:

$$f = \underset{x \in R}{\mathrm{argmax}}(y) \tag{14}$$

where x is the set of normalized input variables, and y is the percent of the hydrogen production.

Table 3 presents the optimal parameters and corresponding hydrogen production using CFD, RSM method and the HBA. The integration between ANFIS and HBA boosted the hydrogen yield from 23.8 L to 25.52 L, increasing by 7.22%.

Table 3. Achieved best parameters using considered approaches.

Method	pH	Flow Rate 10^{-5} m^3/s	Impeller Speed (rpm)	Hydrogen Production (L)
CFD [38]	6	2	120	23.8
RSM [38]	6.2	2.4	115	24.09
Proposed	0.8732 [N]	0.7859 [N]	0.9324 [N]	25.52
	6.1124	3.1436	186.48	

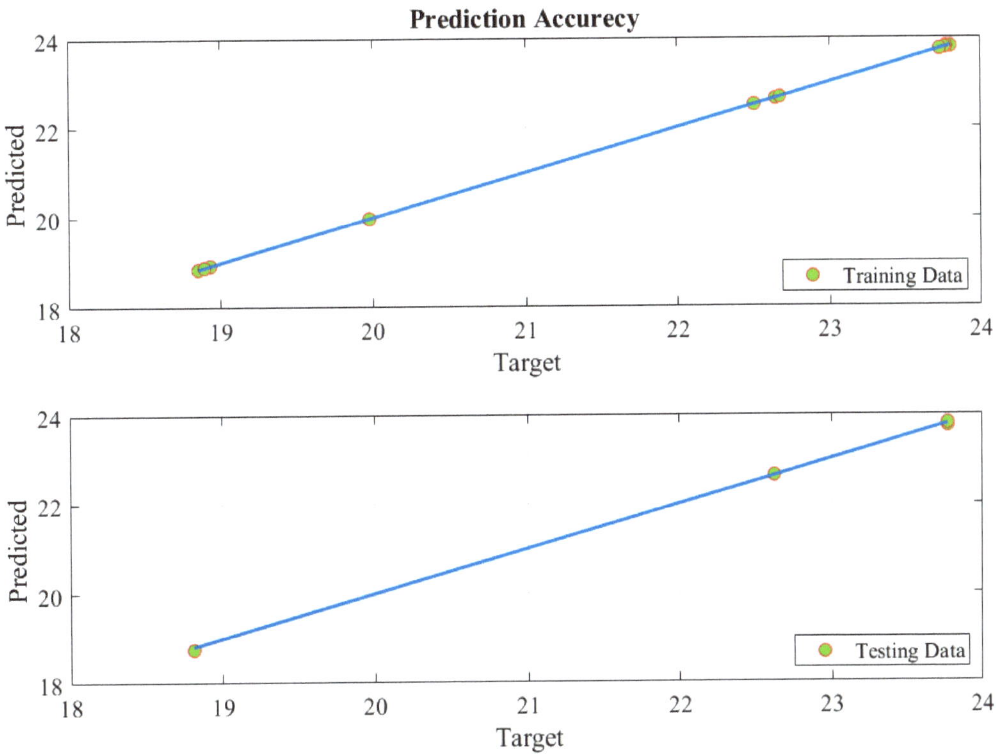

Figure 5. Prediction accuracy of ANFIS model.

Figure 6 presents the variation of particles during the identification procedure. The optimal normalized values are 0.8732, 0.7859 and 0.9324, respectively, for pH, flow rate and impeller speed. The maximum produced hydrogen is 25.52 L. To confirm that this value not achieved arbitrarily, the HBO executed 30 times. The minimum, maximum, mean and STD values are 24.77 L, 25.52 L, 25.314 L and 0.33.

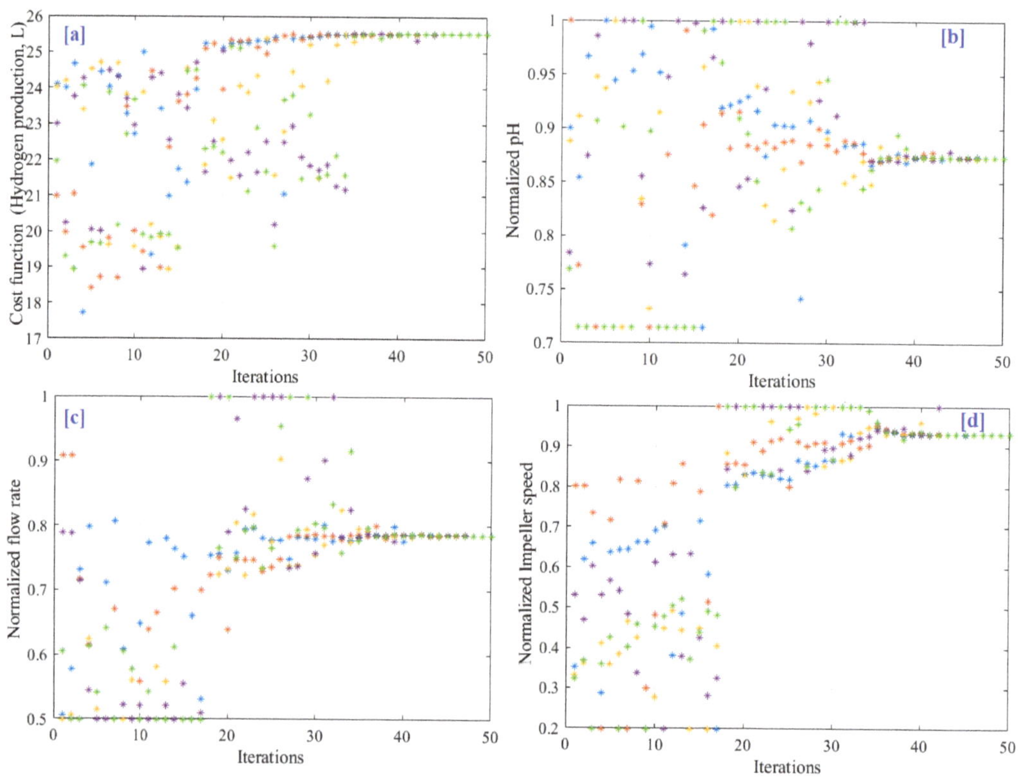

Figure 6. Particle movement during the identification procedure (**a**) hydrogen production, (**b**) normalized pH, (**c**) normalized flow rate and (**d**) normalized impeller speed.

5. Conclusions

The primary target of the paper is modeling and optimizing the performance of the membrane bioreactor. Such a goal has been achieved using artificial intelligence and modern optimization. The suggested method incorporates the ANFIS-based modeling and parameter identification by the honey badger algorithm. The biohydrogen productivity has been boosted through optimal tuning of the impeller speed, pH and gas flow rate. Using ANFIS, the RMSE values are 0.0057 and 0.0322 for training and testing data. The Squared-*R* values are 1.00 and 0.9999 for training and testing. Compared with RMS, thanks to ANFIS, the RMSE decreased from 2.89 using ANOVA to 0.0183 using ANFIS. The low RMSE and the high Squared-*R* using ANFIS demonstrate an effective modeling stage. This demonstrated the robustness of ANFIS modeling. Next, using HBA for the optimal parameters, the optimal normalized values are 6.1124, $3.1436 * 10^{-5}$ m^3/s and 186.48 rpm, respectively, for pH, flow rate and impeller speed. Under the optimal condition, the integration between ANFIS and HBA boosted the hydrogen yield from 23.8 L to 25.52 L, an increased of 7.22%. It is expected that the scaling up of the bioreactor and the type of organic material will affect the overall performance in biohydrogen productivity. Therefore, it is recommended to test various organics and a large-scale bioreactor. Furthermore, future work will consider other recent optimization algorithms with different modeling techniques.

Author Contributions: Conceptualization, H.R., A.G.O., M.A.A. and E.T.S.; Methodology, A.H.A.; Software, H.R.; Validation, A.H.A.; Formal analysis, M.A.A.; Investigation, E.T.S.; Data curation, A.H.A.; Writing – original draft, H.R., A.G.O., M.A.A., A.H.A. and E.T.S.; Writing—review & editing,

H.R., A.G.O., M.A.A., A.H.A. and E.T.S.; Supervision, A.G.O. and E.T.S. All authors have read and agreed to the published version of the manuscript.

Funding: This research was funded by the Deputyship for Research & Innovation, Ministry of Education in Saudi Arabia under grant number [IF2/PSAU/2022/01/20906].

Institutional Review Board Statement: Not applicable.

Informed Consent Statement: Not applicable.

Data Availability Statement: Not applicable.

Acknowledgments: The authors extend their appreciation to the Deputyship for Research & Innovation, Ministry of Education in Saudi Arabia for funding this research work through the project number (IF2/PSAU/2022/01/20906).

Conflicts of Interest: The authors declare no conflict of interest.

References

1. Tan, R.R.; Belmonte, B.A.; Benjamin, M.F.D.; Andiappan, V.; Aviso, K.B. Optimization of enhanced weathering networks with alternative transportation modes. *Carbon Resour. Convers.* **2022**, *5*, 167–176. [CrossRef]
2. Olabi, A.; Wilberforce, T.; Elsaid, K.; Sayed, E.T.; Maghrabie, H.M.; Abdelkareem, M.A. Large scale application of carbon capture to process industries—A review. *J. Clean. Prod.* **2022**, *362*, 132300. [CrossRef]
3. Parvez, I.; Shen, J.; Hassan, I.; Zhang, N. Generation of hydro energy by using data mining algorithm for cascaded hydropower plant. *Energies* **2021**, *14*, 298. [CrossRef]
4. Singh, U.; Rizwan, M.; Malik, H.; García Márquez, F.P. Wind energy scenario, success and initiatives towards renewable energy in India—A review. *Energies* **2022**, *15*, 2291. [CrossRef]
5. Alizadeh Zolbin, M.; Tahouni, N.; Panjeshahi, M.H. Total site integration considering wind/solar energy with supply/demand variation. *Energy* **2022**, *252*, 123928. [CrossRef]
6. Barbarelli, S.; Nastasi, B. Tides and Tidal Currents—Guidelines for Site and Energy Resource Assessment. *Energies* **2021**, *14*, 6123. [CrossRef]
7. Hayat, M.B.; Ali, D.; Monyake, K.C.; Alagha, L.; Ahmed, N. Solar energy—A look into power generation, challenges, and a solar-powered future. *Int. J. Energy Res.* **2019**, *43*, 1049–1067. [CrossRef]
8. Chaudhary, B.; Kumar, V. Emerging Technological Frameworks for the Sustainable Agriculture and Environmental Management. *Sustain. Horiz.* **2022**, *3*, 100026. [CrossRef]
9. Ambaye, T.G.; Vaccari, M.; Bonilla-Petriciolet, A.; Prasad, S.; van Hullebusch, E.D.; Rtimi, S. Emerging technologies for biofuel production: A critical review on recent progress, challenges and perspectives. *J. Environ. Manag.* **2021**, *290*, 112627. [CrossRef]
10. Rasul, M.; Hazrat, M.; Sattar, M.; Jahirul, M.; Shearer, M. The future of hydrogen: Challenges on production, storage and applications. *Energy Convers. Manag.* **2022**, *272*, 116326. [CrossRef]
11. Ishaq, H.; Dincer, I.; Crawford, C. A review on hydrogen production and utilization: Challenges and opportunities. *Int. J. Hydrog. Energy* **2022**, *47*, 26238–26264. [CrossRef]
12. Ahmed, S.F.; Rafa, N.; Mofijur, M.; Badruddin, I.A.; Inayat, A.; Ali, M.S.; Farrok, O.; Yunus Khan, T. Biohydrogen production from biomass sources: Metabolic pathways and economic analysis. *Front. Energy Res.* **2021**, *9*, 753878. [CrossRef]
13. Lee, J.E.; Shafiq, I.; Hussain, M.; Lam, S.S.; Rhee, G.H.; Park, Y.-K. A review on integrated thermochemical hydrogen production from water. *Int. J. Hydrog. Energy* **2021**, *47*, 4346–4356. [CrossRef]
14. Patel, S.K.; Das, D.; Kim, S.C.; Cho, B.-K.; Kalia, V.C.; Lee, J.-K. Integrating strategies for sustainable conversion of waste biomass into dark-fermentative hydrogen and value-added products. *Renew. Sustain. Energy Rev.* **2021**, *150*, 111491. [CrossRef]
15. Olabi, A.G.; Wilberforce, T.; Sayed, E.T.; Elsaid, K.; Rezk, H.; Abdelkareem, M.A. Recent progress of graphene based nanomaterials in bioelectrochemical systems. *Sci. Total Environ.* **2020**, *749*, 141225. [CrossRef]
16. Lama, G.; Meijide, J.; Sanromán, A.; Pazos, M. Heterogeneous Advanced Oxidation Processes: Current Approaches for Wastewater Treatment. *Catalysts* **2022**, *12*, 344. [CrossRef]
17. Kamal, A.; Makhatova, A.; Yergali, B.; Baidullayeva, A.; Satayeva, A.; Kim, J.; Inglezakis, V.J.; Poulopoulos, S.G.; Arkhangelsky, E. Biological treatment, advanced oxidation and membrane separation for landfill leachate treatment: A review. *Sustainability* **2022**, *14*, 14427. [CrossRef]
18. Bhattacharyya, A.; Liu, L.; Lee, K.; Miao, J. Review of Biological Processes in a Membrane Bioreactor (MBR): Effects of Wastewater Characteristics and Operational Parameters on Biodegradation Efficiency When Treating Industrial Oily Wastewater. *J. Mar. Sci. Eng.* **2022**, *10*, 1229. [CrossRef]
19. Singh, H.; Tomar, S.; Qureshi, K.A.; Jaremko, M.; Rai, P.K. Recent advances in biomass pretreatment technologies for biohydrogen production. *Energies* **2022**, *15*, 999. [CrossRef]
20. Salameh, T.; Sayed, E.T.; Olabi, A.G.; Hdaib, I.I.; Allan, Y.; Alkasrawi, M.; Abdelkareem, M.A. Adaptive Network Fuzzy Inference System and Particle Swarm Optimization of Biohydrogen Production Process. *Fermentation* **2022**, *8*, 483. [CrossRef]

21. Pérez, B.J.L.; Jiménez, J.A.M.; Bhardwaj, R.; Goetheer, E.; van Sint Annaland, M.; Gallucci, F. Methane pyrolysis in a molten gallium bubble column reactor for sustainable hydrogen production: Proof of concept & techno-economic assessment. *Int. J. Hydrog. Energy* **2021**, *46*, 4917–4935.
22. Wieland, S.; Melin, T.; Lamm, A. Membrane reactors for hydrogen production. *Chem. Eng. Sci.* **2002**, *57*, 1571–1576. [CrossRef]
23. Ri, P.-C.; Kim, J.-S.; Kim, T.-R.; Pang, C.-H.; Mun, H.-G.; Pak, G.-C.; Ren, N.-Q. Effect of hydraulic retention time on the hydrogen production in a horizontal and vertical continuous stirred-tank reactor. *Int. J. Hydrog. Energy* **2019**, *44*, 17742–17749. [CrossRef]
24. Ladewig, B.; Al-Shaeli, M.N.Z. *Fundamentals of Membrane Bioreactors*; Springer: Singapore, 2017.
25. Iorhemen, O.T.; Hamza, R.A.; Tay, J.H. Membrane bioreactor (MBR) technology for wastewater treatment and reclamation: Membrane fouling. *Membranes* **2016**, *6*, 33. [CrossRef] [PubMed]
26. Robles, Á.; Ruano, M.V.; Charfi, A.; Lesage, G.; Heran, M.; Harmand, J.; Seco, A.; Steyer, J.-P.; Batstone, D.J.; Kim, J. A review on anaerobic membrane bioreactors (AnMBRs) focused on modelling and control aspects. *Bioresour. Technol.* **2018**, *270*, 612–626. [CrossRef]
27. Sattar, A.; Arslan, C.; Ji, C.; Sattar, S.; Ali Mari, I.; Rashid, H.; Ilyas, F. Comparing the bio-hydrogen production potential of pretreated rice straw co-digested with seeded sludge using an anaerobic bioreactor under mesophilic thermophilic conditions. *Energies* **2016**, *9*, 198. [CrossRef]
28. Bakonyi, P.; Kumar, G.; Bélafi-Bakó, K.; Kim, S.-H.; Koter, S.; Kujawski, W.; Nemestóthy, N.; Peter, J.; Pientka, Z. A review of the innovative gas separation membrane bioreactor with mechanisms for integrated production and purification of biohydrogen. *Bioresour. Technol.* **2018**, *270*, 643–655. [CrossRef]
29. Chu, C.-Y.; Zheng, J.-L.; Chen, T.-H.; Bhuyar, P. High performance of biohydrogen production in packed-filter bioreactor via optimizing packed-filter position. *Int. J. Environ. Res. Public Health* **2021**, *18*, 7462. [CrossRef]
30. Wilberforce, T.; Olabi, A.; Rezk, H.; Abdelaziz, A.Y.; Abdelkareem, M.A.; Sayed, E.T. Boosting the output power of PEM fuel cells by identifying best-operating conditions. *Energy Convers. Manag.* **2022**, *270*, 116205. [CrossRef]
31. Rezk, H.; Sayed, E.T.; Abdelkareem, M.A.; Olabi, A. Performance improvement of co-culture inoculated microbial fuel cell using fuzzy modelling and Harris hawks optimization. *Int. J. Energy Res.* **2022**, *46*, 14396–14407. [CrossRef]
32. Yahaya, E.; Lim, S.W.; Yeo, W.S.; Nandong, J. A review on process modeling and design of biohydrogen. *Int. J. Hydrog. Energy* **2022**, *47*, 30404–30427. [CrossRef]
33. Zúñiga, I.T.; Villa-Leyva, A.; Vargas, A.; Buitrón, G. Experimental validation of online monitoring and optimization strategies applied to a biohydrogen production dark fermenter. *Chem. Eng. Sci.* **2018**, *190*, 48–59. [CrossRef]
34. Zhao, C.; Zhang, N.; Zheng, H.; Zhu, Q.; Utsumi, M.; Yang, Y. Effective and long-term continuous bio-hydrogen production by optimizing fixed-bed material in the bioreactor. *Process Biochem.* **2019**, *83*, 55–63. [CrossRef]
35. Ali Rothan, Y.; Ali, F.F.; Issakhov, A.; Selim, M.M.; Li, Z. Optimization analysis of hydrogen production using ammonia decomposition. *J. Mol. Liq.* **2021**, *335*, 116190. [CrossRef]
36. Liu, N.; Jiang, J.; Yan, F.; Xu, Y.; Yang, M.; Gao, Y.; Aihemaiti, A.; Zou, Q. Optimization of simultaneous production of volatile fatty acids and bio-hydrogen from food waste using response surface methodology. *RSC Adv.* **2018**, *8*, 10457–10464. [CrossRef] [PubMed]
37. Mu, Y.; Yu, H.-Q.; Wang, G. A kinetic approach to anaerobic hydrogen-producing process. *Water Res.* **2007**, *41*, 1152–1160. [CrossRef]
38. Jabbari, B.; Jalilnejad, E.; Ghasemzadeh, K.; Iulianelli, A. Modeling and optimization of a membrane gas separation based bioreactor plant for biohydrogen production by CFD–RSM combined method. *J. Water Process Eng.* **2021**, *43*, 102288. [CrossRef]
39. Rezk, H.; Mohammed, R.H.; Rashad, E.; Nassef, A.M. ANFIS-based accurate modeling of silica gel adsorption cooling cycle. *Sustain. Energy Technol. Assess.* **2022**, *50*, 101793. [CrossRef]
40. Olabi, A.; Rezk, H.; Sayed, E.T.; Ghoniem, R.M.; Abdelkareem, M.A. Boosting carbon dioxide adsorption capacity applying Jellyfish optimization and ANFIS-based modelling. *Ain Shams Eng. J.* **2022**, *14*, 101931. [CrossRef]
41. Hashim, F.A.; Houssein, E.H.; Hussain, K.; Mabrouk, M.S.; Al-Atabany, W. Honey Badger Algorithm: New metaheuristic algorithm for solving optimization problems. *Math. Comput. Simul.* **2022**, *192*, 84–110. [CrossRef]
42. Liu, I.C.; Whang, L.-M.; Ren, W.-J.; Lin, P.-Y. The effect of pH on the production of biohydrogen by clostridia: Thermodynamic and metabolic considerations. *Int. J. Hydrog. Energy* **2011**, *36*, 439–449. [CrossRef]
43. Liu, Y.; Yu, P.; Song, X.; Qu, Y. Hydrogen production from cellulose by co-culture of Clostridium thermocellum JN4 and Thermoanaerobacterium thermosaccharolyticum GD17. *Int. J. Hydrog. Energy* **2008**, *33*, 2927–2933. [CrossRef]
44. Ngo, T.A.; Nguyen, T.H.; Bui, H.T.V. Thermophilic fermentative hydrogen production from xylose by Thermotoga neapolitana DSM 4359. *Renew. Energy* **2012**, *37*, 174–179. [CrossRef]
45. Salehpour, R.; Jalilnejad, E.; Nalband, M.; Ghasemzadeh, K. Hydrodynamic behavior of an airlift reactor with net draft tube with different configurations: Numerical evaluation using CFD technique. *Particuology* **2020**, *51*, 91–108. [CrossRef]
46. Trad, Z.; Vial, C.; Fontaine, J.-P.; Larroche, C. Mixing and liquid-to-gas mass transfer under digester operating conditions. *Chem. Eng. Sci.* **2017**, *170*, 606–627. [CrossRef]

47. Ding, J.; Wang, X.; Zhou, X.-F.; Ren, N.-Q.; Guo, W.-Q. CFD optimization of continuous stirred-tank (CSTR) reactor for biohydrogen production. *Bioresour. Technol.* **2010**, *101*, 7005–7013. [CrossRef]
48. Chezeau, B.; Vial, C. Combined effects of digestate viscosity and agitation conditions on the fermentative biohydrogen production. *Biochem. Eng. J.* **2019**, *142*, 105–116. [CrossRef]

sustainability

MDPI

Article

Hybrid ANFIS-PI-Based Optimization for Improved Power Conversion in DFIG Wind Turbine

Farhat Nasim [1], Shahida Khatoon [1], Ibraheem [1], Shabana Urooj [2,*], Mohammad Shahid [3], Asmaa Ali [4] and Nidal Nasser [5]

1 Department of Electrical Engineering, Faculty of Engineering and Technology, Jamia Millia Islamia, New Delhi 110025, India; g.fnasim@jmi.ac.in (F.N.); skhatoon@jmi.ac.in (S.K.); ibraheem@jmi.ac.in (I.)
2 Department of Electrical Engineering, College of Engineering, Princess Nourah bint Abdulrahman University, P.O. Box 84428, Riyadh 11671, Saudi Arabia
3 Department of Electrical Engineering, Galgotias College of Engineering and Technology, Greater Noida 201306, India; eems.j87@gmail.com
4 Optimized Computing and Communications Research Laboratory, Western University, London, ON N6A 3K7, Canada; eng.asmaa.ali@gmail.com
5 College of Engineering, Alfaisal University, Riyadh 11533, Saudi Arabia; nnasser@alfaisal.edu
* Correspondence: smurooj@pnu.edu.sa

Abstract: Wind energy is essential for promoting sustainability and renewable power solutions. However, ensuring stability and consistent performance in DFIG-based wind turbine systems (WTSs) remains challenging due to rapid wind speed variations, grid disturbances, and parameter uncertainties. These fluctuations result in power instability, increased overshoot, and prolonged settling times, negatively impacting grid compliance and system efficiency. Conventional proportional-integral (PI) controllers are simple and effective in steady-state conditions, but they lack adaptability in dynamic situations. Similarly, artificial intelligence (AI)-based controllers, such as fuzzy logic controllers (FLCs) and artificial neural networks (ANNs), improve adaptability but suffer from high computational demands and training complexity. To address these limitations, this paper presents a hybrid adaptive neuro-fuzzy inference system (ANFIS)-PI controller for DFIG-based WTS. The proposed controller integrates fuzzy logic adaptability with neural network-based learning, allowing real-time optimization of control parameters. Implemented within the rotor-side converter (RSC) and grid-side converter (GSC), ANFIS enhances reactive power management, grid compliance, and overall system stability. The system was tested under a step wind speed signal varying from 10 m/s to 12 m/s to evaluate its robustness. The simulation results confirmed that the ANFIS-PI controller significantly improved performance compared with the conventional PI controller. Specifically, it reduced rotor speed overshoot by 3%, torque overshoot by 12.5%, active power overshoot by 2%, and DC link voltage overshoot by 20%. Additionally, the ANFIS-PI controller shortened settling time by 50% for rotor speed, by 25% for torque, by 33% for active power, and by 16.7% for DC link voltage, ensuring faster stabilization, enhanced dynamic response, and greater efficiency. These improvements establish the ANFIS-PI controller as an advanced, computationally efficient, and scalable solution for enhancing the reliability of DFIG-based WTS, facilitating seamless integration of wind energy into modern power grids.

Keywords: adaptive neuro-fuzzy inference system (ANFIS); wind turbine system (WTS); rotor-side converter (RSC); grid-side converter (GSC); proportional-integral (PI) controller; power stability; overshoot reduction; settling time improvement

https://doi.org/10.3390/su17062454

1. Introduction

1.1. Research Background

Wind power has emerged as a key contributor to sustainable electricity generation, driven by the rising demand for eco-friendly energy solutions. Recognized for its availability and accessibility and as a sustainable substitute for non-renewable fuels, wind energy plays a significant role in lowering greenhouse gas emissions and combating environmental change. As of 2024, global wind energy capacity surpassed 117 GW [1], highlighting its growing significance in the energy sector. This capacity will expand further as nations pursue ambitious climate and renewable energy targets. Electricity is generated from wind energy using the wind turbine system (WTS), which operates through three fundamental stages: aerodynamic, mechanical, and electrical. Each stage is essential for the efficient conversion of wind energy into electricity.

1. Aerodynamic stage: This stage captures the wind's dynamic energy and converts it into rotational motion through turbine blades. The blades are aerodynamically optimized to maximize power extraction under different atmospheric conditions.
2. Mechanical stage: The rotational energy is transmitted through a gear mechanism that optimizes speed for generator efficiency.
3. Electrical stage: The generated energy is converted and conditioned for grid integration using advanced power electronic components, including AC–DC–AC converters and filters.

The most widely used generators for wind turbines with variable-speed operation employ doubly fed induction generators (DFIGs), permanent magnet synchronous generators (PMGSs), and synchronous machines with wound rotors [2]. Among the various generator technologies available for variable-speed wind turbines, the doubly fed induction generator (DFIG) is a preferred choice for its efficiency, versatility, and economic feasibility.

The DFIG offers several advantages over other generator types:

- It can regulate real and reactive power, improving grid compatibility and electrical performance.
- It minimizes mechanical load on turbine parts, prolonging their lifespan.
- Its partial-scale power electronic converter design allows operation over a broad range of wind speeds, making it suitable for regions with moderate wind conditions.
- It provides a cost-effective balance between performance and production costs compared with other advanced generator systems [3].

The DFIG system operates using two power electronic converters:

1. The rotor-side converter (RSC) controls the rotor current and manages reactive power.
2. The grid-side converter (GSC) maintains the DC link voltage stability and ensures compliance with grid requirements [4].

Despite its advantages, wind energy conversion faces challenges due to wind's inherently variable nature, which affects the stability, efficiency, and reliability of power generation [5].

1.2. Motivation for the Research

DFIG-based WTSs are preferred for renewable energy due to their efficiency, cost-effectiveness, and power regulation capabilities. However, wind speed variability causes voltage fluctuations, frequency instability, and power quality issues, impacting grid reliability. While simple and effective in steady-state conditions, traditional PI controllers lack adaptability under dynamic changes. AI-driven controllers, such as fuzzy logic controllers (FLCs) and artificial neural networks (ANNs), offer improved adaptability but have limitations—FLCs require expert-defined rules, while ANNs demand extensive training data. To overcome these challenges, a hybrid intelligent control strategy integrating fuzzy reasoning with artificial neural adaptability is needed to improve dynamic response, stability, and real-time adaptability while maintaining computational efficiency.

1.3. Literature Review

Effective control of WTSs is essential for maximizing power extraction and maintaining grid reliability. Traditional PI (proportional-integral) regulators have historically been the primary choice for managing DFIG systems. PI controllers are praised for their simplicity, cost-effectiveness, and ability to eliminate steady-state deviations in controlled environments. These controllers rely on fixed tuning parameters to regulate system dynamics, making them relatively easy to implement and widely adopted in industrial applications [6–9]. However, the reliance on fixed parameters also serves as their primary limitation. In dynamic operating environments, such as those experienced in wind energy systems, fixed-parameter PI controllers struggle to adapt to external disturbances, grid variability, and fluctuating wind speeds. This can lead to overshoot, instability, and prolonged settling times, undermining the efficiency and dependability of the system. To overcome these limitations of conventional PI controllers in DFIG-driven wind turbine systems (WTSs), meta-heuristic optimization techniques have been applied for automatic calibration and adaptive tuning, enhancing system output response and robustness. These methods include the jellyfish search algorithm (JSA), beetle antennae search (BAS), ant colony optimization (ACO) [10], thermal exchange optimization (TEO), grasshopper optimization algorithm (GOA), water cycle algorithm (WCA) [11], cuckoo search algorithm (CS), dragonfly algorithm (DA), modified flower pollination algorithm (MFPA) [12], hunger games search (HGS), grey wolf optimizer (GWO), artificial bee colony (ABC), hybrid meta-heuristics [13], whale optimization algorithm (WOA), hawks optimization (HHO) [14], bat algorithm (BA), squirrel search algorithm (SSA) [15], and artificial fish-swarm algorithm (AFSA) [16]. Despite numerous advantages, meta-heuristic techniques have notable drawbacks. They often involve high computational costs due to iterative processes and complex algorithms, limiting real-time applicability in WTSs. Additionally, these methods may require extensive parameter tuning and initialization, and their performance can sometimes depend on problem-specific customization. In some cases, convergence to global optima is not guaranteed, and there is a risk of premature convergence, especially in dynamic and nonlinear systems like DFIG-based WTSs [17,18].

To further address these challenges, fuzzy logic controllers (FLCs) have gained popularity. Fuzzy logic controllers (FLCs) are increasingly utilized in DFIG-based wind turbine systems for their ability to handle nonlinearities and uncertainties without requiring exact mathematical models. Their rule-based decision-making enables adaptability to dynamic operating conditions, enhancing system stability and control precision. However, their performance depends on carefully designing membership functions and rules, which can be subjective and demand expert knowledge. Studies highlight their superior responsiveness and disturbance rejection compared with conventional PI [19]. Fuzzy logic-based control strategies have been extensively utilized in wind energy systems, particularly in DFIG-based systems, due to their ability to handle uncertainties and nonlinearities without requiring precise mathematical models. However, when compared with artificial neural networks (ANNs), fuzzy controllers present several limitations. One significant drawback is that FLCs heavily rely on expert knowledge to design membership functions and define rules, which can be subjective and time-consuming. Additionally, fuzzy controllers are typically less adaptive to changing dynamic conditions when compared with ANNs, which can learn and adjust their parameters based on real-time data through training.

ANNs, on the other hand, excel in handling large and complex datasets, can generalize well, and can adapt continuously to variations in operating conditions. Their learning capabilities make them more robust in optimizing the control of dynamic systems, such as DFIG-based wind power systems, under varying environmental and operational conditions. Moreover, ANNs can avoid the need for predefined rules, which can make them more flexible and scalable than fuzzy controllers in systems with uncertain or complex behavior [20]. However, while ANNs have these advantages, they come with challenges. They require substantial data for training and may be computationally intensive, especially when deployed in real-time applications. Furthermore, the training process can be lengthy, resulting in model overfitting if not adequately controlled. Neural networks contribute dynamic learning and adaptability, allowing for self-tuning capabilities in changing environments. Still, they demand significant training data and computational power, and their "black-box" nature can complicate interpretability [21]. Recent advancements in intelligent control techniques for DFIG-based WTSs have focused on improving power quality, grid compliance, and system efficiency. Researchers in [22] proposed a fuzzy logic-based neural network controller for power quality enhancement but required large datasets for training, leading to high computational overhead. The proposed ANFIS-PI controller minimizes the need for extensive training data while ensuring superior adaptability. The authors in [23] developed a direct power control method using fuzzy logic, achieving grid stability but encountering longer response times. The ANFIS-PI approach reduces settling time by 25%, ensuring faster and smoother responses. Furthermore, the authors of [24] implemented a model predictive direct torque control (MP-DTC) algorithm, which improved torque precision but suffered from high computational demand. The literature summary cited in Table 1 concludes that the hybrid ANFIS-PI controller surpasses traditional PI controllers by significantly reducing overshoot, improving settling time, and enhancing dynamic adaptability.

Table 1. Summary of the latest literature review on meta-heuristic techniques.

S#	Ref.	Control Technique Applied	Parameters Applied	Signal Applied	Merits	Demerits
1	[19]	PI controller	DC link voltage, wind turbine speed	Variable wind speed	Simple implementation, eliminates steady-state errors	Sudden overshoot, longer settling time
2	[3,4]	Particle swarm optimization (PSO)	Stator current, DC link voltage, active/reactive powers	Short circuit faults, variable wind speed	Enhances performance under faults, provides pitch angle regulation	High settling time, DC voltage instability under fluctuations
3	[5,6]	Genetic algorithm (GA)	Active/reactive powers	Rotor resistance variation	Smooth, reliable performance under variations	Difficulty in stopping criteria
4	[10]	Jellyfish search algorithm (JSA)	Power regulation parameters	Grid variations	Enhances stability, intense global search	High computational complexity
5	[11]	Grasshopper optimization algorithm (GOA)	DC link voltage, power factors	Grid variations	Fast execution, adaptive capabilities	Poor exploitation ability

Table 1. *Cont.*

S#	Ref.	Control Technique Applied	Parameters Applied	Signal Applied	Merits	Demerits
6	[12]	Cuckoo search algorithm (CS)	Frequency and voltage control	Grid fluctuations	Strong global search ability	May get stuck in local optima
7	[13]	Grey wolf optimizer (GWO)	Power system stabilization	Grid disturbances	Adaptive behavior	Sensitive to initialization parameters
8	[14]	Whale optimization algorithm (WOA)	Reactive power optimization	Variable wind speed	Effective in multi-objective problems	Slower convergence
9	[15]	Bat algorithm (BA)	Stator and rotor control	Variable wind speed	Good exploration ability	High dependency on parameter tuning
10	[16]	Artificial fish-swarm algorithm (AFSA)	Grid synchronization parameters	Variable wind speed	Strong global optimization	Computational complexity
11	[22]	Fuzzy logic-based neural network	Power quality parameters	Grid disturbances	Handles nonlinearities, improves stability	Requires large datasets
12	[23]	Direct power control using fuzzy logic	Power control variables	Grid disturbances	Ensures grid stability	Longer response times
13	[24]	Model predictive direct torque control (MP-DTC)	Torque and flux parameters	Dynamic grid Variations	Improved torque precision	High computational demand
14	**Proposed**	**Hybrid ANFIS-PI controller**	**RSC and GSC control parameters**	**Grid Variations**	**Superior adaptability with fuzzy learning and neural network-based optimization, reducing overshoot by 20% and improving settling time by 50%**	**Minimizes need for large datasets while maintaining real-time computational feasibility**

Compared with meta-heuristic techniques, it provides a balanced trade-off between real-time adaptability and computational efficiency. Unlike purely heuristic methods, which often face convergence and tuning challenges, the hybrid ANFIS-PI controller integrates fuzzy logic adaptability with neural learning for superior robustness in dynamic wind energy systems.

1.4. Problem Statement

Despite advancements in PI, AI, and predictive control techniques, existing methods have significant limitations in real-time adaptability, computational efficiency, and stability under dynamic wind conditions. Conventional PI controllers fail to handle rapid wind fluctuations, while AI-driven methods often demand excessive computational resources. There is a need for a hybrid intelligent control system that balances adaptability, computational feasibility, and system robustness.

1.5. Aim of the Research

This research aimed to develop a hybrid ANFIS-PI controller that integrated fuzzy logic-based inference with neural network adaptability. The goal was to optimize DFIG-based WTS operations by improving real-time adaptability, reducing computational overhead, and ensuring stable power generation.

1.6. Objectives of the Research

The primary objectives of this research were as follows:

1. Develop a hybrid ANFIS-PI controller to enhance power stability and grid compliance in DFIG-based WTSs.
2. Reduce overshoot and improve settling time, ensuring better dynamic response under fluctuating wind speeds.
3. Optimize RSC and GSC simultaneously, unlike previous works focusing on only one converter.
4. Validate system performance by comparing it with conventional PI controllers and existing AI-based techniques.

1.7. Contribution

The significant contributions of this research are as follows:

1. Introduction of a novel hybrid ANFIS-PI controller, integrating fuzzy reasoning adaptability with neuro-computational learning.
2. Dual converter optimization, enhancing both RSC and GSC control mechanisms, ensuring improved power stability and grid compliance.
3. Reduction in computational complexity, minimizing the need for large training datasets while maintaining superior real-time adaptability.
4. Validated performance improvements, demonstrating lower overshoot (20%) and improved settling time (50%) over traditional PI controllers.

1.8. Novelty and Advantages of the Proposed Method

1. Hybrid control approach: The research introduced a novel adaptive neuro-fuzzy inference system (ANFIS)-PI hybrid controller for the rotor-side converter (RSC) and grid-side converter (GSC) in a doubly fed induction generator (DFIG)-based wind turbine system (WTS). Unlike traditional fixed-parameter controllers, this hybrid structure combined fuzzy logic inference with neural network adaptability, allowing dynamic parameter tuning.
2. Improved system adaptability: Unlike conventional proportional-integral (PI) controllers, which struggle with real-time adaptation to fluctuating wind speeds and grid disturbances, the ANFIS-based controller dynamically learned from operational conditions and adjusted the control parameters accordingly.
3. Optimized computational efficiency: The model achieved lower training errors (~0.034) in just 25 training epochs, compared with the conventional AI-based controllers requiring double the dataset size and 50 epochs for similar results. This highlights superior computational efficiency without compromising accuracy.
4. Dual-converter optimization: Most previous studies optimized the RSC or the GSC. This research enhanced both converters, ensuring improved rotor current regulation, DC link voltage stability, and active/reactive power management for grid compliance.

5. Real-time feasibility: Unlike meta-heuristic algorithms (such as genetic algorithms, particle swarm optimization, and ant colony optimization), which are computationally expensive and may suffer from slow convergence, ANFIS ensures fast and adaptive control decisions, making real-time applications more feasible.

The proposed hybrid ANFIS-PI controller represents a DFIG-based wind turbine control breakthrough by combining AI-driven adaptability with classical control robustness. Its ability to reduce overshoot, enhance stability, minimize computational overhead, and improve grid compliance makes it a superior alternative to conventional PI and AI-based controllers.

The layout of this paper is structured as follows: Section 2 offers an overview of the system configuration and presents detailed mathematical modeling of its components. Section 3 discusses the development framework of the proposed ANFIS controller, highlighting its integration into the wind turbine system. Section 4 demonstrates the simulation outcomes, offering a comparative analysis of the proposed controller's performance against conventional PI controllers and other advanced methods. Last, Section 5 concludes the paper by outlining the significant insights and contributions and suggesting possible directions for future research.

2. Proposed System Configuration and Modeling of WTS Components

The configuration and modeling of the hybrid ANFIS-PI controller, integrated within MATLAB R2023b/Simulation, along with various components of a horizontal axis wind turbine (HAWT) and double-fed induction generator (DFIG), are described as follows.

2.1. System Configuration

The wind turbine system is essential in driving the transition toward sustainable energy solutions, offering a clean alternative to fossil fuels. Through advanced mechanical and electrical techniques, WTSs harness wind energy to generate electricity efficiently and reliably [25]. These systems integrate aerodynamic modeling, advanced power converters, and grid connection strategies to maximize energy output and ensure stable operation. The designed system featured a variable-speed, direct-drive wind turbine with a power rating of 1.5 MW. It employed a doubly fed induction generator (DFIG). The DFIG model was designed in Simulink with parameter ratings as depicted in Table 2 [26].

Table 2. Parameters of DFIG-based wind turbine.

Notation	Parameter	Magnitude and Unit
P	Nominal power	1.5 MVA
V	Nominal L-L voltage	575
F	Grid frequency	60
p	No. of pole pairs	40
V_{dc}	DC link voltage	1.15 KV
$\mathbb{L}$	Line inductance	0.31 pu
$\mathcal{R}$	Line resistance	0.0032 pu

Figure 1 illustrates a DFIG-based wind turbine system (WTS) integrated with the grid. In this configuration, the stator is directly connected to the grid for direct power injection. At the same time, the rotor is linked to the grid via a power electronic converter, allowing variable-speed operation and bidirectional power flow, enhancing efficiency. The back-to-back power converter system includes the rotor-side converter (RSC), which regulates rotor currents and controls generator speed, and the grid-side converter (GSC), which maintains a constant DC link voltage and ensures smooth, stable power injection into the grid [27]. This paper employed a hybrid controller to enhance the control effectiveness of these B2B converters integrated within WECS.

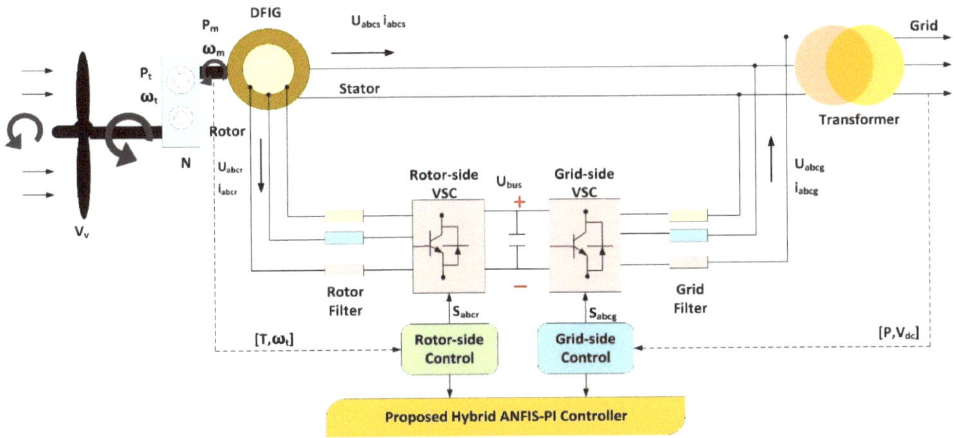

Figure 1. The architecture of the proposed grid-tied DFIG-based WECS.

2.2. *Mathematical Representation of Wind Turbine*

A turbine was created in the last century to transform aerodynamic power into electrical power. The wind's kinetic energy is transformed into rotational energy to drive an electric generator. Although wind turbines can have two or three blades, three-bladed wind turbines are the most widely utilized on land and water. Equations can express the power output and rotational force exerted by a wind turbine. The mechanical power (P_m) generated by the wind turbine is represented by Equation (1).

$$P_m = \frac{1}{2}\rho\pi\mathcal{R}^2V^3C_p(\lambda,\beta) \tag{1}$$

Here, ρ denotes the atmospheric density in kg/m^3, while A signifies the rotor-swept region of the wind turbine blades; this swept area is equivalent to $\pi\mathcal{R}^2$, where $\mathcal{R}$ denotes wind rotor radius, V stands for the wind speed, and C_p stands for the power coefficient of WT, and it is a function of λ and β as given in Equations (2) and (3). Here, β represents the blade pitch angle in degrees, which impacts the aerodynamic efficiency of the wind turbine and affects the power coefficient, whereas λ is described by Equation (4).

$$C_p = f\ (\lambda,\beta) \tag{2}$$

$$C_p = C_1(\frac{C_2}{\lambda_i} - C_3\beta - C_4)e^{-(\frac{C5}{\lambda_i})} + C_6\lambda_i \tag{3}$$

Here, $(\frac{1}{\lambda_i}) = (\frac{1}{\lambda} + 0.008\beta) - (\frac{0.035}{\beta^3} + 1)$, with empirical constants [28], while C_p depends on V_{wind} and the rotational speed of the shaft ω_t. This characteristic of the wind turbine can be depicted through the C_p curves presented in Figure 2.

$$\lambda = \frac{\omega_t\mathcal{R}_T}{V_{wind}} \tag{4}$$

Since the shaft of WT and the generator are directly connected, there is just one state variable in the system. The mechanical equation for the wind turbine is represented as

$$J\frac{d\omega_m}{dt} = T_m - T_{gen} - B_{\omega_m} \quad (5)$$

Here, J represents the total moment of inertia in kilogram-square meters (kg m^2) for the WT and generator. T_m denotes the mechanical torque, T_{gen} represents the electromagnetic torque, and the coefficient of viscous friction is defined as B and measured in kilogram-square meters per second (kg m^2/s), which can be neglected in the case of small-scale WTs. Finally, ω_m represents the mechanical rotor speed. In this model, a 1.5 MW WT is connected with a DFIG that supplies alternating current (AC) power to the grid.

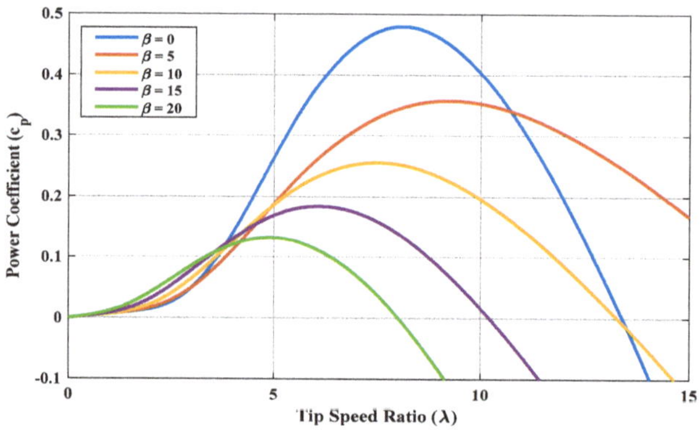

Figure 2. Power coefficient (Cp) versus tip speed ratio (λ) for WTS [29].

The mechanical power from WT can be controlled in the specified operating range of wind speed. The operating regions of WTS are classified into four regions [30] as shown in Figure 3. The WT operating range begins from cut-in speed to cut-out speed. The WTS does not produce any power in regions 1 and 4 because in region 1 the wind speed is too low for the turbine to generate power, while region 4 is the period of too-high wind speed, where the power of the wind is so high that it could be harmful to the turbine, so the turbine shuts down.

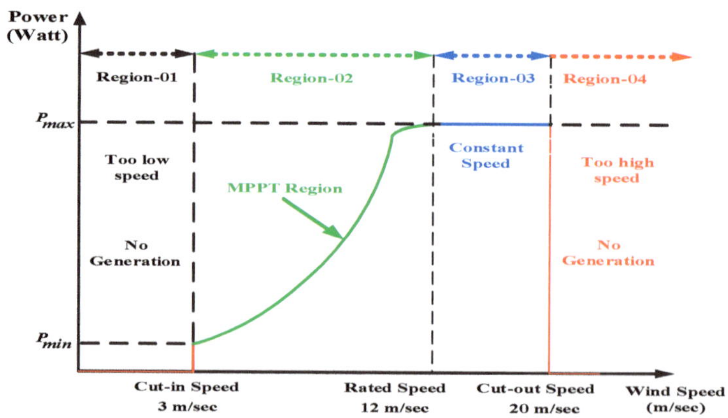

Figure 3. Power operating regions of WTS.

In region 2, the wind is needed such that the power output Cp is maximized for each wind speed. Meanwhile, in region 3, the WTS achieves its peak power generation capacity due to high wind speeds. This shows that the turbine has reached its maximum output capacity and cannot produce additional power, regardless of any further increase in wind speed. The power output becomes limited by the design capabilities of the turbine itself, not by the wind availability [31]. Therefore, it is evident that regions 2 and region 3 are extremely important. In these two regions, robust, rigorous, stable, fast, and accurate controllers are required for smooth and efficient operation of WTS.

2.3. Mathematical Model of DFIG

The DFIG consists of stator and rotor windings and slip rings. The stator has three-phase insulated windings linked to the grid via a three-phase transformer. Similarly, the rotor contains three-phase insulated windings, mirroring the design of the stator. A system of slip rings and brushes establishes a connection between the rotor windings and an external stationary circuit. These components allow for either injection into or absorption from the rotor windings of the control rotor current [32]. The dynamic model of the DFIG is represented using direct and inverse transformations. By applying space vector theory, the three-phase windings of the stator and rotor can be simplified into two equivalent windings: the stationary ab reference frame for the stator and the rotating dq reference frame for the rotor. The voltage vectors for both the stator and rotor are formulated as follows:

$$\vec{\sqcap}_{ʃ} \Longrightarrow \begin{cases} \sqcap_{ds} = \mathcal{R}_s i_{ds} + \frac{d\psi ds}{dt} - \omega_s \psi qs \\ \sqcap_{qs} = \mathcal{R}_s i_{qs} + \frac{d\psi qs}{dt} + \omega_s \psi ds \end{cases} \tag{6}$$

$$\vec{\sqcap}_{\mathrm{r}} \Longrightarrow \begin{cases} \sqcap_{dr} = \mathcal{R}_r i_{dr} + \frac{d\psi dr}{dt} - \omega_r \psi qr \\ \sqcap_{qr} = \mathcal{R}_r i_{qr} + \frac{d\psi qr}{dt} + \omega_r \psi dr \end{cases} \tag{7}$$

where $\sqcap_{ds}$,$\sqcap_{qs}$,$\sqcap_{dr}$, and $\sqcap_{qr}$ are the stator and rotor voltages and are represented in the reference frame of dq. Similarly, i_{ds}, i_{qs}, i_{dr}, and i_{qr} denote the stator and rotor currents in the dq frame. $\mathcal{R}_r$, $\mathcal{R}_s$, ω_s, and ω_r correspond to the stator and rotor phase resistances and angular velocities, respectively. Based on Equations (6) and (7), the equivalent dq electric circuit is illustrated in Figure 4.

The stator and rotor flux vectors are represented in Equations (8) and (9), respectively:

$$\vec{\psi}_{ʃ} \Longrightarrow \begin{cases} \psi_{ds} = \mathbb{L}_s i_{ds} + \mathbb{L}_m i_{dr} \\ \psi_{qs} = \mathbb{L}_s i_{qs} + \mathbb{L}_m i_{qr} \end{cases} \tag{8}$$

$$\vec{\psi}_{\mathrm{r}} \Longrightarrow \begin{cases} \psi_{dr} = \mathbb{L}_m i_{ds} + \mathbb{L}_r i_{dr} \\ \psi_{qr} = \mathbb{L}_m i_{qs} + \mathbb{L}_r i_{qr} \end{cases} \tag{9}$$

where $\vec{\psi}_{ʃ}$ and$\vec{\psi}_{\mathrm{r}}$ represent the flux vectors of the stator and rotor, respectively. ψ_{ds} and ψ_{qs} denote the flux components along the d–q axis of the stator, while ψ_{dr} and ψ_{qr} correspond to the flux components along the d–q axis of the rotor. $\mathbb{L}_s$ and $\mathbb{L}_r$ represent the leakage inductances in the stator and rotor phases; $\mathbb{L}_m$ is the mutual inductance between the stator and the rotor; and p is the generator pole pair count [33].

The mathematical formulation of electromagnetic torque is given in the following equation:

$$T_{em} - \frac{3}{2} p \frac{\mathbb{L}_m}{\mathbb{L}_s} (\psi_{qs} i_{dr} - \psi_{ds} i_{qr}) \tag{10}$$

The equations for active and reactive power in the stator and rotor are presented in Equations (11) and (12), respectively:

$$P_s = \frac{3}{2}(\sqcap_{ds} i_{ds} + \sqcap_{qs} i_{qs}) \tag{11}$$

$$Q_s = \frac{3}{2}(\sqcap_{qs} i_{ds} + \sqcap_{ds} i_{qs})$$

$$P_r = \frac{3}{2}(\sqcap_{dr} i_{dr} + \sqcap_{qr} i_{qr}) \tag{12}$$

$$Q_r = \frac{3}{2}(\sqcap_{qr} i_{dr} + \sqcap_{dr} i_{qr})$$

Here, P_s and Q_s represent the stator's active and reactive power, respectively, while P_r and Q_r denote the rotor's active and reactive power. T_{em} refers to the electromagnetic torque.

The fundamental torque equation is provided in Equation (13):

$$T_{em} - T_{load} = J\frac{d\omega_m}{dt} \tag{13}$$

where J represents the rotor's inertia, T_{load} denotes the load torque applied to the shaft, and ω_m refers to the rotor speed [34].

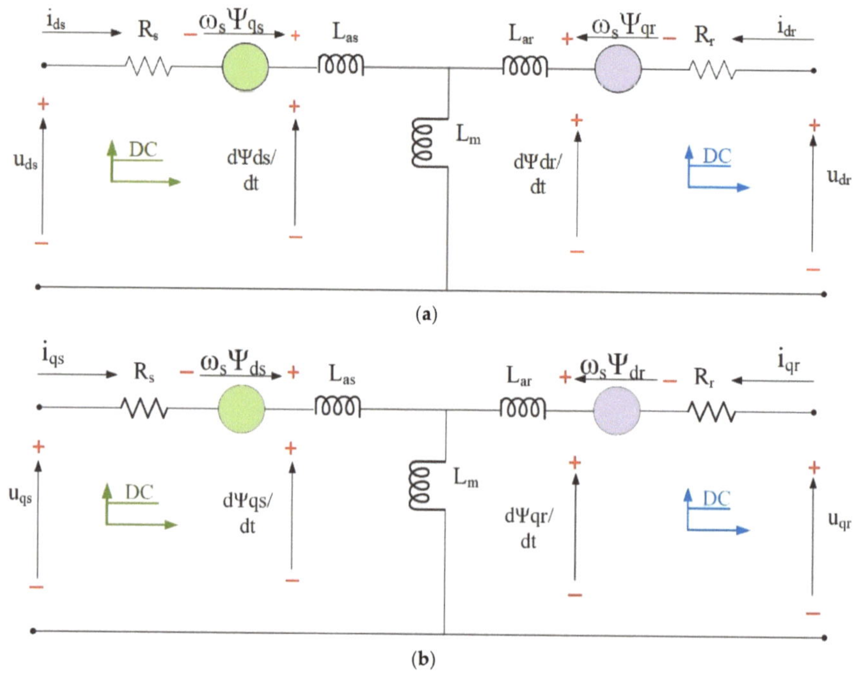

Figure 4. (**a**,**b**) DFIG dq equivalent circuit.

3. Proposed Hybrid ANFIS-PI Controller

ANFIS demonstrates exceptional efficiency in modeling complex and nonlinear systems, even with limited input and output training data. It surpasses other artificial intelligent control strategies like fuzzy logic controllers and artificial neural networks (ANNs) due to its hybrid structure. By combining the strengths of fuzzy inference systems (FISs) and ANNs into a unified model, ANFIS ensures faster convergence and is particularly suitable for modeling complex and unpredictable systems without relying on fully precise mathematical formulations. The ANFIS technique leverages system data and formulates rules directly from the data using ANN methods. To enhance system performance by minimizing overshoot and improving settling time, ANFIS was implemented in the wind turbine (WT) model, and its performance was evaluated in MATLAB/Simulink, using percentage overshoot and settling time as key metrics. The robustness of the ANFIS-based model was then compared with a conventional PI-based model.

ANFIS is an advanced intelligent control system built on the Takagi–Sugeno fuzzy inference framework. It integrates fuzzy logic concepts with neural network methodologies to form a hybrid model capable of learning and adapting to varying conditions. Using fuzzy logic rules, ANFIS generates input variables, which are then processed by a neural network to produce the corresponding control output. This capability allows it to learn from system data and progressively enhance its performance. Unlike the PI controller, ANFIS excels in managing dynamic system behavior and addressing nonlinearities, resulting in superior control outcomes. The ANFIS structure is composed of five core layers, as illustrated in Figure 5.

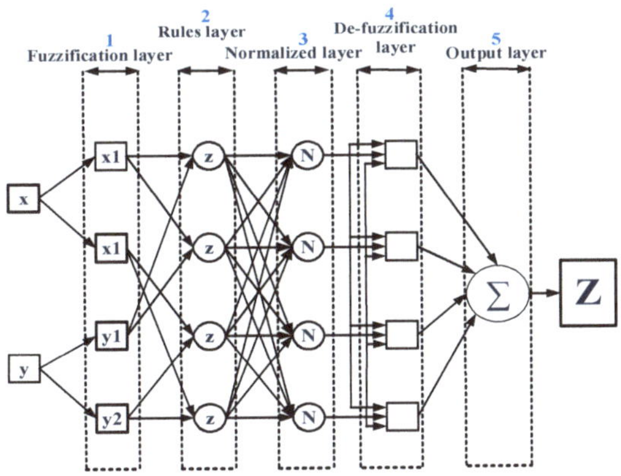

Figure 5. Basic structure of ANFIS.

The initial layer, referred to as the input layer, allocates membership degrees to input data by mapping it into membership functions (MFs). In ANFIS, variables x and y represent the inputs, where x_1 and x_2 correspond to the membership functions of x, and y_1 and y_2 correspond to the membership functions of y. The second layer implements fuzzy logic rules to define the relationships between inputs and outputs. The third layer standardizes these outputs and forwards them to the fourth layer, where the output data are processed to generate the output membership function. Finally, the fifth layer aggregates these outputs into a single value. ANFIS is capable of managing multiple inputs. The flow chart in Figure 6 illustrates the implementation process of the hybrid controller. Training data are collected from the PI-controller-based model, using the error signal as the input and the control signal as the output. Fuzzification maps these inputs to fuzzy sets using membership functions. ANFIS is then trained using historical or simulated data to optimize membership functions and rule weights. As shown in Figure 5, the fuzzification layer converts crisp inputs into fuzzy terms. The rule base then connects these fuzzy inputs to the output through a set of if–then rules.

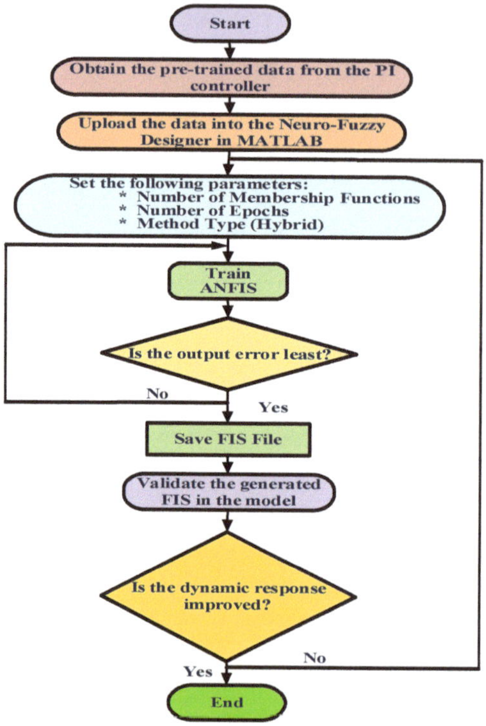

Figure 6. Flow chart of designing the ANFIS controller.

The flow chart incorporates two validation steps: one to ensure minimal output error and another to verify an improved dynamic response. Following the ANFIS training, the first check evaluates the output error. If the error does not meet the desired minimal threshold, the process loops back to retrain ANFIS with further iterations to reduce the error. Conversely, if the output error is minimized to the desired level, the process moves forward to export and saves the FIS (fuzzy inference system) file, which contains the optimized ANFIS model. The saved FIS undergoes a second evaluation to determine whether the system's dynamic response has improved. If the response meets the required criteria, the process is finalized. However, if the dynamic response is inadequate, the process restarts at the relevant step in the flow chart.

In this case, adjustments are made to the membership functions and training epochs, and the system is re-evaluated. This iterative process continues until the desired dynamic response is achieved. The parameters utilized in the designed simulation model are outlined in Table 3.

In wind turbine systems (WTSs), a hierarchical control framework is commonly utilized for both the rotor-side converter (RSC) and the grid-side converter (GSC). This hierarchical control strategy divides the task into multiple control loops, improving the control precision, stability, and response time by decoupling objectives. The cascade control system consists of two primary loops: the machine control loop, which features an inner loop for torque regulation and an outer loop for rotor speed control, and the grid control loop, which includes an inner loop for active power management and an outer loop for DC link voltage regulation. These control loops ensure stable rotor speed, regulated current flow, and proper voltage levels, thereby facilitating smooth wind turbine integration and efficient operation. Ultimately, this hierarchical control approach enhances the system's reliability, performance, and robustness.

Table 3. Parameters configured in the proposed simulation model.

Parameters	Value
Proportion constant, K_p (RSC)	2.25
Integral constant, K_i (RSC)	14
Proportion constant, K_p (GSC)	0.73
Integral constant, K_i (GSC)	2.85
Dataset (input–output pairs)	4,000,001
Test signal applied (Vwind) at 10 s	From 10 m/s to 12 m/s
No. of inputs	01
No. of outputs	01
No. of membership functions	08
Type of membership functions	Constant
Nos. of epochs/iterations	25

Implementation of Proposed Hybrid Controller in WECS Model

This study introduced a hybrid controller into the cascade control framework, where ANFIS controllers were deployed in the outer loops, and conventional PI controllers were utilized in the inner loops of the RSC and GSC converters in WTS. This method sought to attain greater reliability and efficiency. Figure 7 depicts the integration of the proposed hybrid control approach into the RSC converter. In this setup, the ANFIS controller in the outer loop ensured the desired rotor speed, while the inner loop employed a PI controller to manage mechanical torque, resulting in improved performance of the RSC converter.

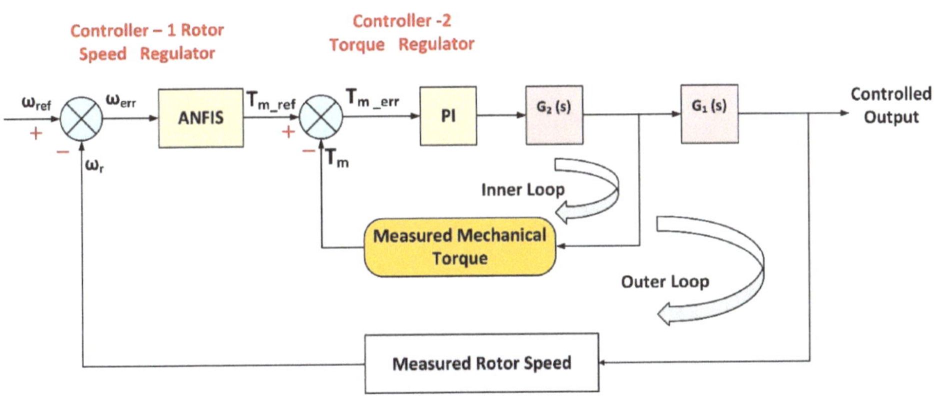

Figure 7. Implementation of ANFIS controller in the rotor-side converter.

Similarly, when the proposed hybrid ANFIS-PI control approach was implemented in the GSC converter, as shown in Figure 8, the ANFIS controller in the outer loop ensured the desired DC voltage level across the capacitor, while the inner loop utilized a PI controller to manage current regulation, resulting in improved performance of the GSC converter.

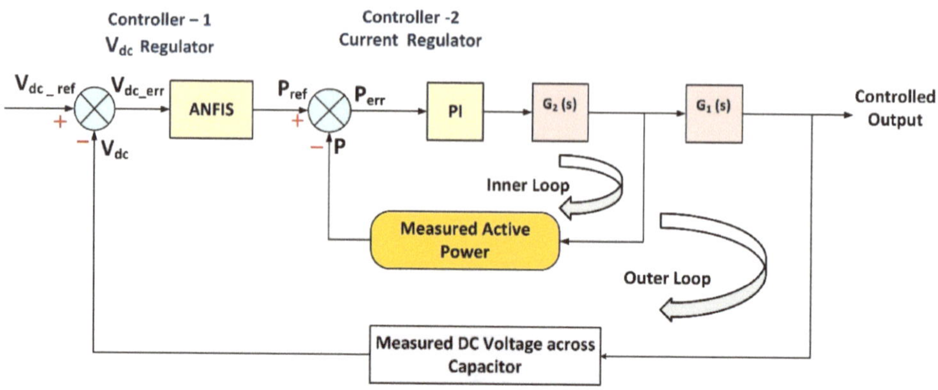

Figure 8. Implementation of the ANFIS controller in the grid-side converter.

Figures 9 and 10 illustrate the error reduction obtained throughout the learning process of the adaptive neuro-fuzzy inference system (ANFIS) applied to both the rotor-side controller (RSC) and the grid-side controller (GSC) in a doubly fed induction generator (DFIG)-based wind turbine system (WTS). The training process, conducted over 25 epochs using the hybrid optimization method, demonstrated a steady decrease in error for both controllers. For the RSC, the error reduced progressively to approximately 0.0657 by the 25th epoch, while the GSC achieved a final error value of approximately 0.0344. The fuzzy inference system was generated using the grid-partitioning approach. The findings confirmed the efficiency of the ANFIS model in achieving precise control, minimizing deviations, and enhancing the overall stability and efficiency of both controllers under varying operating conditions.

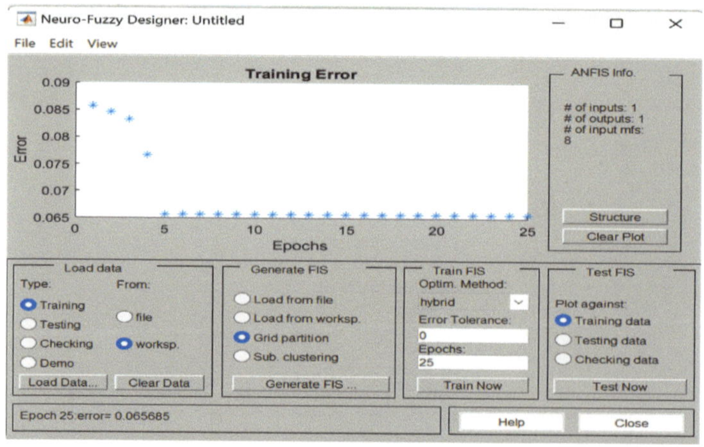

Figure 9. Error reduction in RSC after 25 iterations.

The proposed model demonstrated significant advancements over [2] in terms of computational efficiency, accuracy, and system complexity. With a dataset of 4,000,001 input–output pairs and only 25 training epochs, the model achieved a lower training error (~0.034), as shown in Figure 9, compared with the higher error (~0.094) in [2], which used a larger dataset of 8,000,001 pairs and was trained for 50 epochs. By employing 8 membership functions instead of 10, the proposed model reduced computational complexity while maintaining performance. Additionally, the replacement of PI controllers in both the outer loops of RSC and GSC enhanced the stability of the wind turbine power generation system, whereas [2] modified the controller for only one converter. These improvements underline the proposed model's superior design and practical applicability.

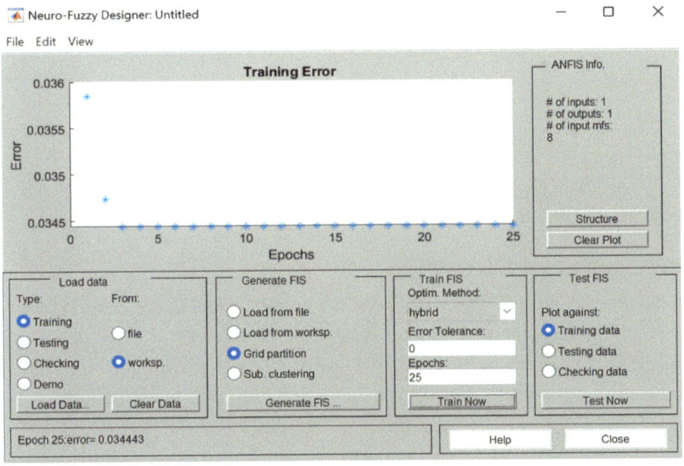

Figure 10. Error reduction in GSC after 25 iterations.

After completing the training, an ANFIS structure was generated for eight membership functions, as shown in Figure 11, which clearly illustrates the distribution of input/output membership functions and the associated rule set. The blue dots represent AND operation while the red dot and green line are associated with hidden layers.

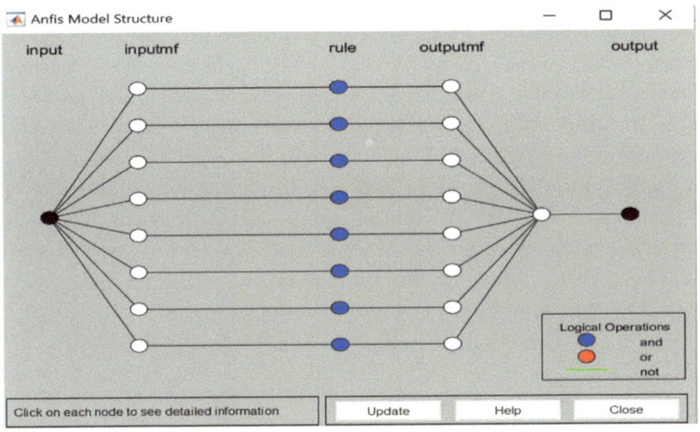

Figure 11. ANFIS model structure.

The test signals of wind speed (V_{wind}) illustrated in Figure 12 are given at 10 s by varying the wind speed from 10 m/s to 12 m/s to evaluate the performance and robustness of the WTS.

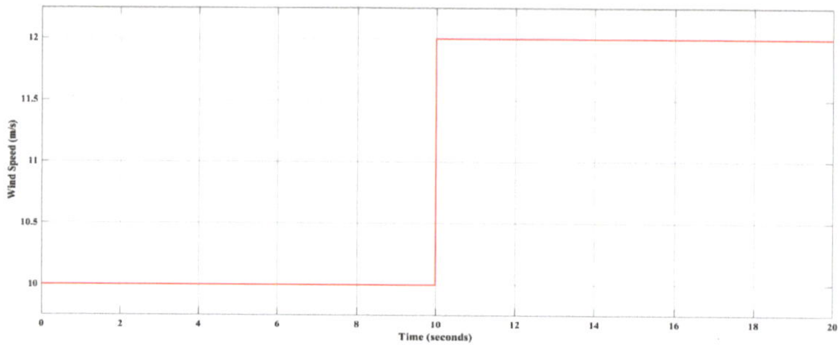

Figure 12. A test signal was applied to the signal builder of the proposed model.

4. Simulation Results and Discussion

The proposed hybrid ANFIS-PI model was designed in MATLAB Simulation by applying a step wind speed signal for simulation conditions, with the wind velocity varying from 10 m/s to 12 m/s, as outlined in Figure 11. This evaluated the system's dynamic response under changing wind regimes. Regarding dataset partitioning, the collected data were divided into three subsets:

1. The training set (70%) was used for training the ANFIS model by adjusting membership functions and rule parameters.
2. The validation set (15%) was applied for refinement and prevent overfitting, ensuring the model generalizes well.
3. The testing set (15%) was used to assess the overall model accuracy and compare it against the PI controller for validation.

By following this structured approach, we ensured that the ANFIS model was trained effectively using a well-defined dataset derived by applying step wind signal in the model as a wind turbine operating condition, enabling robust performance comparisons with conventional PI control strategies. The efficiency of the proposed hybrid ANFIS-PI controller and the conventional PI controller was evaluated under dynamic wind speed conditions, focusing on key performance indicators (KPIs) such as maximum overshoot and settling time, as well as overall system stability. The comparison evaluated electromechanical and power control variables, including the rotor speed, electromagnetic torque, active power, and DC link voltage. The hybrid ANFIS-PI controller consistently demonstrated superior performance across all these parameters, establishing itself as a more robust and adaptable solution for wind turbine control systems. The simulation results of both comparing controllers are illustrated in Figures 13–16.

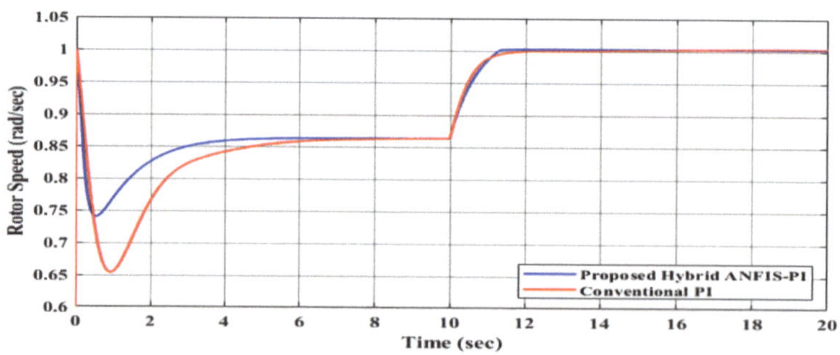

Figure 13. Hybrid and PI controllers' responses for rotor speed.

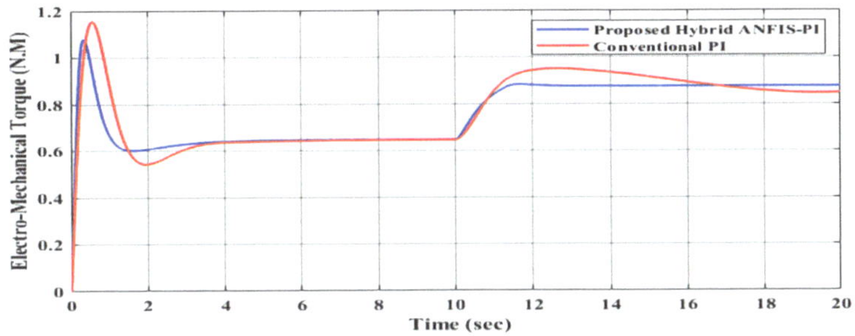

Figure 14. Hybrid and PI controllers' responses for torque.

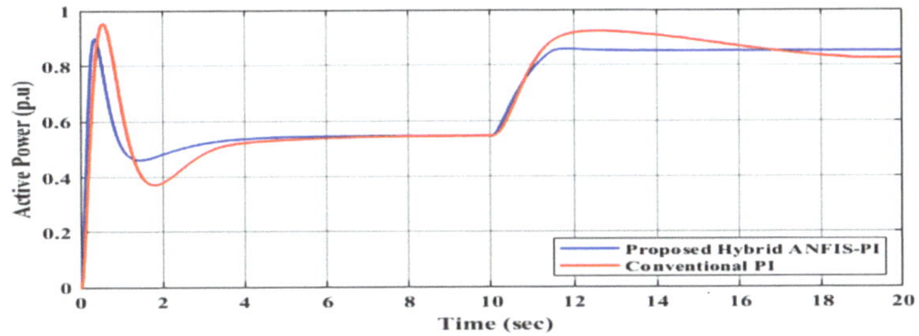

Figure 15. Hybrid and PI controllers' responses for active power.

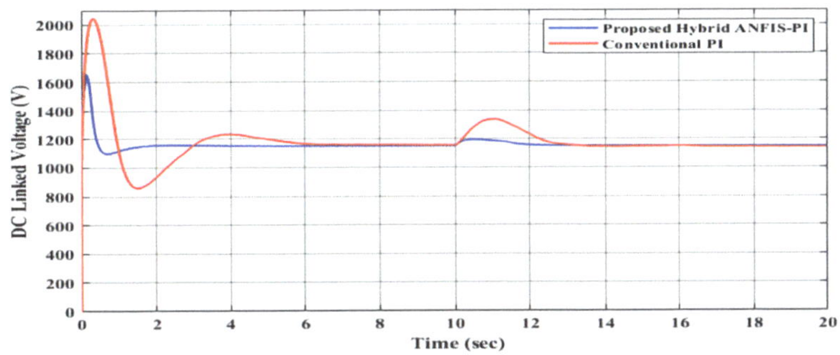

Figure 16. Hybrid and PI controllers' responses for DC link voltage.

In Figure 13, the rotor speed performance of the wind turbine model is presented over time, comparing the responses of the traditional PI controller and the developed hybrid ANFIS-PI controller. The PI controller showed significant undershoot (~0.65 rad/s) and overshoot (1.05 rad/s, 5% deviation), with a settling time of ~2 s and oscillations, indicating less stability. In contrast, the hybrid ANFIS-PI controller achieved a reduced undershoot (0.75 rad/s), minimal overshoot (1.02 rad/s, 2% deviation), and faster stabilization within 1 s, ensuring smoother and more robust performance under dynamic wind conditions.

Similarly, Figure 14 displays the time-dependent response of electro-mechanical torque for both PI and hybrid controllers. The PI controller showed a maximum overshoot of 1.15 N·m (43.75% deviation) with a settling time of 4 s, alongside oscillatory behavior. In contrast, the hybrid ANFIS-PI controller achieved a reduced overshoot of 1.05 N·m (31.25% deviation) and a faster settling time of 3 s, with a smoother and more stable response. This highlights the hybrid ANFIS-PI controller's superior adaptability and robustness, making it a more effective control strategy for wind turbines.

The response of active power (in per unit) over time under the control of both the proposed hybrid ANFIS-PI and the traditional PI controller is depicted in Figure 15. The PI controller showed a larger undershoot (~17.5%) and longer stabilization time (~10 s), with higher overshoots (~9%) and settling times (~4.5 s). In contrast, the hybrid ANFIS-PI controller had a smaller undershoot (~11%), stabilized faster (~6 s), and achieved lower overshoots (~7%) with shorter settling times (~3 s). This highlights the hybrid controller's superior adaptability, stability, and efficiency, making it a more effective wind turbine power generation strategy.

The DC link voltage response, depicted in Figure 16, demonstrated the hybrid controller's enhanced performance in regulating system voltage. The conventional PI controller exhibited significant overshoot, peaking at 1800 V, and took approximately 12 s to stabilize. Such high overshoot and extended settling times could increase the system components' stress, potentially reducing their lifespan and increasing maintenance costs. In contrast, the hybrid controller effectively limited the overshoot to 1600 V and stabilized within 10 s. By maintaining tighter control over the DC link voltage, the hybrid controller reduced the risk of component degradation and enhanced the overall system reliability. The responses of rotor speed, torque, active power, and DC bus voltage parameters when subjected to test signals for both PI and hybrid controllers are presented in Table 4.

Table 4. PI and hybrid ANFIS-PI controllers' responses across parameters.

S. No.	Parameters	Control Scheme	Rise Time (s)	Overshoot (%)	Settling Time (s)	Steady State Error
1	Rotor speed	PI	1.2	5	2	0.88
		ANFIS-PI	**0.8**	**2**	**1**	**0**
2	Torque	PI	1	43.75	4	0.9
		ANFIS-PI	**0.7**	**31.25**	**3**	**0**
3	Active power	PI	0.56	9	4.5	0.83
		ANFIS-PI	**0.55**	**7**	**3**	**0**
4	DC bus Voltage	PI	0.8	50	12	0.9
		ANFIS-PI	**0.7**	**30**	**10**	**0**

As illustrated in Table 3, the results emphasized the notable enhancements introduced by the hybrid ANFIS-PI controller in the simulation. Across all evaluated parameters, the hybrid controller consistently outperformed the conventional PI controller, showcasing its ability to adapt to dynamic wind speed variations with lower overshoot, faster stabilization, and improved stability. These attributes enhanced the overall dynamic response and contributed to robustness and reliability, making the hybrid controller a more effective solution for wind turbine control applications.

Limitations of the Proposed Method

While the proposed hybrid ANFIS-PI controller offers superior adaptability, improved dynamic response, and reduced computational complexity, it has certain drawbacks that should be acknowledged:

1. Initial training complexity: The ANFIS model requires an initial training phase, during which membership functions and rules must be optimized. Although it requires fewer training data than standalone ANN models, the training process still demands careful tuning to ensure stability and accuracy.

2. Dependence on membership function design: The fuzzy logic component in ANFIS relies on well-defined membership functions. Suboptimal membership function selection can lead to degraded performance, requiring expert knowledge or additional optimization techniques for fine-tuning.
3. Computational load during implementation: While ANFIS-PI is computationally more efficient than purely ANN-based or model predictive controllers, it still requires more processing power than traditional PI controllers. Implementing real-time control on low-cost embedded systems may require optimized hardware (e.g., DSP or FPGA) for efficient execution.
4. Limited generalization for different wind turbine configurations: The proposed ANFIS-PI controller was designed for DFIG-based WTS, meaning its direct applicability to other types of wind turbines (e.g., PMSG-based systems) may require modifications in control strategies and parameter tuning.
5. Additional computational cost for real-time adaptation: Since ANFIS continuously adapts control parameters, real-time adaptation in highly dynamic environments increases computational demand, potentially impacting performance in large-scale wind farms without adequate processing resources.

Despite these drawbacks, the proposed ANFIS-PI controller maintains a strong balance between adaptability, efficiency, and real-time feasibility, making it a more practical solution than existing AI-based and conventional control strategies.

5. Conclusions

This study presents a hybrid ANFIS-PI controller for the rotor-side converter (RSC) and grid-side converter (GSC) in wind turbine systems (WTSs), addressing the limitations of conventional PI controllers in handling nonlinear wind variations. The proposed methodology significantly enhances system performance by integrating the adaptability of ANFIS with the effectiveness of PI control. The simulation results confirmed that the ANFIS-PI controller significantly improved performance compared with the conventional PI control. Specifically, it reduced rotor speed overshoot by 3% (from 5% to 2%), torque overshoot by 12.5% (from 43.75% to 31.25%), active power overshoot by 2% (from 9% to 7%), and DC link voltage overshoot by 20% (from 50% to 30%). Additionally, the ANFIS-PI controller shortened settling time by 50% (from 2 s to 1 s) for rotor speed, by 25% (from 4 s to 3 s) for torque, by 33% (from 4.5 s to 3 s) for active power, and by 16.7% (from 12 s to 10 s) for DC link voltage, ensuring faster stabilization, enhanced dynamic response, and greater efficiency. These improvements demonstrated that the hybrid ANFIS-PI controller effectively mitigated the adverse effects of wind fluctuations, ensuring stable and reliable WTS operation. By optimizing power delivery, enhancing reliability, and improving dynamic response, this study contributes to advancing wind power technologies and their efficient assimilation into contemporary electrical networks. The results emphasize the effectiveness of AI-driven control strategies in renewable energy systems, opening new avenues for future research on intelligent and adaptive wind turbine system (WTS) control techniques.

Author Contributions: Conceptualization, F.N. and I.; Methodology, F.N. and S.K.; Software, F.N. and M.S.; Formal analysis, S.K. and M.S.; Investigation, S.K. and I.; Resources, S.U.; Data curation, F.N., I., A.A. and N.N.; Writing—original draft, F.N.; Writing—review & editing, S.U., M.S. and N.N.; Visualization, S.U. and N.N.; Supervision, S.K., I. and M.S.; Project administration, S.U., A.A. and N.N.; Funding acquisition, S.U. All authors have read and agreed to the published version of the manuscript.

Funding: This research was funded by Princess Nourah bint Abdulrahman University Researchers Supporting Project Number (PNURSP2025R79), Princess Nourah bint Abdulrahman University, Riyadh, Saudi Arabia.

Institutional Review Board Statement: Not applicable.

Informed Consent Statement: Not applicable.

Data Availability Statement: The original contributions presented in this study are included in the article. Further inquiries can be directed to the corresponding author.

Acknowledgments: Princess Nourah bint Abdulrahman University Researchers Supporting Project Number (PNURSP2025R79), Princess Nourah bint Abdulrahman University, Riyadh, Saudi Arabia.

Conflicts of Interest: The authors declare no conflict of interest.

References

1. GWEC's Global Wind Report 2024. Available online: https://www.gwec.net/reports/globalwindreport (accessed on 4 March 2025).
2. Ishaque, M.; Laghari, J.A.; Bhayo, M.A.; Chandio, S.; Mahariq, I. Hybrid ANFIS-PI-Based Robust Control of Wind Turbine Power Generation System. *Int. Trans. Electr. Energy Syst.* **2024**, *2024*, 2389751. [CrossRef]
3. Ouhssain, S.; Chojaa, H.; Aljarhizi, Y.; Al Ibrahmi, E.; Maarif, A.; Mossa, M.A. Enhancing the Performance of a Wind Turbine Based DFIG Generation System Using an Effective ANFIS Control Technique. *Int. J. Robot. Control Syst.* **2024**, *4*, 1617–1640. [CrossRef]
4. Martínez, J.C.; Rodríguez Amenedo, J.L.; Arnaltes Gómez, S.; Alonso-Martínez, J. Grid-forming control of doubly-fed induction generators based on the rotor flux orientation. *Renew. Energy* **2023**, *207*, 162–176. [CrossRef]
5. Mossa, M.A.; Echeikh, H.; Iqbal, A. Enhanced control technique for a sensor-less wind driven doubly fed induction generator for energy conversion purpose. *Energy Rep.* **2021**, *7*, 5815–5833. [CrossRef]
6. Boubzizi, S.; Abid, H.; El Hajjaji, A.; Chaabane, M. Comparative study of three types of controllers for DFIG in wind energy conversion system. *Prot. Control Mod. Power Syst.* **2018**, *3*, 21. [CrossRef]
7. Chetouani, E.; Errami, Y.; Obbadi, A.; Sahnoun, S. Optimal tuning of PI controllers using adaptive particle swarm optimization for doubly-fed induction generator connected to the grid during a voltage dip. *Bull. Electr. Eng. Inform.* **2021**, *10*, 2367–2376. [CrossRef]
8. Soomro, M.A.; Memon, Z.A.; Kumar, M.; Baloch, M.H. Wind energy integration: Dynamic modeling and control of DFIG based on super twisting fractional order terminal sliding mode controller. *Energy Rep.* **2021**, *7*, 6031–6043. [CrossRef]
9. Sahito, G.M.; Korai, J.J.; Korai, M.B.; Shaikh, A.R.S. Design of PI Controller for Wind Energy Conversion System Using MATLAB/Simulink. *Int. J. Electr. Eng. Emerg. Technol.* **2022**, *5*, 6–10.
10. Rodríguez, J.A.M. Metaheuristic Algorithm and Laser Projection for Adjusting the Model of the Last Lower Surface to a Footprint. *Biomimetics* **2024**, *9*, 699. [CrossRef]
11. Alhato, M.M.; Bouallègue, S. Direct Power Control Optimization for Doubly Fed Induction Generator Based Wind Turbine Systems. *Math. Comput. Appl.* **2019**, *24*, 77. [CrossRef]
12. Aguilar-Mejía, O.; Minor-Popocatl, H.; Tapia-Olvera, R. Comparison and Ranking of Metaheuristic Techniques for Optimization of PI Controllers in a Machine Drive System. *Appl. Sci.* **2020**, *10*, 6592. [CrossRef]
13. Sunaina, S.; Shivahare, B.D.; Kumar, V.; Gupta, S.K.; Singh, P.; Diwakar, M. Metaheuristic Optimization Algorithms and recent applications: A comprehensive survey. In Proceedings of the 2023 International Conference on Computational Intelligence, Communication Technology and Networking (CICTN), Ghaziabad, India, 20–21 April 2023; pp. 506–511.
14. Alwash, S.M.; Al-Thahab, O.Q.J.; Alwash, S.F. Modeling and Control Strategies for DFIG in Wind Turbines: A Comparative Analysis of SPWM, THIPWM, and SVPWM Techniques. *J. Eur. Des Systèmes Autom.* **2023**, *56*, 963–972. Available online: https://www.academia.edu/119271461/Modeling_and_Control_Strategies_for_DFIG_in_Wind_Turbines_A_Comparative_Analysis_of_SPWM_THIPWM_and_SVPWM_Techniques (accessed on 4 March 2025). [CrossRef]
15. Nassef, A.M.; Abdelkareem, M.A.; Maghrabie, H.M.; Baroutaji, A. Review of Metaheuristic Optimization Algorithms for Power Systems Problems. *Sustainability* **2023**, *15*, 9434. [CrossRef]
16. Tomar, V.; Bansal, M.; Singh, P. Metaheuristic Algorithms for Optimization: A Brief Review. *Eng. Proc.* **2024**, *59*, 238. [CrossRef]
17. Zhou, J.; Lilhore, U.K.; Poongodi, M.; Hai, T.; Simaiya, S.; Jawawi, D.N.A.; Alsekait, D.M.; Ahuja, S.; Biamba, C.; Hamdi, M. Comparative analysis of metaheuristic load balancing algorithms for efficient load balancing in cloud computing. *J. Cloud Comput.* **2023**, *12*, 85. [CrossRef]

18. Altay, E.V.; Altay, O.; Özçevik, Y. A Comparative Study of Metaheuristic Optimization Algorithms for Solving Real-World Engineering Design Problems. *CMES Comput. Model. Eng. Sci.* **2023**, *139*, 1039–1094. [CrossRef]
19. Maafa, A.; Mellah, H.; Benaouicha, K.; Babes, B.; Yahiou, A.; Sahraoui, H. Fuzzy Logic-Based Smart Control of Wind Energy Conversion System Using Cascaded Doubly Fed Induction Generator. *Sustainability* **2024**, *16*, 9333. [CrossRef]
20. Pandey, S.K.; Verma, K.; Kumar, S.; Agnihotri, P. Comparative Analysis of ANFIS and PI Based Control of DFIG for Wind Power Generation with Grid-Integration. In Proceedings of the 2023 9th IEEE India International Conference on Power Electronics (IICPE), Sonipat, India, 28–30 November 2023; pp. 1–7.
21. Behara, R.K.; Saha, A.K. Neural Network Predictive Control for Improved Reliability of Grid-Tied DFIG-Based Wind Energy System under the Three-Phase Fault Condition. *Energies* **2023**, *16*, 4881. [CrossRef]
22. Sayeh, K.F.; Tamalouzt, S.; Ziane, D.; Bekhiti, A.; Belkhier, Y. Utilizing Fuzzy Logic Control and Neural Networks Based on Artificial Intelligence Techniques to Improve Power Quality in Doubly Fed Induction Generator-Based Wind Turbine System. *Int. J. Energy Res.* **2025**, *2025*, 5985904. [CrossRef]
23. Sayeh, K.F.; Tamalouzt, S.; Ziane, D.; Deffaf, B.; Belaid, S.L.; Belkhier, Y.; Bajaj, M.; Blazek, V. Fuzzy Logic-Enhanced Direct Power Control for Wind Turbines with Doubly Fed Induction Generators. *Results Eng.* **2024**, *24*, 103557. [CrossRef]
24. Ouari, K.; Belkhier, Y. Model predictive direct torque algorithm for coordinated electrical grid operation of wind energy conversion system-based doubly fed induction generator. *Int. J. Model. Simul.* **2024**, 1–14. [CrossRef]
25. Chojaa, H.; Derouich, A.; Chehaidia, S.E.; Taoussi, M.; Zamzoum, O.; Mosaad, M.I.; Benbouhenni, H.; Yessef, M. Advanced Control Techniques for Doubly-Fed Induction Generators Based Wind Energy Conversion Systems. In Proceedings of the 2022 Global Energy Conference (GEC), Batman, Turkey, 26–29 October 2022; pp. 282–287.
26. 5MW DFIG Wind Turbine with Variable Speed Operation Connected to a Grid—File Exchange—MATLAB Central. Available online: https://www.mathworks.com/matlabcentral/fileexchange/58202-1-5mw-dfig-wind-turbine-with-variable-speed-operation-connected-to-a-grid (accessed on 4 March 2025).
27. Abad, G.; Lopez, J.; Rodriguez, M.; Marroyo, L.; Iwanski, G. *Doubly Fed Induction Machine: Modeling and Control for Wind Energy Generation*; John Wiley & Sons: Hoboken, NJ, USA, 2011.
28. Hete, R.R.; Shrivastava, T.; Dash, R.; Anupallavi, L.; Fathima, M.; Reddy, K.J.; Dhanamjayalu, C.; Mohammad, F.; Khan, B. Design and development of PI controller for DFIG grid integration using neural tuning method ensembled with dense plexus terminals. *Sci. Rep.* **2024**, *14*, 7916. [CrossRef]
29. Bhayo, M.A.; Mirsaeidi, S.; Koondhar, M.A.; Chandio, S.; Tunio, M.A.; Allasi, H.L.; Aziz, M.J.A.; Idris, N.R.N. An Experimental Hybrid Control Approach for Wind Turbine Emulator. *IEEE Access* **2023**, *11*, 58064–58077. [CrossRef]
30. Liu, J.; Gao, Y.; Geng, S.; Wu, L. Nonlinear Control of Variable Speed Wind Turbines via Fuzzy Techniques. *IEEE Access* **2017**, *5*, 27–34. [CrossRef]
31. Nasim, F.; Khatoon, S.; Ibraheem; Shahid, M. Implementation of crowbar protection in DFIG. In *Advances in AI for Biomedical Instrumentation, Electronics and Computing*; CRC Press: Boca Raton, FL, USA, 2024.
32. Farrag, M.E.A.; Putrus, G.A. Design of an Adaptive Neurofuzzy Inference Control System for the Unified Power-Flow Controller. *IEEE Trans. Power Deliv.* **2012**, *27*, 53–61. [CrossRef]
33. Htet, T.Z.; Zhao, Z.; Gu, Q. Design analysis of direct-driven PMSG in wind turbine application. In Proceedings of the 2016 International Conference on System Reliability and Science (ICSRS), Paris, France, 15–18 November 2016; pp. 7–11.
34. Ukoba, K.; Onisuru, O.R.; Jen, T.-C. Harnessing machine learning for sustainable futures: Advancements in renewable energy and climate change mitigation. *Bull. Natl. Res. Cent.* **2024**, *48*, 99. [CrossRef]

sustainability

MDPI

Review

Wind Energy Contribution to the Sustainable Development Goals: Case Study on London Array

A. G. Olabi [1,2,*], Khaled Obaideen [1], Mohammad Ali Abdelkareem [1,3,*], Maryam Nooman AlMallahi [4], Nabila Shehata [5], Abdul Hai Alami [1], Ayman Mdallal [1], Asma Ali Murah Hassan [1] and Enas Taha Sayed [3]

1 Sustainable Energy & Power Systems Research Centre, RISE, University of Sharjah, Sharjah P.O. Box 27272, United Arab Emirates
2 Mechanical Engineering and Design, School of Engineering and Applied Science, Aston University, Aston Triangle, Birmingham B4 7ET, UK
3 Chemical Engineering Department, Faculty of Engineering, Minia University, Al Minya 61519, Egypt
4 Department of Mechanical and Aerospace Engineering, United Arab Emirates University, Al Ain City P.O. Box 15551, United Arab Emirates
5 Environmental Science and Industrial Development Department, Faculty of Postgraduate Studies for Advanced Sciences, Beni-Suef University, Beni-Suef 65211, Egypt
* Correspondence: aolabi@sharjah.ac.ae (A.G.O.); mabdulkareem@sharjah.ac.ae (M.A.A.)

Abstract: Clean and safe energy sources are essential for the long-term growth of society. Wind energy is rapidly expanding and contributes to many countries' efforts to decrease greenhouse gas emissions. In terms of sustainable development goals (SDGs), renewable energy development promotes energy security while also facilitating community development and environmental conservation on a global scale. In this context, the current article aims to investigate wind energy's role within the SDGs. Furthermore, the present study highlights the role of the London Array wind farm in achieving the SDGs. Indeed, deploying clean and economical energy sources in place of conventional fossil fuel power plants provides vital insights into environmental impacts. The London Array operation is saving approximately 1 million tons of carbon dioxide (CO_2) equivalent. Furthermore, the London Array contributes to the achievement of multiple SDGs, including SDG 8: decent employment and economic growth; SDG 9: industry, innovation, and infrastructure; SDG 11: sustainable cities and communities; and SDG 15: life on land. To enhance the London Array's contribution to the SDGs, a total of 77 indicators (key performance indicators) were proposed and compared to the current measurements that have been carried out. The results showed that the London Array used most of the suggested indicators without classifying them from the SDGs' perspective. The proposed indicators will help cut operation costs, mitigate climate change and environmental damage, improve employee engagement and morale, reduce learning gaps, set goals and plans, and use resources efficiently.

Keywords: renewable energy; wind energy; climate change; sustainable development goals; indicators

1. Introduction

Currently, the global energy demand surpasses the available generation capacity. As a result, future energy requirements must be effectively and economically fulfilled. The use of renewable energy sources should be prioritized in the implementation of energy solutions. At the moment, renewable energy's contribution to primary energy on a global scale is insufficient to fulfil primary energy and power demand [1]. Many developing nations appear to be attempting to enhance their energy sectors. However, bringing new inventions to reality appears to be difficult [2]. There is always the issue of sustainability when it comes to energy resources. Resources must offer enough energy to suit our demands, such as heating homes, powering cities, and driving automobiles. However, it is equally critical to evaluate how these resources will be used in the long run. Renewable resources will never be depleted. Renewable resources also generate clean energy, which means reduced pollution and greenhouse gas emissions, both of which contribute to climate change [3,4]. Renewable energy sources are natural resources that can be transformed into clean, usable energy, such as biomass, geothermal resources, solar, hydropower, ocean, and wind. Table 1 summarizes the benefits and drawbacks of various renewable energy sources.

Table 1. Renewable energy sources comparison.

Source	Advantages	Disadvantages
Solar	• More predictable energy output • Can be installed on the roof • No noise	• Variable source and power output • High initial costs • Land requirement for large-scale projects
Wind	• Can produce energy during the day and night • Releases less carbon dioxide (CO_2) emissions • Produces more energy (high efficiency)	• Environmental impact • Noise pollution • Variable source and power output
Geothermal	• Requires small spaces • No noise • Long-lasting power plants • Does not require frequent maintenance	• Location restrictions • Environmental side effects • High costs
Biomass	• High availability • Reduces waste • Being carbon neutral	• Requires space • Low efficiency • Requires water
Tidal	• Environmentally friendly • High energy density • Low operation and maintenance costs	• Limited locations • Variable intensity of tides • High construction costs
Hydropower	• Safe • Low emissions • Can operate all day (reliable)	• Expensive to construct • Limited locations and reserves • Environmental side effects

The United Nations developed the Sustainable Development Goals (SDGs) to create a more sustainable future for all. The goals were established in 2015, and extensive implementation is scheduled to be achieved by 2030 [5]. The SDGs' framework, as displayed in Figure 1, allows for the tracking of implementation and progress towards the Agenda 2030 [6].

Figure 1. The United Nations Sustainable Development Goals (SDGs). (Open Access, https://www.un.org/development/desa/disabilities/envision2030.html, accessed on 1 November 2022).

In terms of the SDGs, renewable energy development enables energy security in transportation [7], the environment [8], construction [9], mechanical work, industry, economy [10,11], and the internet of things (IoT) [12,13]. Renewable energies such as solar [14], wind [15], hydro [16], tidal [17], geothermal [18], biomass [19], and green hydrogen [20] assist supply energy needs while also facilitating community development and environmental protection on a worldwide scale. In recent decades, clean energy has been advocated as a viable solution to the energy dilemma while avoiding negative climatic and environmental implications [21,22]. Compared to other renewable energy sources, wind power generation has an advantage because of its modern technology, dependable infrastructure, and cheap overall cost [23]. Wind energy is likely to play a larger part in the UK's energy supply in future years. The trends in the world's wind energy capacity are shown in Figure 2.

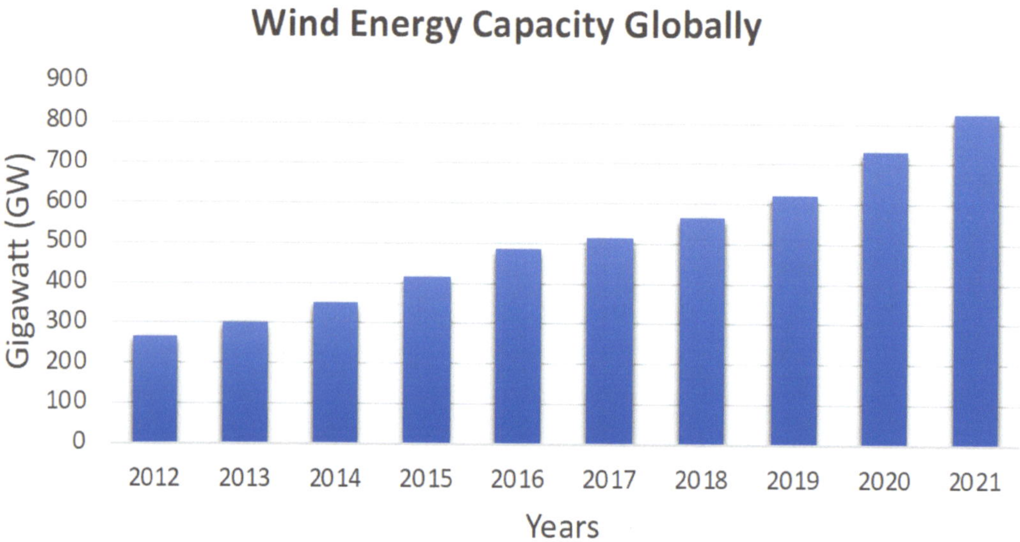

Figure 2. Wind energy capacity globally, data extracted from IRENA 2022 (https://www.irena.org/-/media/Files/IRENA/Agency/Publication/2022/Apr/IRENA_RE_Capacity_Statistics_2022.pdf, accessed on 20 January 2022).

This study investigates the significance of increasing wind energy production to meet the SDGs. The role that wind energy, which is abundant worldwide, plays in contributing to the achievement of each of the 17 SDGs is the subject of specific research. In order to illustrate the significance of wind power to the various SDGs, a case study was conducted using the London Array. Then, in order to ensure that wind energy fully meets the requirements of the SDGs, we proposed a total of 77 indicators. This study provides an exciting opportunity to advance our knowledge of the relationship between wind energy and the SDGs.

2. Wind Energy

Wind energy is an essential component of renewable energy, it also offers several advantages, including being cost-effective, reliable, and environmentally friendly. It is now used to produce electricity [24]. Wind energy consumption is crucial to investigate because of wind's enormous potential. Furthermore, wind turbine technology has advanced significantly in recent years. Turbines' efficiency and reliability have increased, allowing them to produce more power. Moreover, wind turbines are also relatively safe to operate and have much lower environmental impacts than other power plants [25,26]. Such an increase in efficiency will reduce the used areas, meaning that more land might be used for purposes other than turbines, such as farming or forestry [27,28].

Wind power has been utilized for various applications, such as milling grains, pumping water, and sawing wood, for many years [29,30]. The drawback of wind energy is that it can sometimes be intermittent and cannot power a whole city or country. Denmark, for instance, cannot rely only on wind power since there is too much demand during the winter months when the winds are weak [31]. An experiment was carried out in the country to determine the power density of wind turbines at various heights, and the findings revealed that as the height increased, so did the wind velocity, which increased the power density (W/m^2) received from the wind blade. The horizontal axial wind turbine was chosen because it required less wind speed to begin generating power. Masdar city was classified as a bad wind zone with high turbulence intensity, reaching 19.82% at a height of 10 m [32].

Mankind has been harnessing the wind's energy for various reasons, such as sailing ships, grinding grain, using windmills, and recently, generating electricity [29,33]. Wind energy may be harnessed using two distinct types of wind turbines, depending on the direction of the revolving shaft. Wind turbine efficiency is the electrical power obtained by the wind turbine over the available power in the wind. According to Albert Betz, obtaining more than 59.3% efficiency (theoretical efficiency) is unattainable, and the actual efficiency ranges between 25 and 45% [34]. A wind farm can include a few to hundreds of wind turbines spread across a broad region. Wind farms might be built onshore or offshore [35]. Wind farms do not pollute the air or water since no fuel is burnt [36,37]. As a result, wind energy will enhance energy security while simultaneously helping to reduce carbon dioxide emissions. The generator is coupled to a shaft on which the wind blades are rotating. A gearbox connects the slow-rotating turbine shaft to the fast-rotating generator shaft; these gears raise the turbine shaft's rotational speed from 30–60 rpm to 1200–1500 rpm in the generator shaft, resulting in the spinning of copper coils inside a generator-housed magnet. This magnet excites the wire's electrons, creating electricity. The number of copper coils and the speed at which the shaft spins in the magnetic field determine how much power is produced [38].

To achieve sustainability, it is critical to employ renewable energy sources, such as wind energy, which can be harnessed without causing harm to future generations. Wind energy is clean and environmentally friendly since it emits no carbon dioxide or other greenhouse gases into the atmosphere [39]. Wind power is a sustainable and environmentally friendly form of electricity generation since it does not pollute or harm the environment while producing electricity. Wind power also contributes towards a higher quality of life by supplying us with clean air, clean water, and protection from natural disasters, such as hurricanes and floods (associated with global warming) [40,41]. There are several challenges associated with using wind energy, such as the need for more land to operate a wind farm and the difficulty in locating a place with enough wind to provide the optimal efficiency and power [42,43]. Most onshore wind farms are typically located in rural areas, distant from cities that demand power, which necessitates the construction of transmission lines to transport the electricity from the wind farm to the city, eventually increasing the cost. Other disadvantages include the influence on nearby animals; there have been multiple occasions where birds and bats have died due to accidents with spinning turbine blades and turbine obstructions in their flight routes [44].

It is important to note that continuing research is being conducted to discover a solution to the mentioned limitations and any potential major obstacles [45,46].

- Environmental impacts

Several studies have found wind energy to be one of the cleanest, infinitely sustainable, and ecologically beneficial of all the renewable technologies available today. It is considered the most compatible energy source for humans and animals [47,48]. Despite this, several experts have detected small environmental repercussions, notably on animals. Wind farms require a huge amount of land with enough wind to function efficiently, which will undoubtedly impact the area's environment; nevertheless, the ground beneath each turbine may still be utilized for animal grazing or cultivation [40,49]. Furthermore, wind farms directly impact the lives of birds and bats. Birds are among the most common global victims of wind turbine accidents [50]. It is critical to investigate the effects of wind turbine use and wind farms; however, when weighed against the beneficial impact, wind energy has done more benefit than harm to the environment [51]. Wind energy will reduce or eliminate the usage of the fossil fuels that are producing harmful CO_2, NO_x, and SO_x emissions that are the reason behind acid rain and global warming, which eventually cause the greenhouse gas effect, a rise in sea level, and fluctuating weather conditions [52,53].

- Social impact

The primary concerns for every country's growth are social issues. One of the most significant social benefits of wind power expansion is the creation of job opportunities; the rapid growth that the wind energy sector has shown in recent years is promising; according to the Office of Energy Efficiency and Renewable Energy, the field of wind energy employed over 100,000 people in the United States in 2016 [54]. Furthermore, wind farms can meet citizens' demands, allowing them to carry out their activities without fear of blackouts [55].

- Economic impact

In addition to job opportunities, the wind energy industry supplies residents with low-cost power. It is regarded as one of the most affordable energy sources, with prices for the most competitive projects as low as USD 0.030/kWh [56]. A wind farm may be considered an addition to tourism in regions where tourism is a significant component of the local economy for individuals interested in technology.

3. Sustainable Development Goals

The SDGs are a set of 17 goals established by the United Nations to end poverty and hunger, achieve gender equality and empower women, ensure healthy lives and promote well-being for all ages, reduce inequality within and among countries, ensure sustainable consumption and production patterns, and slow or stop climate change and its consequences [57]. The United Nations established the SDGs, also known as the Global Goals, in 2015 as a call to action to eradicate poverty, protect the environment, and promote peace and prosperity for all by 2030 [58]. The Brundtland Commission issued the Brundtland Commission Report in 1987, which defined sustainable development as "development that fulfills existing requirements without harming future generations." The four components of sustainable development are society, environment, culture, and economics [59]. An integrated approach to sustainability takes into account environmental, social, and economic considerations, as well as the desire for a higher quality of life in the future. For example, affluent civilizations may offer food, resources, and clean air to their citizens by maintaining a healthy environment. Sustainability is commonly thought of as an end goal, but sustainable development is the method by which it is attained. All 17 SDGs are interrelated, reflecting the reality that the actions of one zone affect the outcomes in other zones, requiring adjustments to social, financial, and ecological sustainability. Nations have agreed to coordinate advancements for those who are most hampered.

The SDGs also aim to dispense with destitution, starvation, and sexual orientation separation against females. Nowadays, the joined together Countries Office of Financial and Social Issues (UNDESA) Division for Maintainable Advancement Objectives (DSDG) gives substantive support and capacity-building for the SDGs and related topical issues, such as water, vitality, climate, seas, urbanization, transportation, science and innovation, the Worldwide Economic Advancement Report (GSDR). The DSDG plays a critical part in assessing the UN system-wide execution of the 2030 plan and in promotion and outreach activities related to the SDGs. To make the 2030 plan a reality, the far-reaching proprietorship of the SDGs must translate into a solid commitment to execute the worldwide objectives by all partners.

3.1. Renewable Energy and SDGs

The SDGs provide a solid foundation for international collaboration in achieving a sustainable future for the planet. Without access to energy services, it is hard to live a dignified life. Not only technology is capable of producing and supplying low-cost, sustainable energy [60,61]. It is not only a goal in its own right but also a means to attain other goals, and the implementation of renewable energy ensures that no one is left behind. With the correct financing mechanism, energy systems can be deployed independently to produce local value [62]. Renewables are easy to install and manage due to their modularity, flexibility, and ease of installation. Every individual or community has the potential to become an energy producer, contributing to the growth of new industries, fostering innovation and environmentally responsible industrialization, and opening up prospects for employment. Renewable energies can reduce the inequalities between urban and rural populations [63]. Renewable energy makes it possible to pay attention to the requirements of women and children, who typically suffer the most from a lack of essential services in their homes. Cooking with renewable power can help preserve our forests and wildlife. Additionally, utilizing renewable energy results in lower levels of air pollution and emissions of greenhouse gases; therefore, it is one of the most important methods" might be better for combating climate change [64,65].

The same is true for life on land as it is for life in the water: renewables can alleviate the effects of ocean acidification and maintain marine ecosystems [66]. Because geopolitical tensions and military confrontations are linked to access to fossil fuel resources and infrastructure in many regions of the world, a world powered by renewable energy is a more peaceful, secure, and equitable environment for everybody [67]. It is regarded as one of the purest and most environmentally friendly energy sources, which aids in achieving sustainable development goals; similar to nuclear power plants, it also does not create any waste. It is also one of the least expensive renewable energy sources accessible to consumers.

3.2. Wind Energy Role in Achieving the Sustainable Development Goals (SDGs)

The United Nations SDGs aim to expand the share of renewable energy in the global electricity mix to at least 70% by 2030. Wind energy does not contribute to the pollution of the environment, does not run out, and lessens the need for fossil fuels, which are the primary contributors to the greenhouse effect [47,68]. Wind energy has many benefits, such as reducing carbon emissions and creating jobs, but it also has some challenges, such as the potential to harm wildlife if not properly managed. Wind energy also helps to minimize energy imports while providing income and local jobs. Wind energy generation and good usage contribute to sustainable development [15,69]. Figure 3 shows the wind's advantages and disadvantages in terms of the SDGs. The benefits of wind energy are as follows: it is a renewable energy source; it is inexhaustible; it cuts down on the usage of fossil fuels; it cuts down on energy imports; it creates wealth and local employment opportunities. Figure 4 summarizes the contributions of wind to the three pillars of sustainable development.

Wind energy emits no toxic elements or poisons into the atmosphere, which might be exceedingly destructive to both the environment and humans. Air pollution has been related to several health issues, including cancer, cardiovascular disease, and respiratory diseases such as asthma. Wind energy operates in the same way as windmills do, by harnessing the power of the wind to turn a blade. However, offshore usage has its own set of challenges [70]. Achieving this will help us prevent climate change and other environmental issues that are caused by pollution. Table 2 summarizes the contributions of wind energy to the SDGs.

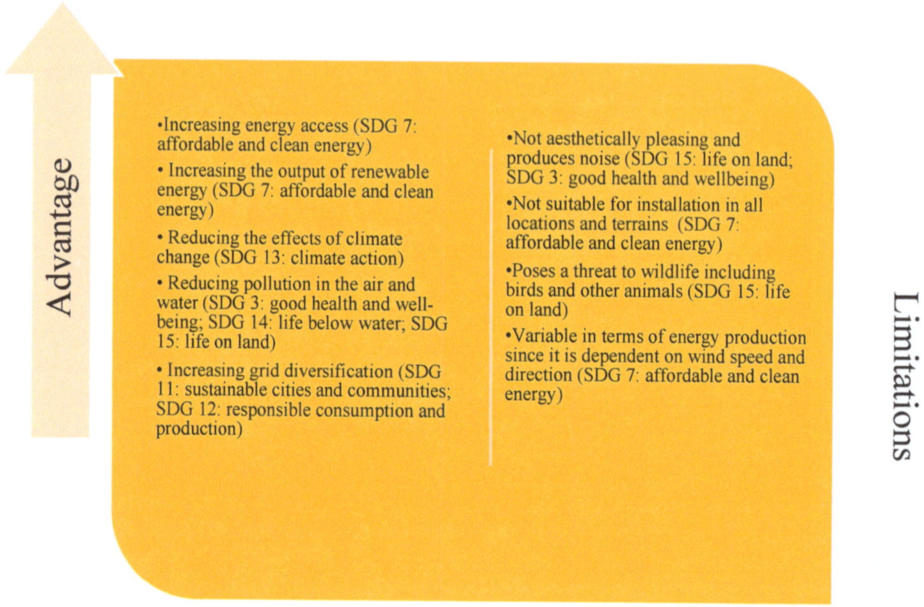

Figure 3. Advantages and disadvantages of wind in the context of the SDGs.

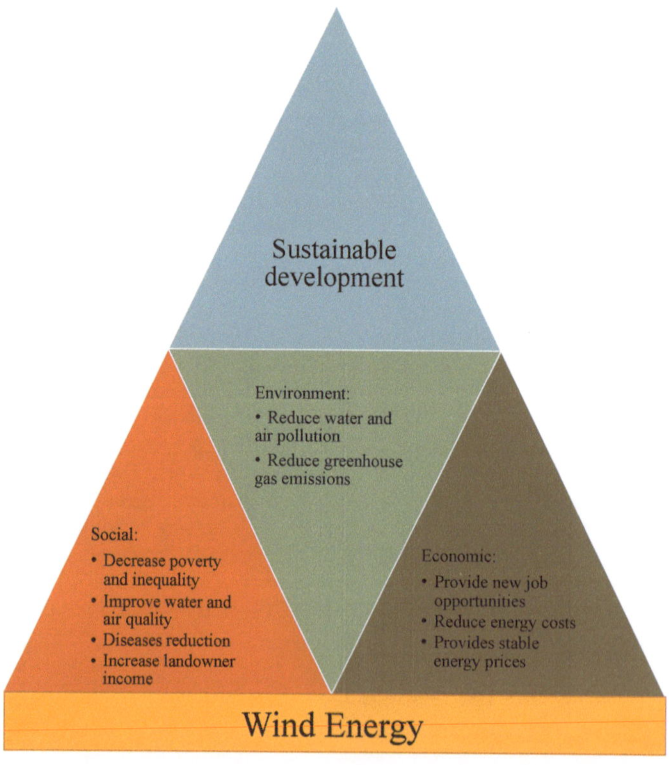

Figure 4. Contributions of wind energy in terms of social, economic, and environmental aspects.

Table 2. Summary of the contributions of wind energy to the SDGs.

SDG	Wind Farms' Contribution to the SDGs
SDG 1: No poverty	• Increases access to energy and creates more jobs.
SDG 3: Good health and wellbeing	• Has the potential to contribute to the improvement of public health by lowering levels of air pollution and emissions of greenhouse gases, both of which have been linked to a variety of adverse health effects.
SDG 6: Clean water and sanitation	• Requires no water consumption, unlike conventional technologies.
SDG 7: Affordable and clean energy	• A clean, renewable source of energy that does not produce greenhouse gas emissions. As a result, it contributes to the goal of increasing the use of clean energy sources, which is to increase the use of clean energy sources.
SDG 8: Decent work and economic growth	• Creates new jobs and improves economic growth.
SDG 9: Industry, innovation, and infrastructure	• Improves energy security and reduces the overall carbon footprint of the industry.
SDG 11: Sustainable cities and communities	• Reduces reliance on traditional power plants, resulting in less pollution.
SDG 12: Responsible consumption and production	• Reduces air and water pollution by supplying cleaner and safer energy.
SDG 13: Climate action	• Reduces GHG emissions by supplying a low-emission energy source. • Prevents climate change.
SDG 14: Life below water	• Avoids the negative effects on ecosystems of conventional resources.
SDG 15: Life on land	• Utilizes unused areas and regions for sustainable energy production.

Several indicators can be utilized to determine whether or not wind energy is sustainable. Some examples include [22,71]:

- Impact on the environment: Wind energy has several impacts on the environment, including habitat destruction, noise pollution, and aesthetic degradation. The number of birds and bats that are killed by wind turbines, the distance between turbines and homes or other sensitive receptors, and the amount of greenhouse gases emitted during the manufacture and transportation of wind energy components are some of the indicators that can be used to evaluate these factors.
- Economic viability: The economic viability of wind energy can be evaluated by taking into consideration a number of factors, including the cost of the energy that is produced, the amount of subsidies that are required, and the return on investment.
- Acceptability to society: The level of community involvement in wind energy projects, public attitudes toward wind energy, and the potential for conflict with other land uses are some of the factors that can be used to evaluate the acceptability of wind energy to society.

- Maturity in terms of technology: The technological maturity of wind energy can be evaluated by considering a number of factors, including the dependability of wind turbines, the effectiveness of energy production, and the possibility of further technological advancements in the near future.

3.3. Case Study: London Array Wind Farm

The London Array was initially designed in four phases and is located 20 km between the counties of Kent and Essex, with the cable path heading southwest towards Cleve Hill in north Kent [72]. It has 175 turbines generating 630 megawatts (MW) of power. The first phase of this project comprised 175 wind turbines capable of powering half a million UK households and reducing damaging carbon dioxide emissions by around 900,000 tons per year. Construction began on March 2011 and was completed in 2013 [73]. Due to concerns about the impact on marine birds, a second phase of the project involving building 166 more turbines (raising the total output to 1000 MW) was denied planning clearance in 2014 [74]. The project was funded in part by E.ON (UK), DONG (Denmark), Masdar (Abu Dhabi), and La Caisse (Canada). Figure 5 shows the London Array project's contribution to sustainable development. Moreover, the London Array can be considered one of Europe's largest wind farms in terms of power generation.

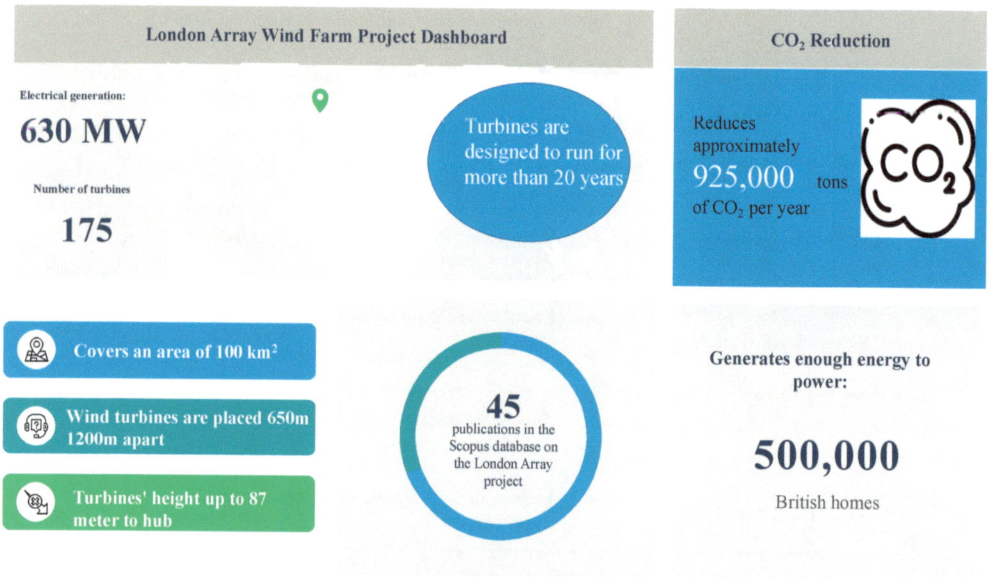

Figure 5. London Array project contributions to sustainable development.

Advantages of the London Array project:

- No waste or pollution generated from the plant.
- Relatively inexpensive to create.
- Renewable.
- Significant energy produced.
- Reduce carbon emissions by a huge amount in terms of carbon tons.
- Assists the UK in attaining the EU's 15% renewable energy goal by 2020.
- Local work prospects during the plant's development and life cycle.

Disadvantages of the London Array project:

- A massive number of turbines are required to generate a considerable amount of electricity.
- The unpredictable output of the wind turbines, which are affected by wind speed.

- Disruption to marine life.
- Disruption to seabird existence in that location.
- Disturbance to local fishing grounds.

In terms of the London Array's sustainability, they were able to effectively combat the negative impacts by collaborating with a variety of stakeholders and a variety of connected agencies, including the Maritime and Coastguard Agency. In addition, they have implemented every conceivable preventative precaution to lessen the severity of negative effects. Numerous environmental impact assessments (EIA) were conducted before any building commenced, as with any new development, but notably with the planning requirements for a renewable energy plant [75]. When creating a renewable energy plant, it is critical to evaluate its influence on habitats, migration habits, feeding places, breeding grounds, and ecosystem health, as well as other potential implications [49]. Based on the EIA report provided by London Array, available at https://londonarray.com/wp-content/uploads/2020/07/Non-technical-summary.pdf (accessed on 2 January 2023), in terms of the physical environment, it was anticipated that the construction, operation, or decommissioning of the facility would not have any significant effects on the surrounding physical environment.

In terms of the biological environment, wind farms provide a considerable risk to birds due to the possibility of accidents involving wind turbines. There is also concern that wind turbines would shift populations or lead birds to avoid the farms entirely due to the perceived risk [50]. Larger wind turbines and quicker rotor speeds, which are more prevalent in offshore turbines, increase the chance of collision. The displacement of birds from the surrounding region as a result of visual intrusion or noise from wind farms can result in significant habitat loss. Birds may also alter their migratory or local flyways to avoid the wind farm, resulting in increased energy expenditure and the disruption of feeding and roosting habits and breeding regions [76]. Some of the implications are favorable, such as the elimination of emissions caused by alternative energy sources. However, certain repercussions are extremely damaging, such as probable rotor blade migration impedance or breeding ground displacement due to onshore substation building [77].

The economic benefits of the wind project, such as the creation of jobs during the construction and operation of the facility, as well as the contribution of the wind project to the local economy through taxes and other economic activity, are included in the definition of the social impact that London Array has. It is possible that the wind farm will have a beneficial societal impact as a result of the provision of clean, renewable energy. This has assisted in the reduction of emissions of greenhouse gases and the amelioration of the effects of climate change.

In terms of its community involvement, London Array has been participating in activities such as donating money to local charities and organizations, organizing events for the community, and collaborating with local educational institutions, such as schools and universities, to advance educational opportunities and professional growth. A social effect has also been caused by the London Array employment and diversity policies. These practices include its attempts to recruit and retain a diverse staff and its efforts to promote equal opportunities and inclusion in the workplace.

In terms of supporting the SDGs, the London Array offshore wind farm can be seen as contributing to several of these. The following are some of the ways that London Array has contributed:

- SDG 1: No poverty. The building and ongoing operation of the London Array wind farm has created employment and other economic possibilities, both of which can contribute to the regional economy's expansion and growth. In addition, the project has offered financing for community projects via its community fund, which has funded efforts such as education about renewable energy and energy efficiency upgrades for local homes. Moreover, the project has provided funding for community projects. These activities may enhance the economic and social well-being of disadvantaged communities, which can contribute to decreasing poverty in such places.

- SDG 3: Good health and well-being. London Array may assist in minimizing the effects of climate change on human health by lowering the emissions of greenhouse gases. Human health is susceptible to being negatively impacted by climate change in various ways, including increasing temperatures, severe weather events, and other effects. In addition, the decrease in air pollution that results from using wind energy may have a beneficial effect on human health. This is especially true for more susceptible groups, such as children, the elderly, and those who already have one or more preexisting health concern.
- SDG 7: Affordable and clean energy. London Array generates electricity from a renewable energy source, wind, which is considered to be more sustainable and less polluting than fossil fuels such as coal and oil. By providing clean electricity, the London Array helps to reduce greenhouse gas emissions and combat climate change, which is a key target of this goal.
- SDG 8: Decent work and economic growth. According to London Array's website, the wind farm has supported the creation of over 3000 jobs, with over 60% of the work going to UK-based companies.
- SDG 9: Industry, innovation, and infrastructure: The London Array is a large-scale infrastructure project that requires significant investment and innovation in designing and constructing the wind turbines and offshore substation. The project has also supported the development of the offshore wind industry in the United Kingdom.

Overall, London Array's focus on research and development demonstrates its commitment to improving offshore wind energy's efficiency and sustainability and advancing the state of knowledge in this field.

- SDG 13: Climate action. As a large offshore wind farm, the London Array is expected to offset approximately 1.9 million tons of carbon dioxide emissions per year, making it a significant contributor to climate action.
- SDG 14: Life below water. London Array generates electricity from a renewable energy source, wind, which is considered to be more sustainable and less polluting than fossil fuels such as coal and oil. By reducing greenhouse gas emissions, London Array can help mitigate climate change's impacts on the oceans, which is a major threat to marine life. In addition, London Array has supported offshore wind energy development, which may help reduce the demand for fossil fuels and potentially reduce the negative impacts of offshore oil and gas development on marine ecosystems.
- SDG 15: Life on land: London Array can help mitigate the effects of climate change on land-based ecosystems and biodiversity by reducing the emissions of greenhouse gases. These ecosystems and species are in danger because of rising temperatures, extreme weather events, and other effects of climate change. In addition, London Array has contributed to the development of offshore wind energy, which may help to reduce the demand for fossil fuels and may help to reduce the negative impacts of the extraction and use of fossil fuels on land-based ecosystems.

Figure 6 provides a summary of how London Array will be able to contribute to the achievement of the SDGs. The purpose of Figure 6 is to provide a visual summary of the potential impact that the London Array project could have on the SDGs, highlighting the ways in which the project is aligned with global efforts to create a sustainable and equitable future. It is possible for stakeholders to gain a better understanding of the potential benefits of the project and how the project contributes to the larger goals of sustainable development if this information is presented in a manner that is both clear and concise.

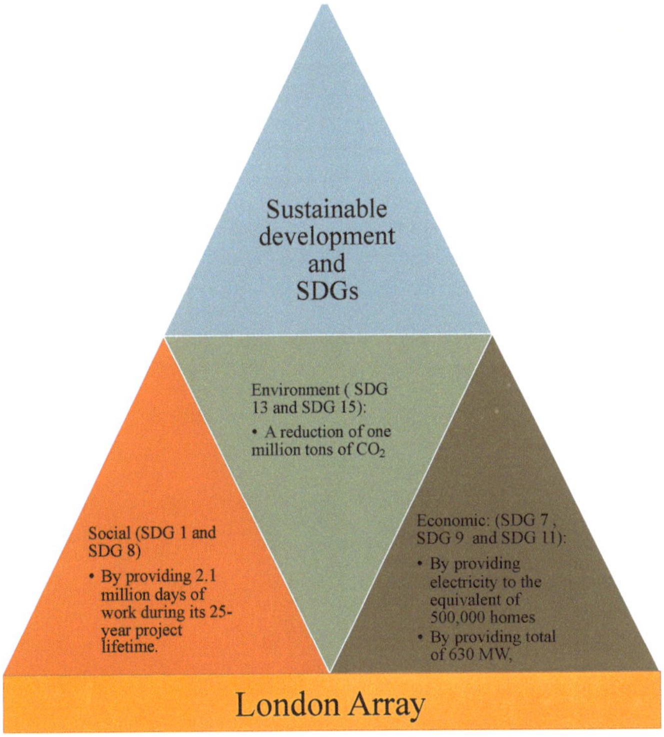

Figure 6. London Array's contribution to the SDGs.

3.4. Sustainability Indicators

Businesses are essential to achieving the SDGs because they are the primary drivers of economic growth and possess the resources and influence necessary to have a constructive effect on society and the environment. Linking business to the SDGs can have a number of benefits, including [53,78–80]:

- Improving the reputation of the company and the image of its brand. Businesses can show their commitment to social and environmental responsibility, as well as attract consumers and stakeholders that place a value on sustainability, by aligning their business practices with the SDGs.
- Companies that address the risks and opportunities presented by sustainability are better prepared to withstand economic and social challenges such as natural disasters, resource constraints, and shifting consumer preferences. These companies have taken steps to reduce risks and increase their resilience.
- SDGs provide a framework that businesses can use to identify and address unmet needs in areas such as healthcare, education, and renewable energy, which can lead to the development of new products and services. One of the benefits of the SDGs is that they create new business opportunities.

In summary, by aligning with the SDGs, organizations and individuals can ensure that their work is not only having a positive impact on the local community but also contributing to the global effort to create a more sustainable and equitable world for all.

Key performance indicators are critical (key) measures of progress towards an anticipated objective. KPIs help center attention on what matters most, offer an analytical foundation for decision-making, and give a focus on strategic and operational development. KPIs are the go-to tool for every expert since they serve as measures for developing plans, monitoring performance, and establishing objectives [81]. They facilitate decision-making by emphasizing goals and providing a framework for examination. KPIs are one foundation that helps firms go on the right path. KPIs have long been used in management [82–84]. One of the reasons KPIs are so popular is because of all the advantages that come with employing them. In this paper, a total of 77 indicators, shown in Table 3, were developed based on the study presented above and several other published publications [78,85–87]. The benefit of these indicators is to verify that the wind project contributes to the achievement of the SDGs by providing a balanced approach to each of the SDGs. In addition to the primary advantages that the suggested indicators provide, there are several further advantages, which may be summed up in Figure 7 [78,83–85].

Table 3. Wind SDGs indicators and sustainability guideline.

SDG 1: No poverty
1. Sum paid in taxes. 2. Total workers' income. 3. Sum of employee salary. 4. Training program offered. 5. Paying reasonable costs, particularly micro, small, and medium-sized businesses. 6. Reducing low-income residents' utility expenditures.
SDG 2: Zero hunger
7. Type of land occupied. 8. Policy on catastrophe risk management available. 9. Assisting smallholders by lowering utility prices.
SDG 3: Good health and well-being
10. Ensuring that employees receive full healthcare benefits. 11. Calculating the possibility of human toxicity, particulate matter, and photochemical ozone formation. 12. Ensuring that a policy on workplace health and safety is available. 13. Decreasing the number of accidents that occur as a result of project completion. 14. Ensuring that noise is controlled and reduced.
SDG 4: Quality education
15. Offering a program to measure employees' skill levels. 16. Ensuring zero tolerance of child labor.
SDG 5: Gender equality
17. Proportion of women hired and recruited. 18. Total money paid to women. 19. Percentage of women in positions of leadership.
SDG 6: Clean water and sanitation
20. Impact on water quality. 21. Calculating the potential for eutrophication and ecotoxicity in freshwater aquatic systems. 22. Improving the efficiency of water usage.
SDG 7 Affordable and clean energy
23. Making an energy balance analysis. 24. Measuring the capacity factor. 25. Decreases in energy usage. 26. Total energy generated overall by the plant. 27. Total amount invested on energy programs. 28. Measuring production outcome. 29. Increasing the amount of energy stored. 30. Measuring availability time.
SDG 8: Decent work and economic growth
31. Increasing gross domestic product through decreasing the cost of energy, a resource required by all kinds of industry. 32. Reducing operating costs by recycling industrial effluent. 33. Research and development expenditures for clean technologies. 34. Total number of new jobs. 35. Maximizing the use of resources. 36. Costs of capital and electricity. 37. Total amount spent on research and development. 38. Integration of employee rights policies. 39. Total amount paid to the community as compensation for resources used in the project. 40. Purchasing all products and services locally.

Table 3. *Cont.*

SDG 9: Industry, innovation, and infrastructure
41. Evaluating social, economic, and environmental effects. 42. Using the circular business model. 43. Tracking and reporting on the effects of climate change. 44. Overall value added. 45. Encouraging industrialization that is inclusive and sustainable. 46. Refitting factories with clean technology that uses few resources. 47. Lowering the amount of CO_2 released per unit of value added. 48. Lessening the carbon footprint of materials.
SDG 10: Reduced inequalities
49. Sharing training programs for underrepresented groups. 50. Total fee differences across various working classes. 51. Total pay disparities across several occupational groupings. 52. Diversity and inclusion level.
SDG 11: Sustainable cities and communities
53. Energy circulation to lessen the effects of urbanization on the environment. 54. Sustainability strategies for resources. 55. Plans for preparing for disasters are offered to vulnerable groups. 56. Ensuring that sustainability regulations are followed.
SDG 12: Responsible consumption and production
57. Enhancing the system for recycling waste products. 58. Total amount of essential resources and materials. 59. Total amount of waste generated. 60. Keeping track of pollution, such as greenhouse gas emissions and methane emissions.
SDG 13: Climate action
61. Recording and reporting of all forms of pollution. 62. The total amount of research focused on climate change. 63. Putting a number on the ecotoxicity of maritime environments.
SDG 14: Life below water
64. Reducing marine pollution. 65. Sustainably managing marine ecosystems. 66. Addressing ocean acidification. 67. Eliminating harmful fisheries subsidies. 68. Increasing research and development in marine technology.
SDG 15: Life on land
69. Biodiversity effect quantification 70. Using the life cycle analysis to measure environmental effects. 71. Brownfield and previously developed land projects. 72. Contributing to research and landscaping.
SDG 16: Peace, justice, and strong institution
73. Engagement of project stakeholders. 74. Disclosure of project regulations and compliance.
SDG 17: Partnerships for the goals
75. Integration of the SDGs into business strategies. 76. Cooperation between the various government organizations. 77. Collaboration with a variety of different groups.

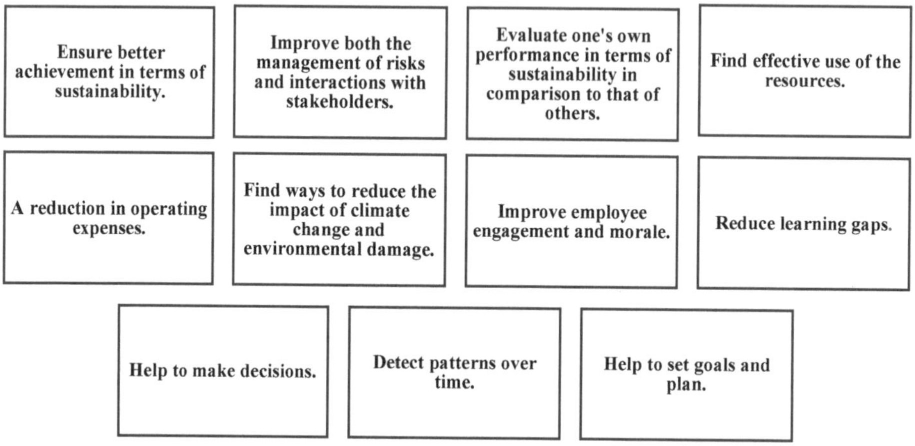

Figure 7. Benefits of adopting the KPIs.

The indicators provided in Table 3 are used to measure the progress that various wind energy projects and initiatives are making towards achieving the SDGs. For example, under SDG 1: No poverty, indicators such as the sum of taxes paid, the total number of workers, employee salaries, and the total training program offered provide information on the economic status of individuals and businesses, and the efforts made to reduce poverty. Similarly, under SDG 2: Zero hunger, indicators such as the type of land occupied and the availability of a policy on catastrophe risk management provide information on food security and the efforts to mitigate the impact of natural disasters. While under SDG 3: Good health and wellbeing, indicators such as employee health benefits, a workplace health and safety policy, and control of noise provide information on the health and safety of workers. Moreover, under SDG 4: Quality education, indicators such as a skills measurement program and a zero tolerance policy for child labor provide information on the education and wellbeing of children. Similarly, under SDG 5: Gender equality, indicators such as the proportion of women hired, the money paid to women, and the percentage of women in leadership positions provide information on gender equality in the workplace. Additionally, the indicators listed under the other SDGs, such as 'Clean water and sanitation', 'Affordable and clean energy', 'Decent work and economic growth', and so on, provide information on various aspects of sustainability, including environmental protection, economic growth, and social equality. Overall, Table 3 provides a comprehensive overview of the various indicators used to measure the progress towards achieving the SDGs, and highlights the importance of considering multiple aspects of sustainability in wind energy-related projects.

The London Array was a large-scale offshore wind farm that was constructed in the waters off the coast of Kent in the UK. Based on EIA and EMMP, the following is a list of some of the KPIs that were utilized during the process of developing the London Array:

- Capacity factor: This is a measure of the efficiency of the wind turbines and was an important KPI for the London Array project because it determined the overall energy output of the wind farm. The capacity factor can be calculated by dividing the total energy output of the wind farm by the total number of wind turbines.
- Accessibility: This metric quantifies the proportion of time that the wind turbines were online and able to produce electricity. In order to achieve the highest possible energy output from the London Array, it was essential to guarantee high availability.
- Maintenance expenses: The London Array required regular maintenance to ensure its continued operation and shorten the amount of time it was offline. As one of the KPIs for the project, the costs of maintenance were closely monitored.

- Production costs: The cost of making electricity was an important KPI because it had a direct effect on the project's ability to make money. The goal was to maintain a high level of production efficiency while minimizing the costs of production.
- Impact on the environment: The London Array was developed to have as little of an effect as possible on the local wildlife and the surrounding environment. The environmental impact of the project was measured and monitored with the help of KPIs.
- Health and safety: It was of the utmost importance that the health and safety of the workers involved in the construction and maintenance of the wind farm be kept in the highest regard. KPIs were utilized in the process of monitoring and ensuring the workers' safety.
- Timeline for the project: The London Array was a big, complicated job, and the timeline for finishing it was carefully monitored and managed. KPIs were utilized in order to monitor progress and guarantee that the project would be finished on schedule.

These KPIs were utilized frequently throughout the development of the London Array wind farm project, and their contribution to the overall success of the project cannot be overstated.

As mentioned above, based on the available information and data, London Array has used most of the above KPIs, to measure and track the performance and efficiency of their business and the overall operation of the wind farm. Some examples of KPIs, with their value, if publicly available, that were used in the London Array include:

- SDG 7: Affordable and clean energy:
    - Capacity factor: The ratio of the actual electricity generated by the wind farm to the theoretical maximum output if the wind turbines were operating at full capacity all the time. The capacity factor of the London Array is equal to 38.3%, (from https://energynumbers.info/uk-offshore-wind-capacity-factors, accessed on 2 January 2023).
    - Availability: The percentage of time that the wind turbines are able to generate electricity, taking into account planned and unplanned maintenance, equipment failures, and other factors that could cause the turbines to be offline
    - Production: The total amount of electricity generated by the wind farm over a specific period of time, such as a month or a year. The nameplate capacity of the London Array is equal to 630 MW and the annual net output is 3100 GWh, (from https://londonarray.com/, accessed on 2 January 2023).
- SDG 8: Decent work and economic growth
    - Revenue: The income generated by the sale of electricity. London Array's estimated annual revenue is currently $8.2M per year, (from https://growjo.com/company/London_Array, accessed on 2 January 2023).
- SDG 13: Climate action and SDG 15: Life on land
    - Carbon emissions avoided: The reduction in greenhouse gas emissions that is achieved by generating electricity from the wind farm instead of fossil fuel sources. The total carbon emission avoided was 925,000 tons of CO2.

Moreover, London Array have conducted an environmental impact assessment (EIA) and an ecological mitigation and management plan (EMMP), both available from https://londonarray.com, accessed on 2 January 2023. The EIA considered the potential impacts of the project on the environment, including the impact on birds, marine mammals, and fish, as well as the potential visual impact of the wind turbines. Measures were put in place to mitigate any negative impacts, such as the installation of radar systems to help detect and avoid bird and bat collisions. The London Array was also designed to operate in a way that minimizes any potential negative impacts on the local community and the surrounding environment. An EMMP is a plan that outlines the measures that will be taken to minimize the potential negative impacts of a project on the environment and to protect and enhance the ecological value of an area. An EMMP is typically developed as part of the EIA process for a project, and it outlines the measures that will be taken to avoid, minimize, and compensate for any negative impacts on the environment. It is not unusual for wind projects such as the London Array to have an EMMP in place to ensure that the project is designed and maintained in a manner that minimizes the negative effects it has on the environment. Table 4 shows the KPIs that were covered in the EIA and EMMP from our proposed KPIs. Table 4 also contains a list of the various KPIs that are associated with the SDGs and that can be tracked in order to evaluate the impact of a project. Each SDG has a corresponding set of KPIs that were used to assess the extent to which the London Array project was able to help in achieving that SDG. It is important to highlight that additional KPIs could have been addressed, but they are not currently being announced publicly, as is the case with the majority of the companies. This is mostly due to concerns about maintaining secrecy.

Table 4. KPIs covered in the EIA and EMMP from the above proposed KPIs.

SDG 3: Good health and wellbeing
1. Calculating the possibility for human toxicity, particulate matter, and photochemical ozone formation. 2. Ensuring that a policy on workplace health and safety is available. 3. Decreasing the number of accidents that occur as a result of project completion. 4. Ensuring that noise is controlled and reduced.
SDG 6: Clean water and sanitation
5. Impact on water quality. 6. Calculating the potential for eutrophication and ecotoxicity in freshwater aquatic systems. 7. Improving the efficiency of water usage.
SDG 7: Affordable and clean energy
8. Making an energy balance analysis. 9. Measuring the capacity factor. 10. Decreases in energy usage. 11. Total energy generated overall by the plant. 12. Total amount invested on energy programs. 13. Measuring the production outcome. 14. Increasing the amount of energy stored. 15. Measuring the availability time.
SDG 8: Decent work and economic growth
16. Total number of new jobs. 17. Maximizing the use of resources. 18. Costs of capital and electricity. 19. Total amount spent on research and development. 20. Integration of employee rights policies.
SDG 9: Industry, innovation, and infrastructure
21. Evaluating the social, economic, and environmental effects. 22. Using the circular business model. 23. Tracking and reporting on the effects on climate change. 24. Overall value added. 25. Lowering the amount of CO_2 released per unit of value added. 26. Lessening the carbon footprint of materials.
SDG 11: Sustainable cities and communities
27. Energy circulation to lessen the effects of urbanization on the environment. 28. Sustainability strategies of resources. 29. Plans for preparing for disasters offered to vulnerable groups. 30. Resource and sustainability regulations.

Table 4. *Cont.*

SDG 12: Responsible consumption and production
31. Enhancing the system for recycling waste products.
32. Total amount of essential resources and materials.
33. Total amount of waste generated.
34. Keeping track of pollution, such as greenhouse gas emissions and methane emissions.
SDG 13: Climate action
35. Recording and reporting of all forms of pollution.
36. Total amount of research that is focused on climate change.
37. Putting a number on the ecotoxicity of maritime environments.
SDG 14: Life below water
38. Reducing marine pollution.
39. Sustainably managing marine ecosystems.
40. Addressing the acidification of the ocean.
41. Eliminating harmful fisheries subsidies.
42. Increasing research and development in marine technology.
SDG 15: Life on land
43. Biodiversity effect quantification.
44. Using the life cycle analysis to measure environmental effects.
45. Brownfield and previously developed land projects.
46. Contribution to research and landscaping.

4. Conclusions

The SDGs depend on the transition to clean, renewable energy, therefore wind energy companies must connect with the SDGs. Wind energy companies can show their commitment to sustainability and social responsibility and help achieve the SDGs by linking their business to them. This could boost the company's reputation and brand image and open new business opportunities due to the rising demand for renewable energy. Wind power can support most SDGs. In our case study of London Array, the main benefits were a 925,000-ton reduction in greenhouse gas emissions (Goal 13: Climate action and Goal 15: Life on land), the creation of new jobs (Goal 1: No poverty and Goal 8: Decent work and economic growth), an increase in GDP by 0.8% (Goal 8), and a 630-MW increase in electrical power (Goal 7: Affordable and clean energy, Goal 11: Sustainable development, and Goal 12: Responsible consumption and production). The London Array, one of the world's most successful offshore wind farms, was the world's first. Large-scale wind farms such as the London Array help to achieve long-term goals. The London Array helps reduce greenhouse gas emissions and generate renewable electricity. The London Array, a large-scale renewable energy project, provides clean electricity, reduces greenhouse gas emissions, and boosts regional economic and social development. This research discusses wind energy projects and energy sector strategies to achieve the SDGs. Wind energy is aiding the SDGs in terms of scientific and economic development. Only publicly available data and the SDGs were used to assess London Array's contribution.

This study shows how wind energy can help achieve the SDGs. This research can also be used to analyze how different renewable energy projects contribute to the SDGs and reduce SDG trade-offs. Future research may use different indicators to measure wind energy's impact on the SDGs. During EIA preparation, the SDGs can be considered and a clear link between the different elements and results can be shown.

Author Contributions: Conceptualization, A.G.O., K.O., M.A.A. and E.T.S.; methodology K.O., M.N.A., N.S., A.H.A., A.M., A.A.M.H. and E.T.S.; formal analysis, M.A.A., M.N.A. and N.S.; investigation, A.G.O., M.A.A., A.H.A. and E.T.S.; resources, A.G.O., M.A.A. and A.H.A.; data curation, M.N.A., N.S., A.H.A. and A.M.; writing—original draft preparation, A.G.O., K.O., M.A.A., M.N.A., N.S., A.H.A., A.M., A.A.M.H. and E.T.S.; writing—review and editing, A.G.O., K.O., M.A.A., M.N.A., N.S., A.H.A., A.M., A.A.M.H. and E.T.S.; supervision, A.G.O. and E.T.S.; project administration, M.A.A. and A.H.A. All authors have read and agreed to the published version of the manuscript.

Funding: This work was supported by the University of Sharjah, project no. 19020406129.

Conflicts of Interest: The authors declare no conflict of interest.

References

1. Rahman, M.M.; Khan, I.; Field, D.L.; Techato, K.; Alameh, K. Powering agriculture: Present status, future potential, and challenges of renewable energy applications. *Renew. Energy* **2022**, *188*, 731–749. [CrossRef]
2. Leal Filho, W.; Balogun, A.-L.; Olayide, O.E.; Azeiteiro, U.M.; Ayal, D.Y.; Muñoz, P.D.C.; Nagy, G.J.; Bynoe, P.; Oguge, O.; Toamukum, N.Y. Assessing the impacts of climate change in cities and their adaptive capacity: Towards transformative approaches to climate change adaptation and poverty reduction in urban areas in a set of developing countries. *Sci. Total Environ.* **2019**, *692*, 1175–1190. [CrossRef]
3. Olabi, A.G.; Abdelkareem, M.A. Renewable energy and climate change. *Renew. Sustain. Energy Rev.* **2022**, *158*, 112111. [CrossRef]
4. Olabi, A.G.; Wilberforce, T.; Elsaid, K.; Sayed, E.T.; Maghrabie, H.M.; Abdelkareem, M.A. Large scale application of carbon capture to process industries—A review. *J. Clean. Prod.* **2022**, *362*, 132300. [CrossRef]
5. Kumar, S.; Kumar, N.; Vivekadhish, S. Millennium development goals (MDGS) to sustainable development goals (SDGS): Addressing unfinished agenda and strengthening sustainable development and partnership. *Indian J. Community Med.* **2016**, *41*, 1–4. [CrossRef] [PubMed]
6. Walsh, P.P.; Murphy, E.; Horan, D. The role of science, technology and innovation in the UN 2030 agenda. *Technol. Forecast. Soc. Change* **2020**, *154*, 119957. [CrossRef]
7. Buonocore, J.J.; Choma, E.; Villavicencio, A.H.; Spengler, J.D.; Koehler, D.A.; Evans, J.S.; Lelieveld, J.; Klop, P.; Sanchez-Pina, R. Metrics for the sustainable development goals: Renewable energy and transportation. *Palgrave Commun.* **2019**, *5*, 136. [CrossRef]
8. Güney, T. Renewable energy, non-renewable energy and sustainable development. *Int. J. Sustain. Dev. World Ecol.* **2019**, *26*, 389–397. [CrossRef]
9. Fei, W.; Opoku, A.; Agyekum, K.; Oppon, J.A.; Ahmed, V.; Chen, C.; Lok, K.L. The critical role of the construction industry in achieving the sustainable development goals (SDGs): Delivering projects for the common good. *Sustainability* **2021**, *13*, 9112. [CrossRef]
10. Franco, I.B.; Arduz, F.G.; Buitrago, J.A. SDG 9 Industry, Innovation, and Infrastructure. In *Actioning the Global Goals for Local Impact*; Springer: Berlin/Heidelberg, Germany, 2020; pp. 135–151.
11. Kynčlová, P.; Upadhyaya, S.; Nice, T. Composite index as a measure on achieving Sustainable Development Goal 9 (SDG-9) industry-related targets: The SDG-9 index. *Appl. Energy* **2020**, *265*, 114755. [CrossRef]
12. Salam, A. Internet of things for sustainable community development: Introduction and overview. In *Internet of Things for Sustainable Community Development*; Springer: Berlin/Heidelberg, Germany, 2020; pp. 1–31.
13. Obaideen, K.; Yousef, B.A.; AlMallahi, M.N.; Tan, Y.C.; Mahmoud, M.; Jaber, H.; Ramadan, M. An overview of smart irrigation systems using IoT. *Energy Nexus* **2022**, *7*, 100124. [CrossRef]
14. Obaideen, K.; AlMallahi, M.N.; Alami, A.H.; Ramadan, M.; Abdelkareem, M.A.; Shehata, N.; Olabi, A. On the contribution of solar energy to sustainable developments goals: Case study on Mohammed bin Rashid Al Maktoum Solar Park. *Int. J.* **2021**, *12*, 100123. [CrossRef]
15. Adeyeye, K.; Ijumba, N.; Colton, J. Exploring the environmental and economic impacts of wind energy: A cost-benefit perspective. *Int. J. Sustain. Dev. World Ecol.* **2020**, *27*, 718–731. [CrossRef]
16. Alnaqbi, S.A.; Alasad, S.; Aljaghoub, H.; Alami, A.H.; Abdelkareem, M.A.; Olabi, A.G. Applicability of Hydropower Generation and Pumped Hydro Energy Storage in the Middle East and North Africa. *Energies* **2022**, *15*, 2412. [CrossRef]
17. Ramachandran, R.; Kularathna, A.; Matsuda, H.; Takagi, K. Information flow to increase support for tidal energy development in remote islands of a developing country: Agent-based simulation of information flow in Flores Timur Regency, Indonesia. *Energy Sustain. Soc.* **2021**, *11*, 26. [CrossRef]
18. Al Radi, M.; Adil Al-Isawi, O.; Abdelghafar, A.A.; Fayez Abu Qiyas, A.; AlMallahi, M.; Khanafer, K.; Assad, M.E.H. Recent progress, economic potential, and environmental benefits of mineral recovery geothermal brine treatment systems. *Arab. J. Geosci.* **2022**, *15*, 832. [CrossRef]
19. Destek, M.A.; Sarkodie, S.A.; Asamoah, E.F. Does biomass energy drive environmental sustainability? An SDG perspective for top five biomass consuming countries. *Biomass Bioenergy* **2021**, *149*, 106076. [CrossRef]
20. Abdelkareem, M.A.; Soudan, B.; Mahmoud, M.S.; Sayed, E.T.; AlMallahi, M.N.; Inayat, A.; Al Radi, M.; Olabi, A.G. Progress of artificial neural networks applications in hydrogen production. *Chem. Eng. Res. Des.* **2022**, *182*, 66–86. [CrossRef]

21. Obaideen, K.; Abdelkareem, M.A.; Wilberforce, T.; Elsaid, K.; Sayed, E.T.; Maghrabie, H.M.; Olabi, A.G. Biogas role in achievement of the sustainable development goals: Evaluation, Challenges, and Guidelines. *J. Taiwan Inst. Chem. Eng.* **2022**, *131*, 104207. [CrossRef]
22. Olabi, A.; Wilberforce, T.; Elsaid, K.; Salameh, T.; Sayed, E.T.; Husain, K.S.; Abdelkareem, M.A. Selection guidelines for wind energy technologies. *Energies* **2021**, *14*, 3244. [CrossRef]
23. Barra, P.; de Carvalho, W.; Menezes, T.; Fernandes, R.; Coury, D. A review on wind power smoothing using high-power energy storage systems. *Renew. Sustain. Energy Rev.* **2021**, *137*, 110455. [CrossRef]
24. Sadorsky, P. Wind energy for sustainable development: Driving factors and future outlook. *J. Clean. Prod.* **2021**, *289*, 125779. [CrossRef]
25. Tran, T.T.; Smith, A.D. fEvaluation of renewable energy technologies and their potential for technical integration and cost-effective use within the US energy sector. *Renew. Sustain. Energy Rev.* **2017**, *80*, 1372–1388. [CrossRef]
26. Sayed, E.T.; Wilberforce, T.; Elsaid, K.; Rabaia, M.K.H.; Abdelkareem, M.A.; Chae, K.-J.; Olabi, A.G. A critical review on environmental impacts of renewable energy systems and mitigation strategies: Wind, hydro, biomass and geothermal. *Sci. Total Environ.* **2021**, *766*, 144505. [CrossRef]
27. Abdelkhalig, A.; Elgendi, M.; Selim, M.Y. Review on validation techniques of blade element momentum method implemented in wind turbines. In *IOP Conference Series: Earth and Environmental Science*; IOP Publishing: Bristol, UK, 2022; Available online: https://iopscience.iop.org/article/10.1088/1755-1315/1074/1/012008/meta (accessed on 1 November 2022).
28. Cheng, S.; Elgendi, M.; Lu, F.; Chamorro, L.P. On the Wind Turbine Wake and Forest Terrain Interaction. *Energies* **2021**, *14*, 7204. [CrossRef]
29. Ragheb, M. History of harnessing wind power. In *Wind Energy Engineering*; Elsevier: Amsterdam, The Netherlands, 2017; pp. 127–143.
30. Pain, S. Power through the ages. *Nature* **2017**, *551*, S134. [CrossRef] [PubMed]
31. Staffell, I.; Pfenninger, S. Using bias-corrected reanalysis to simulate current and future wind power output. *Energy* **2016**, *114*, 1224–1239. [CrossRef]
32. El Ramahi, M. Case study: Masdar renewable energy water desalination program. In *The Water, Energy, and Food Security Nexus in the Arab Region*; Springer: Berlin/Heidelberg, Germany, 2017; pp. 209–221.
33. Pérez-Collazo, C.; Greaves, D.; Iglesias, G. A review of combined wave and offshore wind energy. *Renew. Sustain. Energy Rev.* **2015**, *42*, 141–153. [CrossRef]
34. Gasch, R.; Twele, J. *Wind Power Plants: Fundamentals, Design, Construction and Operation*; Springer Science & Business Media: Berlin/Heidelberg, Germany, 2011.
35. Wen, Q.; He, X.; Lu, Z.; Streiter, R.; Otto, T. A comprehensive review of miniatured wind energy harvesters. *Nano Mater. Sci.* **2021**, *3*, 170–185. [CrossRef]
36. Ghenai, C.; Rasheed, M.A.; Alshamsi, M.J.; Alkamali, M.A.; Ahmad, F.F.; Inayat, A. Design of hybrid solar photovoltaics/shrouded wind turbine power system for thermal pyrolysis of plastic waste. *Case Stud. Therm. Eng.* **2020**, *22*, 100773. [CrossRef]
37. Gomaa, M.R.; Rezk, H.; Mustafa, R.J.; Al-Dhaifallah, M. Evaluating the environmental impacts and energy performance of a wind farm system utilizing the life-cycle assessment method: A practical case study. *Energies* **2019**, *12*, 3263. [CrossRef]
38. Agency, E.P. Renewable Energy Fact Sheet: Wind Turbines. Available online: https://www.epa.gov/sites/default/files/2019-08/documents/wind_turbines_fact_sheet_p100il8k.pdf (accessed on 1 November 2022).
39. Mendecka, B.; Lombardi, L. Life cycle environmental impacts of wind energy technologies: A review of simplified models and harmonization of the results. *Renew. Sustain. Energy Rev.* **2019**, *111*, 462–480. [CrossRef]
40. Nazir, M.S.; Mahdi, A.J.; Bilal, M.; Sohail, H.M.; Ali, N.; Iqbal, H.M. Environmental impact and pollution-related challenges of renewable wind energy paradigm–a review. *Sci. Total Environ.* **2019**, *683*, 436–444. [CrossRef] [PubMed]
41. Schippers, P.; Buij, R.; Schotman, A.; Verboom, J.; van der Jeugd, H.; Jongejans, E. Mortality limits used in wind energy impact assessment underestimate impacts of wind farms on bird populations. *Ecol. Evol.* **2020**, *10*, 6274–6287. [CrossRef] [PubMed]
42. Lucena, J.d.A.Y.; Lucena, K.Â.A. Wind energy in Brazil: An overview and perspectives under the triple bottom line. *Clean Energy* **2019**, *3*, 69–84. [CrossRef]
43. Majid, M. Wind energy programme in India: Emerging energy alternatives for sustainable growth. *Energy Environ.* **2019**, *30*, 1135.
44. Marques, A.T.; Batalha, H.; Rodrigues, S.; Costa, H.; Pereira, M.J.R.; Fonseca, C.; Mascarenhas, M.; Bernardino, J. Understanding bird collisions at wind farms: An updated review on the causes and possible mitigation strategies. *Biol. Conserv.* **2014**, *179*, 40–52. [CrossRef]
45. Smeraldo, S.; Bosso, L.; Fraissinet, M.; Bordignon, L.; Brunelli, M.; Ancillotto, L.; Russo, D. Modelling risks posed by wind turbines and power lines to soaring birds: The black stork (Ciconia nigra) in Italy as a case study. *Biodivers. Conserv.* **2020**, *29*, 1959–1976. [CrossRef]
46. Arnett, E.B.; Baerwald, E.F.; Mathews, F.; Rodrigues, L.; Rodríguez-Durán, A.; Rydell, J.; Villegas-Patraca, R.; Voigt, C.C. Impacts of wind energy development on bats: A global perspective. In *Bats in the Anthropocene: Conservation of Bats in a Changing World*; Springer: Cham, Switzerland, 2016; pp. 295–323.
47. Nazir, M.S.; Bilal, M.; Sohail, H.M.; Liu, B.; Chen, W.; Iqbal, H.M. Impacts of renewable energy atlas: Reaping the benefits of renewables and biodiversity threats. *Int. J. Hydrogen Energy* **2020**, *45*, 22113–22124. [CrossRef]

48. Ellabban, O.; Abu-Rub, H.; Blaabjerg, F. Renewable energy resources: Current status, future prospects and their enabling technology. *Renew. Sustain. Energy Rev.* **2014**, *39*, 748–764. [CrossRef]
49. Dai, K.; Bergot, A.; Liang, C.; Xiang, W.-N.; Huang, Z. Environmental issues associated with wind energy–A review. *Renew. Energy* **2015**, *75*, 911–921. [CrossRef]
50. Laranjeiro, T.; May, R.; Verones, F. Impacts of onshore wind energy production on birds and bats: Recommendations for future life cycle impact assessment developments. *Int. J. Life Cycle Assess.* **2018**, *23*, 2007–2023. [CrossRef]
51. Nazir, M.S.; Ali, N.; Bilal, M.; Iqbal, H.M. Potential environmental impacts of wind energy development: A global perspective. *Curr. Opin. Environ. Sci. Health* **2020**, *13*, 85–90. [CrossRef]
52. Singh, R.L.; Singh, P.K. Global environmental problems. In *Principles and Applications of Environmental Biotechnology for a Sustainable Future*; Springer: Berlin/Heidelberg, Germany, 2017; pp. 13–41.
53. Olabi, A.G.; Obaideen, K.; Elsaid, K.; Wilberforce, T.; Sayed, E.T.; Maghrabie, H.M.; Abdelkareem, M.A. Assessment of the pre-combustion carbon capture contribution into sustainable development goals SDGs using novel indicators. *Renew. Sustain. Energy Rev.* **2022**, *153*, 111710. [CrossRef]
54. Dennis, B. Climate and Environment (The U.S. Wind Industry Now Employs More than 100,000 People). In *The Washington Post*; 2017; Available online: https://www.washingtonpost.com/news/energy-environment/wp/2017/04/19/the-u-s-wind-industry-now-employs-more-than-100000-people/ (accessed on 1 January 2023).
55. Enevoldsen, P.; Sovacool, B.K. Examining the social acceptance of wind energy: Practical guidelines for onshore wind project development in France. *Renew. Sustain. Energy Rev.* **2016**, *53*, 178–184. [CrossRef]
56. Durakovic, F. Levelized Cost of Electricity of Renewable Energy Technologies as a Criterion for Project Prioritization. 2021. Available online: https://lutpub.lut.fi/handle/10024/163284 (accessed on 1 January 2023).
57. Armin Razmjoo, A.; Sumper, A.; Davarpanah, A. Energy sustainability analysis based on SDGs for developing countries. *Energy Sources Part A Recovery Util. Environ. Eff.* **2020**, *42*, 1041–1056. [CrossRef]
58. Howard-Grenville, J.; Davis, G.F.; Dyllick, T.; Miller, C.C.; Thau, S.; Tsui, A.S. Sustainable development for a better world: Contributions of leadership, management, and organizations. *Acad. Manag. Discov.* **2019**, *5*, 355–366. [CrossRef]
59. Fatimah, Y.A.; Govindan, K.; Murniningsih, R.; Setiawan, A. Industry 4.0 based sustainable circular economy approach for smart waste management system to achieve sustainable development goals: A case study of Indonesia. *J. Clean. Prod.* **2020**, *269*, 122263. [CrossRef]
60. Mustafa, M.; Ghazali, F.; Yuniza, M.E. Governance interventions in providing affordable, reliable, sustainable, and modern energy in the ASEAN region (Goal 7). In *Good Governance and the Sustainable Development Goals in Southeast Asia*; Routledge: Abingdon, UK, 2022; Available online: https://www.taylorfrancis.com/chapters/edit/10.4324/9781003230724-8/governance-interventions-providing-affordable-reliable-sustainable-modern-energy-asean-region-goal-7-maizatun-mustafa-farahdilah-ghazali-mailinda-eka-yuniza (accessed on 1 November 2022).
61. McCollum, D.; Gomez Echeverri, L.; Riahi, K.; Parkinson, S. Sdg7: Ensure Access to Affordable, Reliable, Sustainable and Modern Energy for All. Available online: https://www.un.org/en/chronicle/article/goal-7-ensure-access-affordable-reliable-sustainable-and-modern-energy-all (accessed on 1 November 2022).
62. Siksnelyte-Butkiene, I.; Zavadskas, E.K.; Streimikiene, D. Multi-criteria decision-making (MCDM) for the assessment of renewable energy technologies in a household: A review. *Energies* **2020**, *13*, 1164. [CrossRef]
63. Carley, M.; Spapens, P. *Sharing the World: Sustainable Living and Global Equity in the 21st Century*; Routledge: Abingdon, UK, 2017; Available online: https://www.taylorfrancis.com/books/mono/10.4324/9781315087993/sharing-world-phillipe-spapens-michael-carley (accessed on 1 November 2022).
64. Van Vuuren, D.P.; Stehfest, E.; Gernaat, D.E.; Doelman, J.C.; Van den Berg, M.; Harmsen, M.; de Boer, H.S.; Bouwman, L.F.; Daioglou, V.; Edelenbosch, O.Y. Energy, land-use and greenhouse gas emissions trajectories under a green growth paradigm. *Glob. Environ. Change* **2017**, *42*, 237–250. [CrossRef]
65. Alharthi, M.; Hanif, I.; Alamoudi, H. Impact of environmental pollution on human health and financial status of households in MENA countries: Future of using renewable energy to eliminate the environmental pollution. *Renew. Energy* **2022**, *190*, 338–346. [CrossRef]
66. Albright, R.; Cooley, S. A review of interventions proposed to abate impacts of ocean acidification on coral reefs. *Reg. Stud. Mar. Sci.* **2019**, *29*, 100612. [CrossRef]
67. Veers, P.; Dykes, K.; Lantz, E.; Barth, S.; Bottasso, C.L.; Carlson, O.; Clifton, A.; Green, J.; Green, P.; Holttinen, H. Grand challenges in the science of wind energy. *Science* **2019**, *366*, eaau2027. [CrossRef] [PubMed]
68. Shahsavari, A.; Akbari, M. Potential of solar energy in developing countries for reducing energy-related emissions. *Renew. Sustain. Energy Rev.* **2018**, *90*, 275–291. [CrossRef]
69. Wilberforce, T.; Olabi, A.G.; Sayed, E.T.; Alalmi, A.H.; Abdelkareem, M.A. Wind turbine concepts for domestic wind power generation at low wind quality sites. *J. Clean. Prod.* **2023**, *394*, 136137. [CrossRef]
70. Rahman, A.; Farrok, O.; Haque, M.M. Environmental impact of renewable energy source based electrical power plants: Solar, wind, hydroelectric, biomass, geothermal, tidal, ocean, and osmotic. *Renew. Sustain. Energy Rev.* **2022**, *161*, 112279. [CrossRef]
71. Huang, J.; Huang, X.; Song, N.; Ma, Y.; Wei, D. Evaluation of the Spatial Suitability of Offshore Wind Farm—A Case Study of the Sea Area of Liaoning Province. *Sustainability* **2022**, *14*, 449. [CrossRef]
72. Giebel, G.; Hasager, C.B. An overview of offshore wind farm design. *MARE-WINT* **2016**, 337–346.

73. Löhning, T.; Voßbeck, M.; Kelm, M. Analysis of grouted connections for offshore wind turbines. *Proc. Inst. Civ. Eng.-Energy* **2013**, *166*, 153–161. [CrossRef]
74. Aleem, M.; Bhattacharya, S.; Cui, L.; Amani, S.; Salem, A.R.; Jalbi, S. Load utilisation (LU) ratio of monopiles supporting offshore wind turbines: Formulation and examples from European Wind Farms. *Ocean. Eng.* **2022**, *248*, 110798. [CrossRef]
75. Verfuss, U.K.; Sparling, C.E.; Arnot, C.; Judd, A.; Coyle, M. Review of offshore wind farm impact monitoring and mitigation with regard to marine mammals. *Eff. Noise Aquat. Life II* **2016**, 1175–1182.
76. Lamb, J.S.; Newstead, D.J.; Koczur, L.M.; Ballard, B.M.; Green, M.C.; Jodice, P.G. A bridge between oceans: Overland migration of marine birds in a wind energy corridor. *J. Avian Biol.* **2018**, *49*, jav-01474. [CrossRef]
77. Herbert-Acero, J.F.; Probst, O.; Réthoré, P.-E.; Larsen, G.C.; Castillo-Villar, K.K. A review of methodological approaches for the design and optimization of wind farms. *Energies* **2014**, *7*, 6930–7016. [CrossRef]
78. Shehata, N.; Obaideen, K.; Sayed, E.T.; Abdelkareem, M.A.; Mahmoud, M.S.; El-Salamony, A.-H.R.; Mahmoud, H.M.; Olabi, A. Role of Refuse-derived fuel in circular economy and sustainable development goals. *Process Saf. Environ. Prot.* **2022**, *163*, 558–573. [CrossRef]
79. Cordova, M.F.; Celone, A. SDGs and innovation in the business context literature review. *Sustainability* **2019**, *11*, 7043. [CrossRef]
80. Blinova, E.; Ponomarenko, T.; Knysh, V. Analyzing the Concept of Corporate Sustainability in the Context of Sustainable Business Development in the Mining Sector with Elements of Circular Economy. *Sustainability* **2022**, *14*, 8163. [CrossRef]
81. Hinderks, A.; Schrepp, M.; Mayo, F.J.D.; Escalona, M.J.; Thomaschewski, J. Developing a UX KPI based on the user experience questionnaire. *Comput. Stand. Interfaces* **2019**, *65*, 38–44. [CrossRef]
82. Velimirović, D.; Velimirović, M.; Stanković, R. Role and importance of key performance indicators measurement. *Serb. J. Manag.* **2011**, *6*, 63–72. [CrossRef]
83. ElAlfy, A.; Palaschuk, N.; El-Bassiouny, D.; Wilson, J.; Weber, O. Scoping the evolution of corporate social responsibility (CSR) research in the sustainable development goals (SDGs) era. *Sustainability* **2020**, *12*, 5544. [CrossRef]
84. Hristov, I.; Chirico, A. The role of sustainability key performance indicators (KPIs) in implementing sustainable strategies. *Sustainability* **2019**, *11*, 5742. [CrossRef]
85. Bae, H.; Smardon, R.S. Indicators of sustainable business practices. *Environ. Manag. Pract.* **2011**, 177. [CrossRef]
86. Szennay, Á.; Szigeti, C.; Kovács, N.; Szabó, D.R. Through the blurry looking glass—SDGs in the GRI reports. *Resources* **2019**, *8*, 101. [CrossRef]
87. Dumay, J.; Guthrie, J.; Farneti, F. GRI sustainability reporting guidelines for public and third sector organizations: A critical review. *Public Manag. Rev.* **2010**, *12*, 531–548. [CrossRef]

sustainability

MDPI

Article

Optimization of Solar-Assisted CCHP Systems: Enhancing Efficiency and Reducing Emissions Through Harris Hawks-Based Mathematical Modeling

Uchechi Ukaegbu [1], Lagouge Tartibu [1,*] and C. W. Lim [1,2]

1 Department of Mechanical Engineering, Faculty of Engineering and the Built Environment, University of Johannesburg, Johannesburg P.O. Box 2028, South Africa; ukaegbuuchechi531@gmail.com (U.U.); bccwlim@cityu.edu.hk (C.W.L.)
2 Department of Architecture and Civil Engineering, City University of Hong Kong, Hong Kong, China
* Correspondence: ltartibu@uj.ac.za

Abstract: The increasing demand for energy, driven by technological advances, population growth, and economic expansion, has intensified the focus on efficient energy management. Tri-generation systems, such as Combined Cooling, Heating, and Power (CCHP) systems, are of particular interest due to their efficiency and sustainability. Integrating renewable energy sources like solar power with traditional fossil fuels further optimizes CCHP systems. This study presents a novel method for enhancing the CCHP system efficiency by identifying the optimal design parameters and assisting decision makers in selecting the best geometric configurations. A mathematical programming model using the Harris Hawks optimizer was developed to maximize the net power and exergy efficiency while minimizing CO_2 emissions in a solar-assisted CCHP system. The optimization resulted in 100 Pareto optimal solutions, offering various choices for performance improvement. This method achieved a higher net power output, satisfactory exergy efficiency, and lower CO_2 emissions compared to similar studies. The study shows that the maximum net power and exergy efficiency, with reduced CO_2 emissions, can be achieved with a system having a low compression ratio and low combustion chamber inlet temperature. The proposed approach surpassed the response surface method, achieving at least a 4.2% reduction in CO_2 emissions and improved exergy values.

Keywords: CCHP; CO_2 emissions; exergy efficiency; net power; tri-generation system; solar energy

1. Introduction

Energy is a fundamental factor in attaining the economic, social, and environmental growth of any country. A survey by Ramachandran, Mourad and Hamed [1] revealed that the rise in energy consumption in recent times, typically instilled by economic development and population growth, has exacerbated the emission of GHGs, especially CO_2. This emission poses serious environmental and public health risks, contributing significantly to climate change. The environmental degradation resulting from greenhouse gas emissions from power plants poses a substantial hazard to society due to the effects of global warming. As reported by the International Energy Agency (IEA) [2], global CO_2 emissions have increased by 0.9%, reaching an unprecedented high of approximately 36.8 Gigatons. Hence, it is crucial to employ fundamental techniques geared towards improving the design of power generation systems to increase their efficiency and reduce the pollution of the environment.

Multi-generation systems are born out of the need for more sustainable generation of one or more forms of energy as the output. They are mainly categorized, according to the number of outputs, into co-generation and tri-generation systems. Co-generation, or a combined heat and power system, involves the use of a power-generating unit to produce both power and thermal energy simultaneously. In the conventional separate power generation systems, the heat energy would have been dissipated immediately to the atmosphere but is recovered in CHP systems to provide heating power thus enabling savings in fuel and GHG emission reduction [3]. Irrespective of the fact that CHP systems have glaring advantages over conventional separate heat and power generation systems, they are susceptible to fluctuations in their system's overall efficiency, especially during the summer seasons when the heating loads are at a minimum [4]. A stable and constant power and thermal energy demand profile is essential to achieve the maximum overall system efficiency all year round. Therefore, it is ideal to incorporate cooling systems into a CHP system to boost its efficiency, hence the concept of CCHP systems.

Tri-generation systems, often referred to as CCHP systems, are an expansion of CHP systems in which a single energy source is used to create electricity, cooling, and heating power. It typically comprises a power generation unit/prime mover, cooling system, and heat recovery systems [5]. The prime mover generates mechanical power which is used to rotate the shaft of an electric motor for the generation of electricity. The thermal energy generated during power generation is utilized to operate heating and cooling systems, such as the absorption chiller and heat recovery generation unit, or to generate more power through the organic Rankine or Kalina cycle.

Renewable sources, especially solar thermal technologies, have been extensively researched as visible solar harvesting options for power generation in a bid to reduce the fossil fuel consumption. Solar thermal collectors are predominantly employed to operate side-by-side with fossil fuels because of their relatively lower cost and higher conversion efficiency when compared to solar photovoltaic technologies [6]. Solar-assisted CCHP systems offer a sustainable approach to energy production by reducing the carbon emissions and operational costs, while also enhancing the system's overall energy efficiency. These systems ensure the reliability and availability of various energy sources [7]. However, its performance can be further improved through optimization, and hence the search for even better thermodynamic performance metrics is underway.

The optimization domain has applied several use cases in solving the challenge associated with the rapid advancement in industrial technology [8]. According to Hansen, Muller, and Koumoutsakos [9], there are generally two types of optimizations: deterministic and metaheuristic algorithms. In the deterministic/heuristic approach, the gradient of the objective function is required, and a constant output is obtained for a given input. The metaheuristic, on the other hand, can solve several types of optimization problems by tuning its parameters, and it does not yield a constant output for a given input. Two common types of metaheuristic optimization are the evolutionary and swarm optimization techniques [10]. The evolutionary algorithm is driven by the principles of mutation, crossover, and survival of the fittest, as proposed by Charles Darwin. On the other hand, the swarm algorithm aims to mimic the foraging behavior of a specific group of animals in order to achieve a certain objective. The genetic algorithm, genetic programming, and differential evolution are valid examples of evolutionary algorithms, and these are presented in Figure 1. Prominent examples of swarm-intelligence algorithms are the Harris Hawks, grey wolf, moth flame, ant colony, ant–lion optimization, etc.

Typically, practical engineering scenarios involve many objectives that often conflict with each other, as indicated by [11]. Multi-objective problems can be resolved by either decomposing them into a single-objective function or by calculating the Pareto optimal solutions while maintaining the multi-objective formulation. This research study employed the latter approach. Other research studies have described various optimization methods, including the RSM [12], non-dominated sorting genetic algorithm-II (NSGA-II) [13], particle swarm optimization (PSO) [14], HHO [15], grasshopper optimization (GOA) [16], ant–lion optimization [17], moth flame optimization (MFO) [18], and grey wolf optimization (GWO) [19].

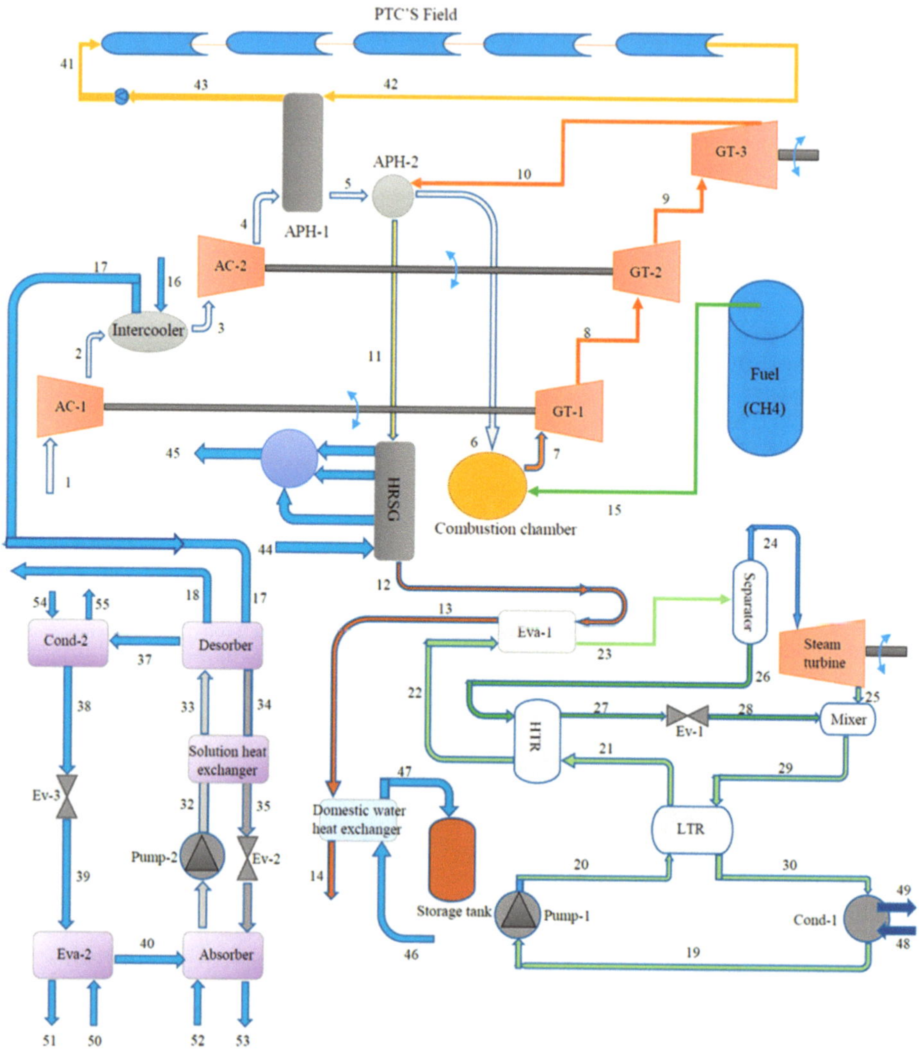

Figure 1. An illustration of the solar-assisted CCHP system.

Prior optimization research has shown that the optimization of CCHP systems primarily depends on three key evaluation categories: exergetic, economic, and environmental variables [20]. These criteria form a comprehensive framework for assessing CCHP system optimization and addressing the efficiency, cost-effectiveness, financial feasibility, and environmental impact. Achieving an optimal CCHP system involves balancing these criteria to achieve efficiency, cost-effectiveness, and environmental sustainability. This study demonstrates the application of HHO in formulating and solving a problem associated with a solar-integrated CCHP system. Based on the information provided, the proposed research seeks to achieve the following goals:

- To develop a mathematical model using the Harris Hawks algorithm on a solar-assisted CCHP system.

- Optimize the net power output and exergy efficiency, while simultaneously limiting CO_2 emissions, in a solar energy-integrated CCHP system utilizing the Harris Hawks optimizer.

2. Literature Review

The pursuit of continual improvement in CCHP systems through the use of different optimization approaches is a forward-thinking trend in energy conservation and management research. A thorough literature study examines the efforts aimed at optimizing specific performance parameters in solar-based CCHP systems.

A comprehensive examination revealed that a substantial proportion of researchers have employed the genetic algorithm to optimize the CCHP system assisted with solar energy. For example, Cao et al. [21] presented a revised solar-powered CCHP system that utilized the genetic algorithm to enhance the electricity generation, exergy performance, and overall cost per unit exergy. Their parametric study investigated the influence of decision variables, such as the proportion of oil mass, the inlet pressure of the Rankine cycle, and operating temperature, on the performance metrics. This technique exhibited enhanced results across several performance metrics, surpassing the conventional methodologies.

Wang et al. [22] performed a thermodynamic assessment and efficiency improvement of a CCHP system that integrates solar and natural gas. They utilized a genetic optimization technique to maximize the system's energetic and exergetic capacities. Furthermore, Wang et al. [23] devised a multi-objective optimization model that utilizes the genetic algorithm to improve the energy and cost efficiency, reduce CO_2 emissions, and boost the grid integration of the CCHP system. Although their suggested operational flexibility strategy enhanced the ability to respond to changing situations, it resulted in decreased levels of grid contact, as well as reduced exergetic efficiency, cost-effectiveness, and environmental impacts.

Song, Liu, and Lin [24] utilized the NSGA-II algorithm to conduct the multi-objective optimization of a solar-powered CCHP system across three different operational configurations. The decision variables considered were the capacity of the gas turbine, the area of the photovoltaic (PV) system, and the size of the solar collector. The objective was to optimize the ratio of cost saving and energy saving. The findings highlighted the significant impact of energy costs and the performance of the PV panels, solar collectors, and gas turbines on the CCHP system.

The multi-objective Harris Hawks method has been utilized to optimize a variety of multi-generation systems. Abba et al. [25] suggested a hybrid optimization strategy that used Harris Hawks and particle swarm optimization, along with support vector regression (SVR), to estimate the load requirements and optimize the sizing of a hybrid renewable energy system, with the goal of minimizing the annual cost. The findings demonstrated that the suggested HHO-based optimization technique surpassed the single SVR in regard to several performance indicators, such as the coefficient of correlation, coefficient of determination, mean squared error, and root mean squared error.

Yousri, Babu, and Fathy [26] introduced the Harmony Search algorithm as a means to optimize the parameters of a proportional-integral controller within a renewable energy-based system. The optimization of the system focuses on minimizing the integral time absolute error and maximizing the tie-line power. The system includes thermal plants, photovoltaic plants, wind turbines, and other components. The resulting results were validated using the grey wolf optimizer, multi-verse optimizer, and sine cosine algorithm.

Zhao et al. [27] described the enhancement of a fuel cell-based CCHP system by optimizing the use of HHO (hydrogen and oxygen) with the introduction of the singer mechanism. The CCHP system was optimized to maximize the exergy performance, system performance, greenhouse gas reduction, and cost efficiency. The computed results were assessed using both the conventional HHO and the NSGA-II algorithms. The updated HHO approach exhibited a reduced exergy efficiency and increased greenhouse gas reduction.

Song, Tan, and Mizzi [28] introduced an adapted HHO method to optimize a proton exchange membrane fuel cell. The HHO method underwent modifications by incorporating the quasi-oppositional concept and a logistic map mechanism. These modifications were aimed at enhancing the system's speed and robustness. This approach was implemented in three case studies and assessed by comparing it with the krill herd method, grass fibrous root optimization, and a conventional HHO algorithm. The objective of reducing the squared deviation between the output voltage and predicted data was successfully accomplished as the proposed approach yielded satisfactory outcomes.

In addition, Pandey and Jadoun [29] investigated the optimal scheduling strategies for a virtual power plant consisting of photovoltaic solar collectors, a wind energy source, a fuel cell, and a CCHP system. A modified HHO system was used to reduce both the financial gains and emissions. The optimization technique was enhanced by augmenting the step duration of the Hawks and incorporating a chaotic mapping mechanism to amplify the algorithm's diversity. The acquired solutions were verified using data from the particle swarm optimization (PSO) and mixed linear programming.

Mahalekshmi and Maruthupandi [30] utilized the Harmony Search algorithm to address the issues of economic and emission dispatch in power systems. The proposed method utilized a fuzzy-based technique to control the size of the Pareto optimal solutions. The overall system was then optimized for emission dispatch and total cost. The acquired findings were evaluated for quality using set spacing, hyper volume, and CM, and were compared with the results derived from the PSO technique.

Mahdavi, Mojaver, and Khalilarya [31] developed an innovative solar-powered CCHP system. The researchers utilized the RSM to perform multi-objective optimization, specifically targeting the net power output, CO_2 emission reduction, and exergy efficiency. In this system, the waste heat generated between the compression stages is captured via a cooling unit to drive a chiller system. By analyzing the combined impacts of several key parameters, a set of six optimal configurations were identified. The TOPSIS approach was utilized to determine the most favorable solution, yielding ideal outcomes for the net power output, CO_2 emission reduction, and exergy efficiency, respectively.

This section has examined literary works, summarized in Table 1, that have showcased the efficacy of solar-integrated CCHP systems. Furthermore, this literature analysis emphasized the increasing enthusiasm for utilizing these optimization techniques in multi-generation systems, such as the CCHP. Furthermore, the author found no evidence of the HHO being utilized in a solar-assisted CCHP system, to the best of their knowledge. Also, related studies have produced relatively high CO_2 emission rates, hence the need for improvement. Given this, this study proposes a new approach based on the HHO. Therefore, the objective of this research was to utilize the HHO to optimize a solar-assisted CCHP system, based on the insights gained from previous studies.

Table 1. Overview of relevant studies.

S/N	Source	Methods for Optimization	Renewable Energy Employed	Metrics for Performance Improvement	System Analyzed
1	Cao et al. [21]	Genetic algorithm	Solar	Exergy efficiency, net output electricity and cost per unit exergy	Tri-generation
2	Wang et al. [22]	Genetic algorithm	Solar, natural gas	Energy efficiency, exergy efficiency	Tri-generation
3	Wang et al. [23]	Genetic algorithm	Solar	Primary energy saving ratio, CO_2 emission reduction ratio, annual cost saving rate	Tri-generation
4	Song Liu and Lin [24]	NSGA-II	Solar	Annual cost saving ratio, primary energy saving ratio	Tri-generation
5	Abba et al. [25]	HHO, PSO & SVR	Solar	Annual capital and maintenance cost	Power generation

Table 1. *Cont.*

S/N	Source	Methods for Optimization	Renewable Energy Employed	Metrics for Performance Improvement	System Analyzed
6	Yousri, Babu, and Fathy [26]	HHO	Solar, wind	Integral time absolute error, tie-line power	Power generation
7	Zhao et al. [27]	HHO	N/A	GHG emission reduction, exergy efficiency, annual cost rate	Tri-generation
8	Song, Tan, and Mizzi [28]	HHO	N/A	Squared deviation between the actual and true output voltage	Power generation
9	Pandey and Jadoun [29]	HHO	N/A	Net profit, GHG emission reduction	Co-generation
10	Mahalekshmi and Maruthupandi [30]	HHO	Wind	Total cost, emission dispatch	Power generation

3. Materials and Methods

3.1. Description of the System

This study specifically examines the solar-assisted CCHP system described in the research conducted by Mahdavi et al. [31]. The system depicted in Figure 1 comprises three gas turbines, two compressors, a Kalina cycle, an absorption chiller, solar collectors, an HRSG, and a hot water generator.

The system presented functions using a specific sequence that includes multiple components. The air, at 298.15 K and 1.013 bar, is compressed using two-stage compressors with a cooling unit positioned between them. The waste heat generated by the cooling unit is utilized to power an absorption chiller, which aids in the heat transfer process within a desorber. Afterward, the warmed air, which has been heated by both solar energy and the exhaust gases from a gas turbine in air preheaters up to a temperature of 850 K, combines with methane fuel injected at a temperature and pressure of 298.15 K and 12 bar into the combustion chamber to produce high-temperature gases up to 1520 K. These gases are utilized to operate three gas turbines, with the third turbine being responsible for driving a generator that produces electrical power. In addition, the waste gases from the third gas turbine are used to heat the HRSG, the Kalina cycle via an evaporator (Eva-1), and a household water heat exchanger before being discharged into the ambient surroundings.

In the Kalina cycle, a mixture of ammonia and water, in a saturated liquid state, is pumped to increase its pressure. It then flows through two heat exchangers, namely the low-temperature recuperator and the high-temperature recuperator, to increase its heat content. This process helps to reduce the amount of energy needed for Eva-1. The resultant mixture is divided into saturated vapor and liquid states, with the vapor utilized to provide the work output in a steam turbine. Simultaneously, the fully saturated liquid is sent back to the HTR in order to recover heat energy, prior to being directed through Expansion Valve-1 (Ev-1), which reduces its pressure. Afterwards, the low-pressure saturated liquid at a temperature of 375 K is mixed with the steam turbine output that comprises the liquid–vapor states. The mixture is then transported through the LTR to disperse its energy before being expelled into the environment in a liquid state at a temperature of 303.15 K through Condenser-1 (cond-1), thus concluding the cycle.

The single-effect absorption chiller utilizes a series of operations to extract energy at the desorber by utilizing the discarded gases from the intercooler. The obtained energy is subsequently used to generate a cooling load at the corresponding evaporator (Eva-2). Pump-2 is applied to the weak LiBr-H_2O solution which then absorbs thermal energy from the rich LiBr-H_2O working fluid contained in the desorber at state 34. The working fluid, in a gaseous state, flows from the desorber into the condenser (cond-2) after coming into contact with the subcooled liquid from Eva-2 (state 40). In cond-2, the vapor undergoes condensation and transforms into a saturated liquid. Afterwards, it flows via the expansion valve (Ev-3) and enters Eva-2 where sub-cooling occurs. In Eva-2, the saturated liquid is transformed to cooled saturated vapor which is absorbed (state 40) by the rich LiBr-H_2O solution thus reverting to its original saturated liquid state at the start of the cycle. States 50 to 51 indicate the cooled water stream provided by Eva-2 which is also utilized in the absorber (52 to 53) to reduce the temperature of the rich LiBr-H_2O. The water stream with an increased temperature (after coming into contact with the rich LiBr-H_2O) is used in states 53 to 54 to condense the incoming water vapor from state 37.

The system's operation was based on several assumptions, including constant operating conditions, minimal fluctuations in the kinetic and potential energy, and insignificant thermal losses from the components other than the combustion chamber. The arrows used in Figure 1 indicate the direction of the flow of the combustion fluids.

3.2. Response Surface Method

The RSM is an analytical approach used to optimize and model complex systems by examining the relationship between the input variables and output results [32]. The RSM involves creating regression models to describe the link between a specific outcome and its influencing factors. The main procedures include the following:

- Experimental design: This stage entails establishing the circumstances for the experiment. It involves choosing pertinent input factors that may influence the response variable. Constraints are subsequently identified to evaluate the design variables during the experiment.
- Experimental testing: This involves the execution of tests according to a predetermined experimental strategy. Data pertaining to the outcome variable are gathered across different combinations of design factor levels. These tests are carefully performed to minimize the influence of less significant design variables.
- Regression model fitting: This involves the development of models utilizing the data collected from experiments. This is accomplished using techniques such as the least squares or maximum likelihood estimation. The resulting regression models are evaluated to determine their level of conformity to the initial model assumptions.
- Validation of the regression model: This occurs after successful fitting of the model by predicting the outcomes using new, unused data from further experimental tests. This stage guarantees the dependability and precision of the established model.

Mahdavi, Mojaver, and Khalilarya [31] employed the response surface methodology to develop regression models that can estimate the net power, CO_2 emission, and exergy efficiency of the system presented in Section 3.1. Table 2 and Equations (10)–(12) display the decision variables, constraints, and objectives.

Table 2. Decision variables.

Symbol	Decision variable	Constraints
X_1	Compression ratio	$10 \leq X_1 \leq 15$
X_2	Pinch point temperature differential	$10 \leq X_2 \leq 30$
X_3	Turbine inlet temperature	$1420 \leq X_3 \leq 1520$
X_4	Combustion chamber inlet temperature	$850 \leq X_4 \leq 950$

This study entails the examination of three separate objective functions:

1. Net power output: This refers to the combined power created by the gas turbine and the Kalina cycle, taking into account the work generated from both systems. The mathematical expression can be represented using the formula proposed by Mahdavi, Mojaver, and Khalilarya [31].

$$\dot{P}_{net} = \dot{P}_{net,\, GT} + \dot{P}_{net,\, KC} \quad (1)$$

$$\dot{P}_{net,\, GT} = \left(\dot{P}_{GT\text{-}1} + \dot{P}_{GT\text{-}2} + \dot{P}_{GT\text{-}3}\right) - \left(\dot{P}_{AC\text{-}1} - \dot{P}_{AC\text{-}2}\right) \quad (2)$$

$$\dot{P}_{net,\, KC} = \dot{P}_{ST} - \dot{P}_{pump\text{-}2} \quad (3)$$

$$\dot{P}_{GT\text{-}1} = \dot{m}_p(h_7 - h_8) = \dot{P}_{AC\text{-}1} \quad (4)$$

$$\dot{P}_{GT\text{-}2} = \dot{m}_p(h_8 - h_9) = \dot{P}_{AC\text{-}2} \quad (5)$$

$$\dot{P}_{GT\text{-}3} = \dot{m}_p(h_9 - h_{10}) \quad (6)$$

where

$\dot{P}_{net}$ = net power output;
$\dot{P}_{net,\, GT}$, $\dot{P}_{net,\, KC}$ = net power generated by gas turbines and the Kalina cycle respectively;
$\dot{P}_{GT\text{-}1}$, $\dot{P}_{GT\text{-}2}$, $\dot{P}_{GT\text{-}3}$ = net power generated by the three gas turbines;
$\dot{P}_{AC\text{-}1}$, $\dot{P}_{AC\text{-}2}$ = net power delivered to the two gas turbines;
$\dot{P}_{ST}$, $\dot{P}_{pump\text{-}2}$ = net power generated by the steam turbine and pump2;
h_7, h_8, h_9, h_{10} = specific enthalpies at the respective states;
$\dot{m}_p$ = mass flow rate of the combustion gases exiting the combustion chamber.

2. Exergy efficiency: This is a useful and effective measure used to evaluate the extent, size, and positions of the irreversibilities in a thermodynamic system [33]. The efficiency is mathematically defined as the quotient of the output exergy divided by the input exergy.

$$\varepsilon = \frac{\dot{P}_{net} + \left(\dot{E}_{45} - \dot{E}_{44}\right) + \left(\dot{E}_{50} - \dot{E}_{51}\right) + \left(\dot{E}_{47} - \dot{E}_{46}\right)}{\dot{E}_{in}} \quad (7)$$

$$\dot{E}_{in} = \dot{E}_{fuel} + \dot{E}_{coll} \quad (8)$$

where

ε = Exergy efficiency;
$\dot{E}_{44}$, $\dot{E}_{45}$, $\dot{E}_{46}$, $\dot{E}_{47}$, $\dot{E}_{50}$, $\dot{E}_{51}$ = Exergy at state 44, 45, 46, 47, 50, and 51;
$\dot{E}_{in}$ = Input exergy;
$\dot{E}_{fuel}$ = Exergy of fuel;
$\dot{E}_{coll}$ = Exergy of solar collector.

3. CO_2 Emission: The discharge of CO_2 into the atmosphere poses adverse environmental impacts, emphasizing the ongoing need for its mitigation within thermal energy systems. Emission, a metric assessing the level of CO_2 production, is defined as the ratio of the mass flow rate of CO_2 to the overall output energy, as defined by Mahdavi and Khalilarya [31].

$$\text{Emission} = \frac{\dot{m}_{CO_2}}{\dot{P}_{net} + \dot{Q}_{heating} + \dot{Q}_{cooling}} \quad (9)$$

where

$\dot{Q}_{heating}$ and $\dot{Q}_{cooling}$ = Heating and cooling loads of the CCHP system;

$$\begin{aligned}\dot{P}_{net}(MW) = {} & 62.19 + 0.4573.X_1 + 0.0259.X_2 - 0.02421.X_3 + 0.03638.X_4 - 0.010867.X_1.X_1 \\ & -0.000029.X_2.X_2 + 0.000005.X_3.X_3 - 0.000009.X_4.X_4 - 0.0003.X_1.X_2 + 0.00022.X_1.X_3 \\ & -0.00042.X_1.X_4 - 0.000005.X_2.X_3 - 0.000005.X_2.X_4 - 0.000003.X_3.X_4;\end{aligned} \quad (10)$$

$$\text{Emission}\left({}^{\text{gr}}/_{\text{MJ}}\right) = 13.1 + 3.722.X_1 + 0.2003.X_2 - 0.0122.X_3 + 0.0451.X_4 - 0.03047.X_1.X_1 + 0.0000296.X_2.X_2 + 0.00004.X_3.X_3 + 0.000052.X_4.X_4 + 0.0049.X_1.X_2 - 0.00294.X_1.X_3 + 0.00286.X_1.X_4 - 0.000285.X_2.X_3 + 0.000285.X_2.X_4 - 0.000099.X_3.X_4; \quad (11)$$

$$\varepsilon(\%) = -29 - 0.36.X_1 + 0.287.X_2 + 0.0659.X_3 + 0.0133.X_4 - 0.01807.X_1.X_1 - 0.000029.X_2.X_2 - 0.000011.X_3.X_3 - 0.000009.X_4.X_4 - 0.0125.X_1.X_2 + 0.0003.X_1.X_3 + 0.00086.X_1.X_4 - 0.000205.X_2.X_3 + 0.000195.X_2.X_4 - 0.000009.X_3.X_4. \quad (12)$$

3.3. Harris Hawks Optimization

This optimization technique is inspired by one of the smartest birds, usually located in South Arizona USA, known as Harris's hawk [34]. These hawks have a peculiar behavior of carrying out coordinated exploration, attacking, and feeding off their prey while in groups. They mainly employ the "surprise pounce" or "seven kills" tactics to catch their prey. This involves a situation where a group of hawks, emerging from different angles, simultaneously converge on the fleeing prey.

The HHO, proposed by Heidari et al. [35], is a common nature-inspired meta-heuristic algorithm that mimics the exploration and attacking behavior of the Harris's hawk. The stages of the HHO in capturing the prey and can be modeled using the following equations.

$$A = 2A_0\left(1 - \frac{r}{R}\right) \quad (13)$$

$$A_0 = 2t_1 - 1 \quad (14)$$

where

A = flight energy of the prey;
r = number of initial repetitions;
R = number of maximum repetitions;
A_0 = initial flight energy of the prey and is a random number in $[-1, 1]$;
t_1 = random value in $[0, 1]$.

The value of A steadily reduces with each iteration since it is correlated with the prey's flight energy. The prey is said to be physically exhausted when A_0 reduces from 0 to -1 and strengthened when A_0 is increased from 0 to 1. Hence, when $A \geq 1$, the hawks are still in the process of searching in various areas for the prey's whereabouts, and this is called the exploration phase. This process is transitioned to the exploitation phase when $A < 1$, indicating that the hawks are prepared to capture the prey. The subsequent section will provide a more detailed explanation of these stages.

3.3.1. Exploration Stage

This is the first stage of the hunting process whereby the hawks (also known as candidate solutions) monitor, track, and locate prey (also known as the target) using their powerful eyes. If an equal likelihood, t, for each waiting strategy is considered, $m < 0.5$ depicts the condition for the first waiting method while $m \geq 0.5$ depicts the condition for the second waiting method. Both methods are modeled in Equation (15).

$$Y(r+1) = \begin{cases} Yrand(r) - t_2|Yrand(r) - 2t_3Y(r)| \; m \geq 0.5 \\ (Y_{prey}(r) - Y_n(r)) - t_4(LB + t_5(UB - LB)) \; m < 0.5 \end{cases} \quad (15)$$

where

$Y(r+1)$ = new position of the hawks based on r repetitions;
Y_{prey} = position of the prey;
Y_n = average position of the current hawks' population;

$Yrand(r)$ = arbitrarily chosen hawk from the hawks' population;
$Y(r)$ = the present position of the hawk;
t_2, t_3, t_4, t_5, m = random values in (0, 1) updated in each repetition, r;
LB = lower bound of the variable;
UB = upper bound of the variable.

3.3.2. Exploitation Stage

This step is known as the "surprise pounce" phase, in which the hawks employ tactical maneuvers to capture the prey that they have spotted after the exploring phase. During the exploitation stage, the HHO models four different methods based on the prey's avoidance behavior and the assault method of the Harris's hawks. These tactics rely on the likelihood of finding the desired solution and the energy of the prey. The options consist of the soft besiege, hard besiege, soft besiege with progressive rapid dives, and hard besiege with progressive rapid dives.

3.3.3. Single-Objective Harris Hawks Optimization

Based on the above-mentioned equations, a mathematical model was developed, and it followed the following methodology:

1. Problem definition: the objective functions and the HHO parameters such as upper and lower boundaries of the decision variables, maximum number of iterations, etc. were defined.
2. Initialization: The initial population size of solutions was initialized. The initialization function was created to ensure that the initialized values do not exceed the defined boundaries of the decision variables.
3. Evaluation: The prey's energy, calculated using Equations (13) and (14), determined if the HHO algorithm employed exploration or exploitation. For any of the two methods employed, the prey's energy was employed to calculate the fitness value of the hawks (objective value) while the prey's location was regarded as the solution for the decision variable. The prey's location and hawks' fitness value were updated at each iteration until the maximum number of iterations was attained. The prey's location at the final iteration became the optimal decision variable solution returned by the algorithm and the corresponding objective value was obtained.

3.3.4. Multi-Objective Harris Hawks Optimization

The multi-objective Harris Hawks optimization is inspired by the multi-objective grey wolf technique to perform organized hunting by the Harris hawks. The MOHHO utilizes an archive repository and a roulette wheel approach to store non-dominated Pareto solutions in the archive. During each iteration, newly discovered non-dominated solutions are compared with the existing ones before updating the archive. Once the archive reaches its maximum capacity, the most densely populated solutions are discarded to make space for new entries. The likelihood of removing a solution from the archive is governed by:

$$P_i = \frac{N_i}{c}, \; c > 1 \tag{16}$$

where
c = constant value;
N_i = number of solutions around the ith solutions.

To develop the mathematical model for the multi-objective HHO, additional components were integrated into the model for single-objective HHO. The model comprised the following:

1. Problem definition: All three objective functions as well as the upper and lower boundaries of the decision variables were defined. Additionally, parameters such as the inertia weight, global learning coefficient, inflation rate, leader selection pressure, mutation rate, etc., were defined.

2. Initialization: the initial population size, number of iterations, and archive/repository size were initialized to store the non-dominated solutions.
3. Iteration improvement: The leader selection approach was employed in this algorithm to maintain the diversity of the repository. The leader selection pressure parameter (beta) and the non-dominated solution were passed as arguments to the "leader" function and yielded the fittest non-dominated solution as the output. This output was subsequently employed to guide the algorithm in performing exploration and exploitation. The roulette wheel was employed to select the leader of the MOHHO where each solution's probability of being selected as a leader depended on their fitness value.
4. Evaluation: During the evaluation process, the Pareto dominance was used to determine the domination of the solutions. During each iteration, a solution was included in the archive if it outperformed one or more of the existing solutions in the archive and if neither the new solution nor the existing solutions in the archive outperformed each other. However, a recently acquired solution was excluded from the archive if it was surpassed by at least one of the existing solutions. When a solution was added to the archive, a resident solution was deleted to maintain the pre-defined archive size. Within each iteration, non-dominated solutions and objective values were obtained and updated until the maximum m number of iterations was attained. The Pareto fronts were used to represent the optimal solutions and their accompanying objective values.

The developed models were implemented on MATLAB R2023a for single- and multi-objective optimization. Table 3 shows the MOHHO parameters employed. Figures 2 and 3 display the flowcharts for single- and multi-objective optimization using Harris Hawks.

Table 3. Hyperparameter values for MOHHO.

S/N	Parameter	Value
1	Maximum Iterations (MaxIt)	100
2	Population Count (nPop)	100
3	Repository Capacity (nRep)	100
4	Inertia Weight	0.5
5	Inertia Weight Damping Factor	0.99
6	Personal Learning Factor	1
7	Global Learning Factor	2
8	Grids per Dimension (nGrid)	4
9	Inflation Factor (alpha)	0.1
10	Leader Selection Pressure (beta)	2
11	Deletion Selection Pressure (gamma)	2
12	Mutation Rate (mu)	0.1

3.4. Mathematical Formulation

The mathematical expression for optimizing the net power is as follows:

$$\text{Maximize } \mathcal{F} = \{P_{net}(X_1, X_2, X_3, X_4)\} \tag{17}$$

Similarly, the process of optimizing CO_2 emission with a single target can be expressed as follows:

$$\text{Maximize } \mathcal{F} = \{(-\text{emission }(X_1, X_2, X_3, X_4)\} \tag{18}$$

The single-objective optimization of exergy efficiency can be expressed as follows:

$$\text{Maximize } \mathcal{F} = \{\varepsilon\,(X_1, X_2, X_3, X_4)\} \tag{19}$$

The multi-objective optimization problem is defined as follows:

$$\text{Maximize } \mathcal{F} = \{P_{net}(X_1, X_2, X_3, X_4), \varepsilon(X_1, X_2, X_3, X_4), -\text{emission}(X_1, X_2, X_3, X_4)\} \tag{20}$$

Equations (17)–(20) are subject to the constraints stated in Table 2.

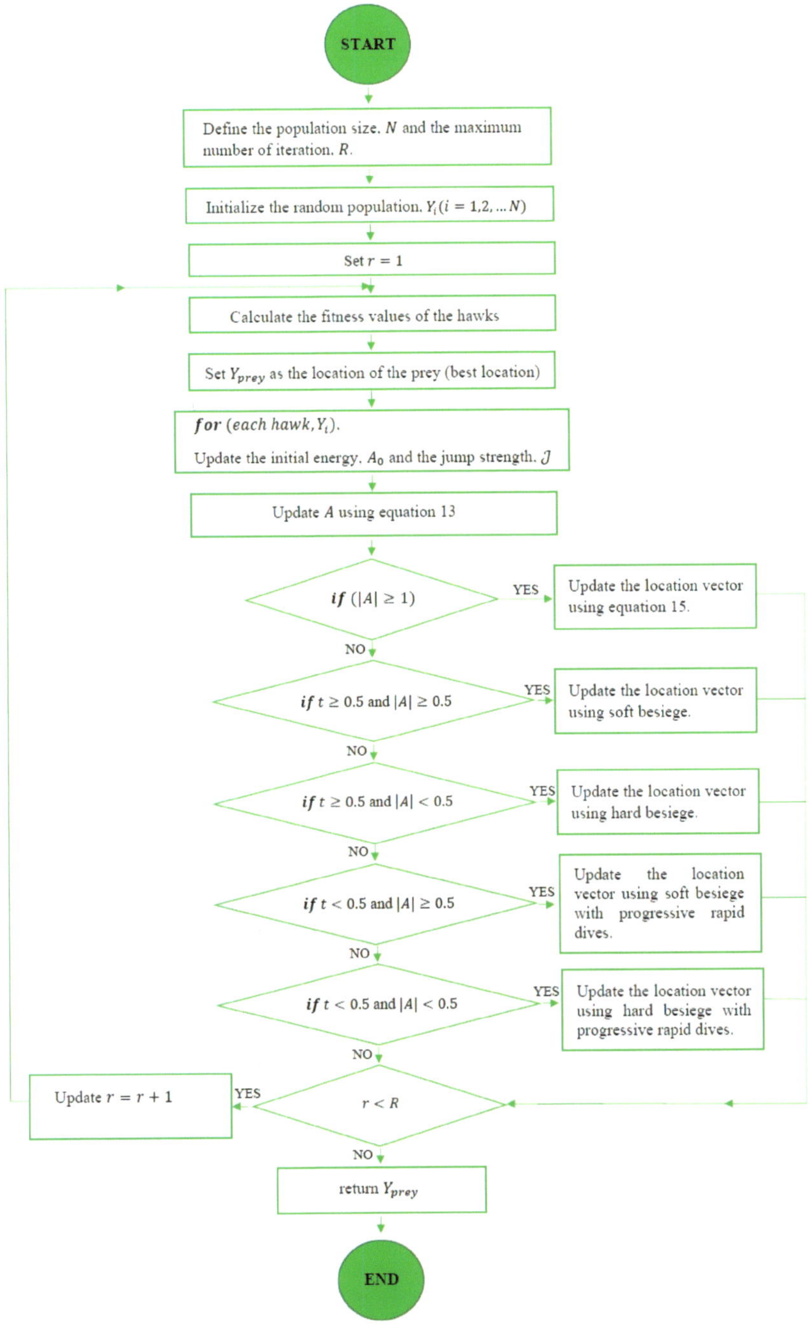

Figure 2. Flowchart for single-objective Harris Hawks optimization.

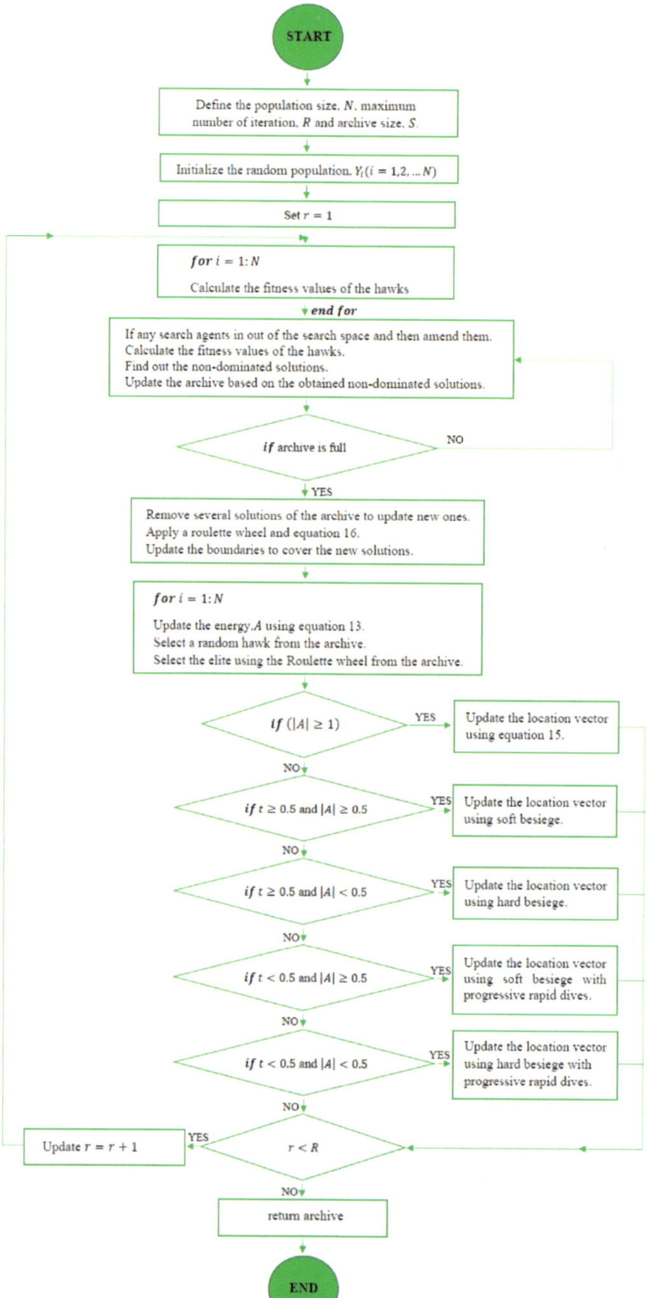

Figure 3. Flowchart for multi-objective Harris Hawks optimization.

4. Results and Discussion

4.1. Single-Objective Optimization

A single-objective HHO algorithm was executed on each objective function for 100 iterations, employing a search agent population size of 100. The MATLAB codes were run on a laptop that has an Intel Core i3-5005U CPU running at a speed of 2.00 GHz and 8 GB of RAM.

4.1.1. Net Power, CO_2 Emission, Exergy Efficiency, and Optimization

The solutions for Equations (10)–(12) were derived utilizing the decision variables and the respective values listed in Table 2, utilizing the HHO algorithm. The outcomes, shown in Table 4, highlight the necessary steps to enhance the net power and exergy efficiency while simultaneously reducing the CO_2 emissions. The key optimization steps for the net power and CO_2 emission are as follows:

- Lower the compression ratio, pinch point temperature difference, and the combustion chamber inlet temperature.
- Increase the turbine inlet temperature.

Table 4. Optimal solutions optimizing the net power, CO_2 emission, and exergy efficiency.

Objective	X_1	X_2	X_3	X_4	Optimal Objective Value
Maximum Power (MW)	10	10	1520	850	60.8552
Minimum CO_2 (gr/MJ)	10	10	1520	850	49.3771
Maximum Exergy Efficiency (%)	10	10	1420	950	42.3507

To enhance the exergy efficiency, the implementation of the following procedures is necessary:

- Decrease the compression ratio, pinch point temperature difference, and turbine inlet temperature.
- Increase the combustion chamber inlet temperature to its highest possible level.

4.1.2. Analysis of the Single-Objective Optimization Outcomes

The results obtained in Table 4 have been summarized in Table 5, which shows the behavior of the optimal solutions. As shown in Table 5, an upward trend is represented by $\uparrow$, a downward trend is represented by $\downarrow$, while a conflict between the parameters is indicated by $\neq$. It could be seen that for decision variables X_1 and X_2, all three objective functions are in harmony. However, for the X_3 and X_4 decision variables, the exergy efficiency conflicts with the net power and CO_2 emission. Thus, a simultaneous optimization of the three objectives is required to determine the Pareto optimal solutions.

Table 5. Illustration of the trend of the objective functions.

Decision Variable	Net Power	CO_2 Emission	Exergy Efficiency
X_1	$\downarrow$	$\downarrow$	$\downarrow$
X_2	$\downarrow$	$\downarrow$	$\downarrow$
X_3	$\uparrow$	$\uparrow$	$\downarrow\neq$
X_4	$\downarrow$	$\downarrow$	$\uparrow\neq$

4.2. Multi-Objective Optimization

The optimization algorithm has produced a hundred non-dominated solution sets, in the Pareto front as shown in Figure 4. The ensuing optimal solution sets provide the best feasible trade-offs despite the inherent conflicts between the three objectives. Each solution on the Pareto front represents a potential optimal, with the final selection determined by the judgment of the decision-maker.

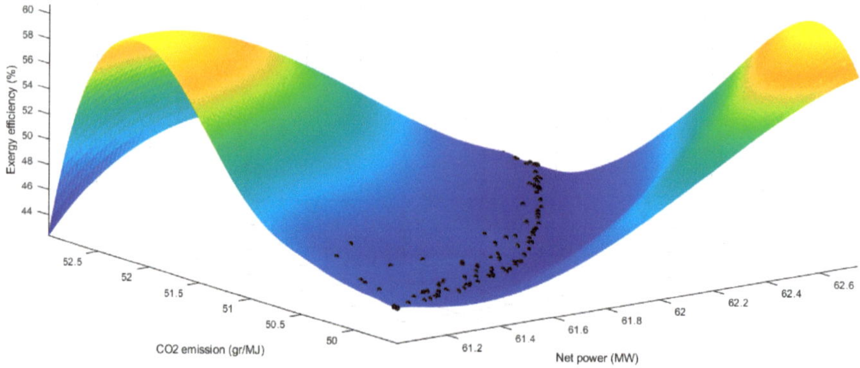

Figure 4. Pareto front illustrating the relationship between CO_2 emissions, net power, and exergy efficiency.

An analysis was conducted on three scenarios to have a deeper comprehension of the correlation between each objective function. This study examined the correlations between the CO_2 emissions and net power output, exergy efficiency and net power output, and exergy efficiency and CO_2 emissions in various situations. For clarity, the overall outcomes have been condensed into three two-dimensional Pareto fronts that emphasize crucial aspects. Figure 5 shows the Pareto front, highlighting the trade-off between the net power and CO_2 emissions during the decision-making process as the CO_2 emission rate increased from 50.5 to 51 gr/MJ following an increase in net power from 61.85 to 62.1 MW. This increase is largely due to the higher fuel consumption and intensified combustion processes needed to achieve the additional power output, which naturally leads to more significant carbon emissions. Thus, maximizing the net power will unavoidably result in increasing CO_2 emissions, which may have detrimental consequences for the environment. This emphasizes the need for a balance between maximizing power and minimizing the emissions within the CCHP system. In the CCHP system under study, the net power output was significantly influenced by the power generated by the GT-3 and then the Kalina cycle. To mitigate the CO_2 emissions without substantially compromising the net power output, the system can be configured to ensure that the combustion gases entering from APH-2 reach the unfired HRSG at a higher temperature than when they enter the combustion chamber. Furthermore, it is feasible to simultaneously reduce the rate of fuel consumption and improve the output of the photovoltaic thermal collectors. These adjustments enhance the energy conversion efficiency of the HRSG, resulting in lower fuel use per unit of power generated. The principal power source of the system, the GT-3, produces less net power as a result of these modifications, which are intended to minimize the CO_2 emissions. However, there will be an increase in the cleaner, non-combustible power output from the Kalina cycle. Even if in the end, this method lowers the net power generated, it is optimized to minimize the CO_2 emissions.

Figure 6 demonstrates the relationship between the exergy efficiency and CO_2 emissions as increasing the exergy efficiency value from 42.75 to 44.5 which reduced the CO_2 emissions by about 2.5%, thus indicating a compatibility between the two. An increased exergy efficiency is correlated with a lower CO_2 emission value. This correlation can be attributed to the improved energy conversion processes within the CCHP system, which reduce waste energy and require less fuel input. By achieving a higher exergy efficiency, the system utilizes a greater portion of the available energy, thereby decreasing the CO_2 emissions through lower fuel consumption. This was seen to be in alignment with similar studies that analyzed a double-effect absorption chiller [36] and a solar-assisted energy storage CCHP system [37]. Figure 6 also shows that in order to significantly boost the energy efficiency, the CO_2 emissions must be reduced by a certain extent. Additionally, the study reveals that by reducing the fuel use, there would be a corresponding drop in the input exergy of the fuel. Consequently, this would enhance the exergy outputs and usable work produced by the Kalina cycle, water heat exchanger, and absorption chiller, so greatly optimizing the exergy efficiency. As a result, there would be minimal CO_2 emissions, and this improves the sustainability of the CCHP system.

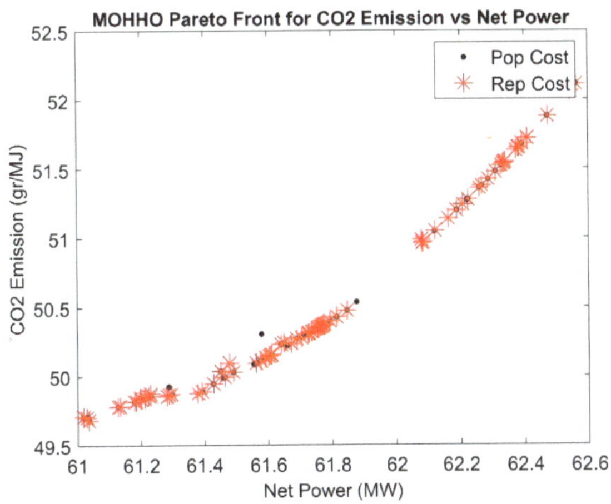

Figure 5. Pareto front showing the trade-off between CO_2 emissions and net power.

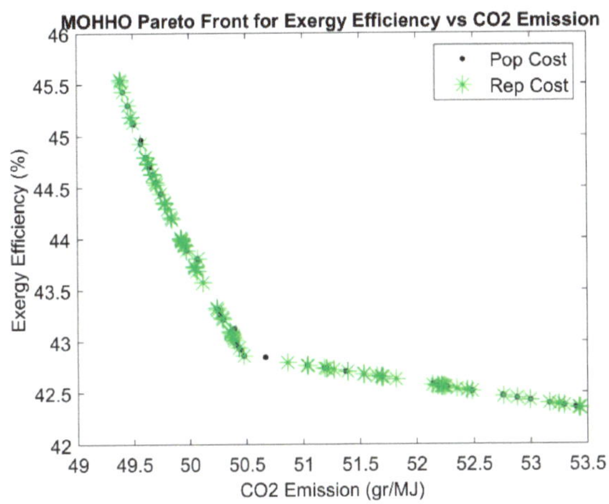

Figure 6. Pareto front showing the trade-off between exergy efficiency and CO_2 emissions.

Additionally, Figure 7 illustrates a negative correlation between the exergy efficiency and net power as increasing the net power from 61 to 62.75 MW which led to a 6% reduction in the exergy efficiency. This indicates that a system can have a high net power output yet still lack efficiency. This inverse relationship occurs because additional power output often relies on increased fuel combustion, which inherently has a lower exergy efficiency due to the waste heat and irreversibility in the combustion process. To address this effect, it is recommended to optimize the configuration of bottom cycle systems like the Kalina cycle and absorption chiller. Running these subsystems in a way that maximizes the use of waste heat can improve the system's overall exergy efficiency while reducing the fuel dependency. By focusing on non-combustible power sources such as the Kalina cycle, the system can achieve a cleaner energy output and improved efficiency without the drawbacks of high emissions and energy loss from additional fuel combustion.

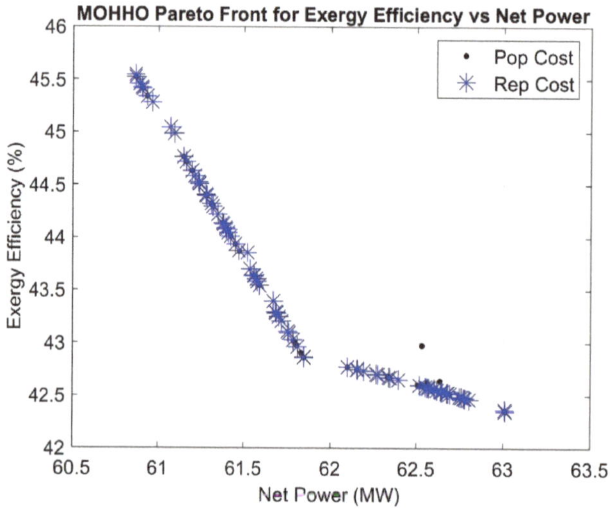

Figure 7. Pareto front showing the trade-off between exergy efficiency and net power.

The constraints placed on the HHO technique and the convergence of each choice variable are shown in Figure 8's scatter plot distribution. The results show that lower compression ratios and pinch point temperatures are preferred by the suggested HHO strategy. However, values in the range between the minimum and the midpoint were selected, by the optimization method, for the combustion chamber inlet temperature. The scatter plot for the gas turbine inlet temperature is distributed between the constraint's upper and lower bounds.

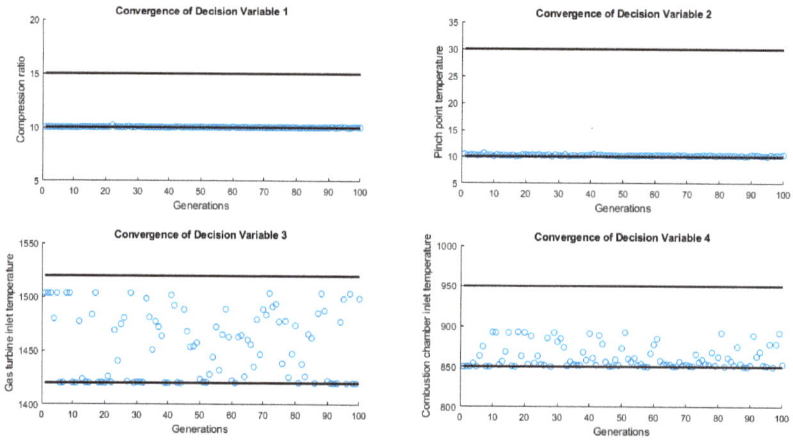

Figure 8. Scatter pattern of decision variables.

The MOHHO was designed to produce six ideal solutions, which were then compared with the outcomes of similar research that utilized the RSM. The data gathered, detailed in Table 6, indicate that the MOHHO generally produced CO_2 emissions ranging from 50.33 gr/MJ to 50.66 gr/MJ, which is at least 4.2% lower than the CO_2 emission values produced by the RSM. The data in Table 6 yielded a high net power of 61.91 MW and low values of X_1, X_2, X_3, and X_4. This, however, resulted in high CO_2 emission and low exergy efficiency values. Better values of CO_2 emissions and exergy efficiency were obtained at relatively higher X_3 and X_4 values although at the reduced net power of 61.52 MW. Overall, the MOHHO yielded decision variable configurations that produced higher net power (61.91 MW), higher exergy efficiency (45.28%), and a relatively lower CO_2 emission (50.33 gr/MJ) as highlighted in Table 6. Therefore, the proposed MOHHO outperformed the RSM in optimizing the net power, CO_2 emission, and exergy efficiency.

Table 6. Comparison of optimization results with a related study.

Optimization Technique	Best-Performing Decision Variables				Optimal Performance Metrics		
	X_1	X_2	X_3	X_4	Net Power	CO_2 Emission	Exergy Efficiency
MOHHO	10	10	1420	855.34	61.91	50.61	42.84
	10	10	1422.67	850	61.82	50.44	42.94
	10.28	11.14	1514.57	896.17	61.52	50.66	45.28
	10	10	1426.26	850	61.78	50.38	43.04
	10.01	10.77	1469.76	870.77	61.60	50.33	44.15
	10.28	11.14	1514.31	896.01	61.52	50.66	45.28
RSM (Mahdavi, Mojaver, and Khalilarya [29])	11.66	11.96	1470	900	61.73	52.87	44.22
	11.11	20.00	1470	900	61.73	52.99	44.09
	11.98	20.00	1470	890	61.75	53.84	44.12
	12.50	16.10	1484	900	61.75	54.07	44.58
	12.50	15.72	1470	891	61.79	54.07	44.10
	12.50	20.00	1468	882	61.75	54.28	44.30

5. Conclusions

This study aimed to maximize the efficiency of a solar-assisted CCHP system through multi-objective optimization using the Harris Hawks approach. The optimization sought to minimize the greenhouse gas emissions, which have detrimental effects on human health and contribute to global warming while increasing the net power and energy efficiency. This study emphasizes how important it is to slow down the rate at which fossil fuels are depleting as demonstrated by the optimized solar-assisted CCHP system. In addition, this work introduces a new optimization strategy, whereas previous studies have utilized different methods to achieve the multi-objective optimization of the solar-powered CCHP system being studied. This method was employed due to its dynamic balance of various exploration and exploitation, its robustness across different kinds of objectives, and its fast convergence time. However, it had the drawback of premature convergence as seen with the inlet gas turbine temperature. The research's results and findings are described in the subsequent section.

- The most effective parameter configuration for the thermodynamic design of the solar-powered CCHP system is identified using a multi-objective optimization technique with decision variables including the compression ratio, pinch point temperature difference, turbine inlet temperature, and combustion chamber inlet temperature.
- The net power, CO_2 emissions, and exergy efficiency serve as objective functions for evaluating the efficacy of the CCHP system by assessing how effectively the decision variables meet the restrictions.

- The Harris Hawks technique is utilized to accomplish multi-objective optimization by generating non-dominated Pareto optimum solutions. This study calculates a collection of Pareto optimal solutions, which are then presented as alternative choices to the decision-maker. The objective is to empower the decision-makers to exercise their judgment and improve the efficiency of the CCHP system. The convergence of the individual decisions streamlines the decision-making process.
- Interestingly, the mutual reliance among the four decision variables, within the CCHP system under study, signifies that alterations to a single variable will invariably influence the remaining three. Consequently, employing a multi-objective optimization methodology is recommended for the comprehensive assessment of the system's operational efficacy.
- The research identified a conflict within the decision-making processes concerning the optimization of the net power and CO_2 emissions. Maximizing the net power tends to correlate with increased CO_2 emissions, presenting a trade-off as a higher net power often leads to elevated CO_2 emissions, which is undesirable.
- Another significant observation pertains to the correlation between exergy efficiency and CO_2 emissions. It suggests that a system characterized by minimal CO_2 emissions tends to exhibit high exergy efficiency. This indicates compatibility wherein a system with minimal CO_2 emissions tends to be highly efficient. The study's findings challenge the assumption that a system with high net power production invariably guarantees a high efficiency. This assertion is supported by the inverse relationship observed between the net power and exergy efficiency. Hence, it is important to note that attaining a high level of net power does not always result in a higher efficiency, as there exists a negative correlation between the net power and exergy efficiency.
- The analysis of the convergence of decision variables reveals that optimizing the objective values can be achieved by minimizing the compression ratio, pinch point temperature difference, and inlet combustion chamber temperature. However, it is noteworthy that the ranges for the gas turbine temperature are spread across the upper and lower bounds of the constraints. This dispersion suggests challenges in achieving convergence for this variable, hinting at the potential difficulties in its optimization. Consequently, additional techniques may be necessary to address these challenges and achieve convergence for the gas turbine temperature within the defined constraints. The HHO may be hybridized with the artificial neural network or the ant–lion approach to achieve optimal convergence. Also, adaptive tuning of the HHO parameters during the optimization process can be explored to improve the optimization performance.

As a recommendation, alternative meta-heuristic optimization techniques like genetic algorithms and particle swarm optimization could be explored to enhance the performance of the solar-assisted CCHP system. Additionally, predictive control methods, such as model predictive control and fuzzy logic control, can be combined with meta-heuristic solutions to dynamically adjust the CCHP system's configuration in response to changing environmental conditions and the preferences of users or decision-makers. Moreover, validating the optimal results obtained from these meta-heuristic approaches with experimental data would strengthen the credibility of the proposed solutions in practical applications. Finally, for future research, incorporating other renewable energy sources like wind and biomass, along with renewable energy storage systems, would significantly mitigate the variability in energy supply often encountered when relying solely on solar power.

Supplementary Materials: The following supporting information can be downloaded at: https://www.mdpi.com/article/10.3390/su162310694/s1.

Author Contributions: Conceptualization, U.U. and L.T.; methodology, formal analysis, writing—original draft preparation, U.U.; writing—review and editing, supervision, L.T. and C.W.L. All authors have read and agreed to the published version of the manuscript.

Funding: The APC was funded by the University of Johannesburg.

Informed Consent Statement: Not applicable.

Data Availability Statement: The original contributions presented in the study are included in the article/supplementary material, further inquiries can be directed to the corresponding author.

Acknowledgments: The authors wish to express their gratitude to the University of Johannesburg Research Committee and the Commonwealth Scholarship Fund for their support and sponsorship of this research.

Conflicts of Interest: The authors declare no conflicts of interest.

Nomenclature

CCHP	Combined Cooling Heating and Power
CHP	Combined Heat and Power
GHG	Greenhouse Gas
HRSG	Heat Recovery Steam Generator
GT	Gas turbine
RSM	Response surface method
HHO	Harris Hawks Optimization
MOHHO	Multi-objective Harris Hawks Optimization
HTR	High Temperature Recuperator
LTR	Low Temperature Recuperator
SHE	Solution Heat Exchanger
APH	Air-preheater

References

1. Ramachandran, T.; Mourad, A.H.I.; Hamed, F. A Review on Solar Energy Utilization and Projects: Development in and around the UAE. *Energies* **2022**, *15*, 3754. [CrossRef]
2. BP Energy Outlook. 2023. Available online: https://www.bp.com/en/global/corporate/energy-economics/energy-outlook.html (accessed on 31 May 2023).
3. Ortiga, J.; Bruno, J.C.; Coronas, A. Operational optimization of a complex trigeneration system connected to a district heating and cooling network. *Appl. Therm. Eng.* **2013**, *50*, 1536–1542. [CrossRef]
4. Lozano, M.A.; Carvalho, M.; Serra, L.M. Allocation of economic costs in trigeneration systems at variable load conditions. *Energy Build.* **2011**, *43*, 2869–2881. [CrossRef]
5. Wang, L.; Lu, J.; Wang, W.; Ding, J. Energy, environmental and economic evaluation of the CCHP systems for a remote island in south of China. *Appl. Energy* **2016**, *183*, 874–883. [CrossRef]
6. Sharma, M.K.; Bhattacharya, J. Finding optimal operating point for advection-cooled concentrated photovoltaic system. *Sustain. Energy Technol. Assess.* **2022**, *49*, 101769. [CrossRef]
7. Dincer, I.; Zamfirescu, C. Renewable-energy-based multigeneration systems. *Int. J. Energy Res.* **2012**, *36*, 1403–1415. [CrossRef]
8. Nadimi-Shahraki, M.H.; Taghian, S.; Mirjalili, S.; Zamani, H.; Bahreininejad, A. GGWO: Gaze cues learning-based grey wolf optimizer and its applications for solving engineering problems. *J. Comput. Sci.* **2022**, *61*, 101636. [CrossRef]
9. Hansen, N.; Muller, S.D.; Koumoutsakos, P. Reducing the time complexity of the derandomized evolution strategy with covariance matrix adaptation (CMA-ES), *Evol. Comput.* **2003**, *11*, 1–18.
10. Abualigah, L. Multi-verse optimizer algorithm: A comprehensive survey of its results, variants, and applications. *Neural Comput. Appl.* **2020**, *32*, 12381–12401. [CrossRef]
11. Mirjalili, S.; Saremi, S.; Mirjalili, S.M.; Coelho, L.D.S. Multi-objective grey wolf optimizer: A novel algorithm for multi-criterion optimization. *Expert Syst. Appl.* **2016**, *47*, 106–119. [CrossRef]
12. Hasanzadeh, R.; Mojaver, M.; Azdast, T.; Park, C.B. A novel systematic multi-objective optimization to achieve high-efficiency and low-emission waste polymeric foam gasification using response surface methodology and TOPSIS method. *Chem. Eng. J.* **2022**, *430*, 132958. [CrossRef]
13. Ren, F.; Wang, J.; Zhu, S.; Chen, Y. Multi-objective optimization of combined cooling, heating and power system integrated with solar and geothermal energies. *Energy Convers. Manag.* **2019**, *197*, 111866. [CrossRef]
14. Azaza, M.; Wallin, F. Multi objective particle swarm optimization of hybrid micro-grid system: A case study in Sweden. *Energy* **2017**, *123*, 108–118. [CrossRef]
15. Shehab, M.; Mashal, I.; Momani, Z.; Shambour, M.K.Y.; Al-Badareen, A.; Al-Dabet, S.; Bataina, N.; Alsoud, A.R.; Abualigah, L. Harris Hawks Optimization Algorithm: Variants and Applications. *Arch. Comput. Methods Eng.* **2022**, *29*, 5579–5603. [CrossRef]

16. Sharifian, Y.; Abdi, H. Solving multi-zone combined heat and power economic emission dispatch problem considering wind uncertainty by applying grasshopper optimization algorithm. Sustain. *Energy Technol. Assess.* **2022**, *53*, 102512. [CrossRef]
17. Ji, J.; Wang, F.; Zhou, M.; Guo, R.; Ji, R.; Huang, H.; Zhang, J.; Nazir, M.S.; Peng, T.; Zhang, C.; et al. Evaluation Study on a Novel Structure CCHP System with a New Comprehensive Index Using Improved ALO Algorithm. *Sustainability* **2022**, *14*, 15419. [CrossRef]
18. Xu, L.; Luo, X.; Wen, Y.; Wu, T.; Wang, X.; Guan, X. Energy Management of Hybrid Power Ship System Using Adaptive Moth Flame Optimization Based on Multi-Populations. *IEEE Trans. Power Syst.* **2023**, *39*, 1711–1727. [CrossRef]
19. Ukaegbu, U.; Tartibu, L.; Lim, C.W. Multi-Objective Optimization of a Solar-Assisted Combined Cooling, Heating and Power Generation System Using the Greywolf Optimizer. *Algorithms* **2023**, *16*, 463. [CrossRef]
20. Wang, J.; Han, Z.; Guan, Z. Hybrid solar-assisted combined cooling, heating, and power systems: A review. *Renew. Sustain. Energy Rev.* **2020**, *133*, 110256. [CrossRef]
21. Cao, Y.; Dhahad, H.A.; Togun, H.; Haghghi, M.A.; Athari, H.; Mohamed, A.M. Exergetic and economic assessments and multi-objective optimization of a modified solar-powered CCHP system with thermal energy storage. *J. Build. Eng.* **2021**, *43*, 102702. [CrossRef]
22. Wang, J.; Lu, Y.; Yang, Y.; Mao, T. Thermodynamic performance analysis and optimization of a solar-assisted combined cooling, heating and power system. *Energy* **2016**, *115*, 49–59. [CrossRef]
23. Wang, J.; Liu, Y.; Ren, F.; Lu, S. Multi-objective optimization and selection of hybrid combined cooling, heating and power systems considering operational flexibility. *Energy* **2020**, *197*, 117313. [CrossRef]
24. Song, Z.; Liu, T.; Lin, Q. Multi-objective optimization of a solar hybrid CCHP system based on different operation modes. *Energy* **2020**, *206*, 118125. [CrossRef]
25. Abba, S.I.; Rotimi, A.; Musa, B.; Yimen, N.; Kawu, S.J.; Lawan, S.M.; Dagbasi, M. Emerging Harris Hawks Optimization based load demand forecasting and optimal sizing of stand-alone hybrid renewable energy systems—A case study of Kano and Abuja, Nigeria. *Results Eng.* **2021**, *12*, 100260. [CrossRef]
26. Yousri, D.; Babu, T.S.; Fathy, A. Recent methodology based Harris Hawks optimizer for designing load frequency control incorporated in multi-interconnected renewable energy plants. *Sustain. Energy Grids Netw.* **2020**, *22*, 100352. [CrossRef]
27. Zhao, L.; Li, Z.; Chen, H.; Li, J.; Xiao, J.; Yousefi, N. A multi-criteria optimization for a CCHP with the fuel cell as primary mover using modified Harris Hawks optimization. *Energy Sources Part A Recovery Util. Environ. Eff.* **2020**, *46*, 10630–10645. [CrossRef]
28. Song, Y.; Tan, X.; Mizzi, S. Optimal parameter extraction of the proton exchange membrane fuel cells based on a new Harris Hawks Optimization algorithm. *Energy Sources Part Recovery Util. Environ. Eff.* **2020**, *46*, 8718–8735. [CrossRef]
29. Pandey, A.K.; Jadoun, V.K. Real-time and day-ahead risk averse multi-objective operational scheduling of virtual power plant using modified Harris Hawk's optimization. *Electr. Power Syst. Res.* **2023**, *220*, 109285. [CrossRef]
30. Mahalekshmi, T.; Maruthupandi, P. Multiobjective Economic/Environmental Dispatch Using Harris Hawks Optimization Algorithm. *Intell. Autom. Soft Comput.* **2023**, *36*, 445–460. [CrossRef]
31. Mahdavi, N.; Mojaver, P.; Khalilarya, S. Multi-objective optimization of power, CO_2 emission and exergy efficiency of a novel solar-assisted CCHP system using RSM and TOPSIS coupled method. *Renew. Energy* **2022**, *185*, 506–524. [CrossRef]
32. Costa, E.; Almeida, M.F.; Alvim-Ferraz, C.; Dias, J.M. Otimization of Crambe abyssinica enzymatic transesterification using response surface methodology. *Renew. Energy* **2021**, *174*, 444–452. [CrossRef]
33. Kumar, R. A critical review on energy, exergy, exergoeconomic and economic (4-E) analysis of thermal power plants. *Eng. Sci. Technol. Int. J.* **2017**, *20*, 283–292. [CrossRef]
34. Bednarz, J.C. Cooperative hunting Harris' hawks (*Parabuteo unicinctus*). *Science* **1988**, *239*, 1525–1527. [CrossRef] [PubMed]
35. Heidari, A.A.; Mirjalili, S.; Faris, H.; Aljarah, I.; Mafarja, M.; Chen, H. Harris hawks optimization: Algorithm and applications. *Future Gener. Comput. Syst.* **2019**, *97*, 849–872. [CrossRef]
36. Mousafarash, A. Exergy and Exergoenvironmental Analysis of a CCHP System Based on a Parallel Flow Double-Effect Absorption Chiller. *Int. J. Chem. Eng.* **2016**, *1*, 2370305. [CrossRef]
37. Zhang, D.; Yang, X.; Li, H.; Jia, Z.; Zhang, S.; Tang, S.; Liu, D.; Wu, X. 4E analysis and parameter study of a solar-thermochemical energy storage CCHP system. *Energy Convers. Manag.* **2024**, *301*, 118002. [CrossRef]

sustainability

MDPI

Article

Decarbonizing Public Transportation: A Multi-Criteria Comparative Analysis of Battery Electric Buses and Fuel Cell Electric Buses

Afnan Fayez Eliyan *, Mohamed Haouari and Ahmad Sleiti

Department of Mechanical and Industrial Engineering, College of Engineering, Qatar University, Doha 2713, Qatar; mohamed.haouari@qu.edu.qa (M.H.); asleiti@qu.edu.qa (A.S.)

* Correspondence: aeliyan@qu.edu.qa

Abstract: To combat global warming, many industrialized countries have announced plans to ban vehicles powered by fossil fuel in the near future. In alignment with this global initiative, many countries across the globe are committed to decarbonizing their public transportation sector, which significantly contributes to increased greenhouse gas emissions. A promising strategy to achieve this goal is the adoption of electric buses, specifically battery electric buses and fuel cell electric buses. Each technology offers distinct advantages and drawbacks, making the decision-making process complex. This research aims to answer two critical questions: What is the optimal choice for decarbonizing the bus transportation sector—electric battery buses or fuel cell electric buses? And what are the best energy carrier pathways for charging or refueling these buses? We propose a methodological framework based on multi-criteria decision-making to address these questions comprehensively. This framework utilizes the entropy weighting and the Technique for Order Preference by Similarity to the Ideal Solution (TOPSIS) methodologies to rank alternative bus technologies along with energy carrier pathways. The framework evaluates a range of criteria, including economic viability, energy demand, and environmental aspects. To illustrate the framework, we considered Qatar as a case study. Our results indicate that, with respect to economic viability and energy consumption, the operation of battery electric buses is favored over fuel cell electric buses, regardless of the energy pathway utilized during both the energy production and bus operation phases. However, from an environmental perspective, operating both bus alternatives using energy from green sources provides superior performance compared to when these buses are powered by natural gas sources.

Keywords: battery electric buses; fuel cell electric buses; multicriteria decision making; public transportation; Shannon entropy; TOPSIS

1. Introduction

The electrification of the transportation sector plays a crucial role in reducing carbon emissions across the energy and mobility systems. Passenger road transportation accounted for about 45% of global transport emissions throughout the years 2000 to 2023 [1]. Additionally, the transportation sector accounts for approximately 34% of global energy consumption, with passenger transportation making up 60% of this sector's total energy consumption [2]. Therefore, the transition from fossil-fuel-powered public buses to cleaner, more sustainable energy sources for public transit systems becomes increasingly essential to ensure the energy transition and to preserve energy sources [3,4].

In the global push for clean transportation, many cities are spearheading the transition away from fossil fuels. This transformation is evident in the surge of electric and hydrogen bus deployments worldwide. Each of these bus technologies can be powered using different energy carrier pathways. For battery electric buses (BEBs), the commonly used method to produce electricity is from natural gas or renewable sources including, but not limited to, wind and solar. For fuel cell electric buses (FCEBs), hydrogen can be produced from the steam reforming of natural gas or through water electrolysis using electricity generated from renewable sources. Despite these advancements, a key question remains unanswered: which bus technology, BEBs or FCEBs, offers a more attractive long-term solution? Moreover, what energy carrier pathway is more preferred?

A holistic assessment that considers the overall cost, energy efficiency, and environmental impact is crucial for making informed decisions as these criteria form the core of the sustainability framework's environmental, economic, and social pillars. Economic viability, assessed through life cycle costing, ensures a cost-effective and financially sustainable solution over time. The environmental impact, measured in terms of global warming potential through CO_2 equivalents, addresses the need for low-impact solutions that align with global sustainability goals. We selected CO_2 emissions as the primary indicator of environmental efficiency because decarbonizing public transportation is our core objective. To clarify, the authors made no claim that CO_2 directly contributes to global warming. However, in academic and scientific contexts, global warming potential is commonly measured by the equivalent CO_2 emissions, as referenced in numerous studies, including but not limited to the work of [5]. Energy demand was also considered due to its direct influence on greenhouse gas emissions and resource consumption.

While a comprehensive evaluation should ideally cover the entire life cycle of bus technologies—including design, manufacturing, operation, and disposal—we focused on the well-to-wheel phase. This decision was based on the fact that this phase contributes the largest share of energy consumption and emissions compared to other stages, as demonstrated in the studies [6] and [5], respectively.

It is noteworthy to mention that this study focuses exclusively on zero-emission public transportation alternatives, such as BEBs and FCEBs, which produce no emissions during operation. Therefore, the assessment of hybrid electric vehicles or plug-in buses falls beyond the scope of this research. However, readers interested in these alternatives can refer to the works of [7,8], who provide extensive research on hybrid and plug-in electric bus technologies.

Research Contributions

This study contributes to the literature by introducing an integrated MCDM framework to evaluate electric bus alternatives and their associated energy carrier pathways. Specifically, the framework assesses these alternatives in terms of three key dimensions: economic viability, energy demand, and environmental performance during the well-to-wheel (WTW) phase, which includes both WTT and TTW stages.

The novelty of this study lies in its two-phase methodological approach. First, the Shannon entropy method is employed to objectively assign weights to the evaluation criteria, ensuring that each criterion's influence is proportionate to its variability. Second, the Technique for Order Preference by Similarity to the Ideal Solution (TOPSIS) is applied to rank the bus alternatives based on their proximity to the ideal solution. By integrating these two methods, the study offers a rigorous and systematic evaluation framework, providing valuable insights into the comparative performance of BEBs and FCEBs and their respective energy carrier pathways. This approach enhances decision-making for sustainable public transportation systems.

The remainder of the paper is organized as follows. Section 2 provides the background and related literature, contextualizing the research within the existing body of work. Section 3 demonstrates the methodological framework. Section 4 presents the results and discusses them in relation to existing literature, and Section 5 offers the conclusions and future perspectives.

2. Background and Related Works

This section aims to provide an overview of global efforts and methodologies related to the decarbonization of public transportation in Section 2.1. Additionally, in Section 2.2, we provide a comprehensive analysis of the existing methodologies for evaluating bus alternatives, focusing on the use of multi-criteria decision-making (MCDM) tools to compare the economic, energy, and environmental impacts of BEBs and FCEBs.

2.1. Global Advances in Decarbonizing Public Transportation

The global push for decarbonizing public transportation has seen varying levels of progress across different regions and countries. Recent data reflect this trend, with a significant increase in electrically chargeable bus sales in Europe, rising by 78.7% in 2021 and capturing a market share of 10.6% [9]. Moreover, London recently announced plans to transition its entire double-decker bus fleet to electric by 2030, joining cities like Oslo, Norway, which boasts the world's largest electric bus fleet [10]. Meanwhile, Groningen, Netherlands, is pioneering the use of hydrogen buses, aiming for a completely zero-emission public transport system by 2030 [11].

Moreover, the European Union has developed long-term strategies aimed at reducing transport emissions, with a target to achieve carbon neutrality by 2050 [12]. The European Green Deal emphasizes the electrification of public transport systems, particularly through BEBs and FCEBs. Similarly, countries in Latin America and the Caribbean, including Mexico, Brazil, and Colombia, have embarked on initiatives to transition to cleaner public transportation systems. Mexico City has integrated electric buses into its public transit system, while Ecuador and Colombia have set goals for electric mobility by 2040 [13].

Countries like Japan, China, and South Africa are also at the forefront of this transition. China, for example, has become a global leader in electric bus deployment, with over 400,000 electric buses in operation [14]. The Balearic and Canary Islands in Spain, as well as island nations such as the Galapagos Islands, Cuba, and American territories, are adopting green mobility to mitigate the unique challenges they face due to limited resources and environmental fragility.

2.2. Methodologies for Assessing Bus Alternatives

Various methodologies in the literature exist for ranking alternative bus technologies from the economic perspective [15]. The methodologies span from performing cost-benefit analysis as in the work of [16–18], performing total cost of ownership (TCO) analysis as in the work of [19,20], or life cycle costing (LCC) methodologies [21]. It is worth mentioning that LCC provides a more comprehensive overview of the alternative costs compared to TCO. This is because LCC considers the entire life span of the asset, including all phases from design and manufacturing to disposal or recycling.

Another stream of research focused on assessing alternative bus technologies based on their energy demand and environmental impact. These studies evaluate the alternatives with regard to energy consumption and environmental effects across the energy carrier and material cycles [22]. The energy carrier cycle includes two phases: the well-to-tank (WTT) and the tank-to-wheel phase (TTW). The WTT stage represents the upstream phase of energy production, including material extraction, processing, and transportation till finally reaching the tank, as in the work of [6,20,22,23]. The TTW evaluates the downstream phase, which includes the actual energy consumption needed for bus operation as in the work of [24,25]. The energy and emissions considered in the bus cycle, however, include bus manufacturing, maintenance, and end-of-life processes (energy needed for bus disassembly and disposal) and can be extended to include battery manufacturing [5]. It was demonstrated that energy demand and greenhouse gas emissions throughout the entire life cycle are predominantly concentrated in the energy carrier cycle [6]. This highlights the need to focus on optimizing the energy carrier pathways to achieve better environmental and energy performance for electric bus technologies.

Ranking or evaluating alternative bus technologies cannot be solely based on economic, energy, or environmental efficiencies independently from each other. Therefore, Ref. [20] proposed an integrated index methodology comprising three indices to assess well-to-wheel energy use, global warming potential equivalent emissions, and the total cost of ownership. Additionally, other researchers have proposed a multi-criteria decision-making (MCDM) approach to evaluate and rank bus technologies by considering two or more performance measures as in the work of [21,26,27]. It is worth mentioning that the integrated index methodology simplifies the analysis by combining multiple parameters into a single metric, which may not adequately capture the trade-offs and conflicting criteria involved in complex decision-making processes. MCDM is more suited for making informed decisions involving multiple, often conflicting, criteria.

Table 1 summarizes the recently published research works that emerged in evaluating and assessing electric bus technologies. In particular, the first column presents the cited work; the second column indicates what electric bus technology was under study. The next three columns specify the scope of the study. In particular, it indicates whether the alternative buses were evaluated against the economic viability, energy consumption, and environmental impact, respectively. The last column indicates whether the study included any ranking of the considered alternatives.

Table 1. Boundaries of the studies of bus technologies.

Reference	EB Technology		Economic	Energy Carrier Cycle		Environmental	Ranking Methodology
	BEBs	FCEBs		WTT	TTW		
[28]	✓	✗	✓	✗	✓	✓	✗-simulation approach
[6]	✓	✓	✗	✓	✓	✓	✗
[29]	✓	✓	✗	✓	✓	✓	✗-LCA
[2]	✓	✓	✓	✓	✓	✓	✗
[23]	✓	✗	✓	✗	✓	✓	✓
[30]	✗	✓	✓	✓	✓	✓	Best-worst method and MARCOS
[17]	✓	✓	✓	✗	✗	✓	✗-CBA
[31]	✗	✓	✓	✗	✗	✗	✗-LCC
[32]	✗	✓	✗	✗	✗	✗	✗-LCC
[22]	✗	✗	✓	✓	✓	✓	✗
[27]	✓	✗	✗	✗	✗	✗	✓AHP-TOPSIS
[26]	✓	✗	✓	✗	✗	✓	✓
This study	✓	✓	✓	✓	✓	✓	✓-MCDM

3. Methodology

This study delves into the optimal strategy for decarbonizing public transportation by leveraging electric buses. Our focus lies in identifying the most efficient electric bus technology and its corresponding energy pathway. By analyzing economic viability, energy demand, and environmental impact, we aim to unlock the most sustainable and cost-effective path toward a zero-emission public transport system.

Therefore, in this section, we describe the overall methodology adopted in this research to answer the research question. First, we describe the energy carrier pathways considered to charge the alternative bus technologies. Then, we detail the methodologies needed to compute the economic viability, energy demand, and environmental performance of the considered alternatives. The alternatives considered are

- BEBs powered by electricity produced from natural gas.
- BEBs powered by electricity produced from solar energy.
- FCEBs powered by liquid or gaseous hydrogen generated from steam reforming of natural gas.
- FCEBs powered by liquid or gaseous hydrogen generated through water electrolysis using electricity generated from renewable sources.

In practice, energy supplied to charge bus batteries usually comes from a mix of sources. However, in the specific case considered in this paper, the grid is almost entirely powered by natural gas, making it a more straightforward scenario to evaluate BEBs charged solely by natural gas. In what follows, we shall consider charging BEBs directly from photovoltaic sources to compare the environmental impacts of renewable energy. One potential strategy to ensure charging from a specific source, like photovoltaics, is the use of battery swap technology. With this approach, pre-charged batteries, charged exclusively using solar energy or natural gas-powered systems, can be swapped with depleted batteries. This method also reduces downtime for charging and can improve the operational efficiency of BEBs.

Lastly, we describe the entropy-based weighting and TOPSIS alternative ranking to rank the bus technologies and the corresponding energy carrier pathways. To illustrate our methodology, we considered Qatar as a case study. However, the proposed methodology can be straightforwardly applied to any country or region.

The overall methodological framework is summarized in Figure 1.

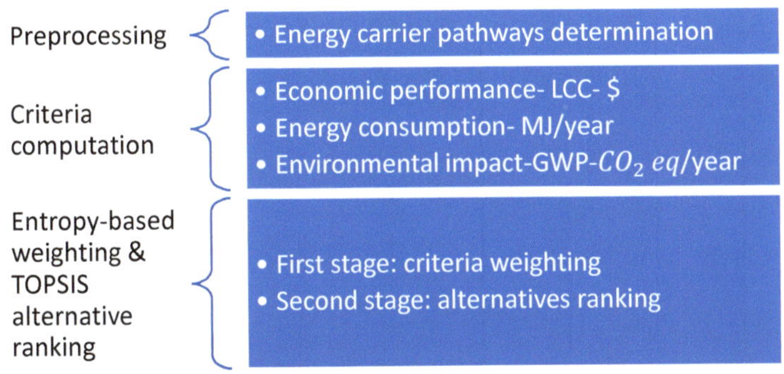

Figure 1. Methodological framework.

3.1. *Energy Carrier Pathways*

In the case of Qatar, we consider the electricity production from natural gas and photovoltaic panels for powering BEBs. For powering FCEBs, we consider hydrogen production through two methods: steam reforming of the natural gas while using carbon capture and storage (CSS) technology (blue hydrogen) or electrolysis of water using electricity from renewable sources (green hydrogen). The assumption of using CCS technology in our study was based on Qatar's commitment, as well as that of many other countries, to reduce CO_2 emissions as part of global efforts to combat climate change under international climate agreements, such as the Paris Agreement. For example, by 2022, Qatar had captured and sequestered nearly 5 million tons of CO_2 equivalent through CCS since 2019 [33].

The consideration of natural gas as a source of electricity and hydrogen production is justified as Qatar possesses substantial reserves of natural gas, which is utilized extensively for electricity generation. Renewable sources of electricity are considered for two reasons: first, according to Qatar National Vision 2030, Qatar aims to generate 20% of its electricity from renewable sources by 2030, as well as achieving a carbon-zero footprint by 2050. To support this goal, Qatar launched its first major solar energy plant, Al Kharsaah, in October 2022. This plant features over 1.8 million solar panels and is expected to produce approximately 2 TWh of electricity annually. The second reason is that steam reforming of natural gas remains the most cost-effective and commonly used method for hydrogen production [34].

In the following sections, we describe the electricity and hydrogen pathways.

3.1.1. Electricity Pathways

The electricity production sources considered in this study are natural gas and solar panels. To generate electricity from natural gas, a combined-cycle power plant uses both a gas and a steam turbine together while utilizing carbon capture technology that captures up to 60% of the emitted CO_2 emissions. Generating electricity from solar panels involves converting sunlight into electricity using photovoltaic cells, which are then converted into usable electricity and distributed to users.

3.1.2. Hydrogen Pathways

For hydrogen pathways, gaseous and liquid forms are considered. The blue hydrogen pathway uses natural gas steam reforming, while green hydrogen is generated from water electrolysis.

For blue hydrogen, the process begins when the majority of natural gas is introduced into the pre-heating system, where it undergoes a reforming reaction with steam under high pressure and temperature to generate H_2 and CO as per Equation (1).

$$CH_4 + H_2O \rightarrow CO + 3H_2 \tag{1}$$

Following this, CO reacts with steam using shift catalysts to generate more hydrogen in the water–gas shift reaction, as described in Equation (2).

$$CO + H_2O \rightarrow CO_2 + H_2 \tag{2}$$

The last step involves removing CO_2 and other impurities to obtain pure hydrogen.

For green hydrogen generation from water electrolysis, the process is presented in Equation (3) as in [35].

$$H_2O(l) \rightarrow H_2(g) + 1/2O_2(g) \tag{3}$$

Hydrogen is typically transported and delivered as either a gas or a liquid, using high-pressure gas hydrogen tube trailers for the gaseous form and liquid hydrogen tanks for the liquid form.

In its gaseous state, hydrogen must be compressed at least three times during transportation. For liquid hydrogen, however, gaseous hydrogen first needs to be liquefied. Once liquefied, the hydrogen can be transported to terminals and delivered to tank trucks before reaching the liquid hydrogen refueling station. Unlike gaseous hydrogen refueling stations, liquid hydrogen refueling stations use a liquid pump instead of a gaseous hydrogen compressor to increase the liquid hydrogen pressure to the desired level. The liquid hydrogen is then vaporized and transferred to high-pressure hydrogen storage tanks [35].

The calculation of the compressor energy demand is based on the design principles described in [36]. For liquid hydrogen, the energy requirements for liquefaction, pre-cooling, and pumping are determined based on previous studies [37,38].

Below, we refer to hydrogen production from natural gas as blue hydrogen, while hydrogen production generated from water electrolysis by electricity from renewable sources is referred to as green hydrogen.

3.2. *Criteria Determination: Computation of Economic, Energy, and Environmental Performance*

This section outlines the process for calculating economic viability, energy demand, and environmental impact, including the sources and references required for data collection for each of the analyzed perspectives.

3.2.1. Computation of Economic Aspect

To measure the total cost of each alternative, we use the LCC method. It considers all the relevant costs of each alternative bus technology over its entire lifespan, starting from the initial purchase price and including operating costs, maintenance, and disposal expenses. In this study, the LCC analysis covers the purchasing, operating, and maintenance costs for each alternative. The cost elements for the alternative bus technologies are summarized in Table 2.

To calculate the LCC, cost estimates for each cost element (including purchase, operation, and maintenance costs) are first determined for each bus alternative while considering the interest rates and the alternative life expectancy. Then, the present worth analysis is employed to determine, on an annual basis, the present values of each cost element. Then, for each alternative, the LCC is calculated by summing the present value of each cost element. The bus technologies and energy carrier pathways are compared, and the alternative with the lowest LCC is chosen. By doing so, the alternative with the lowest ownership cost based on a present worth analysis is selected.

Table 2. Cost components for each bus technology alternative.

Life Cycle Costs	BEB	FCEB
Purchase costs (K QAR/bus)	1700 [39]	2146 [39]
Maintenance costs (QAR/km)	0.420 [39]	1.30 [39]
Power costs (QAR/functional unit)	0.149/kWh [21]	9.540/kg green hydrogen [40]
	0.506/kWh [41]	4.880/kg blue hydrogen [42]

1 USD = 3.65 QAR (Qatari Riyal).

3.2.2. Computation of Energy Consumption.

Energy consumption is evaluated across two phases: WTT and TTW. Energy consumption in the WTT phase encompasses all the processes involved in producing and delivering fuel to the vehicle's tank. It includes feedstock extraction, processing, production, and delivery to the distribution centers and ultimately to the vehicle's tank. In TTW analysis, we analyze the energy needed to move the bus [43].

Annual TTW energy consumption (E_{TTW}) is computed as per the work of [44]. It is computed by multiplying the average traveled distance per year by the calorific value divided by the fuel economy, as indicated in Equation (4).

$$E_{TTW} = \frac{\text{distance per year} \times \text{calorific value}}{\text{fuel economy}} \tag{4}$$

WTT energy consumption (E_{WTT}) is computed by the multiplication of (E_{TTW}) by the fuel production rate as in Equation (5)

$$E_{WTT} = \text{fuel production rate} \times E_{TTW} \tag{5}$$

In the WTT energy consumption analysis, data regarding the fuel production rate and the calorific value of electricity from natural gas were obtained from [45] and [46], respectively. Data for the hydrogen production rate from various sources were adopted from [47] work.

In the TTW energy consumption analysis, data regarding the energy consumption of BEBs were gathered from a very recent study originating from operating 741 BEBs in Qatar during the World Cup in 2022 [24]. Data regarding the energy consumption due to the operation of FCEBs were gathered from [48] work.

Table 3 summarizes the collected data with the corresponding reference for calculating the energy demand in the WTT and TTW phases. In Table 3 and throughout the remainder of the document, we refer to electricity produced from natural gas and solar energy as E-NG and E-SE, respectively.

Table 3. Energy demand of various energy carrier pathways.

Energy Carrier	Calorific Value *	Fuel Economy **	Production Rate (MJ/MJ) ***
E-NG	3.598 MJ/kWh	0.889 km/kWh [24]	1.933
E-SE			0.20
Blue H_2	141.7 MJ/kg	9.756 km /kg H_2 [48]	G-1.437 L-1.605
Green H_2			G-1.550 L-1.718

Notes. *: how much energy is stored in one kg or in kWh. **: the rating of how far a vehicle can travel on a specific amount of fuel. ***: the ratio of the energy needed to deliver one megajoule of usable energy to the vehicle.

3.2.3. Computation of Environmental Efficiency

The environmental performance of the various energy carrier pathways is measured in terms of CO_2 equivalent emissions per year (kg CO_2-eq/year). This metric quantifies the total greenhouse gas emissions, including carbon dioxide (CO_2), methane (CH_4), and nitrous oxide (N_2O), expressed as the amount of CO_2 that would have the same global warming potential (GWP) over a specific period.

Electric buses operate solely on electric power, emitting no greenhouse gases or other pollutants during operation, resulting in effectively zero TTW emissions. Therefore, in calculating the environmental effect of electric bus alternatives, we measure the CO_2 equivalent emissions per year in the WTT phase. Table 4 summarizes the emissions generated per energy carrier during the WTT phase along with the data sources.

Table 4. Emissions of various energy carrier pathways.

Energy Carrier	Emissions (g CO_2-eq/MJ)	Reference
E-NG	122.218	[49]
E-SE	11.389	[45]
Blue H_2-G	54.410	[47]
Blue H_2-L	57.710	
Green H_2-G	9.010	
Green H_2-L	12.310	

3.3. Entropy-Based Weighting and TOPSIS Alternative Ranking

3.3.1. First Stage: Determining Criteria Weights Using the Entropy Method

In multi-criteria decision-making (MCDM), there are two main approaches for determining criteria weights: subjective and objective methods. The subjective method relies on the judgments of decision-makers, often using pairwise comparisons of both qualitative and quantitative data. However, obtaining reliable judgments from the decision-makers can be challenging sometimes, making it important to consider objective criteria weighting methods as well.

Objective criteria weighing is derived from an algorithm that does not incorporate the decision-maker's judgments. The most commonly used method is the Shannon entropy method [50]. The entropy method is based on the concept of entropy from information theory, which measures the amount of diversity within a dataset. The entropy method is performed as follows:

Step 1. Construct the initial decision matrix (DM) according to Equation (6)

$$DM_{ij} = \begin{bmatrix} a_{11} & a_{12} & \cdots & a_{1n} \\ a_{21} & a_{22} & \cdots & a_{2n} \\ \cdot & \cdot & \cdot & \cdot \\ \cdot & \cdot & \cdot & \cdot \\ \cdot & \cdot & \cdot & \cdot \\ a_{m1} & a_{m2} & \cdots & a_{mn} \end{bmatrix} \quad (6)$$

This initial matrix includes the performance data of each alternative across all criteria, a_{ij}, where $i = 1, 2, \ldots, m$ (m is the number of alternatives) and $j = 1, 2, \ldots, n$ (n is the number of criteria).

Step 2. Normalize the decision matrix

In this step, the performance values of each alternative against each criterion are normalized to ensure comparability. Therefore, for a given decision matrix, DM_{ij}, this step involves normalizing the matrix to transform all criteria values into a comparable scale as indicated in Equation (7):

$$r_{ij} = \frac{a_{ij}}{\sum_{i=1}^{m} a_{ij}} \quad (7)$$

where r_{ij} is the normalized value of a_{ij}.

Step 3. Calculate the entropy value

For each criterion, the entropy value, e_j, is calculated using the normalized values. This quantifies the dispersion of the performance data collected by each criterion as indicated in Equation (8)

$$e_j = -k \sum_{i=1}^{m} r_{ij} ln(r_{ij}) \tag{8}$$

where k is a constant defined as $k = \frac{1}{ln(m)}$ to ensure that e_j is within the range [0, 1].

Step 4: Determine the degree of diversification

The diversification for each criterion j is calculated to reflect the distinctiveness of the criterion's information. This is calculated as indicated in Equation (9)

$$d_j = 1 - e_j. \tag{9}$$

Step 5. Find the criteria weights

The weights w_j of each criterion j are derived by normalizing the degree of diversification values. Then, the criteria with higher dispersion and more unique information receive higher weights. The criteria weights are calculated as per Equation (10)

$$w_j = \frac{d_j}{\sum_{j=1}^{n} d_j} \tag{10}$$

3.3.2. Second Stage: Alternative Ranking Using TOPSIS

After determining the criteria weights, the alternatives should be ranked to identify the best one with the highest score. We adopt an approach for ranking the alternatives based on their relative closeness to an ideal solution, which was proposed by [51].

This widely recognized technique is commonly used for solving MCDM problems. It operates on the principle that the chosen alternative should be as close as possible to the ideal solution (which minimizes cost criteria and maximizes benefit criteria) and as far as possible from the worst solution.

The TOPSIS method involves the construction of the decision matrix, normalization, weighting of criteria, calculation of the best and worst solutions, and finding the distance measures used for ranking alternatives. Finally, the closeness coefficient to the ideal solution.

The next section offers a detailed explanation of the TOPSIS process.

The process works as follows:

Step 6. Construct the initial decision matrix (DM)

$$DM_{ij} = \begin{bmatrix} a_{11} & a_{12} & \cdots & a_{1n} \\ a_{21} & a_{22} & \cdots & a_{2n} \\ . & . & . & . \\ . & . & . & . \\ . & . & . & . \\ a_{m1} & a_{m2} & \cdots & a_{mn} \end{bmatrix} \tag{11}$$

This step involves constructing an initial DM by the decision maker as shown in Equation (11).

Step 7. Normalize the decision matrix

For the same decision matrix DM_{ij}, the normalized value r_{ij} of a_{ij} is obtained as per Equation (12):

$$r_{ij} = \frac{a_{ij}}{\sqrt{\sum_{i=1}^{m} a_{ij}^2}} \tag{12}$$

Step 8. Construct the weighted normalized decision matrix

The weighted normalized decision matrix, v_{ij}, is obtained by multiplying the weights of the criteria w_j (obtained in **Step 5**) by the normalized decision matrix, as indicated in Equation (13).

$$v_{ij} = w_j \times r_{ij} \tag{13}$$

Step 9. Determine the best and worst solutions

Identify the positive solution S^+ and the negative solution S^-, as per Equations (14) and (15), respectively.

$$S^+ = \{v_1^+, v_2^+, \ldots, v_n^+\}, \tag{14}$$

$$S^- = \{v_1^-, v_2^-, \ldots, v_n^-\}, \tag{15}$$

where

$$v_j^+ = \max(v_{ij}) \text{ for benefit criteria and } \min(v_{ij}) \text{ for cost criteria,}$$

$$v_j^- = \min(v_{ij}) \text{ for benefit criteria and } \max(v_{ij}) \text{ for cost criteria.}$$

Step 10: Calculate Euclidean distance from the ideal best and worst as per Equations (16) and (17), respectively.

$$S_i^+ = \sqrt{\sum_{j=1}^{n} (v_{ij} - v_j^+)^2}, (1 \le i \le m, 1 \le j \le n), \tag{16}$$

$$S_i^- = \sqrt{\sum_{j=1}^{n} (v_{ij} - v_j^-)^2}, (1 \le i \le m, 1 \le j \le n), \tag{17}$$

where S_i^+ is the separation from the best solution and S_i^- is the separation from the worst-solution.

Step 11. Calculate the closeness coefficient to the ideal solution

For each alternative, we calculate the closeness coefficient CC_i^*, after which the alternative with the highest value of this coefficient is considered the best choice. The CC_i^* is calculated as per Equation (18)

$$CC_i^* = \frac{S_i^-}{S_i^+ + S_i^-} \tag{18}$$

where $CC_i^* \in [0, 1]$. When the value gets closer to 1, it indicates an approach to the ideal solution; when it gets closer to 0, it indicates an approach to the non-ideal solution. In the last step of the TOPSIS methodology, alternatives are ranked based on the closeness value. The alternative with the highest closeness value receives the highest rank.

4. Results and Discussion

In this section, we first detail the economic, energy, and environmental performance of the considered alternative bus technologies with the associated energy carrier cycles. Then, we demonstrate the output of the ranking approach.

4.1. *Economic, Energy, and Environmental Performance*

4.1.1. Economic Performance

Figure 2 compares the total ownership costs of the different bus alternatives, calculated using LCC analysis. The analysis assumes an average annual distance of 79,200 km (240 km/day for 330 days/year) over a 10-year life expectancy as per the work of [24]. The LCC model factors in the purchase price, fuel costs (electricity from the grid, electricity from solar energy, blue hydrogen, and green hydrogen), maintenance, and salvage value, with an interest rate considered.

The overall findings indicate that the BEBs being charged from solar energy have the lowest cost, followed by BEBs being charged using the grid. This is followed by

FCEBs being charged with blue hydrogen, and finally, the FCEBs being charged with green hydrogen. The wide variation in fuel costs between different energy pathways significantly impacts the overall economic viability of these options. For example, BEBs charged from the grid incur a cost of 0.149 QAR/km, while the cost rises to 0.506 QAR/km when using solar energy. Similarly, the fuel costs for FCEBs vary greatly depending on the type of hydrogen, with blue hydrogen costing 4.88 QAR/kg and green hydrogen costing 9.54 QAR/kg.

While initial purchase costs for FCEBs are higher (2146 k QAR compared to 1700 k QAR for BEBs), the long-term operational costs, especially fuel, are the key determinants of overall economic performance. Furthermore, the minimal variation in maintenance costs across alternatives suggests that fuel consumption is the more critical factor to consider when evaluating total ownership costs.

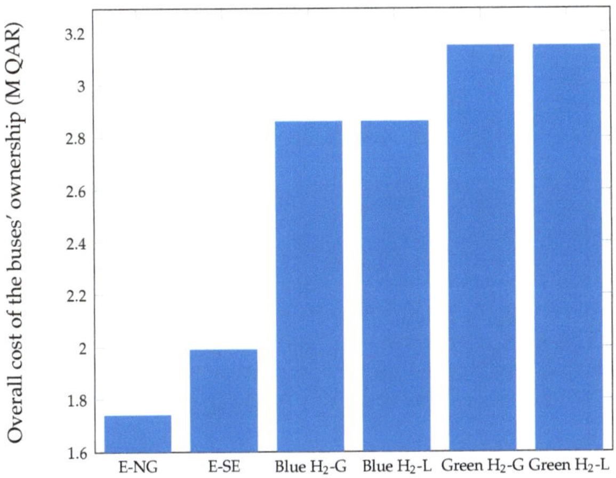

Figure 2. Economic performance of the considered alternatives.

To provide greater insight into the cost dynamics, we performed a sensitivity analysis on fuel price fluctuations. Our findings indicate that a 10% increase in hydrogen costs results in a proportional rise in total ownership costs for FCEBs, while the impact on BEBs is less sensitive due to lower and more stable electricity prices, particularly when charged from renewable sources. Moreover, future reductions in hydrogen production costs, particularly for green hydrogen, could shift the economic balance in favor of FCEBs.

4.1.2. Energy Consumption

The results are based on an annual distance of 79,200 km. The analysis assumes the use of 12-meter-long BEBs and FCEBs.

Figure 3 displays the WTW energy consumption for each alternative. The analysis includes both the WTT and TTW phases, with the WTT phase accounting for over 60% of the total energy consumption for most pathways, except for electricity from solar energy, where it represents only 18% of the total energy consumption.

Among the options, BEBs demonstrate the lowest WTW energy consumption, regardless of whether the electricity comes from the grid or solar energy. This is followed by blue hydrogen production (both liquid and gas) and finally by green hydrogen.

To provide a comparative understanding of the energy consumption of the alternatives, BEBs powered by grid electricity serve as a reference point. Solar-powered BEBs reduce energy consumption by 60% compared to grid-powered BEBs. On the other hand, FCEBs exhibit much higher energy demand, with consumption increasing by 198% to 233% relative to grid-powered BEBs. The process of liquefying hydrogen, especially green hydrogen, further exacerbates energy consumption, with liquid hydrogen consuming 10–12% more energy than gaseous hydrogen due to the liquefaction process.

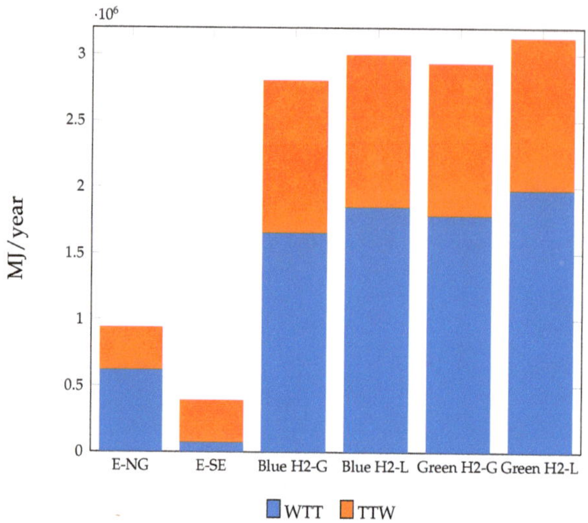

Figure 3. WTW energy consumption.

Examining the hydrogen pathway in more detail, liquid hydrogen (whether blue or green) requires relatively higher energy than gaseous hydrogen. Figure 3 shows that liquid hydrogen consumes 10% to 12% more energy than gaseous hydrogen, primarily because of the energy and emissions involved in the liquefaction process. Additionally, green hydrogen consumes 7% to 8% more energy due to the lower power generation efficiency from renewable sources.

During the WTT phase, charging BEBs from the grid increases energy consumption by about 167% to 219%. In contrast, charging BEBs using electricity from solar sources decreases energy consumption by about 90%.

The TTW energy consumption of BEBs and FCEBs strongly depends on operating conditions such as driving speed, elevation, and operating temperature [24,52]. Additionally, factors like HVAC systems, passenger loads, and auxiliary systems consume considerable energy [52,53]. The low energy consumption of BEBs is attributed to their highly efficient powertrain, which benefits from energy recovery during braking [24]. Given that the lower heating value of hydrogen is 141 MJ/kg H_2, the operational efficiency of FCEVs is less than half that of BEVs.

4.1.3. Global Warming Potential

Emissions associated with various energy carrier pathways vary significantly between these processes, as shown in Figure 4. While electricity and hydrogen production from natural gas contributes to increased GWP, green hydrogen is considered a renewable and environmentally friendly process, as it utilizes solar energy and produces hydrogen with significantly lower emissions.

To gain a comparable perspective of the results, we refer to the electricity from the grid as the reference case. A decrease of about 99% of emissions is observed when electricity is generated by solar energy. However, an increase of about 19% to 41% of the emissions is observed when charging FCEBs with blue hydrogen. A decrease of about 68–80% of the emissions is observed when charging FCEBs with green hydrogen.

Therefore, it can be concluded that in the absence of solar energy for electricity generation, green hydrogen is the preferred energy carrier among the alternatives considered.

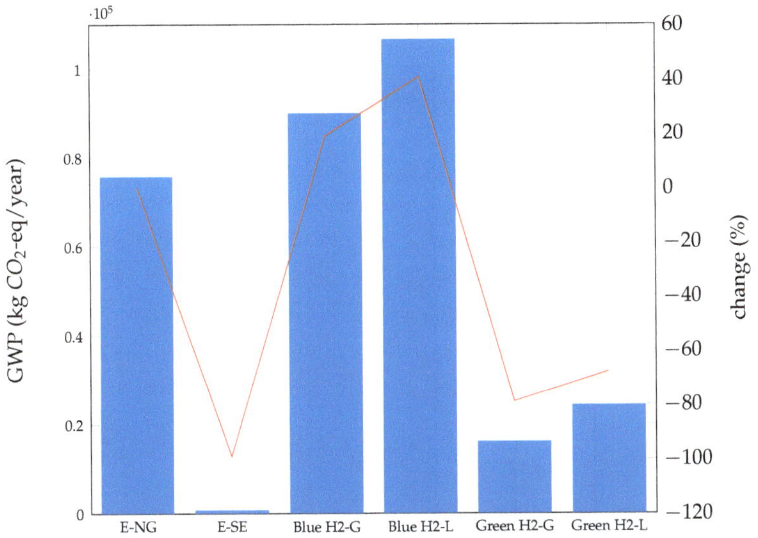

Figure 4. GWP (bar chart—left axis) and potential to change GWP (line plot—right axis) compared to the BEBs-NG baseline.

4.2. Results of the Ranking Approach

4.2.1. First Stage: Determine Criteria Weights Using Entropy Method

Step 1. Table 5 exhibits the matrix based on collected data from the literature, as shown previously.

Table 5. Initial decision matrix.

Energy Carrier	Economic ($)	Energy (MJ/year)	GWP (kg CO_2-eq/year)
E-NG	1,743,871	940,149	76
E-SE	1,991,439	391,061	1
Blue H_2-G	2,855,361	2,803,145	90
Blue H_2-L	2,855,361	2,997,166	107
Green H_2-G	3,150,620	2,933,033	16
Green H_2-L	3,150,620	3,127,054	24

Step 2. Using Equation (7), the normalized matrix is displayed in Table 6.

Table 6. Normalized decision matrix.

Energy Carrier	Economic ($)	Energy (MJ/year)	GWP (kg CO_2-eq/year)
E-NG	0.111	0.071	0.242
E-SE	0.126	0.030	0.003
Blue H_2-G	0.181	0.212	0.287
Blue H_2-L	0.181	0.227	0.340
Green H_2-G	0.200	0.222	0.051
Green H_2-L	0.200	0.237	0.078

Step 3. Using Equation (8), we calculate the entropy value.

Step 4. The degree to which each criterion is distinct is calculated using Equation (9).

Step 5. By normalizing the degree of diversity values using Equation (10) and employing the entropy method, we determine the criterion weights for economic factors, energy consumption, and GWP. The economic criteria received the lowest weight of 0.008, followed by energy consumption with a weight of 0.304, and GWP with the highest weight of 0.688. This exhibits that the economic factors exhibit the least variability, while GWP shows the most variability among the criteria considered.

The results of Steps 3–5 are shown in Table 7.

Table 7. Steps 3–5 for criteria weighting.

Energy Carrier	Economic ($)	Energy (MJ/year)	GWP (kg CO_2-eq/year)
E-NG	−0.244	−0.188	−0.343
E-SE	−0.262	−0.104	−0.015
Blue H_2-G	−0.310	−0.329	−0.358
Blue H_2-L	−0.310	−0.337	−0.367
Green H_2-G	−0.322	−0.334	−0.152
Green H_2-L	−0.322	−0.341	−0.198
sum	−1.768	−1.634	−1.434
Step 3	0.987	0.912	0.800
Step 4	0.013	0.088	0.200
Step 5 (Entropy)	0.043	0.293	0.663

4.2.2. Second Stage: Alternative Ranking Using TOPSIS

Step 6. We use the same decision matrix in Step 1.

Step 7. The normalized matrix using Equation (12) is shown in Table 8.

Table 8. Normalized decision matrix.

Energy Carrier	Economic ($)	Energy (MJ/year)	GWP (kg CO_2-eq/year)
E- NG	0.265	0.156	0.469
E- SE	0.303	0.065	0.005
Blue H_2-G	0.435	0.466	0.557
Blue H_2-L	0.435	0.498	0.661
Green H_2-G	0.480	0.487	0.100
Green H_2-L	0.480	0.519	0.151

Step 8. The weighted normalized decision matrix is calculated by multiplying the criterion weights obtained from **Step 5** by the normalized matrix obtained from the previous step. The normalized weighted matrix is shown in Table 9.

Table 9. Normalized weighted matrix.

Energy Carrier	Economic ($)	Energy (MJ/year)	GWP (kg CO_2-eq/year)
E-NG	0.002	0.044	0.334
E-SE	0.003	0.018	0.004
Blue H_2-G	0.004	0.130	0.397
Blue H_2-L	0.004	0.139	0.470
Green H_2-G	0.004	0.136	0.071
Green H_2-L	0.004	0.145	0.107

Step 9. Using Equations (14) and (15), the best and worst solution for each criterion is determined, respectively. It is noteworthy to mention that economic, energy, and environmental aspects are considered as cost criteria. This means the lower the value of the cost, i.e., the energy consumption, the more likely it is that GWP is the best solution. The best and the worst solutions for each criterion are shown in Table 10.

Table 10. Ideal best and worst solutions for each criterion.

Energy Carrier	Economic ($)	Energy (MJ/year)	GWP (kg CO_2-eq/year)
V+	0.002	0.018	0.004
V−	0.004	0.145	0.470

Step 10. After determining the best and worst solutions, we apply Equations (16) and (17) to measure the Euclidean distance to the positive ideal solution (shown in column 2 in Table 11) and to the negative solution (shown in column 3 in Table 11).

Table 11. Steps 9–10.

Energy Carrier	Si+	Si−	CC_i^*	Rank
E-NG	0.332	0.170	0.339	4
E-SE	0.000	0.484	0.999	1
Blue H_2-G	0.409	0.075	0.155	5
Blue H_2-L	0.482	0.006	0.012	6
Green H_2-G	0.136	0.400	0.746	2
Green H_2-L	0.164	0.363	0.689	3

Step 11. Using Equation (18), the closeness scores for the considered alternatives are determined (shown in column 4 in Table 11. Finally, we rank the alternatives based on their closeness score, as shown in the last column in Table 11.

Our analysis employed two primary methodologies: the entropy method and the TOPSIS methodology. Initially, the entropy method was used to objectively assign weights to three key criteria—economic viability, energy consumption, and environmental impact—across six different bus alternatives, each utilizing various energy pathways. Among these criteria, the economic performance exhibited the least variation among the alternatives, leading to the assignment of the lowest weight (0.008). Energy consumption was assigned a higher weight of 0.304, while the environmental impact, measured through GWP, received the highest weight of 0.663.

Subsequently, the TOPSIS methodology was applied to rank the bus alternatives based on these weighted criteria. The results, presented in Table 11, indicate that bus alternatives powered by green energy sources outperform others, regardless of the bus technology used, underscoring the significance of the energy source in powering electric vehicles. In scenarios where green energy is unavailable, BEBs powered by electricity from natural gas emerged as the most favorable option, followed by FCEBs powered by gaseous hydrogen and then those powered by liquid hydrogen.

The significant weighting assigned to environmental impact in this analysis reflects global priorities surrounding climate change and emissions reduction. While BEBs currently outperform FCEBs across most criteria, ongoing improvements in hydrogen production technologies could shift these rankings in the future. The analysis also highlights the critical role of energy sources in shaping the overall sustainability of public transport systems.

4.3. Policy Implications

The results show that the proposed method can effectively be applied in real-world scenarios, particularly for decision-making in the public transportation sector. By incorporating energy source variations and their respective impacts, our method can help policymakers and industry professionals evaluate the best alternatives for decarbonizing transportation systems.

The case study's findings suggest that bus alternatives powered by green energy sources perform best, and policies should prioritize developing and integrating renewable energy infrastructure. Governments could incentivize the adoption of green energy in public transportation systems by offering subsidies or tax breaks for using renewable energy.

For Qatar, specific policies should focus on energy diversification to decrease dependence on a single energy type and enhance the resilience of public transportation systems. Additionally, policies should aim to lower the carbon footprint of public transport by implementing stricter emission standards and directing funds toward projects that minimize GWP. These policy implications can help shape a more sustainable, efficient, and environmentally friendly public transportation system.

5. Conclusions and Future Work

Qatar's commitment to decarbonizing its public transportation necessitates a thorough evaluation of BEBs and FCEBs. To make an informed decision on the optimal bus technology and energy carrier pathway, we propose a comprehensive multi-modal decision-making framework. This framework assesses bus technologies and energy carrier pathways in terms of economic viability, energy consumption, and environmental impact, ensuring the selection of the most sustainable urban transportation mode.

The framework utilizes the Shannon entropy approach for criteria weighting, followed by ranking the alternative bus technologies to minimize the distance from the ideal best solution and maximize the distance to the worst solution. The entropy method is the most suitable approach for criteria weighting when making objective decisions with the presence of accurate data. This methodology can be employed in policy-making at the governmental level to rank and select the best alternative. Once an alternative is selected, necessary policy actions can be taken to support the favorable option. These actions may include infrastructure development for electricity and hydrogen recharging/refueling, incentives, and subsidies.

This study has employed several methodologies to come to an informed decision as follows: (1) Economic, energy demand, and environmental indicators were systematically evaluated for a selection of sustainable bus technologies and their corresponding energy carrier pathways, ensuring a holistic assessment of each alternative. (2) The evaluation scope of the alternatives was extended to include the energy carrier pathway to assess its contribution to the total energy demand and GWP. (3) The Shanon entropy methodology for criteria weighting and the TOPSIS methodology were integrated for for the ranking of alternatives in a systematic way to reach an informed alternative ranking.

The main findings of this study are as follows.

- Each bus technology and energy carrier pathway has its advantages and drawbacks when evaluated against the various criteria, making the multi-criteria decision-making approach essential for ranking alternatives with conflicting objectives. For instance, BEBs charged with the electricity grid (NG) exhibit the lowest costs but a relatively high GWP. While FCEBs charged by green hydrogen demonstrate the lowest GWP but incur the highest costs.
- Energy consumption and GWP are highly dependent on the energy carrier pathways. Therefore, this factor must be considered when selecting an alternative sustainable bus technology.
- BEBs charged by solar energy or the grid are preferred over all FCEBs in the LCC criterion due to their lower purchase and fuel costs compared to FCEBs and hydrogen production. With advancements in hydrogen production efficiency, FCEBs may become more favorable in the future.
- In terms of energy consumption, BEBs charged by solar energy or the grid are preferred over all FCEBs, regardless of the energy carrier pathways. The high energy consumption in blue and green hydrogen production is attributed to its less efficient process compared to electricity generation.
- Environmentally, BEBs charged with solar energy, followed by FCEBs charged with green hydrogen, are the most favorable alternatives. However, in the absence of renewable energy sources as an option, BEBs charged by the grid perform the best due to the carbon capture technology deployed in Qatar.

- Liquid hydrogen production requires more energy compared to gaseous hydrogen due to the need for multiple compression operations to achieve the liquid state.
- Using the integrated approach of entropy weighting and TOPSIS ranking methodology, BEBs powered by solar are the preferred alternative, while the liquid blue hydrogen for fueling FCEBs is the least preferred alternative.

It should be noted that although the proposed framework is applied to Qatar as a case, it is scalable and can be applied to any other country.

For future research, this study can be expanded to include a broader range of bus types, such as various configurations of battery technologies for electric buses and to encompass additional evaluation criteria. Potential criteria for further consideration could include social acceptance, operational reliability, and technological maturity. Moreover, the incorporation of experts' and decision-makers' judgments into the criteria weighting process is essential. This can be achieved by employing methodologies such as Fuzzy Entropy [54] and Fuzzy TOPSIS, as demonstrated in the works of [55,56].

Additionally, several specific areas warrant further investigation:

1. Life cycle assessment of batteries: Our assessment focuses primarily on the operational phase of electric buses. However, a more comprehensive approach would encompass the environmental impact of battery production, disposal, and potential recycling. Additionally, the replacement cost of batteries over the lifespan of the buses is a significant economic factor that deserves further consideration.
2. The maturity of green hydrogen technology: The reliance on green hydrogen as an energy source for hydrogen buses introduces another layer of complexity. Electrolysis technology for green hydrogen production is still undergoing development, and the current cost of electrolysis remains a hurdle. Future research should explore the impact of potential cost reductions in electrolysis on the overall viability of hydrogen buses.
3. Additional considerations: Beyond the limitations mentioned above, here are some other factors that could be explored in future studies:
   (a) Infrastructure Development: Whichever technology prevails, the success of electric or hydrogen buses hinges on the development of a robust charging or refueling infrastructure. The cost and time required to build this infrastructure need to be factored into the overall assessment.
   (b) Energy Mix Variations: The environmental impact of electric buses heavily depends on the source of the electricity used for charging. Studies should consider the impact of variations in the energy mix across different regions.
   (c) Safety Considerations: While hydrogen fuel cells offer promising clean energy potential, safety concerns regarding hydrogen storage and refueling infrastructure need to be addressed. Future studies should explore safety protocols and regulations for hydrogen bus operations to ensure public confidence. On the other hand, while lithium-ion batteries offer high energy density for electric buses, concerns persist regarding thermal runaway and fire risks in the case of battery failure. Future studies should explore advancements in battery safety technology, fire suppression systems for electric buses, and stricter regulations for battery manufacturing and disposal to mitigate these risks. Recent fire incidents involving electric buses highlight the need for prioritizing safety measures [12].
   (d) Reliability Considerations: The reliability of battery performance and range in electric buses, particularly in extreme weather conditions, is an important factor for public transportation operators. Further research should explore advancements in battery technology that can enhance reliability and minimize operational disruptions. Furthermore, while fuel cells offer a promising clean energy solution for hydrogen buses, their long-term durability and degradation characteristics require further investigation. Studies should explore real-world data on fuel cell performance over extended periods to assess their reliability for public transportation applications. Additionally, advancements in fuel cell

technology that enhance durability and minimize maintenance requirements are crucial for wider adoption.

(e) Social and Operational Factors: The adoption of new technologies often comes with social and operational challenges. Future studies could explore factors like public acceptance, driver training needs, and maintenance considerations for both electric and hydrogen buses.

By acknowledging these limitations and exploring additional considerations, future research can provide a more nuanced understanding of the most suitable approach for decarbonizing public transportation with electric or hydrogen buses.

Author Contributions: Conceptualization, A.F.E. and M.H.; methodology, A.F.E. and M.H.; software, A.F.E. and M.H.; validation, M.H. and A.S.; resources, A.F.E. and M.H.; data curation, A.F.E., M.H. and A.S.; writing—original draft preparation, A.F.E. and M.H.; writing—review and editing, A.F.E., M.H. and A.S. All authors have read and agreed to the published version of the manuscript.

Funding: This research received no external funding.

Institutional Review Board Statement: Not applicable.

Informed Consent Statement: Not applicable.

Data Availability Statement: Data are contained within the article.

Conflicts of Interest: The authors declare no conflicts of interest.

Abbreviations

The following abbreviations are used in this manuscript:

Nomenclature	
BEBs	Battery electric buses
DM	Decision matrix
E-NG	Electricity produced from natural gas
E-SE	Electricity produced from solar panels
E_{TTW}	Annual TTW energy consumption
E_{WTT}	WTT energy consumption
FCEBs	Fuel cell electric buses
LCC	Life cycle costing
MCDM	Multi-criteria decision-making
TCO	Total cost of ownership
TOPSIS	Technique for Order Preference by Similarity to the Ideal Solution
TTW	Tank-to-wheel
WTT	Well-to-tank
WTW	Well-to-wheel
Constants	
a_{ij}	Performance data of alternative i against each criterion j
Variables	
CC_i^*	Closeness coefficient for alternative i
e_j	Entropy value for criterion j
r_{ij}	Normalized value of a_{ij}
S_i^+	Euclidean distance for alternative i from the best solution
S_i^-	Euclidean distance for alternative i from the worst solution
v_{ij}	Weighted normalized decision matrix
w_j	Weight value for criterion j
Other symbols	
m	Number of alternatives
n	Number of criteria
S^+	Best solution
S^-	Worst solution

References

1. Transport Sector CO_2 Emissions by Mode in the Sustainable Development Scenario, 2000–2030—Charts—Data & Statistics. Available online: https://www.iea.org/data-and-statistics/charts/transport-sector-co2-emissions-by-mode-in-the-sustainable-development-scenario-2000-2030 (accessed on 31 July 2024).
2. Kuyumcu, A.M.; Bingül, B.; Akar, F.; Yıldız, A. Well-to-wheel carbon footprint and cost analysis of gasoline, diesel, hydrogen ICE, hybrid and fully electric city buses. *Energy* **2024**, *301*, 131685. [CrossRef]
3. Nanaki, E.A.; Koroneos, C.J. Climate change mitigation and deployment of electric vehicles in urban areas. *Renew. Energy* **2016**, *99*, 1153–1160. [CrossRef]
4. Rietmann, N.; Hügler, B.; Lieven, T. Forecasting the trajectory of electric vehicle sales and the consequences for worldwide CO_2 emissions. *J. Clean. Prod.* **2020**, *261*, 121038. [CrossRef]
5. Ellingsen, L.A.W.; Thorne, R.J.; Wind, J.; Figenbaum, E.; Romare, M.; Nordelöf, A. Corrigendum to "Life cycle assessment of battery electric buses" [Transp. Res. Part D: Transp. Environ. 112 (2022) 103498]. *Transp. Res. Part D Transp. Environ.* **2023**, *116*, 103628. [CrossRef]
6. Grazieschi, G.; Zubaryeva, A.; Sparber, W. Energy and greenhouse gases life cycle assessment of electric and hydrogen buses. A real-world case study in Bolzano Italy. *Energy Rep.* **2023**, *9*, 6295–6310. [CrossRef]
7. Xu, S.; Tian, X.; Wang, C.; Qin, Y.; Lin, X.; Zhu, J.; Sun, X.; Huang, T. A Novel Coordinated Control Strategy for Parallel Hybrid Electric Vehicles during Clutch Slipping Process. *Appl. Sci.* **2022**, *12*, 8317. [CrossRef]
8. Tian, X.; Cai, Y.; Sun, X.; Zhu, Z.; Wang, Y.; Xu, Y. Incorporating Driving Style Recognition Into MPC for Energy Management of Plug-In Hybrid Electric Buses. *IEEE Trans. Transp. Electrif.* **2023**, *9*, 169–181. [CrossRef]
9. Fuel Types of New Buses: Electric 12.7%, Diesel 67.3% Market Share Full-Year 2022. Available online: https://www.acea.auto/fuel-cv/fuel-types-of-new-buses-electric-12-7-diesel-67-3-market-share-full-year-2022/ (accessed on 31 July 2024).
10. Double Decker Fuel Cell Bus Fleet Launched in LONDON. 2024. Available online: https://www.sustainable-bus.com/fuel-cell-bus/double-decker-fuel-cell-bus-fleet-launched-in-london/ (accessed on 8 August 2024).
11. The Future Is Green. Available online: https://ovbureau.nl/en/themas-2030/the-future-is-green/ (accessed on 8 August 2024).
12. Agnellini, S.; Butturi, F.; Gimmelli, A.; Franchini, F. On the Reasons Behind e-Buses (and EVs) Fires and How to Decrease Risks (The Problem is Smaller Than One Might Think). 2022. Available online: https://www.sustainable-bus.com/news/reasons-why-electric-buses-vehicles-fire/ (accessed on 18 August 2024).
13. Planning a Transition to Electrification of Public Transit Systems: Learnings from the Bus Rapid Transit System of Metrobus in Mexico City 2022. 2022. Available online: https://www.apec.org/docs/default-source/publications/2022/10/planning-a-transition-to-electrification-of-public-transit-systems-learnings-from-the-bus-rapid-transit-system-of-metrobus-in-mexico-city/222_ewg_planning-a-transition-to-electrification-of-public-transit-systems.pdf?sfvrsn=587548d3_2 (accessed on 8 August 2024).
14. BBC Future. Climate Change: How China's Electric Vehicle Revolution Began with Buses. BBC Future 2023. Available online: https://www.bbc.com/future/article/20231206-climate-change-how-chinas-electric-vehicle-revolution-began-with-buses (accessed on 10 February 2024).
15. Borghetti, F.; Carra, M.; Besson, C.; Matarrese, E.; Maja, R.; Barabino, B. Evaluating alternative fuels for a bus fleet: An Italian case. *Transp. Policy* **2024**, *154*, 1–15. [CrossRef]
16. Malmgren, I. Quantifying the Societal Benefits of Electric Vehicles. *World Electr. Veh. J.* **2016**, *8*, 996–1007. [CrossRef]
17. Siddiqui, O.; Ishaq, H.; Khan, D.A.; Fazel, H. Social cost-benefit analysis of different types of buses for sustainable public transportation. *J. Clean. Prod.* **2024**, *438*, 140656. [CrossRef]
18. Razy-Yanuv, E.; Meron, N. Early transition to cleaner bus fleets: Benefits, costs, and policy evaluation of alternatives from a life cycle perspective. *Clean. Environ. Syst.* **2024**, *12*, 100172. [CrossRef]
19. Tong, F.; Hendrickson, C.; Biehler, A.; Jaramillo, P.; Seki, S. Life cycle ownership cost and environmental externality of alternative fuel options for transit buses. *Transp. Res. Part D Transp. Environ.* **2017**, *57*, 287–302. [CrossRef]
20. Muñoz, P.; Franceschini, E.A.; Levitan, D.; Rodriguez, C.R.; Humana, T.; Correa Perelmuter, G. Comparative analysis of cost, emissions and fuel consumption of diesel, natural gas, electric and hydrogen urban buses. *Energy Convers. Manag.* **2022**, *257*, 115412. [CrossRef]
21. Cihat Onat, N. How to compare sustainability impacts of alternative fuel Vehicles? *Transp. Res. Part D Transp. Environ.* **2022**, *102*, 103129. [CrossRef]
22. Chi, Y.; Xu, W.; Xiao, M.; Wang, Z.; Zhang, X.; Chen, Y. Fuel-cycle based environmental and economic assessment of hydrogen fuel cell vehicles in China. *Energy* **2023**, *282*, 128773. [CrossRef]
23. Elagouz, N.; Onat, N.C.; Kucukvar, M.; Ayvaz, B.; Kutty, A.A.; Osman Kusakci, A. Integrated modelling for sustainability assessment and decision making of alternative fuel buses. *Transp. Res. Part D Transp. Environ.* **2023**, *117*, 103656. [CrossRef]
24. Alarrouqi, R.A.; Bayhan, S.; Al-Fagih, L. Battery Electric Buses Operating in Qatar: A Comprehensive Energy Consumption Analysis. In Proceedings of the 2022 IEEE Transportation Electrification Conference & Expo (ITEC), Anaheim, CA, USA, 15–17 June 2022; pp. 1236–1241. [CrossRef]
25. Sun, S.; Batista, S.; Menéndez, M.; Wang, Y.; Zhang, S. Powering up urban mobility: A comparative study of energy efficiency in electric and diesel buses across various lane configurations. *Sustain. Cities Soc.* **2024**, *101*, 105086. [CrossRef]

26. Büyüközkan, G.; Feyzioğlu, O.; Göçer, F. Selection of sustainable urban transportation alternatives using an integrated intuitionistic fuzzy Choquet integral approach. *Transp. Res. Part D Transp. Environ.* **2018**, *58*, 186–207. [CrossRef]
27. Hamurcu, M.; Eren, T. Electric Bus Selection with Multicriteria Decision Analysis for Green Transportation. *Sustainability* **2020**, *12*, 2777. [CrossRef]
28. Dulce, F.; Murillo-Hoyos, J.; Caicedo, E. Comparative analysis of the performance, environmental impact, and costs of electric, combustion, and gas buses in an operating context of a mid-sized city of an emerging country. *Transp. Res. Interdiscip. Perspect.* **2024**, *25*, 101113. [CrossRef]
29. Lubecki, A.; Szczurowski, J.; Zarebska, K. A comparative environmental Life Cycle Assessment study of hydrogen fuel, electricity and diesel fuel for public buses. *Appl. Energy* **2023**, *350*, 121766. [CrossRef]
30. Pamucar, D.; Iordache, M.; Deveci, M.; Schitea, D.; Iordache, I. A new hybrid fuzzy multi-criteria decision methodology model for prioritizing the alternatives of the hydrogen bus development: A case study from Romania. *Int. J. Hydrogen Energy* **2021**, *46*, 29616–29637. [CrossRef]
31. Ayetor, G.; Mbonigaba, I.; Sunnu, A.K.; Nyantekyi-Kwakye, B. Impact of replacing ICE bus fleet with electric bus fleet in Africa: A lifetime assessment. *Energy* **2021**, *221*, 119852. [CrossRef]
32. Ally, J.; Pryor, T. Life cycle costing of diesel, natural gas, hybrid and hydrogen fuel cell bus systems: An Australian case study. *Energy Policy* **2016**, *94*, 285–294. [CrossRef]
33. Haider, U. Your Energy Transition Partner—Sustainability Report 2022. Available online: https://www.qatarenergy.qa/en/MediaCenter/Publications/QatarEnergy%202022%20Sustainability%20Report.pdf (accessed on 23 July 2024).
34. Hydrogen Production: How Is Hydrogen Made and What Is It? 2024. Available online: https://www.atlascopco.com/en-uk/compressors/air-compressor-blog/what-is-hydrogen-and-how-is-it-produced (accessed on 23 June 2024).
35. Gu, Y.; Chen, Q.; Xue, J.; Tang, Z.; Sun, Y.; Wu, Q. Comparative techno-economic study of solar energy integrated hydrogen supply pathways for hydrogen refueling stations in China. *Energy Convers. Manag.* **2020**, *223*, 113240. [CrossRef]
36. Yu, Y.; Jiang, P.; Sun, S. Compressor Engineering Manual 2012. Available online: https://www.irantpm.ir/wp-content/uploads/2014/12/Compressor-Handbook.pdf (accessed on 16 June 2024).
37. Elgowainy, A.; Reddi, K.; Lee, D.Y.; Rustagi, N.; Gupta, E. Techno-economic and thermodynamic analysis of pre-cooling systems at gaseous hydrogen refueling stations. *Int. J. Hydrogen Energy* **2017**, *42*, 29067–29079. [CrossRef]
38. Petitpas, G.; Aceves, S.M. Liquid hydrogen pump performance and durability testing through repeated cryogenic vessel filling to 700 bar. *Int. J. Hydrogen Energy* **2018**, *43*, 18403–18420. [CrossRef]
39. Borghetti, F.; Longo, M.; Bonera, M.; Libretti, M.; Somaschini, C.; Martinelli, V.; Medeghini, M.; Mazzoncini, R. Battery Electric Buses or Fuel Cell Electric Buses? A Decarbonization Case Study in the City of Brescia, Italy. *Infrastructures* **2023**, *8*, 178. [CrossRef]
40. Qatar Leads Countries That Produce Green Hydrogen Cheaper Than Blue. 2024. Available online: https://inspenet.com/en/noticias/qatar-leads-countries-that-produce-green-hydrogen-cheaper-than-blue/ (accessed on 31 July 2024).
41. KAHRAMAA: A Successful Financial Close of the Siraj-1 Solar Power Plant Project. 2020. Available online: https://www.km.qa/MediaCenter/pages/NewsDetails.aspx?ItemID=357 (accessed on 20 June 2024).
42. The Future of Blue Hydrogen in the Middle East. 2024. Available online: https://www.lexology.com/library/detail.aspx?g=6de2656e-e00b-4212-81db-3a76b5d68d29 (accessed on 31 July 2024).
43. Correa, G.; Muñoz, P.; Rodriguez, C. A comparative energy and environmental analysis of a diesel, hybrid, hydrogen and electric urban bus. *Energy* **2019**, *187*, 115906. [CrossRef]
44. Kosai, S.; Nakanishi, M.; Yamasue, E. Vehicle energy efficiency evaluation from well-to-wheel lifecycle perspective. *Transp. Res. Part D Transp. Environ.* **2018**, *65*, 355–367. [CrossRef]
45. Administration, U.E.I. How Much Coal, Natural Gas, or Petroleum Is Used to Generate a Kilowatthour of Electricity? 2024. Available online: https://www.eia.gov/tools/faqs/faq.php?id=667&t=6 (accessed on 23 June 2024).
46. Europe, C. Fuel Calorific Values. 2024. Available online: https://cngeurope.com/fuel-calorific-values/ (accessed on 23 June 2024).
47. Cluzel, C.; Baltac, S.; Nock, W.; Robinson, O.; Drake, W. Low Carbon Hydrogen Well-to-Tank Pathways Study. August 2021. Available online: https://www.zemo.org.uk/assets/reports/Zemo%20Low%20Carbon%20Hydrogen%20WTT%20Pathways%20-%20Summary%20(2).pdf (accessed on 14 June 2024).
48. Pederzoli, D.; Carnevali, C.; Genova, R.; Mazzucchelli, M.; Del Borghi, A.; Gallo, M.; Moreschi, L. Life cycle assessment of hydrogen-powered city buses in the High V.LO-City project: Integrating vehicle operation and refuelling infrastructure. *SN Appl. Sci.* **2022**, *4*, 57. [CrossRef]
49. Administration, U.E.I. How Much Carbon Dioxide Is Produced per Kilowatthour of U.S. Electricity Generation? 2024. Available online: https://www.energybot.com/energy-faq/how-much-carbon-dioxide-is-produced-per-kilowatthour-of-u-s-electricity-generation.html#:~:text=On%20average%200.855%20pounds%20of%20CO2%20emissions%20are%20produced%20per%20kWh%20generated (accessed on 16 June 2024).
50. Shannon, C.E. A mathematical theory of communication. *Bell Syst. Tech. J.* **1948**, *27*, 379–423. [CrossRef]
51. Hwang, C.L.; Yoon, K. Technique for Order of Preference by Similarity to Ideal Solution. In *Multiple Attribute Decision Making: Methods and Applications—A State-of-the-Art Survey*; Springer: Berlin/Heidelberg, Germany, 1981; pp. 128–136.
52. Zhou, B.; Wu, Y.; Zhou, B.; Wang, R.; Ke, W.; Zhang, S.; Hao, J. Real-world performance of battery electric buses and their life-cycle benefits with respect to energy consumption and carbon dioxide emissions. *Energy* **2016**, *96*, 603–613. [CrossRef]

53. Hjelkrem, O.A.; Lervåg, K.Y.; Babri, S.; Lu, C.; Södersten, C.J. A battery electric bus energy consumption model for strategic purposes: Validation of a proposed model structure with data from bus fleets in China and Norway. *Transp. Res. Part D Transp. Environ.* **2021**, *94*, 102804. [CrossRef]
54. Saraswat, S.; Digalwar, A.K. Evaluation of energy alternatives for sustainable development of energy sector in India: An integrated Shannon's entropy fuzzy multi-criteria decision approach. *Renew. Energy* **2021**, *171*, 58–74. [CrossRef]
55. Coimbra, E.C.L.; Borges, A.C.; Mounteer, A.H.; Rosa, A.P. Using wastewater treatment performance, biomass and physiological plant characteristics for selection of a floating macrophyte for phytoremediation of swine wastewater through the integrative Entropy-Fuzzy AHP-TOPSIS method. *J. Water Process. Eng.* **2023**, *53*, 103793. [CrossRef]
56. Fahmi, A.; Khan, A.; Abdeljawad, T.; Alqudah, M.A. Natural gas based on combined fuzzy TOPSIS technique and entropy. *Heliyon* **2024**, *10*, e23391. [CrossRef]

sustainability

Article

A Two-Stage Evaluation of China's New Energy Industrial Policy Package

Qiao Wang [1,*], Shiyun Chen [1] and Hongtao Yi [2]

[1] College of Public Administration, Huazhong Agricultural University, Wuhan 430070, China; sychen@webmail.hzau.edu.cn
[2] School of Public Administration and Policy, Renmin University of China, Beijing 100872, China; yihongtao@ruc.edu.cn
* Correspondence: wangqiao0318@163.com; Tel.: +86-18164023580

Abstract: Energy structural transformation plays a strategically important role in achieving the dual-carbon reduction goals. Among the various approaches to carbon reduction, the Chinese government regards the growth of the new energy industry as an essential means. Considering that the government policy support determines the long-term growth of the new energy industry, how to improve and optimize the policy support system has always been the core issue. Based on the fact that policy evaluation is a prerequisite, and the new energy industrial development requires the government to promote solutions in the form of a policy package rather than just individual policies, we investigate whether the implementation of the new energy industry policy package (NEIPP) is effective through an empirical case study of Shanghai. A two-stage evaluation method, which integrates the content analysis method (CAM) and synthetic control method (SCM), was used to empirically evaluate the actual effect of the NEIPP. At Stage One, four policy goals were summarized. SCM was used to identify the pure multi-effect of the NEIPP. The results showed that the NEIPP had a significant positive effect on green economic growth and industrial structure, while having a negative effect on carbon emissions. The NEIPP had no impact on the promotion of technological innovation. Several policy implications were drawn from this study.

Keywords: policy package; new energy industry; two-stage evaluation; content analysis method; synthetic control method; Shanghai

1. Introduction

Under the current grievous situation of the energy crisis, resource shortage, and the increasing enhancement of global warming, how to guarantee sustainable supplies of energy and reduce carbon emissions have become the two core policy issues worldwide [1–3]. Many countries have shifted their environmental policies to focus on promoting energy structural transformation and actively promoting changes in the way energy is produced and used to improve energy efficiency. Among all these government initiatives, developing new energy industries plays a strategically important role in addressing the global energy crisis and climate change, as it not only affects a country's energy structure transformation and energy conservation, but also further influences carbon emission reduction [4]. Accordingly, stimulating the development of the new energy industry to promote energy structural transformation has become a major policy issue.

As one of the largest and most economically significant developing countries in the world, China is currently entering a crucial phase of energy structural transformation since the year of 2020 when the government set the goal of achieving carbon peak by 2030 and carbon neutrality by 2060 (Dual-Carbon Target). Pursuing the Dual-Carbon Target creates an urgent demand for the green-oriented energy transition, the Chinese government regards the growth of the new energy industry, including the solar energy industry, the hydrogen energy industry, and so on [5], as an essential means of energy structure transformation and carbon emission reduction. Through making use of various policy instruments to promote the utilization and development of new energy, the Chinese new energy industry has achieved booming growth in the past few decades [6]. Admittedly, the Chinese government's vigorous development in the new energy industry can solve the national energy puzzle and achieve sustainable economic growth. While limited by factors including enormous capital investment, advanced technology barriers, and pervasive financing risk in the early stages of new energy industrial development, its long-term growth depends heavily on financial, technological, and commercial support from the government. Consequently, how to improve and optimize the policy support system for the new energy industry has always been the focus of academic and policy discussions. It is recognized that policy evaluation is a prerequisite for the optimization of the existing policy system and for the improvement of the effective policy supply. Therefore, there is a pressing need to conduct a comprehensive assessment of the existing new energy policies under the background of carbon peaking and carbon neutrality.

In practice, the new energy industry is characterized by its complexity and high degree of systemic interdependence, which objectively require the government to promote new energy industrial development in the form of a policy package rather than isolated individual policies. Although the term policy package has not been explicitly used in the area of the new energy industry, scholars have employed similar concepts, such as policy mixes, policy group, and policy combination [7,8], to convey the same meaning. Building on previous research, we introduce the concept of a policy package in our analysis of the new energy industry, defining it as a combination of policies implemented in the way that is synergetic and complementary to each other in a given region, which shares a common conceptual output, has a similar orientation, and serves the same strategic objective. Compared to a specific or single policy, a policy package involves various policy texts issued by different levels of government and relevant functional departments; a complete policy package is usually composed of the main policy and a series of sub-policies. Each sub-policy focuses on different directions and contents under the value orientation of the main policy, thus forming a vertical pyramid structure. Additionally, the policy package has specific characteristics, which include the multidimensional and hierarchical objective structure [9]. For instance, when the local government aims to pursue the development of the new energy industry, it always considers regional green economy growth as a long-term goal. This complexity makes it more challenging to analyze, especially to evaluate the new energy industrial policy package (NEIPP).

Although some scholars have drawn attention to the study of new energy industry policies' effectiveness in China [10], in the extant policy evaluation research, most of these emerging studies have mainly focused on using some policy evaluation method to evaluate the influence of new energy policy on singular objectives. There are very few studies that discuss the evaluation of new energy industrial policy packages; especially, empirical studies that exactly assess the actual multi-effectiveness of the NEIPP are extremely insufficient. Given that the use of policy packages has been theoretically proven to be significant during the new energy industry development and that policy packages have the specific characteristics of a multidimensional and hierarchical objective structure, a key question arises: how can scientific methods be used to evaluate their diverse and multidimensional impacts? Therefore, in this exploratory study, we attempt to introduce a two-stage evaluation method that integrates the content analysis method and synthetic control method to empirically evaluate the actual effect of the new energy industrial policy package. In the first stage, we attempt to summarize the diversified policy goals through deconstructing the various new energy industrial policy texts based on the method of content analysis, then decompose the abstract and general strategic objective into observable and measurable specific goals. In the second stage, we take the counterfactual policy evaluation method, namely the synthetic control method, to synthesize virtual control groups on each dimension of specific goals separately and then compare the differences in policy effectiveness between the treatment group and the virtual control group.

2. Literature Review

The existing literature has broadly established a consensus that the development of the new energy industry is highly dependent on government policy support. Accordingly, they have extensively and deeply studied the new energy industry policy, which provides a solid theoretical foundation for our analysis of the NEIPP. Considering that we mainly focus on evaluating the actual effectiveness of the NEIPP, this paper chiefly relates to the literature on policy packages and new energy policy evaluation.

2.1. Literature Review on Policy Packages

Since Rizwanul Islam pioneered the concept of the policy package as early as 1980 in his study of public food distribution in Bangladesh [11], scholars have become increasingly aware that overemphasizing a single policy and ignoring linkages between policies can lead to biased and one-sided policy analysis [12–15]. Accordingly, more attention has been paid to the study of policy packages [16–18]. Among these related research studies, a full understanding of the conceptualization is crucial. Actually, the previous literature has basically reached a consensus on the connotation of the term policy package, and most have regarded it as a set of related policies that are designed and implemented together in order to create a comprehensive and coordinated approach to achieve certain governmental goals [19–21]. In general, the policy package is approximately akin to the comprehensive treatment philosophy of Chinese traditional medicine; even though there are differences in policy instruments and policy components, all have similar guiding principles and a common strategic goal. Through bundling these related policies together, local governments can ensure that they are addressing all aspects of the specific goal in a cohesive and effective manner [18,22]. Compared to a single policy, policy packages are often more effective in coping with realistic and complex issues. Recently, policy packages have been widely used in complicated areas, and studies have explored their application in areas, such as technological innovation [23–25], social security [26,27], targeted poverty alleviation, and rural social assistance [28].

2.2. Literature Review on New Energy Policy Evaluation

To better understand the current research trends, this paper further categorizes the existing evaluations of new energy industry policy evaluation research from the dimensions of policy subjects and objects. One type of policy evaluation can be conducted from the perspective of policy contents. Scholars have evaluated the effectiveness of new energy policy from the perspective of the policy itself, including the following: (i) Evaluations based on the inherent attributes of the policy texts, where most scholars describe and analyze policy texts from aspects such as annual trend changes, issuing institutions, text form, and completeness, using these standards to assess the intensity of new energy industry policies [29,30]; (ii) Deconstructing the content of new energy policies from the perspective of policy tools and evaluating the government's preferences and policy tool selection during the new energy industry development [31–34]. When evaluating new energy industry policy, scholars have always selected the various policy texts as the first material for analysis. Factually, deconstructing policy texts can better help scholars to gain insights into the content, structure, and objectives of policy packages from the inherent normativity of the policy texts since they offer objective evidence recording government intentions, policy objectives, and implementation processes. Therefore, the current academic analysis of new energy industry policy texts can provide certain references for us to explore the multidimensional effects of new energy industrial policies based on the policy text content.

The second type of policy evaluation can be conducted from the perspective of policy objects, mainly paying attention to investigating their actual influence on promoting the development of a new energy industry and regional economic growth. In these evaluations, many scholars used the new energy policy texts as independent variables and the policy results, such as carbon emission reduction [2,35,36], carbon capture technology utilization [37–39], and green economic growth [40–42], as dependent variables, further constructing corresponding econometric models and empirically exploring the relationship between the policy intervention variables and policy outcome variables. Among these studies, the policy objects' changes before and after policy implementation are the focus of scholars' research. Although scientifically rigorous methods of policy effect evaluation can help us to obtain an objective understanding of the new energy industry policy, evaluation of the single new energy industrial policy's effects on its single policy goal often overlook the multidimensional and diversified influence of new energy policy packages.

From the above analysis, current scholars have conducted in-depth and beneficial explorations of new energy policies, which provide a certain degree of literature support for our study. However, the existing literature mainly focused on one specific new energy policy and evaluated its effect on a single policy goal rather than putting all new energy-related policies into the same evaluation framework and investigating the multidimensional and diversified influence of a new energy policy package. Compared with the previous studies, we intend to overcome the limitations of the existing literature in the following aspects: (1) This paper introduces the policy package into the new energy industry area and focuses on empirically exploring the actual effect of the policy package. In the extant literature, the only few existing studies on the NEIPP discussed a similar concept—policy mixes—but only investigated the influence of the mix of feed-in tariffs, carbon emissions trading, and R&D subsidies. We expand the policy mix, and all policies related to developing the new energy industry are included in the policy package. (2) Prior studies have always used counterfactual policy evaluation methods to investigate the policy's actual effect, while these methods are only suitable for the evaluation of a single policy goal. Considering that the NEIPP has the characteristic of a multidimensional and hierarchical objective structure, we attempt to introduce a two-stage evaluation method that integrates the content analysis method and synthetic control method to evaluate the actual effect of the NEIPP empirically from the perspective of multidimensional goals.

3. Methodology for Evaluating the Effect of the NEIPP

As outlined in the introduction part, we adopt a two-stage evaluation method integrating the content analysis method (CAM) and synthetic control method (SCM) to empirically evaluate the actual effect of the NEIPP. We use this method mainly based on the two following considerations: Firstly, a policy package consists of a set of interrelated policies issued by different governmental departments and has the salient features of a multidimensional and hierarchical objective structure that is distinguished from a single policy. These unique features lead the evaluation of policy packages to face the challenge of how to identify multidimensional goals from the various new energy policy texts. Thus, in the first stage of our evaluation, we introduce CAM to clarify the overall and specific objectives of the policy package to be evaluated, and translate the target language into measurable indicators. Secondly, identifying the net policy effect has always been the core issue of policy evaluation. Although some estimation methods, such as DID, PSM, and RD, have advantages in identifying the policy impact, they require multiple untreated and treated units as a prerequisite. Considering that we chose Shanghai as our single evaluation example, which does not meet the requirement of having multiple treated units, SCM is selected as the empirical methodology, because it is a critical methodology widely used to assess policy outcomes with very few treated units. Therefore, in the second stage of our evaluation, we use SCM to identify the multi-effect of the NEIPP. The specific steps of our two-stage evaluation method are as follows.

3.1. Text Analysis Based on the NEIPP

Current policy evaluation mainly focuses on assessing the degree to which policy objectives have been achieved; thus, clarifying the policy objectives is the first step to empirically evaluating the NEIPP. As policy objectives represent the intentions and expectations of policymakers and implementers, they are explicitly articulated in various policy documents. We begin by applying CAM to clarify both the overall and specific policy objectives through interpreting the content, structure, language, etc., of the policy texts. Then, we need to translate the target language into measurable indicators, namely decompose the policy objectives into specific, observable, comparable, and calculable variables, which serve as an outcome variable for the next counterfactual evaluation of the NEIPP. Figure 1 illustrates our refinement process of the NEIPP.

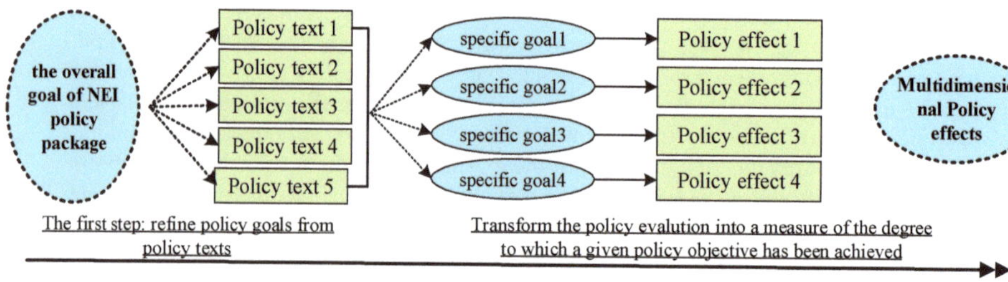

Figure 1. Refinement process of the NEIPP.

3.1.1. The First Step of Text Analysis: Policy Text Collection and Coding

Collecting related policy documents is the first step for our evaluation. In this study, the selection of new energy industry policy should adhere to the following three conditions: Firstly, the NEIPP should have a clearly defined overall objective, typically the strategic goal for new energy development set out by the Central Committee of the Party. Secondly, all policies should be formal and legitimate. Lastly, the title or content of the policy should explicitly correspond to the overall objective of the NEIPP. After collecting all the related new energy industry policies, we have filtered the policy texts through seeking advice from some related experts. Then, we use the method of policy text analysis to encode the relevant policy text, according to the Grounded Theory Approach [43] Generally, the steps for policy coding are as follows:

(1) Open coding: using concise phrases as labels to conceptualize each clause in the policy text, and using the representative key terms in the policy text as much as possible to abstract the policy text into a set of concepts.
(2) Axial coding: classifying and integrating the concepts generated by open coding to form main categories. If the policy text has a clear objective concept, we mark it directly as the main category. If the policy text has concepts related to policy tools and measures, we then extract and aggregate the policy objectives from them to form the main category.
(3) Selective coding: confirming the core category of the NEIPP, namely, the overall objective of the NEIPP. Further, we adjust and optimize the relationship between the different concepts, main categories, and core categories, so that the overall framework has the maximum coverage and explanatory power.

3.1.2. The Second Step of Text Analysis: Establishing Evaluation Index System

Policy text coding provides us with some specific policy objectives, which serve as the outcome variables for the next SCM evaluation. Given that the prerequisite for SCM evaluation is that the outcome variable is measurable and observable, the next crucial step is to further translate the specific policy objectives into observable quantitative indicators and build up a well-defined and hierarchical objective system. The objective system includes three levels of indicators: the overall objective as the first-level indicator, the specific goal as the second-level indicator, and the third-level indicator as the representative policy object directed by the specific policy objective.

3.2. SCM to Identify the Multi-Effect of the NEIPP

In the second stage of our evaluation, we use SCM to identify the multi-effect of the NEIPP. As we all know, SCM has gained popularity in recent years due to its ability to address the challenges of traditional methods, such as selection bias and unobserved confounding variables, and it is particularly useful when a randomized control trial is not feasible or ethical. In contrast with the traditional DID method, the key idea behind SCM is to create a synthetic control group that closely resembles the treated group before the intervention occurred. By combining information from multiple control units, we create a weighted average that closely matches the characteristics of the treated unit; further comparing the outcome of the treated unit with the synthetic control group, we can estimate the causal effect of the treatment. According to the above analysis and some related studies [44], the basic principle of SCM is as follows:

Suppose we have a treated unit (such as a region that implemented a specific policy) and multiple control units (regions that did not implement the policy). The goal is to evaluate the impact of this policy on the treated unit. The policy effect formula is $\alpha_{it} = Y_{it}(1) - Y_{it}(0)$, where 0 represents the region that did not implement the policy, and 1 represents the region that implemented the policy. Accordingly, $Y_{it}(1)$ represents the potential outcome variable of region i in period t implementing the new energy policy package, while $Y_{it}(0)$ represents the potential outcome variable of region i in period t not implementing the policy package. Therefore, the counterfactual outcome variable $Y_{it}(0)$ is constructed for calculation as follows:

$$Y_{i,t} = Y_{i,t}^{N} + D_{i,t} \tag{1}$$

$$Y_{i,t}^{N} = \theta_t Z_i + \lambda_t \mu_i + \varepsilon_{i,t} \tag{2}$$

In Equations (1) and (2), for the policy implementation unit, $i = 0$; for the control group (not affected by policy intervention), $i = 1, 2, \ldots, j$; the interval before policy implementation is $t = 1, 2, \ldots T_0$; and the interval after policy implementation is $t = T_0 + 1, T_0 + 2, \ldots, T$. $Y_{i,t}$ is the actual result. $Y_{i,t}^{N}$ is the expected result of the treatment group (i.e., the region affected by the policy intervention) in the absence of policy intervention. $\varepsilon_{i,t}$ is the disturbance term that varies with individuals and time. θ_t refers to the $(1 \times F)$ dimensional vector of observed common factors. Z_i is the $(F \times 1)$ dimensional vector of observed factor loadings. λ_t is the $(R \times 1)$ dimensional vector of unobserved common factors. μ_i is the $(R \times 1)$ dimensional vector of unobserved factor loadings. $\varepsilon_{i,t}$ is the short-term shock that cannot be observed in each region, with a mean of 0.

The SCM aims to find a set of weights w_i^* such that for any $i \in \{1, 2, \ldots, J\}$, the following approximate relationship holds:

$$Y_{0,t} \approx \sum_{j=1}^{J} w_j^* Y_{j,t} (t \leq T_0) \tag{3}$$

This means that before implementing the policy ($t \leq T0$), a counterfactual synthetic group similar to the treatment group can be constructed by weighting the predictive variables of each sample in the control group. The observed result $Y_{0,t}$ of this synthetic group can be approximated by the weighted result of other groups.

$$Z_{0,t} \approx \sum_{j=1}^{J} w_j^* Z_j \tag{4}$$

This means that the selected weight w_j^* also needs the covariate Z_0 of the treatment group to be approximated by the weighted sum of the covariate Z_j of the control group.

An important step is to find a better set of weights w_j through the constrained optimization problem:

$$\{\hat{w}_1, \hat{w}_2, \ldots, \hat{w}_j\} = \operatorname{argmin}\left\| X_0 - \sum_{j=1}^{J} w_j X_j \right\| \quad (5)$$

There exists a region of $w_j \geq 0$ satisfyingj $\in \partial\{1, 2, \ldots, J\}$ and $\sum_{j=1}^{J} w_j = 1$.

3.3. Robustness Test on the Multi-Effect of the NEIPP

To ensure the robustness and reliability of our SCM evaluation results, we apply the permutation test method, which is widely used in SCM research to conduct a robustness test. The permutation test is a quasi-rank test method similar to the rank test in statistics, with the aim of calculating the probability of other provincial units in the control group exhibiting the same policy effects as the treatment group provincial units. The judgment basis of the ranking test is the distribution difference of the prediction error curves of the treatment group provinces and other provinces. The further the curve is from the gathering place of most treatment group curves, the better the robustness test result is represented.

4. Empirical Evaluation on the NEIPP of Shanghai

4.1. Research Area

We selected Shanghai as our empirical case for evaluating the NEIPP for several reasons: Firstly, as the largest and most developed city in China, Shanghai has faced significant challenges brought by excessively traditional energy use and has been under considerable pressure to transform its energy structure. Thus, there is an extremely urgent need for Shanghai to promote new energy industrial development. Furthermore, Shanghai is a prominent city in China in the field of new energy development with a large market demand, a complete industrial chain, strong innovation capabilities, and a supportive policy environment for new energy industrial development. As a result, Shanghai boasts a wide range of new energy policies that encompass new energy vehicles, new energy industrial parks, and new energy technology innovation, etc. These policies form a comprehensive and cohesive package with the strategic goal of developing new energy as a burgeoning industry. More importantly, Shanghai has undertaken a number of pilot projects related to the development of new energy issued and organized by the central government, such as the private purchase subsidy pilot policy for new energy vehicles and the new energy demonstration city project, among others. To implement central government strategies in the new energy industry, local governments are more active in providing policy support.

4.2. Policy Collection and Text Coding Based on Shanghai's NEIPP

4.2.1. Policy Text Collection

In the history of China's new energy industrial development, 2010 is a landmark year. That year, the Emerging Energy Industry Development Plan from 2011 to 2020 was formulated by the China National Energy Administration, which was the first national strategic plan for large-scale development of the new energy industry over the past decade. Meanwhile, the Notice on Carrying out Subsidy Pilots for the Private Purchase of New Energy Vehicles was jointly issued by various ministries of the government in 2010, which launched a subsidy pilot for the private purchase of new energy vehicles in cities such as Shanghai, Hangzhou, Hefei, Shenzhen, and other cities. These policies not only marked the formal establishment of Shanghai's new energy industrialization policy objectives, but a large number of related supporting policies were also introduced and rapidly developed in the following year. From the perspective of policy changes, this stage shows obvious policy breakpoint characteristics, providing a clear time division node for counterfactual inference. Therefore, we take 2010 as the observation starting point, use keywords such as new energy, emerging strategic industry, photovoltaic, solar energy, etc., and collect relevant policies, regulations, and normative documents from databases, such as the Peking University Law

Database, Laws and Regulations Database of the People's Republic of China, Law Star, and the official website of the local government of Shanghai. After initially collecting 143 policy texts, through careful review and screening, duplicate and non-related texts were excluded, and finally, 57 policy texts closely related to the new energy industry policy were determined, providing a basis for further research. Figure 2 shows the process of NEIPP text collection.

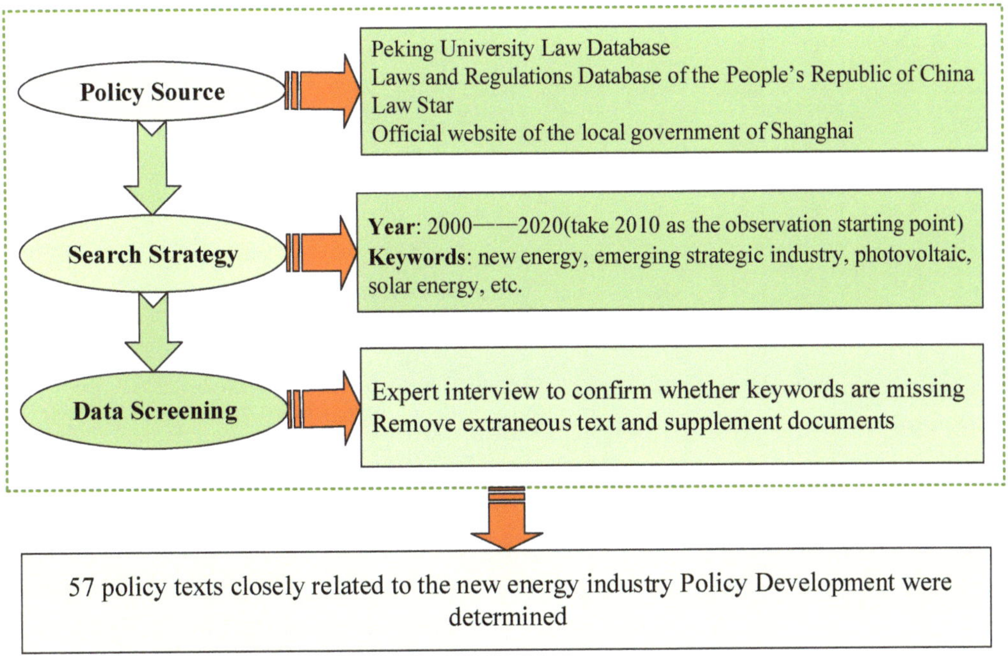

Figure 2. Process of NEI policy text collection.

4.2.2. Policy Text Encoding

This paper utilizes Nvivo 11Pro and Python 3.9 software to assist in the encoding of policy texts. During the open coding stage, we extracted key words from the 57 policy texts, initially refining and summarizing 27 concepts, mainly about the instruments and measures adopted by the government to promote the new energy industrial development. Then, coming to the stage of axial coding, we abstracted these 27 concepts into government objectives in new energy industry development based on the logical relationship between government action and objectives. Four main categories were obtained at this stage, namely, realizing energy saving and emission reduction, promoting industrial restructuring and upgrading, improving regional technological innovation, and achieving sustainable urban economic development. In the last stage of selective coding, we mainly merge and adjust the relationships between concepts, main categories, and core categories according to the principle of genus relationship and maximum difference. The final result is shown in Table 1.

Table 1. Results of policy text encoding.

Core Category	Main Category	Concept
New Energy Industry development	Industrial Structure Optimization	❖ Promote the cross-border integration of energy with transportation, finance, and other industries. ❖ Focus on building a strategic emerging industry and leading industry development system, centered around three major industries, following the '9 + X' model. Promote the transformation, demonstration, and application of scientific and technological achievements from innovative enterprises. ❖ Promote the low-carbon transformation of the industrial structure, aiming to achieve the 'five transformations' in industrial development.
	Sustainable Economic Development	❖ Develop clean and low-carbon new energy industries such as new energy vehicles, wind power, photovoltaics, etc. ❖ Improve energy utilization efficiency and safety. ❖ Reduce energy consumption intensity and carbon emission intensity. ❖ Promote energy structure optimization and transformation and upgrading. ❖ Build a green transportation energy system. ❖ Create an international energy innovation center. ❖ Support the integrated development of the Yangtze River Delta and promote international cooperation. Fully implement energy efficiency benchmarking in major energy-consuming sectors, key industries, and for main energy-consuming products.
	Energy Saving and Emission Reduction	❖ Accelerate the green upgrading of the energy industry and actively promote the development of new energy such as photovoltaics, wind power, hydrogen energy, etc. ❖ Promote the low-carbon transformation of industry. ❖ Encourage green development in urban and rural construction, scaling up the promotion of ultra-low energy consumption buildings and energy-saving renovations of existing buildings. ❖ Build a comprehensive green transportation system, vigorously promoting the new energy transformation of terminal transportation tools. ❖ Improve the energy efficiency of the circular economy industry and carry out the construction of waste material recycling system demonstration cities. ❖ Increase the intensity of scientific and technological innovation and accelerate the basic research and frontier technology layout of carbon neutrality. ❖ Consolidate and enhance the carbon sink capacity and steadily promote the plan of the "Thousand Parks". ❖ Hold the Shanghai International Carbon Neutrality Technology, Product, and Achievement Expo and guide the whole population to participate in low-carbon actions.

Table 1. *Cont.*

Core Category	Main Category	Concept
New Energy Industry development	Green Technological Innovation	❖ Support the research and development of new energy independently innovative products and the construction of industrial bases. ❖ Accelerate the breakthrough of power battery technology and lay out the research and development and industrialization of solid-state batteries and other next-generation products. ❖ Support enterprises to continuously carry out low-carbon/zero-carbon/negative-carbon basic research. ❖ Promote the application of energy-saving new processes, new technologies, new equipment, and new products. ❖ Strengthen support for projects and product technologies in energy-saving environmental protection, new energy, low-carbon transportation, green low-carbon building, and carbon capture and utilization. ❖ Focus on the actual situation of the city's industrial green low-carbon development and regularly publish the green technology catalogue.

4.3. Measurement of Objective Variables

In the above steps, we conclude and extract four specific policy objectives from 57 policy texts. In the following steps, we need to translate the specific policy objectives into observable quantitative indicators. Following the practice adopted in existing studies, we operationalize these variables, such as urban economic sustainable development, industry transformation and upgrade, carbon emission, and technological innovation. Specifically, the variables are as follows:

- Sustainable Economic Development. We constructed a comprehensive index system, naming this indicator Green Total Factor Productivity (GTFP), to measure Shanghai's economic sustainable development. This index system includes input variables and desirable output variables as well as undesirable output variables, and the specific indicators are as follows: the number of employees in the secondary and tertiary industries (labor input); the gross fixed capital stock of each city, which is measured by the perpetual inventory method (capital input) and the total amount of electricity usage of each city (energy input). The output variables contain not only desirable but also undesirable variables. As for the desirable output, we mainly select economic output measured by GDP with a constant price in 2000 and environmental output measured by industrial waste. We adopt the SBM-DDF method to calculate GTFP in China for each provincial unit. Generally, GTFP reflects the production efficiency of each provincial unit while considering environmental factors and is an important indicator for measuring the sustainable development dimension of new energy industry policies.
- Energy Saving and Emission Reduction. We calculate the carbon emissions of each provincial unit based on the IPCC method, using the total consumption of ten types of energy. Carbon emissions reflect the environmental burden of each provincial unit and are an important indicator to measure the energy-saving and emission-reduction dimension of new energy industrial policies.
- Industrial Structure Optimization. We use the ratio of the output value of tertiary industry to secondary industry to measure the industrial structure of each provincial unit. The industrial structure reflects the economic development level and structural transformation degree of each provincial unit. It is an important indicator to measure the industrial structure upgrade of new energy industry policies.
- Green Technological Innovation. We use the number of green invention patent authorizations of each provincial unit to measure its technological innovation ability. It can well reflect the innovative activities of each provincial unit in the field of new energy,

and thus it is an important indicator to measure the technological innovation of new energy industry policies.

To construct a synthetic control group, this article also selects some important factors that may affect local new energy industrialization as predictive control variables, including urbanization level, foreign direct investment, government intervention, transportation infrastructure, social consumption level, and labor level. These variables may have a direct impact on the result variables, representing the early endowments of different regions. Table 2 shows the descriptive statistics of each variable.

Table 2. Statistical description of the selected variables.

Variable Type	Variable Description	Variable Name	Statistical Indicator	Mean	Standard Deviation
Outcome Variables	Optimizing Industrial Structure	Industrial Structure	Tertiary Industry Output/Secondary Industry Output	0.967	0.376
	Sustainable Development	GTFP	GTFP	1.306	0.263
	Energy Saving and Emission Reduction	Per Capita Carbon Emissions	Carbon Emissions (100,000 tons)/Population	9.086	8.123
	Technological Innovation	Technological Innovation	Number of Green Patent Authorizations	2386.141	5300.620
Predictive Control Variables	Regional Social Development Endowment	Urbanization Level	Urban Population/Total Population	0.488	0.142
	Regional Industrial Resource Endowment	Industrialization Level	Industrial Added Value/Regional GDP	0.350	0.082
	Regional Ability to Attract Foreign Investment	Foreign Direct Investment	(Total Amount of Foreign Direct Investment ×Exchange Rate of US Dollar to RMB)/Regional GDP	0.022	0.019
	Degree of Government Intervention	Government Intervention	Fiscal Expenditure/Regional GDP	0.210	0.100
	Regional Transportation Infrastructure Level	Transportation Infrastructure	Natural Logarithm of Highway Mileage	11.491	0.829
	Regional Social Resource Endowment	Social Consumption Level	Total Retail Sales of Social Consumer Goods/GDP	0.358	0.060
	Regional Logistics Transportation Capacity	Logistics Transportation Capacity	Natural Logarithm of Total Freight Volume	11.266	0.929
	Labor Resource Endowment	Labor Level	Natural Logarithm of Employment	7.614	0.807

4.4. Sample Selection and Data Collection

The subject of this study is the new energy industrial policy in Shanghai, mainly empirically examining the effects of this policy in promoting technological innovation, sustainable development, energy saving, and emission reduction and optimizing industrial structure. To evaluate the causal effect of the policy package, this study uses the SCM for cross-time zone analysis, constructs a synthetic Shanghai as a control group, and compares it with the real Shanghai where the policy is implemented. The sample period of this article is $[1, T] = [2000, 2020]$, where the pre-policy intervention period is $[1, T_0 - 1] = [2000, 2009]$, and the post-policy intervention period is $[T_0, T] = [2010, 2020]$.

In the process of constructing the synthetic control group, the following principles should be met: firstly, selecting provincial units with similar economic, social, and technological characteristics to Shanghai before the implementation of the policy as the candidate control group to ensure the comparability of the synthetic control group; secondly, excluding provincial units that implemented similar new energy industry policies in the post-policy implementation period to avoid the interference effect of the policy. Finally, provincial units with missing or incomplete data were excluded to ensure the effectiveness of the synthetic control group. This article excludes Hong Kong, Macao, Taiwan, Tibet, and other regions because the statistical data for these regions are seriously missing. After these steps, the control group pool used by this article to synthetically control Shanghai includes the remaining 26 provincial units nationwide. Based on the principle of objectivity and authority, data for these provincial units were primarily collected from the Provincial Statistical Yearbooks, China Science and Technology Statistical Yearbook, China Labor Statistical Yearbook, city government website, NEA official website, and so on. For those missing data, we use the mean interpolation method to estimate and supplement the gaps.

The results of the fitted means of the predicted control variables compared with the real Shanghai are shown in Table 3. From the table, it can be seen that the difference between the real Shanghai and the synthetic Shanghai is small at the overall level. This indicates that the latter better fits the characteristics of the former before 2015. This further suggests that the synthetic Shanghai, weighted by the corresponding provinces, can be used to assess the policy effects of the real Shanghai on the dimensions for each of the outcome variables included in this study.

Table 3. Comparison of the fitted mean values of the control variables predicted by the synthetic Shanghai and the real Shanghai.

Predictive Control Variables	Treated Shanghai	Synthetic Shanghai			
		Economic Sustainable Development	Energy Saving and Emission Reduction	Industrial Structure Optimization	Green Technological Innovation
Urbanization Level	0.5335	0.468	0.479	0.454	0.444
Industrialization Level	0.428	0.392	0.405	0.398	0.309
Foreign Direct Investment	0.061	0.032	0.039	0.038	0.053
Government Intervention	0.209	0.233	0.0225	0.231	0.227
Transportation Infrastructure	9.287	10.862	11.015	10.866	10.808
Social Consumption Level	0.417	0.379	0.373	0.38	0.382
Transportation Capacity	11.439	11.182	11.353	11.058	10.811
Labor Level	7.091	7.039	7.12	6.893	7.021

5. Empirical Test and Analysis

5.1. Empirical Test Results

This paper selects Shanghai as the research sample and 26 other provinces as the control group. The SCM method is used to evaluate the pure policy effect of the NEIPP. We respectively compare the constructed counterfactual control group synthetic Shanghai with the treatment group real Shanghai in four aspects: the effect of the NEIPP on green technological innovation, economic sustainable development, industrial structure optimization, energy saving, and emission reduction.

5.1.1. The Impact of the NEIPP on Green Technological Innovation

The effect of Shanghai's NEIPP in promoting technological innovation can be observed by comparing the constructed counterfactual control group synthetic Shanghai with the treatment group real Shanghai. When the number of green patent authorizations is used as the outcome variable, Zhejiang (0.359), Jiangsu (0.062), Hainan (0.563), and Ningxia (0.017) constitute the synthetic Shanghai. The result shows that before the implementation of

Shanghai's new energy policy, the performance of the real Shanghai and synthetic Shanghai in terms of technological innovation was basically consistent, indicating that the synthetic Shanghai can effectively replicate the development trajectory of the real Shanghai before the policy implementation. However, after the implementation of the new energy policy, the two results still maintained a basically consistent trend, as shown in Figure 3. As of 2020, the number of green patent authorizations in the real Shanghai is only slightly higher than in the synthetic Shanghai, indicating that the effect of Shanghai's new energy industry policy in promoting technological innovation has not met expectations. This may be due to the fact that the strength and scope of Shanghai's policy in supporting technological innovation are not comprehensive enough. According to the Shanghai Energy Development Twelfth Five-Year Plan, the policy mainly focuses on new energy vehicles, wind energy, solar energy, and other fields, with less support for other new energy fields such as biomass energy, geothermal energy, and marine energy, which may cause Shanghai's new energy industry's technological innovation ability and level to be relatively limited, making it challenging to establish a diversified technological innovation system, which could affect overall development and competitiveness. In addition, Shanghai faces higher innovation costs as one of China's economic and technological innovation centers. The costs of labor, land, equipment, etc., are higher than in other provinces, putting pressure on innovation activities. At the same time, Shanghai pays more attention to the quality of innovation rather than quantity. New energy industry policies may be more inclined to support high-level innovation projects rather than many innovation projects, which may make it difficult to reflect policy effects in terms of quantity. In conclusion, although the policy's effect is not as significant as expected, it is worth noting that the actual Shanghai outperforms the synthetic Shanghai slightly. This suggests that while the policy's impact may not be substantial in terms of quantity, it may have contributed to a slight edge in the quality of technological innovation in Shanghai's new energy industry. This improvement highlights the policy's potential in promoting a sustainable and innovative new energy sector in Shanghai.

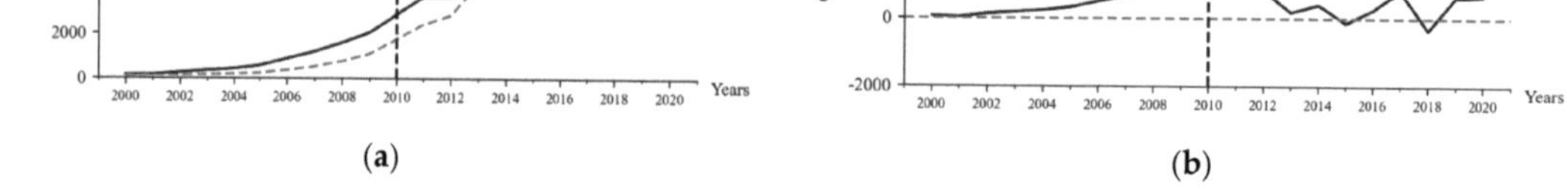

Figure 3. Impact of the NEIPP on the real Shanghai's technological innovation in contrast with the synthetic Shanghai: (**a**) Comparison of green technology innovation between real Shanghai and synthetic Shanghai: (**b**) Mean value gap between real Shanghai and synthetic Shanghai in green technology innovation.

5.1.2. The Impact of the NEIPP on Economic Sustainable Development

To study the impact of Shanghai's NEIPP on economic sustainable development, the real Shanghai can be used as the treatment group, and the counterfactual synthetic Shanghai composed of Zhejiang (0.508), Hainan (0.128), and Ningxia (0.364) can be used as the control group for comparison. When the green total factor productivity (GTFP) was used as the outcome variable, as shown in Figure 4, before the implementation of Shanghai's NEIPP, there was a certain gap between the real Shanghai and the synthetic

Shanghai in terms of economic sustainable development, indicating that the synthetic Shanghai did not fully simulate the development trend of the real Shanghai before the policy implementation, but the trends of the two were basically consistent, and there was a certain reliability of trend prediction. However, after the policy implementation, the GTFP of the real Shanghai was significantly higher than that of the synthetic Shanghai, indicating that the policy had a significant positive effect, which persisted throughout the observation period. As shown in Figure 4 and Table 4, the difference between the real Shanghai and the synthetic Shanghai in GTFP in 2020 was 2.169, indicating that the policy effect was significant. From the whole observation period, the difference showed an upward trend year by year, indicating that the real Shanghai's NEIPP policy had a significant policy effect on improving GTFP. This result may be caused by two reasons: Firstly, Shanghai's NEIPP covers multiple fields such as new energy vehicles, wind energy, solar energy, etc., providing comprehensive support and guarantees for the development of the new energy industry, promoting the scale expansion and technological progress of the new energy industry, as well as improving the resource utilization efficiency and environmental friendliness of the new energy industry. Secondly, Shanghai's NEIPP has strong innovation and foresight, introducing market mechanisms and incentives, stimulating the innovation vitality and competitiveness of new energy enterprises, and enhancing the core competitiveness and economic sustainable development ability of the new energy industry. For example, Shanghai's NEIPP encourages new energy enterprises to cooperate with scientific research institutions and universities, through strengthening technology research and development and thus improving the quality and performance of new energy products.

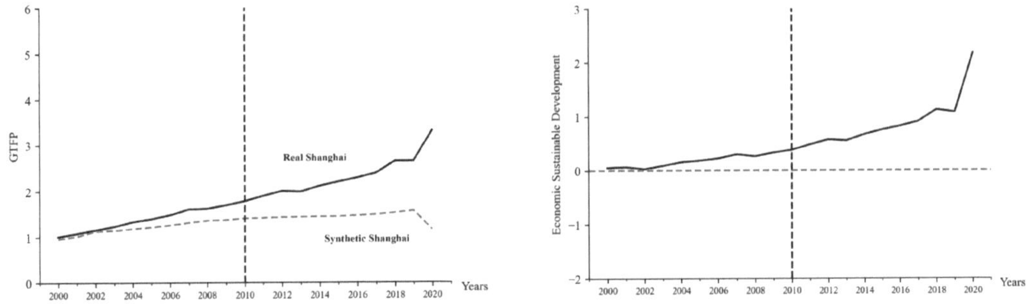

(a) (b)

Figure 4. Impact of the NEIPP on the real Shanghai's eEconomic sSustainable dDevelopment in contrast with the synthetic Shanghai: (**a**) Comparison of economic sustainable development between real Shanghai and synthetic Shanghai; (**b**) Mean value gap between real Shanghai and synthetic Shanghai in economic sustainable development.

Table 4. Policy effect.

Year	The Difference between "Treated Shanghai" and "Synthetic Shanghai"			
	Economic Sustainable Development	Energy Saving and Emission Reduction	Industrial Structure Optimization	Green Technological Innovation
2000	0.051	2.562	0.237	62.470
2001	0.064	4.034	0.222	36.635
2002	0.027	4.180	0.245	132.634
2003	0.087	1.083	0.209	192.478
2004	0.153	1.188	0.230	240.769
2005	0.180	0.702	0.205	335.053

Table 4. *Cont.*

Year	The Difference between "Treated Shanghai" and "Synthetic Shanghai"			
	Economic Sustainable Development	Energy Saving and Emission Reduction	Industrial Structure Optimization	Green Technological Innovation
2006	0.220	−0.247	0.255	517.143
2007	0.293	−0.993	0.357	666.131
2008	0.258	−1.112	0.446	827.276
2009	0.324	−1.950	0.644	960.814
2010	0.375	−2.628	0.538	1091.963
2011	0.480	−4.966	0.600	1216.673
2012	0.570	−5.765	0.722	837.017
2013	0.550	−5.242	0.837	188.590
2014	0.666	−6.339	0.998	411.986
2015	0.753	−6.308	1.198	−124.599
2016	0.821	−6.079	1.366	256.535
2017	0.905	−8.702	1.227	832.846
2018	1.122	−10.946	1.267	−331.226
2019	1.077	−12.505	1.572	635.927
2020	2.169	−15.539	1.561	680.087

5.1.3. The Impact of the NEIPP on Energy Saving and Emission Reduction

To investigate Shanghai's performance in energy conservation and emission reduction under the impact of the NEIPP, we use Shanghai as the intervention group and a synthetic Shanghai composed of Zhejiang (0.244), Liaoning (0.283), Ningxia (0.358), and Fujian (0.114) as the baseline group to conduct difference analysis. When carbon emissions were used as the outcome variable, as shown in Figure 5, before the implementation of Shanghai's NEIPP, there was a certain gap between the real Shanghai and the synthetic Shanghai in terms of carbon emissions, indicating that the synthetic Shanghai did not fully simulate the development trend of the real Shanghai before the policy implementation. Still, the trends of the two were basically the same, and there was a certain reliability of trend prediction. However, after the policy implementation, the carbon emissions of the real Shanghai were significantly lower than those of the synthetic Shanghai, indicating that the policy had a significant negative effect, which persisted throughout the observation period. As shown in Table 4, the difference between the real Shanghai and the synthetic Shanghai in per capita carbon emissions in 2020 was −15.539 (100,000 tons), indicating that the policy effect was significant. From the whole observation period, the difference showed an upward trend year by year, indicating that the real Shanghai's NEIPP had a significant policy effect on reducing carbon emissions. Two key factors may explain this outcome: Firstly, Shanghai's NEIPP has strong innovation and foresight. During the 11th Five-Year Plan period, Shanghai proposed a strategy to prioritize research and development and the industrialization of new energy vehicles. The formulation and implementation of these policies push Shanghai to surpass other provinces in the synthetic control group in terms of the development level and speed of the new energy industry, thus effectively reducing carbon emissions; the effect was already evident in 2010, and the advantages brought by the policy gradually expanded after 2010. Secondly, Shanghai's regional and first-mover advantages in the new energy industry are also important factors for reducing carbon emissions. Shanghai has a number of leading new energy enterprises and research institutions in China, forming a relatively complete new energy industrial chain and innovation system, providing solid support for the development of the new energy industry. At the same time, as China's largest economic center and international metropolis, Shanghai's energy demand and consumption levels are high, providing great potential and space for the application and promotion of new energy. These regional and first-mover advantages make the development effect and the energy saving and emission reduction effect of the new energy industry in Shanghai significantly better than in other provinces in the synthetic control group. Shanghai's new energy vehicle ownership has reached the first in the country, and it also has comprehensive measures in terms of new energy vehicle charging facilities, subsidy policies, and license issuance.

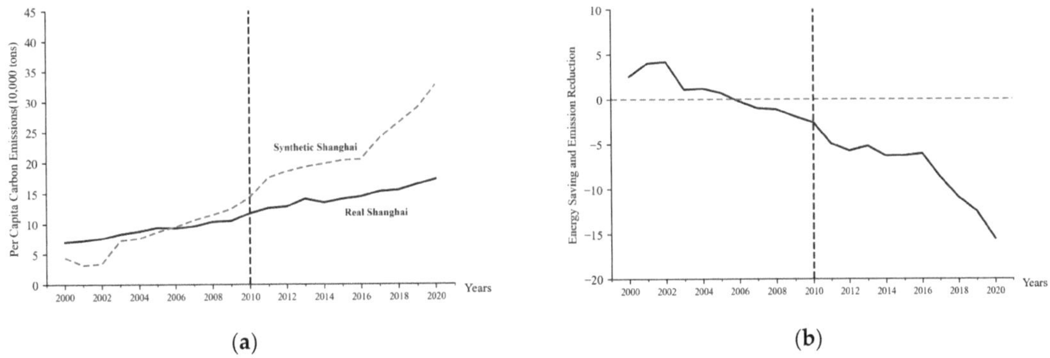

Figure 5. Impact of the NEIPP on the real Shanghai's eEnergy sSaving and eEmission rReduction in contrast with the synthetic Shanghai: (**a**) Comparison of energy saving and emission reduction between real Shanghai and synthetic Shanghai; (**b**) Mean value gap between real Shanghai and synthetic Shanghai in energy saving and emission reduction.

5.1.4. The Impact of the NEIPP on Industrial Structure

To evaluate the impact of Shanghai's NEIPP on the optimization of industrial structure, we compare the real Shanghai with the synthetic Shanghai composed of weighted averages of Zhejiang (0.209), Fujian (0.324), and Ningxia (0.467). We use the ratio of value added in the tertiary industry to that in the secondary industry as an indicator. As shown in Figure 6, before the implementation of the policy, the real Shanghai and the synthetic Shanghai closely mirrored each other on this indicator. Although the synthetic Shanghai did not fully reproduce the historical trajectory of the real Shanghai, their trends were highly similar, so it can be considered that the synthetic Shanghai has a high predictive power. After the implementation of the policy, the difference between the real Shanghai and the synthetic Shanghai gradually increased. As shown in Figure 6 and Table 4, in 2020, the real Shanghai was 1.19 points higher than the synthetic Shanghai on this indicator, indicating that the policy had a significant effect. From the entire observation period, the difference showed an upward trend, indicating that Shanghai's NEPP was important in optimizing the industrial structure. This result can be explained from the following two aspects: Firstly, Shanghai's NEIPP effectively promoted the development of the tertiary industry, especially the innovation and industrialization of new energy vehicles, wind energy, solar energy, and other fields, enhancing the value-added and competitiveness of the tertiary industry, thereby increasing the proportion of the tertiary industry in the total value-added. Second, Shanghai's NEIPP was conducive to the transformation and upgrading of the secondary industry, encouraging the traditional energy industry to shift to the new energy industry, reducing the energy consumption and environmental pollution of the secondary industry, enhancing the sustainability and efficiency of the secondary industry, and thereby reducing the proportion of the secondary industry in the total value-added.

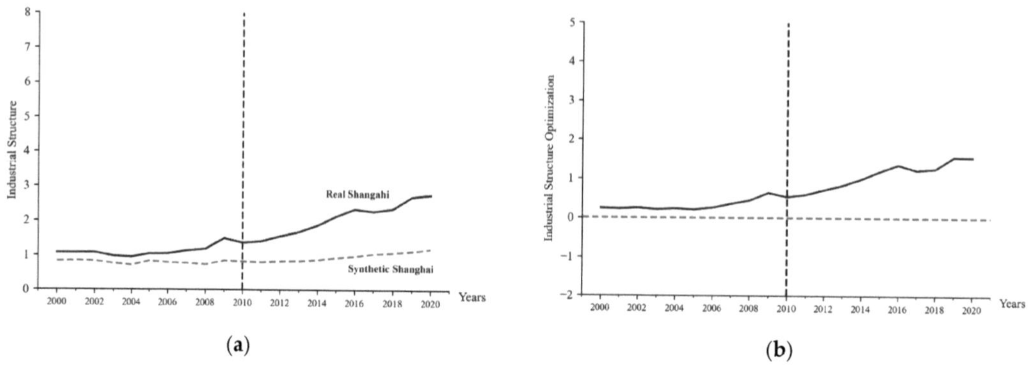

Figure 6. Impact of the NEIPP on the real Shanghai's Industrial Structure Optimization in contrast with the synthetic Shanghai: (**a**) Comparison of industrial structure optimization between real Shanghai and synthetic Shanghai; (**b**) Mean value gap between real Shanghai and synthetic Shanghai in industrial structure organization.

5.2. Robustness Test

Although after considering the estimated bias probably caused by sample self-section and endogenous problems, we have chosen the SCM method to obtain an accurate estimated result. This paper further introduces the method of Abadie et al. (2010) to further verify the robustness of the policy evaluation results [45]. The basic idea of this method is to assume that all control group provinces implement the new energy industrial policy with Shanghai in the same year and then use the SCM to construct a synthetic control unit for each control group province, calculating the difference between the synthetic control unit and the actual province as the effect of the hypothetical policy. Then, from the goal dimensions of optimizing the industrial structure, promoting sustainable development, promoting technological innovation, and energy saving and emission reduction, we compare the actual policy effects of Shanghai and the hypothetical policy effects of the control group provinces. If, on a certain goal dimension, the actual effect of Shanghai is significantly higher than the hypothetical effect of the control group, it indicates that the implementation of the new energy industry policy in Shanghai is effective on this goal dimension, and the conclusion is robust. Conversely, if on a certain goal dimension, the difference in effects between the two is not significant, or the actual effect of Shanghai is lower than the hypothetical effect of the control group, there are two possibilities: first, the implementation of the new energy industry policy is ineffective in this goal dimension, and the conclusion is robust; second, Shanghai's new energy industry policy is effective in this goal dimension, but the conclusion is not robust.

In addition, attention should also be paid to the square root of the average prediction standard deviation (RMSPE, which measures the degree of fit between a province and its synthetic control province). The calculation formula is as follows:

$$\mathrm{RMSPE_{POST}} = \sqrt{\frac{1}{T - T_0}\sum\nolimits_{t=T_0+1}^{T}\left(y_{it} - y_{it}^{*}\right)^2} \quad (6)$$

$$\mathrm{RMSPE_{PRE}} = \sqrt{\frac{1}{T_0}\sum\nolimits_{t=1}^{T_0}\left(y_{it} - y_{it}^{*}\right)^2} \quad (7)$$

If the square root of the average prediction standard deviation of a province before 2010 is relatively large, it indicates that the fitting effect of the province before 2010 is not ideal, which leads to insufficient validity of using the gap of the province after 2010 as a comparison sample. Therefore, when a province's synthetic control unit shows a poor fit before the policy implementation, further analysis of its ranking test is no longer conducted. This study excludes provinces in the control group where the square root of the average prediction standard deviation is more than twice that of the corresponding treatment group.

From Figure 7, it can be seen that after adjusting the prediction error value, the new energy industry policy in Shanghai has effects in promoting sustainable development, optimizing the industrial structure, and energy saving and emission reduction. The robustness of these findings is strong. Combined with the previous policy evaluation results, a robust conclusion can be reached that the new energy industry policy in Shanghai is relatively ineffective in promoting technological innovation. In the ranking test where the industrial structure is used as the result variable, Qinghai is excluded according to the RMSPE before the policy. In the ranking test where per capita carbon emissions are used as the result variable, Inner Mongolia, Guangdong, Guangxi, and Ningxia are excluded according to the RMSPE before the policy.

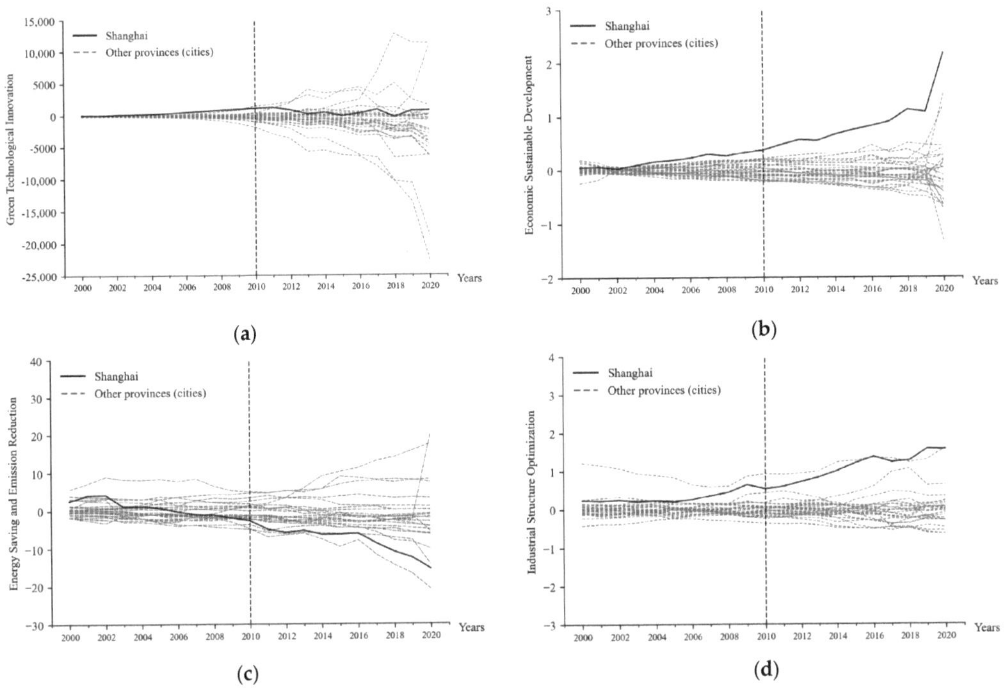

Figure 7. Distribution of differences by province and city: (**a**) Distribution of differences in green technological innovation by province (city); (**b**) Distribution of differences in economic sustainable development by province (city); (**c**) Distribution of differences in energy saving and emission reduction by province (city); (**d**) Distribution of differences in industrial structure optimization by province (city).

6. Conclusions, Policy Implications, and Future Research

Recently, achieving the dual-carbon goal has increasingly become the government policy direction and strategic goal, and it is well-recognized that the new energy industry plays a crucial role during the process of carbon emission reduction. Considering the government policy supports determines the long-term growth of the Chinese new energy industry, how to improve and optimize the policy support system for the new energy industry has always been the one of the core issues for related scholars and policy practitioners. In practice, the new energy industry development requires the government to promote solutions in the form of policy packages rather than just individual policies. We thus introduced the notion of a policy package into our new energy industry policy analysis and defined it as a set of related policies that are designed and implemented together to create a comprehensive and coordinated approach to achieve the new energy industry's long-term growth. Based on the fact that policy evaluation is a prerequisite for the improvement of government policy supports, we turned to investigating whether the implementation of the new energy industry policy package is effective or not and tried to provide new evidence to demonstrate the effectiveness of China's new energy industrial policy package through the empirical case study of Shanghai.

Some conclusions can be drawn as follows. This paper developed a two-stage analytical framework which integrates CAM and SCM to empirically evaluate the actual effect of the new energy industrial policy package. In the first stage, we summarized the diversified policy goals through deconstructing the various new energy industrial policy texts, then we decomposed the abstract and general strategic objectives into observable and measurable specifics. Four specific policy objectives from 57 policy texts were concluded and extracted, namely, urban economic sustainable development, industry transformation and upgrade, technological innovation, energy saving, and emission reduction. In the second stage of our evaluation, we used SCM to identify the multi-effect of the NEIPP, and we selected Shanghai as the research sample to empirically investigate the pure policy effect of the NEIPP. The results showed that the NEIPP had a significant positive effect on Shanghai's GTFP and industrial structure, while it had a significant negative effect on carbon emissions. It is worth noting that the effect of Shanghai's new energy industry policy in promoting technological innovation did not meet expectations. We conducted robustness checks after the SCM analysis, and the results remained robust.

As the evaluation of the NEIPP mainly assessed the degree to which governmental policy goals had been achieved, the objective and reasonable evaluation results can provide an empirical basis for future government policy adjustment, reinforcement, or abolition. Several policy implications can be drawn from this study. Firstly, since the NEIPP plays an important role in improving urban GTFP, optimizing regional industrial structure, and reducing carbon emissions, both central and local governments should strengthen policy support for the development of the new energy industry. Central government should mainly improve the guidance and supervision for urban new energy development, while the local government should spare no effort in promoting the new energy industry development through tax reduction, talent encouragement, and building related infrastructure, especially providing policy support for high-level innovation projects.

Although this paper proposed a two-stage analytical framework to empirically evaluate the actual effect of the NEIPP, which can also be used in other similar policy package evaluations, and presented initial evidence on the average effect of the NEIPP from the perspective of multidimensional objectives, some limitations need to be acknowledged. Due to limited data availability, we only chose Shanghai as a single empirical case study, and whether the NEIPP has the similar impact in other jurisdictions still needs further empirical research. What is more, whether the two-stage evaluation method can be further applied to the evaluation practice of other similar policy packages still needs future research to explore.

Author Contributions: For this research article, different authors performed different work; H.Y. and Q.W. conceived and designed the experiment, Q.W. wrote the paper, and S.C. is responsible for the data. All authors have read and agreed to the published version of the manuscript.

Funding: This research was funded by projects of the Humanity and Social Science Research Funds of Ministry of Education of China (Project No. 22YJC630140); the National Natural Science Foundation of China (Project No. 72204092); and the Fundamental Research Funds for Central Universities (Project No. 510322189).

Institutional Review Board Statement: Ethical review and approval were waived for this study, due to some relevant legal provisions (such as *Personal Information Protection Law of the People's Republic of China and Measures for the Ethical Review of Biomedical Research Involving Humans*), this study only involves expert interviews and does not collect, handle, or disclose personal identification information (such as names or contact details). Furthermore, the information we collected does not pose a risk of identifying individuals. Therefore, an ethical review is not required.

Informed Consent Statement: Informed consent was waived due to some relevant legal provisions (such as *Personal Information Protection Law of the People's Republic of China and Measures for the Ethical Review of Biomedical Research Involving Humans*), this study only involves expert interviews and does not collect, handle, or disclose personal identification information (such as names or contact details). Furthermore, the information we collected does not pose a risk of identifying individuals. Therefore, an ethical review is not required.

Data Availability Statement: The original contributions presented in the study are included in the article, further inquiries can be directed to the corresponding author.

Conflicts of Interest: The authors declare no conflicts of interest.

References

1. Woon, K.S.; Phuang, Z.X.; Taler, J.; Varbanov, P.S.; Chong, C.T.; Klemes, J.J.; Lee, C.T. Recent advances in urban green energy development towards carbon emissions neutrality. *Energy* **2023**, *267*, 126502. [CrossRef]
2. Ma, W.; Liu, X.; Li, C.; Yin, H.; Liu, R.; He, G.; Zhao, X.; Luo, J.; Ding, Y. Rechargeable Al–CO_2 batteries for reversible utilization of CO_2. *Adv. Mater.* **2018**, *30*, 1801152. [CrossRef] [PubMed]
3. Yeung, F.D.; Sammarchi, S.; Wang, E.; Gao, Q.; Li, J. Interdisciplinary challenges in bio-energy carbon capture utilization & storage deployment: A review. *Carbon Capture Sci. Technol.* **2024**, *13*, 100283.
4. Zhao, X.; Wu, L.; Zhao, Y. How to achieve incentive regulation under renewable portfolio standards and carbon tax policy? A China's power market perspective. *Energy Pol.* **2020**, *143*, 111576.
5. Yu, Q.; Hao, Y.; Ali, K.; Hua, Q.; Sun, L. Techno-economic analysis of hydrogen pipeline network in China based on levelized cost of transportation. *Energy Convers. Manag.* **2024**, *301*, 118025. [CrossRef]
6. Zuo, T.; He, L. Comparison Study of the Institutional Connection and Integration of Rural Minimum Living Security and Poverty Alleviation Systems. *Public Adm. Rev.* **2022**, *10*, 7–25+213.
7. Wei, Z.; Yu, S.; Zhang, D. Study on the Poverty Alleviation Policy System from the Perspective of Policy Groups: Evolution Process, Policy Integration, and Path Optimization. *J. Jiangsu Adm. Inst.* **2019**, *1*, 36–43.
8. Guo, W.; Tao, K.; Li, Z. Impact Analysis of Policy Combinations on the Formation of Leading Markets: A Case Study of the New Energy Vehicle Industry. *Sci. Res. Manag.* **2018**, *39*, 30–36.
9. Gomel, D.; Rogge, K.S. Mere deployment of renewables or industry formation, too? Exploring the role of advocacy communities for the Argentinean energy policy mix. *Environ. Innov. Soc. Transit.* **2020**, *36*, 345–371. [CrossRef]
10. Li, P.; Zhang, Z.X. The effects of new energy vehicle subsidies on air quality: Evidence from China. *Energy Econ.* **2023**, *120*, 106624. [CrossRef]
11. Li, Y.; Chiu, Y.-H.; Lu, L.C. New Energy Development and Pollution Emissions in China. *Int. J. Environ. Res. Public Health* **2019**, *16*, 1764. [CrossRef] [PubMed]
12. Islam, R. Foodgrain procurement, input subsidy and the public food distribution system in Bangladesh: An analysis of the policy package. *Bangladesh Dev. Stud.* **1980**, *8*, 89–120.
13. Bouma, J.A.; Verbraak, M.; Dietz, F.; Brouwer, R. Policy mix: Mess or merit? *J. Environ. Econ. Policy* **2019**, *8*, 32–47.
14. Ma, Y.; Sha, Y.; Wang, Z.; Zhang, W. The effect of the policy mix of green credit and government subsidy on environmental innovation. *Energy Econ.* **2023**, *118*, 106512. [CrossRef]
15. Bazzan, G.; Daugbjerg, C.; Tosun, J. Attaining policy integration through the integration of new policy instruments: The case of the Farm to Fork Strategy. *Appl. Econ. Perspect. Policy* **2023**, *45*, 803–818.
16. Caloffi, A.; Freo, M.; Ghinoi, S.; Mariani, M.; Rossi, F. Assessing the effects of a deliberate policy mix: The case of technology and innovation advisory services and innovation vouchers. *Res. Policy* **2022**, *51*, 104535. [CrossRef]
17. Khreis, H.; Sanchez, K.A.; Foster, M.; Burns, J.; Nieuwenhuijsen, M.J.; Jaikumar, R.; Ramani, T.; Zietsman, J. Urban policy interventions to reduce traffic-related emissions and air pollution: A systematic evidence map. *Environ. Int.* **2023**, *172*, 107805. [CrossRef]
18. Wicki, M.; Huber, R.A.; Bernauer, T. Can policy-packaging increase public support for costly policies? Insights from a choice experiment on policies against vehicle emissions. *J. Public Policy* **2020**, *40*, 599–625. [CrossRef]
19. Zhang, Q. Contemporary China's Policy Groups: Conceptual Introduction and Characteristics Analysis. *J. Beijing Adm. Coll.* **2000**, *1*, 13–14.
20. Givoni, M.; Macmillen, J.; Banister, D.; Feitelson, E. From policy measures to policy packages. *Transp. Rev.* **2013**, *33*, 1–20. [CrossRef]
21. Kern, F.; Kivimaa, P.; Martiskainen, M. Policy packaging or policy patching? The development of complex energy efficiency policy mixes. *Energy Res. Soc. Sci.* **2017**, *23*, 11–25. [CrossRef]

22. Howlett, M.; Rayner, J. Patching vs. packaging in policy formulation: Assessing policy portfolio design. *Politics Gov.* **2017**, *1*, 170–182. [CrossRef]
23. Fitriadi, M.; Prawira, M.Y.; Aidi, Z. Economic Policy Package: How Policy Delivery Affects Business Competition. In Proceedings of the Seminar International Academic Network on Competition Policy, Bali, Indonesia, 6 September 2017.
24. Lu, Y.; Zhang, Y.; Cao, X.; Wang, C.; Wang, Y.; Zhang, M.; Ferrier, R.C.; Jenking, A.; Yuan, J.; Beiley, M.J.; et al. Forty years of reform and opening up: China's progress toward a sustainable path. *Sci. Adv.* **2019**, *5*, eaau9413. [CrossRef] [PubMed]
25. Magro, E.; Navarro, M.; Zabala-Iturriagagoitia, J.M. Coordination-mix: The hidden face of STI policy. *Rev. Policy Res.* **2014**, *31*, 367–389. [CrossRef]
26. Meissner, D.; Kergroach, S. Innovation policy mix: Mapping and measurement. *J. Technol. Transf.* **2021**, *46*, 197–222. [CrossRef]
27. Wang, S. Study on the Integration of China's Urban Minimum Living Security System and Employment Assistance System: A Policy Group Theoretical Perspective. *China Labor* **2018**, *8*, 23–29.
28. Dugan, A.; Mayer, J.; Thaller, A.; Bachner, G.; Steininger, K.W. Developing policy packages for low-carbon passenger transport: A mixed methods analysis of trade-offs and synergies. *Ecol. Econ.* **2022**, *193*, 107304. [CrossRef]
29. Meng, W.; Li, Y. Assessment of the Effectiveness of Science and Technology Policy Groups: A Case Study of Shanghai's 'Science and Technology Innovation Center' Policies. *Sci. Sci. Manag. ST* **2021**, *42*, 45–65.
30. Liu, H. *Textual Measurement and Effectiveness Evaluation of New Energy Industry Policies*; Shandong University of Finance and Economics: Jinan, China, 2022.
31. Li, Q.; Zhao, Y.; Liu, B. Quantitative Analysis of New Energy Industry Policies and Their Environmental Effects. *J. Beijing Inst. Technol. (Soc. Sci. Ed.)* **2021**, *23*, 30–39.
32. Cheng, Y.; Mum, D. Study on optimal subsidy strategy in new energy vehicle supply chain based on SD game model. *China Popul. Resour. Environ.* **2018**, *28*, 29–39.
33. Wei, S.; Guo, S. Policy Tool Selection for China's New Energy Vehicle Industry Development. *Sci. Technol. Prog. Policy* **2014**, *31*, 99–103.
34. Wang, H.; Yin, J. Local Industrial Policies and Industry Innovation Development: Empirical Evidence from Texts of New Energy Vehicle Industry Policies. *Econ. Res. J.* **2021**, *47*, 64–78.
35. Guo, B.; Li, J.; Zhang, X. Impact of Policy Synergy on Policy Effectiveness: An Empirical Study Based on 227 Chinese Photovoltaic Industry Policies. *Stud. Sci. Sci.* **2018**, *36*, 790–799.
36. Zhao, M.; Sun, T. Dynamic spatial spillover effect of new energy vehicle industry policies on carbon emission of transportation sector in China. *Energy Policy* **2022**, *165*, 112991. [CrossRef]
37. Zhang, X.; Zhang, Y. Environment-friendly and economical scheduling optimization for integrated energy system considering power-to-gas technology and carbon capture power plant. *J. Clean. Prod.* **2020**, *276*, 123348. [CrossRef]
38. Shao, X.; Zhong, Y.; Liu, W.; Yi, R.; Li, M. Modeling the effect of green technology innovation and renewable energy on carbon neutrality in N-11 countries? Evidence from advance panel estimations. *J. Environ. Manag.* **2021**, *296*, 113189. [CrossRef]
39. Gowd, S.C.; Ganeshan, P.; Vigneswaran, V.S.; Hossain, M.S.; Kumar, D.; Rajendran, K.; Ngo, H.H.; Pugazhendhi, A. Economic perspectives and policy insights on carbon capture, storage, and utilization for sustainable development. *Sci. Total Environ.* **2023**, *883*, 163656. [CrossRef]
40. Xu, H.; Liu, B.; Qiu, L.; Liu, X.; Lin, W.; Liu, B. Does the new energy demonstration cities construction reduce CO_2 emission? Evidence from a quasi-natural experiment in China. *Environ. Sci. Pollut. Res.* **2022**, *29*, 50408–50426. [CrossRef]
41. Wang, Q.; Yi, H. New energy demonstration program and China's urban green economic growth: Do regional characteristics make a difference? *Energy Policy* **2021**, *151*, 112161. [CrossRef]
42. Zeng, J.; Liu, T.; Feiock, R.; Li, F. The impacts of China's provincial energy policies on major air pollutants: A spatial econometric analysis. *Energy Policy* **2019**, *132*, 392–403. [CrossRef]
43. Jia, X.D.; Heng, L. The "jungle", history, and approach road of the grounded theory. *Sci. Res. Manag.* **2020**, *41*, 151–163.
44. Zhang, C.; Kong, X.Z. Administrative boundary adjustment and grain production: Evidence from synthetic controlmethods. *J. Nanjing Agric. Univ. (Soc. Sci. Ed.)* **2017**, *17*, 121–133+159.
45. Abadie, A.; Diamond, A.; Hainmueller, J. Synthetic control methods for comparative case studies: Estimating the effect of California's tobacco control program. *J. Am. Stat. Assoc.* **2010**, *105*, 493–505. [CrossRef]

sustainability

MDPI

Article

Cold Chain Distribution Route Optimization for Mixed Vehicle Types of Fresh Agricultural Products Considering Carbon Emissions: A Study Based on a Survey in China

Shuangli Pan [1,2], Huiyu Liao [1], Guijun Zheng [3,*], Qian Huang [1] and Maozhuo Shan [1]

1 School of Logistics, Central South University of Forestry and Technology, Changsha 410004, China; t20070772@csuft.edu.cn (S.P.); 19727312611@163.com (H.L.); huangqian0826@126.com (Q.H.); mz867265517@126.com (M.S.)
2 Hunan Key Laboratory of Intelligent Logistics Technology, Changsha 410004, China
3 Business School, Central South University of Forestry and Technology, Changsha 410004, China
* Correspondence: t20071329@csuft.edu.cn

Abstract: With the improvement of people's living standards and the widening of circulation channels, the demand for fresh agricultural products continues to increase. The increase in demand will lead to an increase in delivery vehicles, costs, and carbon emissions, among which the increase in carbon emissions will aggravate pollution and is not conducive to sustainable development. Therefore, it is very important to balance economic and environmental benefits in the distribution of fresh agricultural products. Based on the analysis of the distribution characteristics of fresh agricultural products, this paper studies the optimization of the cold chain distribution route of fresh agricultural products considering carbon emission. Firstly, the cold chain distribution route planning of fresh agricultural products was investigated and analyzed by the interview method, and the basis for establishing the model objective and constraint conditions was obtained. Then, taking the minimum total cost including carbon emission cost as the optimization goal, the cold chain distribution route optimization model for mixed vehicle types is established considering electric refrigerated vehicles, gasoline refrigerated vehicles, and so on. Genetic algorithm was used to solve the model, and MATLAB2018b was used to substitute specific case data for simulation analysis. The analysis results show that increasing the consideration of carbon emission and mixed vehicle types in the distribution route of fresh agricultural products can not only reduce the distribution cost but also reduce the carbon emission. To some extent, the research content of this paper can provide a reference for enterprises in planning cold chain distribution routes of fresh agricultural products.

Keywords: distribution; vehicle; carbon; route; cost; optimization; algorithm

1. Introduction

With the improvement of living standards and the upgrading of consumption structure, people's demand for fresh agricultural products is increasing, and the demand for cold chain distribution of fresh agricultural products is rising higher and higher. According to the "China Cold Chain Logistics Development Report (2024 edition)", the total demand for cold chain logistics in China in 2023 is about 350 million tons, of which food (vegetables, fruits, meat, aquatic products, and other fresh agricultural products) accounts for 90%. Due to the particularity of fresh agricultural products, in order to ensure the freshness of agricultural products, it is necessary to maintain a low-temperature distribution environment and efficient distribution efficiency. Compared with ordinary logistics, the cooling demand of vehicles makes the cold chain distribution of fresh agricultural products consume more energy and produce more carbon dioxide (CO_2). Reductions of between 210 and 460 billion tonnes of CO_2 equivalent emissions can be delivered over the next four decades through actions to improve the cooling industry's energy efficiency together with the transition to climate-friendly refrigerants, according to the Cooling Emissions and Policy Synthesis Report from the United Nations Environment Programme (UNEP) and the International Energy Agency (IEA). How to plan the distribution route scientifically and reasonably, so as to improve the distribution efficiency and satisfaction, and reduce the distribution cost and carbon emission, is a very noteworthy problem.

Due to the enhancement of environmental awareness, the use of new energy vehicles is becoming more and more widespread, and many fresh agricultural products cold chain logistics enterprises are in the stage of mixed-use such as gasoline refrigerated trucks and electric refrigerated trucks. China's "14th Five-Year Plan" Cold Chain Logistics Development Plan clearly points out that it is necessary to speed up the elimination of high-emission refrigerated vehicles and encourage new or updated refrigerated vehicles to adopt new energy models in order to meet the needs of urban green distribution development. According to statistics on the website, 3696 new energy refrigerated vehicles were sold in the first five months of 2024 in China (excluding exports), an increase of 310% compared with 2023. In this case, how to give full play to the advantages of various types of vehicles, use vehicles, and plan distribution routes reasonably is a practical problem worth considering. In view of these aspects, this paper comprehensively considers carbon emissions and mixed vehicle types and other factors to study the cold chain distribution route optimization of fresh agricultural products, so as to promote the high-quality development of agricultural cold chain logistics.

2. Literature Review

Since the Vehicle Routing Problem (VRP) was first proposed by Dantzig and Ramser in 1959, it has attracted much attention, and there have been many mature research results. Among them, many scholars have explored the optimization of the distribution route according to the characteristics of the cold chain of fresh agricultural products. For example, Tarantilis and Kiranoudis [1] planned the distribution route for manufacturers producing and selling fresh meat; Zhu et al. [2] optimized the distribution route of fresh agricultural products considering the input of preservation cost. Wang et al. [3] set minimization of distribution cost and maximization of customer satisfaction as the optimization objectives of intelligent distribution of fresh agricultural products. Yang and Tao [4] established a VRP model for cold chain logistics of fresh products with dual objective optimization. With the deepening of research, more and more scholars pay attention to the importance of time in VRP. Marius and Desrosiers [5] introduced the time window limit into vehicle routing research earlier. Ahkamiraad [6] established a mixed-integer linear programming model for a special capacity multi-hop library VRP with pick-up, delivery, and time window to achieve minimal transportation and fixed costs. Liu and Li [7] built a cold chain logistics distribution route optimization model considering the time window requirements. Ho et al. [8], Ding et al. [9], Wu et al. [10], and Wang et al. [11] combined the characteristics of fresh agricultural products to study the vehicle routing optimization problem considering time window constraints. As the cold chain distribution of fresh agricultural products has high timeliness requirements, this paper will consider the time window constraint in exploring the optimization of the cold chain distribution path.

With the increasing awareness of environmental protection, some scholars have begun to pay attention to the problem of carbon emissions in cold chain distribution. Lin et al. [12] took the lead in introducing carbon emission into logistics research, arguing that enterprises should not only consider traditional economic costs when making logistics decisions but also consider their impact on the ecological environment and society. Liu et al. [13] studied the influence of carbon trading mechanisms on cold chain distribution route planning. Ji et al. [14], Zhang et al. [15], and Yang et al. [16] discussed the optimization of low-carbon routes of cold chain logistics combined with traffic conditions. Demir et al. [17], Hariga et al. [18], Zhao et al. [19], Afra and Behnamain [20], Wu et al. [21], Jiang et al. [22], and Ma et al. [23] established a multi-objective cold chain distribution path optimization model by taking factors such as the lowest fuel consumption, the lowest carbon emission, and the lowest environmental pollution into the objective function. Fu et al. [24], Yao et al. [25], Feng et al. [26], and Pérez-Lechuga et al. [27] focused on the problem of green vehicle routing for fresh agricultural products. However, in general, there are relatively few studies on the optimization of the cold chain distribution path of fresh agricultural products considering carbon emissions, and there is a large research space.

In the actual distribution process, there are often different types of vehicles for distribution. Their rated load, fuel/power consumption, driving speed, etc., are different and have a greater impact on the distribution path and cost of logistics enterprises. So, it is of practical significance to consider the problem of mixed vehicles in the model. Golden et al. [28] took multi-vehicle factors into account in the model earlier and established the objective function of minimum total cost. Heiman et al. [29] studied the routing problems of traditional fuel vehicles, plug-in hybrid vehicles, and electric vehicles. Macrina et al. [30] studied the vehicle routing problem of a hybrid fleet consisting of electric vehicles and conventional vehicles. Ene et al. [31] established a heterogeneous vehicle fleet VRP multi-objective model. Kabadurmus and Erdogan [32] also considered allowing different vehicle types when studying the vehicle routing problem. Leung et al. [33] and Kwon et al. [34] studied the route optimization model of multi-type refrigerated vehicles. Chen et al. [35] discussed the cold chain distribution vehicle routing problem under heterogeneous fleet and parking constraints. Chen et al. [36] analyzed how to plan the green vehicle route using mixed fleets for cold chain distribution. Multiple studies have shown that the distribution of many types of vehicles can reduce the total distribution cost and reduce carbon emissions to a certain extent. Therefore, it is necessary to take mixed vehicle types into account in the optimization of cold chain distribution routes for fresh agricultural products.

To sum up, there are many research results on cold chain distribution route optimization, but the focus of attention is different. The comparison of existing studies is shown in Table 1, depending on whether they are specific to fresh agricultural products, and whether time windows, carbon emissions, and multiple vehicle types are considered. It can be seen that time window constraints are basically considered, some studies specifically combine the characteristics of fresh agricultural products, carbon emission factors have attracted attention, and there are relatively few studies on multiple vehicle types. Studies that consider all four factors at the same time are lacking.

Table 1. Comparison of existing studies on cold chain distribution routes.

Document Number	Fresh Agricultural Products	Time Window	Carbon Emission	Multiple Vehicle Types
1–4	√			
5–7		√		
8–11	√	√		
12–13			√	
14–23		√	√	
24–27	√	√	√	
28–32				√
33–35	√	√		√
36		√	√	√

Note: A "√" indicates that this factor was taken into account in the relevant literature.

As the demand for cold chain distribution of fresh agricultural products is strong, and carbon emission and distribution of various vehicle types are realistic problems that have to be considered, this paper intends to comprehensively consider the four factors mentioned above, and explore the optimization of cold chain distribution path of mixed vehicle types of fresh agricultural products considering carbon emission, in order to provide decision-making reference for the sustainable development of cold chain of agricultural products. The main features of this study are as follows: (1) Considering the time window constraint in the optimization of the distribution path. Maximize customer requirements for delivery time and improve customer satisfaction. (2) Consider the carbon emissions in the cold chain distribution of fresh agricultural products. Pursue economic benefits while improving environmental benefits. (3) Consider the impact of the distribution of multiple types of vehicles. Rationally allocate vehicle resources according to realistic conditions.

3. Method

Establishing a mathematical model to guide decision-making is a common method in distribution route planning. Based on the existing decision-making model, this paper intends to establish a new cold chain distribution route optimization model for fresh agricultural products by comprehensively considering factors such as time window, carbon emission, and mixed vehicle types, and select a suitable solution algorithm. Since the factors and ideas considered in decision-making are reflected in the objective function and constraint conditions of the model, this paper, before establishing the model, mastered the core elements of cold chain distribution of fresh agricultural products through literature analysis and investigation, and provided a basis for setting the model's objectives and constraint conditions.

3.1. Investigation and Analysis of Cold Chain Distribution Route Planning for Fresh Agricultural Products

3.1.1. Investigation Plan

Through the analysis of the existing literature, it is found that cost minimization is one of the main goals of cold chain distribution route optimization for fresh agricultural products. Almost all of the research is about how to reduce costs, but different studies focus on different types of costs, among which fixed cost and transportation cost are generally considered, and some involve penalty costs, refrigeration costs, cargo damage costs, etc., followed by what costs are considered by enterprises in reality, and what costs are worthy of attention. This study conducted a survey on cold chain logistics companies of fresh agricultural products in view of these problems; at the same time, through the investigation to understand the actual situation of enterprises and decision-making methods.

(1) Investigation purpose: To understand the concerns and influencing factors considered by cold chain logistics companies of fresh agricultural products in the process of distribution route planning, and to find out what cost components they consider when calculating costs, so as to provide reference for the establishment of multi-objective cold chain logistics distribution route optimization model of fresh agricultural products.
(2) Investigation object: The respondents were eight different fresh agricultural products cold chain logistics companies in Wuhan, China. All of them can provide distribution services of fresh agricultural products such as fruits, aquatic products, meat, eggs, and quick-frozen products within the city, and are equipped with cold storage, refrigerated trucks, distribution centers, and other facilities to ensure the quality of distribution services and the freshness of goods.
(3) Investigation time and location: The survey was conducted from 25 February 2024, to 10 March 2024, over a period of 15 days. The respondents were distributed in 4 districts of Wuhan City, China, namely Hongshan District, Dongxihu District, Jiangxia District, and Hannan District.
(4) Investigation content: The core content is the practice of cold chain logistics companies in planning distribution routes for fresh agricultural products. For example, what are the influencing factors they will consider? What are the delivery costs that they focus on? What types of refrigerated trucks do they use for distribution?
(5) Investigation method: The survey was completed by interview. Through the communication with the company's employees, we understand the influencing factors and distribution costs considered by cold chain logistics enterprises when making distribution route plans. The survey outline is in Appendix A.

3.1.2. Analysis of Cold Chain Distribution Route Planning Objectives

Regarding the goal of cold chain distribution route planning, the survey mainly understands the cost considered by the enterprise. The surveyors know what aspects of costs are mainly considered by enterprises in cost calculation through inquiry and record them. When the interviewees could not answer, the researchers would remind them based on the results of the literature analysis, such as whether the cost of carbon emissions was considered. Through sorting, it is found that the costs considered by the eight companies mainly involve fixed costs, transportation costs, penalty costs, refrigeration costs, cargo loss costs, and carbon emission costs, as shown in Table 2. Three companies accounted for all six costs at the same time, while the others accounted for only some of them.

Table 2. The main costs considered by the surveyed companies.

Company Number	Fixed Cost	Transportation Cost	Penalty Cost	Cargo Loss Cost	Refrigeration Cost	Carbon Emission Cost
1	√	√	√	√	√	√
2	√	√			√	
3	√	√	√	√	√	√
4	√	√	√	√		
5	√	√				
6	√	√	√	√	√	√
7	√	√	√			
8	√	√			√	

Note: A "√" indicates that the company has taken this cost into account.

From the above statistical results, it can be seen that fixed cost and transportation cost are the most important parts in the cost calculation of distribution route planning, and all eight companies have taken them into account, while only a few companies have taken other costs into account due to the difficulty of calculation and quantification. The cost of carbon emissions has attracted the attention of some companies due to increased environmental awareness, but the attention needs to be increased. In fact, the interviewer also found in the process of communication with staff that it is not that companies do not know what should be included in the calculation of the total cost of distribution, but because of the complexity of its calculation, and it is time-consuming and laborious, they choose to ignore it, or convert to other cost calculations, such as refrigeration costs often being converted into fuel costs. In addition, even for companies that fully consider the above six kinds of costs, in addition to transportation and fixed costs, the other costs are simple calculations and are not calculated scientifically, reasonably, and carefully.

3.1.3. Analysis of Constraints in Cold Chain Distribution Route Planning

In the investigation of 8 cold chain logistics companies, it was found that all companies took into account the constraints of vehicle use, drivers, and time windows acceptable to customers when carrying out goods distribution. In addition, in response to the call for low-carbon environmental protection and effectively reducing carbon emissions in the distribution process, four companies that have taken into account the cost of carbon emissions have introduced pure electric refrigerated vehicles and gas–electric dual-use refrigerated vehicles. Two other companies are also planning to introduce gas–electric refrigerated trucks. It can be predicted that the application of electric refrigerated trucks in the cold chain distribution of fresh agricultural products will be more and more extensive. Companies that use electric refrigerated trucks need to consider power constraints when planning distribution routes. Because electric refrigerated trucks are still in the development stage, most companies now have multiple vehicle types at the same time. Therefore, when studying the cold chain distribution of fresh agricultural products, it is of practical significance to consider the joint distribution of mixed vehicle types.

According to the literature analysis and investigation, in order to reduce the carbon emissions in the cold chain distribution of fresh agricultural products, the cost of carbon emissions can be calculated and controlled. At the same time, in order to achieve the overall optimization, the fixed cost, transportation cost, penalty cost, cargo loss cost, refrigeration cost, and carbon emission cost, which account for a relatively large proportion, should be considered comprehensively, and the total cost should be minimized. In terms of constraints, in addition to considering the time required by customers, how to rationally allocate a variety of vehicle resources such as gasoline refrigerated vehicles, pure electric refrigerated vehicles, and gas–electric dual-use refrigerated vehicles is a practical problem worth considering. In order to achieve both economic and environmental benefits and promote the sustainable development of cold chain logistics of fresh agricultural products, it is particularly necessary to consider the time window and mixed vehicle types when planning the distribution route, and to minimize the total cost including carbon emission cost as the optimization goal.

3.2. Model Construction

3.2.1. Problem Description and Model Assumptions

According to the investigation results, this paper aims to minimize the comprehensive distribution cost of fixed cost, transportation cost, penalty cost, cargo loss cost, refrigeration cost, and carbon emission cost, and establishes a cold chain distribution route optimization model for fresh agricultural products considering different vehicle types. Assume that a distribution center needs to provide distribution services for M demand points, and the distribution center can be equipped with K refrigerated trucks for the transportation of fresh agricultural products. The quantity demanded at each demand point is Q_i. The latest time for delivery of fresh agricultural products is l_i, and the earliest time is e_i, so the service time window is $(e_i,\ l_i)$. The staff driving the refrigerated vehicle should deliver the goods to the customer demand point within the prescribed time, and certain penalty costs will be incurred if the time is advanced or delayed. It is necessary to make a reasonable distribution route while meeting the time required by the demand point, so as to reduce the total cost in the distribution process. The assumptions are as follows: (1) Only one distribution center is considered, and the goods transported are fresh agricultural products. (2) There are many types of refrigerated trucks in the distribution center, and the load capacity is known. (3) The quantity of goods in the distribution center is sufficient. (4) The starting point and end point of the refrigerated truck are the distribution center, and each demand point can only be distributed once in a plan. (5) The service time window of each demand point is known. (6) The road conditions between the distribution center and the demand point are complicated, and the linear distance between two points on the map is considered in this paper. (7) The demand of goods will not change in the process of distribution. The relevant variables used in the model are shown in Table 3.

Table 3. Variables in the model.

Symbol	Implication
m	Collection of demand points, where 0 represents the distribution center. $\mathrm{m} = \{0, 1, 2\cdots, \mathrm{M}\}$.
k	Number of refrigerated trucks used for distribution, $\mathrm{k} = \{1, 2\cdots, \mathrm{K}\}$.
r	Refrigerated vehicle types, including gasoline refrigerated vehicles and electric refrigerated vehicles, etc., $\mathrm{r} = \{1, 2\cdots, \mathrm{R}\}$.
K_r	Number of refrigerated vehicles of type r, and the total number of vehicles is K.
i, j	Demand points.
Z_1	Cost objective function.
Z_{11}	Fixed cost.

Table 3. *Cont.*

Symbol	Implication
Z_{12}	Transportation cost.
Z_{13}	Penalty cost.
Z_{14}	Cargo loss cost.
Z_{15}	Refrigeration cost.
Z_{16}	Carbon emission cost.
Z_2	CO_2 emissions generated during the distribution of refrigerated vehicles.
G_r	Full load of type r refrigerated vehicle.
g_r	Load weight of type r refrigerated vehicle.
$V^r_{\max}$	Maximum speed of type r refrigerated vehicle.
E_i	The earliest service time allowed by demand point i.
L_i	The latest service time allowed by demand point i.
e_i	The earliest delivery time required by demand point i.
l_i	The latest delivery time required by demand point i.
d_{ij}	Distance between node i and node j.
$L^r_{\max}$	Maximum driving distance of type r refrigerated vehicle.
Q_i	Quantity of goods demanded at demand point i.
M_0	Minimum battery capacity of electric refrigerated vehicle.
$M_{\max}$	Maximum battery capacity of an electric refrigerated vehicle.
$f(g)_{1i}$	Electricity of electric refrigerated vehicle at node i.
t^k_{ir}	The time when the k refrigerated vehicle of type r is transported from i to j.
w_1	Unit penalty cost of refrigerated vehicle arriving earlier than the specified time.
w_2	Unit penalty cost of refrigerated vehicle being late to the specified time.
x^k_{ijr}	when the k refrigerated vehicle of type r goes from the demand point i to j, $x^k_{ijr} = 1$, otherwise 0.
y^k_{ir}	If the k refrigerated vehicle of type r serves the demand point i, $y^k_{ir} = 1$, otherwise 0.
p	Unit carbon trading price.

3.2.2. Target Analysis

In this paper, the cold chain distribution route of fresh agricultural products is studied, and various costs are analyzed to minimize the comprehensive distribution cost. This paper does not set the weight of all kinds of costs; in the actual application process, enterprises can set reasonable weights according to their own conditions and demand points. The total cost is calculated as follows:

$$Z_1 = Z_{11} + Z_{12} + Z_{13} + Z_{14} + Z_{15} + Z_{16} \tag{1}$$

(1) Fixed cost

The fixed cost Z_{11} is composed of the depreciation cost of refrigerated vehicle, the driver's salary, and so on, generated in the process of distribution. It does not change with the transport distance or load. It is calculated as follows:

$$Z_{11} = \sum_{k=1}^{K_r} \sum_{r=1}^{R} Z_{11r} \times k \tag{2}$$

where Z_{11r} is the fixed cost of type r (gasoline or electric, etc.) refrigerated vehicle.

(2) Transportation cost

The transportation cost Z_{12} is the fuel consumption or power consumption cost generated by the refrigerated vehicle in the process of distributing fresh agricultural products. The longer the distance the refrigerated vehicle travels, the higher the fuel consumption cost and power consumption cost. The calculation function expression of transportation cost is as follows:

$$Z_{12} = \sum_{i=0}^{M}\sum_{j=0}^{M}\sum_{r=1}^{R}\sum_{k=1}^{K_r} x_{ijr}^{k} d_{ij} Z_{12r} \tag{3}$$

where Z_{12r} is the transportation cost per unit distance for type r vehicle.

(3) Penalty cost

If the driver of the vehicle does not deliver the goods within the time window acceptable to the customer, there will be a certain penalty cost. The penalty cost $Z_{13i}(t_i)$ at demand point i is a linear function of vehicle arrival time t_i. There is a penalty if the vehicle arrives within the range of $[E_i, e_i]$ and $[l_i, L_i]$, while there is no penalty if it arrives within the time of $[e_i, l_i]$, and the penalty value beyond the range of $[E_i, L_i]$ is set to an infinite positive number M. The expression for penalty cost is as follows:

$$Z_{13i} = \begin{cases} M(t_i < E_i) \\ (e_i - t_i) \times w_1(E_i \le t_i < e_i) \\ 0(e_i \le t_i < l_i) \\ (l_i - t_i) \times w_2(l_i \le t_i < L_i) \\ M(t_i > L_i) \end{cases} \tag{4}$$

$$Z_{13} = \sum_{i=1}^{M} Z_{13i} \tag{5}$$

(4) Cargo loss cost

Due to the characteristics of fresh agricultural products and their distribution characteristics, the cargo loss cost will be generated in the distribution process. This paper considers the cost caused by natural loss of fresh agricultural products in the process of distribution and the cargo damage caused by the temperature difference between inside and outside the refrigerated vehicle during loading and unloading. The expression of cargo loss cost is as follows:

$$Z_{14} = \sum_{i=0}^{M}\sum_{r=1}^{R}\sum_{k=1}^{K_r} y_{ir}^{k} \times q_i \times \left(1 - e^{-\beta_1 \times t_{ir}^{k}}\right) \times N + \sum_{i=0}^{M}\sum_{r=1}^{R}\sum_{k=1}^{K_r} y_{ir}^{k} \times q_{it} \times \left(1 - e^{-\beta_2 * t_{ir}^{k}}\right) \times N \tag{6}$$

β_1 and β_2 indicate the rate of corruption during closing and loading, respectively. N is the unit value of the goods. q_{it} is the weight at t time.

(5) Refrigeration cost

In the distribution process of fresh agricultural products, refrigerated vehicles are required to maintain a low temperature, which will generate refrigeration costs. The refrigeration cost has two parts, respectively, the cost in the closed state and the cost generated during the loading and unloading process. This paper ignores the cost of refrigeration in the loading and unloading process and only considers the cost of refrigeration in the closed state. The expression of refrigeration cost is as follows:

$$Z_{15} = \sum_{i=0}^{M}\sum_{r=1}^{R}\sum_{k=1}^{K_r} \alpha_r \times t_{ir}^{k} \times y_{ir}^{k} \tag{7}$$

α_r is the refrigeration cost per unit time of type r refrigerated vehicle.

(6) Carbon emission cost

It can be seen from the relevant information that under certain conditions such as road conditions, driving technology, speed, etc., the fuel consumption or power consumption per kilometer traveled by the vehicle has a linear relationship with the load capacity. The linear function is as follows:

$$f(g)_r = \left[f'(G)_r - f(0)_r\right]/G \times g + f(0) \tag{8}$$

In the expression, $f''(G)_r$ represents the fuel consumption or electricity consumption of the refrigerated vehicle when it is fully loaded; $f(0)_r$ represents the fuel consumption or electricity consumption of the refrigerated vehicle without load; g is the load of refrigerated vehicle; G is the full load of the refrigerated vehicle.

Then, based on vehicle fuel consumption or electric energy consumption in the distribution process, carbon dioxide emissions are calculated, and the calculation function expression is as follows:

$$Z_2 = \sum_{i=0}^{M}\sum_{j=0}^{M}\sum_{r=1}^{R}\sum_{k=1}^{K_r} f(g)_r \times d_{ij} \times x_{ij}^{k_r} \times \lambda_r \tag{9}$$

where $f(g)_r$ is the fuel or electricity consumption of the vehicle. λ_r is the carbon dioxide emission coefficient of the refrigerated vehicle. The carbon emission cost expression is:

$$Z_{16} = Z_2 \times p \tag{10}$$

3.2.3. Constraint Analysis

Considering the distribution characteristics of cold chain logistics of fresh agricultural products and the characteristics of vehicle path planning, the following constraints are set for the model.

(1) Constraints of the customer time window. The delivery vehicle should arrive within the time limit required by the customer, otherwise it will be subject to corresponding penalties.
(2) During the delivery process, a demand point is only served by one refrigerated vehicle and is only served once.
(3) Constraints on the number of vehicles in the distribution center. The number of vehicles to complete the distribution task cannot exceed all the vehicles owned by the distribution center, otherwise the scheme cannot be implemented.
(4) All refrigerated vehicles start from the distribution center and return to the distribution center after the end of service.
(5) Each refrigerated vehicle has a load limit, so in the distribution process, it is necessary to ensure that the total weight of fresh products does not exceed the carrying capacity of the refrigerated vehicle.
(6) In order to ensure driving safety, the total length of each refrigerated vehicle driving route shall not exceed its maximum mileage during the distribution process.
(7) In order to ensure safety, during the distribution process, the speed of each refrigerated vehicle should not exceed its maximum speed.
(8) Each electric refrigerated vehicle used for distribution work should not be lower than the minimum allowable power, nor higher than the maximum battery capacity.

3.2.4. Construction of Cold Chain Distribution Path Optimization Model

According to the above analysis, the objective function of the cold chain distribution route optimization model of fresh agricultural products considering mixed vehicle types and carbon emissions is as follows:

$$Z_1 = Min(Z_{11} + Z_{12} + Z_{13} + Z_{14} + Z_{15} + Z_{16}) \tag{11}$$

The constraints are as follows:

$$E_i \le t_{ir}^k \le L_i;\ \forall r \in R, \forall k \in K \tag{12}$$

$$\sum_{i=0}^{M}\sum_{k=1}^{K_r}\sum_{r=1}^{R} x_{ijr}^{k} = 1;\ \forall j \in M \tag{13}$$

$$\sum_{j=0}^{M}\sum_{k=1}^{K_r}\sum_{r=1}^{R} x_{ijr}^{k} = 1;\ \forall j \in M \tag{14}$$

$$\sum_{j=1}^{M}\sum_{k=1}^{K} x_{ijr}^{k} \le K_r;\ i = 0 \tag{15}$$

$$\sum_{i=0}^{M}\sum_{r=1}^{R} x_{ijr}^{k} = \sum_{i=0}^{M}\sum_{r=1}^{R} x_{jir}^{k};\ \forall k \in K, \forall i \in M \tag{16}$$

$$\sum_{i=1}^{M}\sum_{r=1}^{R} Q_i \times y_{ir}^{k} \le G_r;\ \forall k \in K \tag{17}$$

$$\sum_{i=0}^{M}\sum_{j=0}^{M} x_{ijr}^{k} \times d_{ij} \le L_{\max}^{r};\ \forall r \in R, \forall k \in K \tag{18}$$

$$\sum_{i=0}^{M}\sum_{j=0}^{M} d_{ij}/t_{ij} \times x_{ijr}^{k} \le V_{\max}^{r};\ \forall r \in R, \forall k \in K \tag{19}$$

$$x_{ijr}^{k} = \{0,1\};\ \forall r \in R, \forall k \in K \tag{20}$$

$$y_{ir}^{k} = \{0,1\};\ \forall r \in R, \forall k \in K \tag{21}$$

$$M_0 \le f(g)_{1i} \le M_{\max} \tag{22}$$

Formula (11) is the objective function with the lowest cost. The Formula (12) in the constraints indicates that the service time is within the time specified by the customer. Formulas (13) and (14) indicate that each customer point has only one refrigerated vehicle for service, and is only served once. (15) means that the number of distribution vehicles must not exceed the number of vehicles owned by the distribution center. (16) indicates that each vehicle starts from the distribution center and finally returns there. (17) means that the total demand for goods at customer demand points on the distribution path cannot be greater than the maximum load of the distribution refrigerated vehicle. (18) indicates that the total mileage of each refrigerated vehicle in the distribution process is not greater than its maximum allowable mileage. (19) means that the speed of each refrigerated vehicle in the distribution process cannot be greater than the fastest speed it is allowed. (20) and (21) are decision variable functions. (22) represents the power constraint of electric refrigerated vehicle at each node.

3.3. Model Solution Algorithm

3.3.1. Algorithm Selection

There are many common algorithms to solve the vehicle routing problem, such as precise algorithm, ant colony algorithm, genetic algorithm, tabu search algorithm, simulated annealing algorithm, particle swarm algorithm, and so on. Considering the established model, this paper adopts genetic algorithm to solve the established model for the following three reasons: First, the operation of genetic algorithm is simple, with strong robustness, and has a wider range of application. Secondly, it can deal with the objective function and fitness function well and minimize the probability of falling into the local optimal solution. Third, the model constructed in this paper is relatively complex, which is a multi-objective distribution route optimization model considering mixed vehicle types, and genetic algorithm is just suitable for large-scale path planning. Therefore, this paper designs the solution of genetic algorithm and then solves the constructed path optimization model.

3.3.2. Solution Steps

The operation process of the genetic algorithm includes the following steps: Firstly, the coding design is carried out, and then the initial group selection is carried out. Then, it is necessary to calculate according to the fitness of individuals and select appropriate operators for operation. In this process, operator crossover and operator variation are used to optimize the solution, followed by, according to the termination condition, ultimately judging whether the solution is complete, as shown in Figure 1.

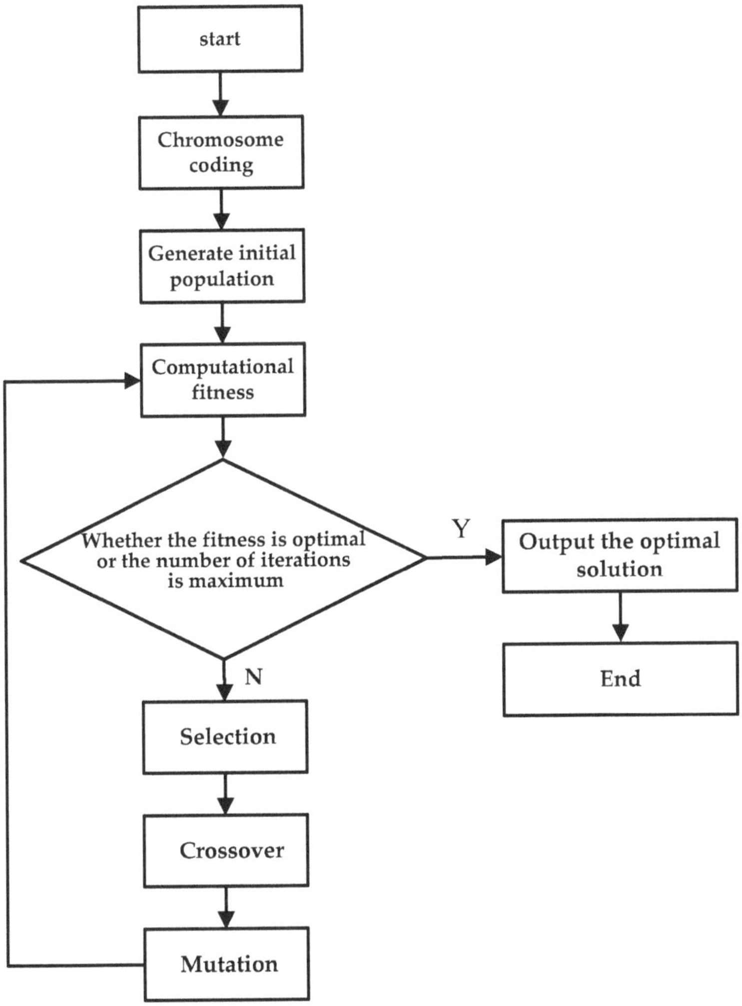

Figure 1. Genetic algorithm operation steps.

3.3.3. Solution Design

(1) Coding Design

Common chromosome coding methods include natural number coding, binary coding, and data structure coding. It can be seen from the literature that most of the coding methods related to the distribution path optimization problem are natural number coding. In the distribution between the distribution center and the customer points, the node code can be continuous, and there is only one distribution to the customer point, so the natural number coding method is also adopted in this paper to encode the chromosome. The coding process is specifically expressed as follows: the number "0" represents the distribution center, and the number "n" represents the customer point, where the code length = the number of customer points + the number of vehicles used $\times 2 - 1$, that is, the code length = $n + 2 \times m - 1$. The problem becomes any natural number from "1" to "$n + 2 \times m - 1$", where a number greater than "n" represents a split point of route information. If the chromosomes that meet the constraints are solved according to the rules, then the chromosomes form an alternative distribution path. The location of "0" is used to record which type of vehicle is selected. Suppose that there are two types of vehicles, before the appearance of the m-th "0", the first type of vehicle is selected, and the second type of vehicle is selected later.

(2) Generate initial population

A population is a collection of all individuals, each representing a solution. The larger the size of the initial population, the greater the number of individuals in the population that can be selected, crossed, and mutated. It can reduce the possibility of falling into the local optimal solution and improve the efficiency of searching for the global optimal solution. Once the number and size of the population are determined, the corresponding number of gene sequences can be generated in a suitable way. For the model in this paper, an individual represents a cold chain distribution path scheme. A variety of constraints are set, and individuals that do not meet the requirements are removed to obtain a high-quality initial population.

(3) Fitness function

The evaluation of a solution by genetic algorithm does not depend on the structure of its solution, but depends on the fitness value of the solution, which reflects the characteristics of genetic algorithm "survival of the fittest". Since the objective of the model established in this paper is to find the minimum cost, the reciprocal of the objective function is used to determine its adaptability.

(4) Selection

Chromosome selection is to select the most dynamic and potential chromosomes in the existing population. The roulette selection method was used in this study. The basic principle of this method is that whether a chromosome is selected has a certain correlation with its fitness function value, that is, the higher the fitness of the chromosome, the greater the probability of being selected. If the population size is i and the fitness function of individuals is $\mathrm{f_i}$, then the probability of the solution of the distribution path optimization model being selected is:

$$p_i = \frac{f_i}{\sum_i^n f_i} \tag{23}$$

(5) Crossover

Crossover is an important method to generate new individuals, which determines the global optimization ability of genetic algorithm. In this paper, the single-point crossover method is chosen, as shown in Figure 2. The greater the crossover probability, the more new solutions can be obtained to ensure the diversity of the population and avoid premature convergence of the algorithm. Therefore, the crossover probability chosen in this paper is 0.9.

Note: The arrow points to the crossover point, and the part after that point is swapped.

(6) Mutation

Mutation is to replace the gene value of a specific site that meets certain conditions in order to overcome premature maturity in crossover and improve the diversity of the population. The algorithm is based on real number coding, which randomly selects chromosome individuals through population mutation. In this paper, the probability of mutation is 0.5.

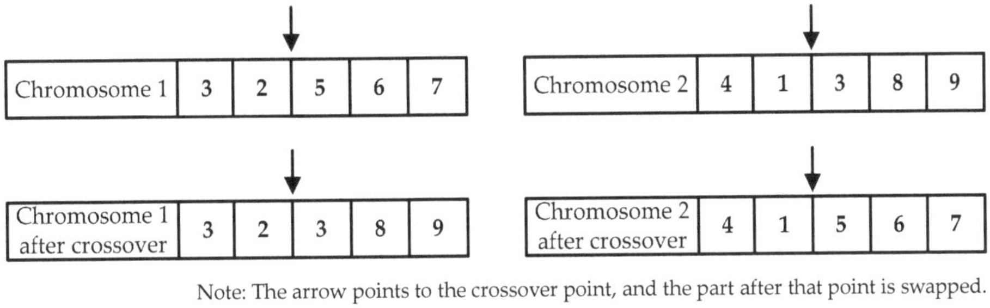

Figure 2. Diagram of single-point crossover.

4. Example Analysis

4.1. Case Background

Company A is one of the cold chain distribution companies of fresh agricultural products in Wuhan, China, investigated in this study. One of its distribution centers provides distribution services for many customers every day. This paper chooses the business of the distribution center and 21 customer demand points around it as the object of analysis. Baidu map is used to query the longitude and latitude coordinates of each demand point and distribution center. The difference between geographic coordinates and plane coordinates is ignored, and the Baidu map query result is taken as plane coordinates. At the same time, through the function of Baidu map, the distance between the demand points and the distance from the distribution center to the demand points are measured. Taking P as the coordinate point of the distribution center, its coordinate is (64, 45). Take M1-M21 as the coordinates of the demand points, and related data are shown in Table 4. The distances between different demand points are shown in Appendix B.

Table 4. Information about the demand points.

Demand Points Number	Demand Point Coordinates		Distance to Distribution Center (km)	Quantity Required (t)	Time Window		Maximum Tolerance Time (h)
	Abscissa	Ordinate					
M1	87	37	19.5	0.98	7:30	9:00	1
M2	77	39	10.1	1.98	7:30	9:00	1
M3	79	36	16.4	3.26	7:30	9:00	1
M4	66	37	18.1	2.76	7:30	9:00	1
M5	75	44	8.8	0.88	7:30	9:00	1
M6	65	39	12.9	2.59	7:30	9:00	1
M7	66	41	15.6	1.5	7:30	9:00	1
M8	68	36	7.5	2.19	7:30	9:00	1
M9	81	39	15.1	3.11	7:30	9:00	1
M10	77	45	9.3	3.48	7:30	9:00	1
M11	70	35	9.4	1.35	7:30	9:00	1
M12	76	41	6.5	2.61	7:30	9:00	1
M13	71	44	4.8	3.41	7:30	9:00	1
M14	83	35	20.7	3.5	7:30	9:00	1
M15	80	38	16.2	2.83	7:30	9:00	1
M16	81	39	15.8	2.05	7:30	9:00	1
M17	73	42	6.5	2	7:30	9:00	1
M18	73	34	14.1	1.28	7:30	9:00	1
M19	87	38	20.2	2.625	7:30	9:00	1
M20	81	32	20.3	2.86	7:30	9:00	1
M21	75	36	12.6	2.1	7:30	9:00	1

According to the survey, the distribution center has a total of 140 refrigerated vehicles, of which the load capacities of 5 t gasoline refrigerated vehicles and 4t electric refrigerated vehicles are 70 each. The relevant data of fuel consumption and power consumption of refrigerated vehicles are shown in Table 5.

Table 5. Data of fuel and electricity consumption of refrigerated vehicles.

Name	Numerical Value
Power consumption per 100 km when the 4t electric refrigerated vehicle is unloaded.	40
Power consumption per 100 km when the 4t electric refrigerated vehicle is fully loaded.	50
Fuel consumption per 100 km when the 5t gasoline refrigerated vehicle is unloaded.	10
Fuel consumption per 100 km when the 5t gasoline refrigerated vehicle is fully loaded.	13

From the investigation, the maximum speed of vehicles in the distribution center is 50 km per hour, while the maximum operating distance of vehicles is set at 110 km. The average sales price of fresh agricultural products is 30 yuan/kg. The fixed cost of a 4t electric refrigerated vehicle is 150 yuan, the transportation cost is 0.3 yuan per kilometer per hour, the maximum battery capacity of electric vehicles is 200 kWh, and the minimum allowable electricity is 30 kWh. The fixed cost of a 5t gasoline refrigerated vehicle is 200 yuan, and the transportation cost is 0.9 yuan per kilometer per hour. The loading and unloading rate of the vehicle is 6 t/h, and the refrigerated vehicle will start the delivery at 7 am. If the vehicle arrives early, the waiting fee is 2 yuan/h (less than 1 h is counted as 1 h), and the late penalty fee is 10 yuan/h (less than 1 h is counted as 1 h). The detailed relevant data are shown in Table 6.

Table 6. Summary of relevant data.

Name	Numerical Value
Vehicle speed	50
Loading and unloading speed	6
Fixed cost of 5t gasoline refrigerated vehicle	200
Unit transportation cost of 5t gasoline refrigerated vehicle	0.9
Fixed cost of 4t electric refrigerated vehicle	150
Unit transportation cost of 4t electric refrigerated vehicle	0.3
Waiting cost	2
Penalty cost	10
Unit refrigeration cost of 4t electric refrigerated vehicle when closed	3.5
Unit refrigeration cost of 5t gasoline refrigerated vehicle when closed	4
Unit refrigeration consumption of 4t electric refrigerated vehicle when closed	0.5
Unit refrigeration consumption of 5t gasoline refrigerated vehicle when closed	0.6
Corruption rate when refrigerated vehicle is not opened	0.008
Corruption rate during loading and unloading of refrigerated vehicle	0.01
Unit carbon trading price	0.6
CO_2 emission coefficient of electric refrigerated vehicle	0.0997
CO_2 emission coefficient of gasoline refrigerated vehicle	0.2925

4.2. Solving Procedure

The relevant data of distribution center and demand points were substituted into the cold chain distribution path optimization model of fresh agricultural products, and MATLAB 2018b was used to solve the problem. The number of demand points is 21, and the distribution center has 4t electric refrigerated vehicles and 5t gasoline refrigerated vehicles. Chromosome coding is performed accordingly. The population size is 750. The fitness is $f = \frac{1}{Z}$. The roulette method is used for selection, any number is generated between the range of [0,1], and it is used as a pointer to determine the number of chromosomes to be retained. The number of chromosomes selected to be retained is 70% of the initial population number. On this basis, a single point method is used to cross. Two chromosomes are randomly generated, and cross operation is performed between the two chromosomes. For chromosomes that do not meet the restriction requirements, the mutation operation is repeated until the chromosome that meets the restriction is obtained.

The number of iterations is set to 750. From the actual situation, with the increase in the number of iterations, the change in the optimal value gradually decreases, tends to be flat after 214 iterations, and reaches a stable and unchanged state after 500 iterations, that is, the optimal solution is reached.

4.3. Optimization Result Analysis

The solution results show that in order to serve 21 customer points, the distribution center needs to send 12 vehicles to form 12 routes, as shown in Table 7. The scheme generated a total distribution cost of 3233.7852 yuan and carbon emissions of 393.409 kg.

Table 7. Optimized delivery path.

Serial Number	Type of Refrigerated Vehicle	Delivery Routes
1	5t gasoline refrigerated vehicle	0→ -> 8 → 6 → 0
2	5t gasoline refrigerated vehicle	0 → 5 → 9 → 0
3	5t gasoline refrigerated vehicle	0 → 15 → 2 → 0
4	5t gasoline refrigerated vehicle	0 → 11 → 14→ 0
5	5t gasoline refrigerated vehicle	0 → 18 → 19 →1 → 0
6	5t gasoline refrigerated vehicle	0 → 3 → 7 → 0
7	5t gasoline refrigerated vehicle	0 → 4 → 16→ 0
8	5t gasoline refrigerated vehicle	0 → 10→ 0
9	5t gasoline refrigerated vehicle	0 → 21 → 20 → 0
10	4t electric refrigerated vehicle	0 → 12 → 0
11	4t electric refrigerated vehicle	0 → 13 → 0
12	4t electric refrigerated vehicle	0 → 17→ 0

In particular, the various costs, the number of vehicles used, and the total mileage of distribution are shown in Table 8. Comparing the data before and after optimization, it can be seen that some changes have occurred.

Table 8. Comparison of results before and after optimization.

Type of Target	Before Optimization	After Optimization	Reduced Amount
Fixed cost	2400	2250	150
Transportation cost	286.2	279.42	6.78
Penalty cost	0	0	0
Cargo loss cost	47.428	39.4231	8.0049
Refrigeration cost	463.59	428.8967	34.6933
Carbon emission	533.37	393.409	139.961
Carbon emission cost	320.02	236.0454	83.9746
Number of vehicles used	13	12	1
Total delivery miles	374.4	334.2	40.2
Total delivery cost	3517.238	3233.7852	283.4528

Although the model is only applied in a small range, the obvious optimization effect can still be seen. The total number of refrigerated vehicles required after optimization is one less than the original distribution route scheme. At the same time, in addition to the same penalty costs, other costs have been reduced by varying degrees, and carbon emissions and total miles of distribution have also been reduced. From these data and changes, it can be seen that the optimized scheme not only saves the distribution cost of fresh agricultural products but also reduces carbon emissions. This also shows that the optimized distribution route reduces some unnecessary transportation in the distribution process and improves the economic and environmental benefits of Company A in the cold chain distribution of fresh agricultural products. It can be seen that the cold chain distribution route optimization model established in this paper is effective and has application value, and can provide reference for the distribution route planning of cold chain logistics companies of fresh agricultural products.

5. Discussion

Reasonable planning of cold chain logistics distribution route is an important measure to save logistics resources, reduce the impact on the environment, and promote the sustainable development of agricultural cold chain logistics. On the basis of field investigation, this paper establishes a new optimization model of cold chain distribution route for fresh agricultural products, and carries out an example analysis. The results show that:

(1) It is necessary and feasible for cold chain distribution route planning of fresh agricultural products to comprehensively consider factors such as carbon emissions and mixed vehicle types. In view of the high timeliness requirements and high energy consumption of cold chain distribution of fresh agricultural products, the planning of distribution routes should not only consider the constraints of the time window but also consider how to rationally use vehicle resources, give play to the advantages of different types of vehicles, in order to reduce distribution costs and reduce carbon emissions. This paper is a helpful supplement to the existing research. In the study of cold chain distribution route of fresh agricultural products, Ho et al. [8], Ding et al. [9], Wu et al. [10], and Wang et al. [11] took the time window constraint into consideration; Fu et al. [24], Yao et al. [25], Feng et al. [26], and Pérez-Lechuga et al. [27] considered the time window and carbon emissions; Leung et al. [33], Kwon et al. [34], and Chen et al. [35] consider the time window and mixed vehicle types. In this paper, the three factors mentioned above are taken into account to establish a decision-making model, which can better balance economic benefits and environmental benefits, and has good application value in the current practical conditions.

(2) The optimal overall cost should be pursued in the cold chain distribution route planning of fresh agricultural products. The cold chain distribution process will produce a variety of costs; only paying attention to a few costs may lead to the increase in other costs, affecting the overall operation effect. Therefore, when planning the distribution route, we should fully understand the possible costs and pursue the minimum comprehensive cost. Previous studies have considered the cost of cold chain distribution of fresh agricultural products from different perspectives. For example, Wang et al. [11] took into account the cost of freshness loss, cold chain refrigeration cost, and delay penalty cost; Jiang et al. [22] took into account the cost of transportation, refrigeration cost, and carbon emission cost; Ma et al. [23] considered fixed transportation cost, variable transportation cost, time penalty cost, and carbon emission cost. This paper comprehensively considers six major costs in the cold chain distribution of fresh agricultural products, including fixed cost, transportation cost, penalty cost, cargo loss cost, refrigeration cost, and carbon emission cost, which is conducive to improving the effectiveness of decision-making.

(3) Genetic algorithm is an effective method to solve the cold chain distribution route optimization model of fresh agricultural products. There are many methods to solve mathematical models, and each has its own advantages. Similar to the studies of Liu and Li [7], Jiang et al. [22], and Chen et al. [35], this paper selects genetic algorithm to solve the problem based on the characteristics of the model. From the algorithm design and practical application, the algorithm can explain the model decision thinking well and help find the optimal solution quickly, so it is an effective solution method. Of course, if the model is further optimized, genetic algorithms mixed with other methods can be tried to improve the ability to solve the problem.

6. Conclusions

In this paper, the cold chain distribution route optimization of fresh agricultural products is taken as the research object, with the minimum total cost including carbon emission cost as the optimization goal. The cold chain distribution route optimization model is constructed considering different types of refrigerated vehicles and is solved with the help of genetic algorithm. Finally, taking the distribution center of Company A as an example, the empirical analysis of the proposed method is carried out, and the results show that the method is effective. The main work completed in this paper includes:

(1) Through the investigation on the distribution route planning of eight fresh agricultural products cold chain logistics companies in Wuhan, China, it is learned that the costs related to decision-making include fixed cost, transportation cost, penalty cost, cargo loss cost, refrigeration cost, carbon emission cost, etc., but most companies do not consider comprehensively, especially that the carbon emission costs are relatively less. In addition, many companies have multiple vehicle types such as gasoline, electric, and hybrid vehicles, which need to be considered when optimizing distribution routes.

(2) Based on the investigation results and the literature, a cold chain distribution route optimization model for fresh agricultural products is established. The optimization goal of the model is to minimize the six total costs including carbon emission costs, and the constraints of the time window are emphasized, and the mixed vehicle types is also taken into account. According to the characteristics of the model, genetic algorithm is chosen as the method to solve the model.

(3) The validity and application value of the model are verified by the case data of a distribution center of Company A. Through the data comparison before and after optimization, it can be seen that the distribution route determined by the new model can effectively utilize vehicle resources and reduce various costs and carbon emissions in the distribution process. This is beneficial to improve the economic benefits of the distribution center and improve the social environmental benefits and can provide a reference for the distribution route planning of cold chain logistics companies of fresh agricultural products.

This paper fully combines the characteristics of cold chain distribution of fresh agricultural products, considers the current situation of coexistence of multiple vehicle types, and comprehensively considers how to rationally plan the cold chain distribution path from the perspective of low-carbon development, which helps to better improve the quality and efficiency of fresh agricultural products distribution. However, there are still some details that have not been carefully considered in this study, such as complex traffic conditions, sudden increase in order demand, etc. Further research can make a more comprehensive consideration around this problem and seek more efficient solution algorithms.

Author Contributions: Conceptualization, S.P., H.L. and M.S.; methodology, S.P. and G.Z.; validation, S.P., G.Z. and Q.H.; investigation, M.S., H.L. and Q.H.; data analysis, S.P., G.Z. and M.S.; writing—original draft preparation, S.P., H.L. and M.S.; writing—review and editing, S.P., H.L., G.Z. and Q.H. All authors have read and agreed to the published version of the manuscript.

Funding: This study was supported by the Scientific Research Project of Hunan Provincial Department of Education, China, grant number 21B0251; supported by the Hunan Key Laboratory of Intelligent Logistics Technology (2019TP1015).

Institutional Review Board Statement: Not applicable.

Informed Consent Statement: Not applicable.

Data Availability Statement: Data are contained within the article.

Conflicts of Interest: The authors declare no conflicts of interest.

Appendix A. The Survey Outline of Cold Chain Distribution Route Planning for Fresh Agricultural Products

Dear Sir/Madam,

We are members of a research team from Central South University of Forestry and Technology. In order to improve the quality and efficiency of cold chain distribution of fresh produce, we are conducting a survey focusing on the consideration of cost and its influencing factors in route planning of fresh produce distribution. We hope to get your help and answer some relevant questions from a professional perspective.

1. What is your company most concerned about in the cold chain distribution process of fresh agricultural products?
2. What do you think is the most important problem to solve in cold chain distribution?
3. Does your company carry out cost calculation and route planning in advance when arranging vehicle distribution routes?
4. If cost calculation is carried out in advance, what aspects of cost are considered by your company?
5. In addition to existing costs, what other costs do you think your company should consider when calculating distribution costs?
6. What do you think are the main reasons why companies currently consider/do not consider the cost of carbon emissions? What do you think about the cost of carbon emissions?
7. What influences does your company take into account when planning distribution routes?
8. What other factors do you think should be considered when planning the distribution routes?
9. Does your company use new energy refrigerated vehicles? What are the current types of distribution vehicles?
10. Do you have any suggestions for cold chain distribution route planning of fresh agricultural products?

Thank you for your cooperation and support!

(Note: We have designed the survey outline, but the actual interview content is not limited to the above questions.)

Appendix B. Distances between Customer Demand Points in the Example

Table A1. Distances between demand points.

Demand Points	M1	M2	M3	M4	M5	M6	M7	M8	M9	M10	M11	M12	M13	M14	M15	M16	M17	M18	M19	M20	M21
M1	0	8	12	4.8	7	6.8	5	7	8	6.2	7	8.2	6.6	8	6	7	8.6	12	4	3.5	8
M2	8	0	10	9	15	6.6	5.7	5.1	13	12	12	16	6.2	14.5	6.4	11	15	6	5.1	9.3	3
M3	12	10	0	16.5	17	5	7	6	15	13	16.2	15	5.1	18	16.2	18.2	13	3.7	8.5	9	6
M4	4.8	9	16.5	0	6.5	10	8.5	11	7	6.2	4	8.5	10	7	4	5	10.5	12.6	8	6.5	9
M5	7	15	17	6.5	0	9	8	13	5	2	7.4	7.2	9	8	8.2	9.3	5.1	11	9.5	4	10.3
M6	6.8	6.6	5	10	9	0	3	4	12	7	11	9	1	13	12	16	9	3	4.3	4	6
M7	5	5.7	7	8.5	8	3	0	4.5	10	9	9	8.2	2	11	8.3	12	9.5	5	2.7	6	5
M8	7	5.1	6	11	13	4	4.5	0	13	12	12	11	4.5	12	10	13	11	5	4.5	8.3	4
M9	8	13	15	7	5	12	10	13	0	5	6	6.1	15	9	11	15	3	11	11	8	16
M10	6.2	12	13	6.2	2	7	9	12	5	0	5	7.2	8	9	9.2	11	5.6	13	8	4.5	18
M11	7	12	16.2	4	7.4	11	9	12	6	5	0	15	11	4	6	3	13	14	4	7.3	10
M12	8.2	16	15	8.5	7.2	9	8.2	11	6.1	7.2	15	0	8	13	10	12	3	8	9	5	21
M13	6.6	6.2	5.1	10	9	1	2	4.5	15	8	11	8	0	11	14	15	8.7	3	5	4	7
M14	8	14.5	18	7	8	13	11	12	9	9	4	13	11	0	6	6	11	13	8	9	16
M15	6	6.4	16.2	4	8.2	12	8.3	10	11	9.2	6	10	14	6	0	3	19	15	5	7	9
M16	7	11	18.2	5	9.3	16	12	13	15	11	3	12	15	6	3	0	17	17	11	13	13
M17	8.6	15	13	10.5	5.1	9	9.5	11	3	5.6	13	3	8.7	11	19	17	0	11	9	7	23
M18	12	6	3.7	12.6	11	3	5	5	11	13	14	8	3	13	15	17	11	0	6	4	13
M19	4	5.1	8.5	8	9.5	4.3	2.7	4.5	11	8	4	9	5	8	5	11	9	6	0	5.4	5.6
M20	3.5	9.3	9	6.5	4	4	6	8.3	8	4.5	7.3	5	4	9	7	13	7	4	5.4	0	8
M21	8	3	6	9	10.3	6	5	4	16	18	10	21	7	16	9	13	23	13	5.6	8	0

References

1. Tarantilis, C.D.; Kiranoudis, C.T. Distribution of fresh meat. *J. Food Eng.* **2002**, *51*, 85–91. [CrossRef]
2. Zhu, S.J.; Fu, H.M.; Li, Y.H. Optimization research on vehicle routing for fresh agricultural products based on the investment of freshness-keeping cost in the distribution process. *Sustainability* **2021**, *13*, 8110. [CrossRef]
3. Wang, H.; Li, W.; Zhao, Z.Z.; Wang, Z.W.; Li, M.H.; Li, D.F. Intelligent distribution of fresh agricultural products in smart city. *IEEE Trans. Ind. Inform.* **2022**, *18*, 1220–1230. [CrossRef]
4. Yang, F.; Tao, F.M. A bi-objective optimization VRP model for cold chain logistics: Enhancing cost efficiency and customer satisfaction. *IEEE Access* **2023**, *11*, 127043–127056. [CrossRef]
5. Marius, M.; Desrosiers, J. Time window constrained routing and scheduling problems. *Transp. Sci.* **1988**, *22*, 1–13.
6. Ahkamiraad, A.; Wang, Y. Capacitated and multiple cross-docked vehicle routing problem with pickup, delivery, and time windows. *Comput. Ind. Eng.* **2018**, *119*, 76–84. [CrossRef]
7. Liu, Z.H.; Li, X.J. Optimization model of cold chain logistics delivery path based on genetic algorithm. *Int. J. Ind. Eng. Theory Appl. Pract.* **2024**, *31*, 152–169.
8. Ho, W.K.; Ang, J.C.; Lim, A. A hybrid search algorithm for the vehicle routing problem with time windows. *Int. J. Artif. Intell. Tools* **2001**, *10*, 431–439. [CrossRef]
9. Ding, Q.L.; Hu, X.P.; Sun, L.J.; Wang, Y. An improved ant colony optimization and its application to vehicle routing problem with time windows. *Neurocomputing* **2012**, *98*, 101–107. [CrossRef]
10. Wu, D.Q.; Li, J.Y.; Cui, J.Y.; Hu, D. Research on the time-dependent vehicle routing problem for fresh agricultural products based on customer value. *Agriculture* **2023**, *13*, 681. [CrossRef]
11. Wang, W.J.; Wen, S.Z.; Gao, S.; Lin, P.Y. A multi-objective dynamic vehicle routing optimization for fresh product distribution: A case study of Shenzhen. *Electron. Res. Arch.* **2024**, *32*, 2897–2920. [CrossRef]
12. Lin, C.; Choy, K.L.; Ho, G.T.; Chung, S.H.; Lam, H.Y. Survey of green vehicle routing problem: Past and future trends. *Expert Syst. Appl.* **2014**, *41*, 1118–1138. [CrossRef]
13. Liu, G.K.; Hu, J.Y.; Yang, Y.; Xia, S.; Ming, K.L. Vehicle routing problem in cold Chain logistics: A joint distribution model with carbon trading mechanisms. *Resour. Conserv. Recycl.* **2020**, *156*, 104715. [CrossRef]
14. Ji, Y.Y.; Ji, S.F.; Ji, T.T.; Meng, F.L. Considering the carbon penalty rates to optimize the urban distribution model in time-varying network. *Complexity* **2022**, *2022*, 1013861. [CrossRef]
15. Zhang, X.; Chen, H.Z.; Hao, Y.C.; Yuan, X.M. A low-carbon route optimization method for cold chain logistics considering traffic status in China. *Comput. Ind. Eng.* **2024**, *193*, 110304. [CrossRef]
16. Yang, L.; Gao, Y.L.; Sun, Y.; Li, J. Two-phase hybrid search algorithm for time-dependent cold chain logistics route considering carbon emission and traffic congestion. *IEEE Access* **2024**, *12*, 95128–95151. [CrossRef]
17. Demir, E.; Bektas, T.; Laporte, G. The bi-objective pollution-routing problem. *Eur. J. Oper. Res.* **2014**, *232*, 464–478. [CrossRef]
18. Hariga, M.; As'Ad, R.; Shamayleh, A. Integrated economic and environmental models for a multi stage cold supply chain under carbon tax regulation. *J. Clean. Prod.* **2017**, *166*, 1357–1371. [CrossRef]
19. Zhao, B.L.; Gui, H.X.; Li, H.Z.; Xue, J. Cold chain logistics path optimization via improved multi-objective ant colony algorithm. *IEEE Access* **2020**, *8*, 142977–142995. [CrossRef]
20. Afra, A.P.; Behnamian, J. Lagrangian heuristic algorithm for green multi-product production routing problem with reverse logistics and remanufacturing. *J. Manuf. Syst.* **2020**, *58*, 33–43. [CrossRef]
21. Wu, D.Q.; Cui, J.Y.; Mansour, D.L.R.F. A new route optimization approach of fresh agricultural logistics distribution. *Intell. Autom. Soft Comput.* **2022**, *34*, 1553–1569. [CrossRef]
22. Jiang, X.Y.; Liu, X.X.; Pan, F.B.; Han, Z.N. Optimizing cold chain distribution routes considering dynamic demand: A low-emission perspective. *Sustainability* **2024**, *16*, 2013. [CrossRef]
23. Ma, Z.C.; Zhang, J.; Wang, H.H.; Gao, S.C. Optimization of sustainable bi-objective cold-chain logistics route considering carbon emissions and customers' immediate demands in China. *Sustainability* **2023**, *15*, 5946. [CrossRef]
24. Fu, Q.M.; Li, J.Y.; Chen, H.H. Resource scheduling method for optimizing the distribution path of fresh agricultural products under low-carbon environmental constraints. *Sci. Program.* **2022**, *2022*, 7692135. [CrossRef]
25. Yao, Q.; Zhu, S.J.; Li, Y.H. Green vehicle-routing problem of fresh agricultural products considering carbon emission. *Int. J. Environ. Res. Public Health* **2022**, *19*, 8675. [CrossRef]
26. Feng, Q.; Zhao, G.; Li, W.J.; Shi, X.J. Distribution path optimization of fresh products in cold storage considering green costs. *Buildings* **2023**, *13*, 2325. [CrossRef]
27. Pérez-Lechuga, G.; Martínez-Sánchez, J.F.; Venegas-Martínez, F.; Madrid-Fernández, K.N. A routing model for the distribution of perishable food in a green cold chain. *Mathematics* **2024**, *12*, 332. [CrossRef]
28. Golden, B.; Assad, A.; Levy, L.; Gheysens, F. The fleet size and mix vehicle routing problem. *Comput. Oper. Res.* **1984**, *11*, 49–66. [CrossRef]
29. Hiermann, G.; Hartl, R.F.; Puchinger, J.; Vidal, T. Routing a mix of conventional, plug-in hybrid, and electric vehicles. *Eur. J. Oper. Res.* **2019**, *272*, 235–248. [CrossRef]
30. Macrina, G.; Laporte, G.; Guerriero, F.; Pugliese, L.D.P. An energy-efficient green-vehicle routing problem with mixed vehicle fleet, partial battery recharging and time windows. *Eur. J. Oper. Res.* **2019**, *276*, 971–982. [CrossRef]

31. Ene, S.; Kucukoglu, I.; Aksoy, A.; Ozturk, N. A hybrid metaheuristic algorithm for the green vehicle routing problem with a heterogeneous fleet. *Int. J. Veh. Des.* **2016**, *71*, 75–102. [CrossRef]
32. Kabadurmus, O.; Erdogan, M.S. A green vehicle routing problem with multi-depot, multi-tour, heterogeneous fleet and split deliveries: A mathematical model and heuristic approach. *J. Comb. Optim.* **2023**, *45*, 89. [CrossRef]
33. Leung, S.C.H.; Zhang, Z.Z.; Zhang, D.F.; Hua, X.; Lim, M.K. A meta-heuristic algorithm for heterogeneous fleet vehicle routing problems with two-dimensional loading constraints. *Eur. J. Oper. Res.* **2013**, *225*, 199–210. [CrossRef]
34. Kwon, Y.J.; Choi, Y.J.; Lee, D.H. Heterogeneous fixed fleet vehicle routing considering carbon emission. *Transp. Res. Part D Transp. Environ.* **2013**, *23*, 81–89. [CrossRef]
35. Chen, Y.Y.; Chen, T.L.; Chiu, C.C.; Wu, Y.J. A multi-trip vehicle routing problem considering time windows and limited duration under a heterogeneous fleet and parking constraints in cold supply chain logistics. *Transp. Plan. Technol.* **2023**, *46*, 335–358. [CrossRef]
36. Chen, W.R.; Zhang, D.Z.; Van Woensel, T.; Xu, G.M.; Guo, J. Green vehicle routing using mixed fleets for cold chain distribution. *Expert Syst. Appl.* **2023**, *233*, 120979. [CrossRef]

MDPI AG
Grosspeteranlage 5
4052 Basel
Switzerland
Tel.: +41 61 683 77 34

Sustainability Editorial Office
E-mail: sustainability@mdpi.com
www.mdpi.com/journal/sustainability

www.ingramcontent.com/pod-product-compliance
Lightning Source LLC
Chambersburg PA
CBHW041426210926
51850CB00013B/3367
* 9 7 8 3 7 2 5 8 4 3 2 7 5 *